About Science, Myself and Others

Editorial team

Managing Editor of Translation:

Dr Maria S Aksent'eva
"Uspekhi Fizicheskikh Nauk"
(Physics–Uspekhi) journal
Leninsky Prospect 15,
119071 Moscow, Russia
E-mail: maria@ufn.ru

Translations:

Part I

Ms M V Tsaplina
Dr K G Postnov

Part II

Dr A V Leonidov
Dr E M Yankovsky
Dr E G Strel'chenko
Ms M A Blagonravova
Dr E N Ragozin

Part III

Dr V I Kisin
Dr Yu V Morozov

Desk Editor: Dr S G Rudnev
Copy-Editor: Ms E A Frimer
Bibliography Editor: Ms E V Zakharova

This book contains material based on chapters published in *The Physics of a Lifetime: Reflections on the Problems and Personalities of 20th Century Physics* by Vitaly L Ginzburg and published by Springer (2001, ISBN: 3-540-67534-5). The author and IOP Publishing are thankful to Springer for the permission to use this material in this manner.

About Science, Myself and Others

Vitaly L Ginzburg

I E Tamm Theory Department,
P N Lebedev Physical Institute,
Moscow,
Russia

CRC Press
Taylor & Francis Group
Boca Raton London New York

CRC Press is an imprint of the
Taylor & Francis Group, an **Informa** business

About Science, Myself and Others

Vitaly L Ginzburg

I E Tamm Theory Department,
P N Lebedev Physical Institute,
Moscow,
Russia

CRC Press
Taylor & Francis Group
Boca Raton London New York

CRC Press is an imprint of the
Taylor & Francis Group, an **Informa** business

CRC Press
Taylor & Francis Group
6000 Broken Sound Parkway NW, Suite 300
Boca Raton, FL 33487-2742

© 2005 by Taylor & Francis Group, LLC
CRC Press is an imprint of Taylor & Francis Group, an Informa business

First issued in Paperback 2019

ISBN 13: 9780750309929 (hbk)
ISBN 13: 9780367393632 (pbk)

**Visit the Taylor & Francis Web site at
http://www.taylorandfrancis.com**

**and the CRC Press Web site at
http://www.crcpress.com**

Contents

Preface to the English edition

The language barrier has always complicated the communication between scientists working in Russia and their foreign colleagues. The many years of the Soviet period saw additional difficulties arising from censorship and the complexity and sometimes even impossibility of publishing papers or their translations into English and other languages. Fortunately, these difficulties are no longer relevant but the language barrier still persists and, hence, the translation of scientific journals is inevitably late and much is not translated at all. The latter particularly applies to non-periodic literature and various articles that are biographic or journalistic in genre. Specifically, my book *O Fizike i Astrofizike* [*On Physics and Astrophysics*] containing a series of articles of this kind was published three times in Russian (in 1985, 1992, and 1995) but was only published in English in 2001—*The Physics of a Lifetime: Reflections on the Problems and Personalities of 20th Century Physics* (Berlin: Springer). The table of contents of that book is given at the end of the present book, which is its continuation in a sense. The first edition, which contained 15 articles, saw the light of day in 1997. Fourteen articles were added in the second edition (2001) and four more in the third one (2003). The English version is, naturally, a translation of the last Russian edition, with only two articles being added ('Once again on the history of the discovery of combinational (Raman) scattering' and 'Russia must not slide into a slough of clericalism').

The translation faithfully follows the third Russian edition, with the exception of the location of some articles. Furthermore, a few comments were made, at the end of articles in particular. Such comments given at the end of the articles are marked with an asterisk (for instance, ^{1*}) to distinguish them from footnotes.

The prefaces to the Russian editions are omitted from the translation, because the story of how this book changed from edition to edition is of little interest. However, it seems to me that it would be appropriate to characterize to some extent the material included in the book. Apart from articles 1 and 6, which are fairly conventional reviews, part I (articles 2–5 and 7) is primarily concerned with my own work. But why have activities in the areas of the radiation of uniformly moving sources, astrophysics, ferroelectric effects, superconductivity, and superfluidity been thereby emphasized? First, I believe that these are precisely the areas in which I obtained results that deserve attention, at least from the viewpoint of the history of physics. Second, it is precisely these articles that were written in due time and turned out to be in the way. For those who would like to familiarize themselves with the general balance of my work, I refer to the article 'A scientific autobiography—an attempt' included in my aforementioned book *The Physics of a Lifetime*.

Part II of the book is compiled of articles, one might say, of biographic and autobiographic nature. Namely, articles 8–16 are dedicated to the memory of I E Tamm, L D Landau, E M Lifshits, D A Kirzhnits, E S Fradkin, V F Shvartsman, V A Sidur, and S I Vavilov. In articles 17–19, I write primarily of myself. Part III contains articles which can most likely be referred to as journalistic or something close to it. Their contents is largely made clear from their titles (see table of contents). It is worth noting that my articles and speeches concerned with the activities of the Russian Academy of Sciences are not included in the book, since this seems inappropriate to the occasion (references to the most of the corresponding publications are given in comment 7 to article 17).

The publication of this translation would have been impossible had this work not been undertaken by Dr M S Aksent'eva and I take advantage of this opportunity to express deep sense of gratitude to her. I also wish to express my appreciation to the large team of translators and editors (listed on the reverse of the title-page) whose labors have made possible the publication of this book in English. I would like to conclude by expressing my thanks to IOP Publishing Managing Director Mr Jerry Cowhig and to the entire staff of the Institute of Physics Publishing.

V L Ginzburg
Moscow, September 2004

PART I

Article 1

What problems of physics and astrophysics seem now to be especially important and interesting at the beginning of the 21st century?[1*]

1.1 Introduction

The rate of development of science nowadays is striking. Great changes in physics, astronomy, biology, and many other fields of science have come about within a period of not more than one to two generations. The readers may even see an example of it in their own families. My father, for instance, was born in 1863 and was a younger contemporary of Maxwell (1831–79). I myself was already 16 when the neutron and positron were discovered in 1932. Before that, only the electron, proton, and photon were known. It is somehow not easy to realize that the electron, X-rays, and radioactivity were discovered only about a hundred years ago and quantum theory was born in 1900. At the same time, one hundred years is such a short period not only compared to the approximately three billion years since life appeared on Earth but also to the age of modern man (*homo sapiens*), which amounts to nearly fifty thousand years! It is also useful to remember that the first great physicists—Aristotle (384–322 BC) and Archimedes (about 287–212 BC)—are separated from us by already more than two thousand years. But further progress in science was comparatively slow and religious dogmatism played not the least part in this. Only since the time of Galileo Galilei (1564–1642) and Kepler (1571–1630) has the development of physics been increasingly rapid. But, incidentally, even Kepler was of the opinion that there exists a sphere of motionless stars which 'consists of ice or a crystal'. The fight of Galileo for acknowledgement of the heliocentric concepts, for which he was convicted by the Inquisition in 1633, is generally known. What a path has been overcome since then in only 300–400 years! The result is contemporary science. It has already freed itself from religious chains and the church today does not at least deny the role of science [4].[1] True, pseudoscientific tendencies and pseudoscience (especially astrology) continue to be propagated, particularly in Russia. But it is only the triumph of totalitarianism (bolshevik communism or fascism) that can radically obstruct the progress of science the most striking example of which was the appearance of Lysenko's 'theories' and their supporters. We shall hope that this will not happen again. In any event, one can expect that, in the 21st century, science will develop no slower than it did in the passing 20th century. The difficulty on this journey, and may be even the largest one, is in my opinion associated with the mammoth increase of the accumulated material and the body of information. Physics is now so much extended and differentiated that 'the wood cannot be seen for the trees' and it is difficult to catch in the mind's eye a picture of modern physics as a whole.

Meanwhile, such a picture does exist and, in spite of all its branches, physics has its pivot. Such a pivot is represented by the fundamental concepts and laws formulated in theoretical physics. The contents of the latter are clearly seen from the *Course* written by L D Landau, E M Lifshitz, and L P Pitaevskii (LLP). The latter author continues the cause begun by his predecessors. An updated edition of the *Course* has been issued, although unfortunately rather slowly. The *LLP Course*, as well as other manuals and monographs, describes the base underlying the work in all fields of physics and in related areas. However, all these books cannot reflect the most recent advances in science and reading them one can hardly, if at all, feel the pulse of scientific life. As is known, seminars serve this purpose. I personally have been head of one such seminar at the P N Lebedev Physical Institute (FIAN) for over 40 years. It is conducted weekly (on Wednesdays) and lasts two hours. The typical agenda covers news from current literature and then two or sometimes one talk is given on various physical and near-physical topics. The 1500th session of the seminar took place on 24 May, 1996 in a form close to a skit and was reflected in the journal *Priroda* [*Nature*] [5]. The 1600th session was held on 13 January, 1999. The seminar is customarily attended by, on average, 100 people—research workers from FIAN and other institutes, as well as a few students from the Moscow Physico-Technical Institute (now the Moscow Institute of Physics and Technology).[2*] With somewhat surprise, I should note that there are very few such many-sided seminars. Highly specialized seminars or, especially abroad, so-called colloquia prevail. The latter last an hour and are devoted to a single review report. But at the same time, such journals as *Nature, Physics Today, Physics World, Contemporary Physics* and some

---

[1] References [1–3] are cited later and in comment 1 at the end of this article.

others containing a lot of news are widespread abroad. Unfortunately, all these journals are now not quite so easily accessible in Russia or appear with some delay. I believe that *Uspekhi Fizicheskikh Nauk* (*Physics–Uspekhi*) is accessible enough and of great benefit.

However, I have long been of the opinion that all this is not enough and I am advocating a 'project' (which is now a popular word) reflected in the title of the present article. I mean a compilation of a certain 'list' of problems which currently seem to be the most important and interesting. These problems should primarily be discussed and commented on in special lectures and papers. The formula 'everything about a particular issue and something about everything' is rather attractive but already unrealistic, for one cannot keep up with everything. At the same time, some subjects, questions and problems are somewhat distinguished for different reasons. The importance of an issue for humanity (to put it in a high-flown manner) may play its role. Such, for example, is the problem of controlled nuclear fusion with the purpose of obtaining energy. Of course, the questions concerning the fundamentals of physics, its leading edge (this field has frequently been referred to as elementary particle physics) is also distinguished. Particular attention is undoubtedly attracted by some problems of astronomy which, as in the times of Galileo, Kepler, and Newton, are now hard (and needless) to separate from physics. Such a list (of course duly updated) constitutes, I believe, a certain physical minimum. It includes issues of which every physicist should have an idea. Less trivial is perhaps the opinion that it is not at all difficult to attain such a goal and not much time and strength is needed for it. But this requires some effort not only on the part of those who learn but also on the part of 'senior fellows'. Namely, one should select problems to constitute the physical minimum, compile the corresponding 'list' and comment on it, explaining and filling it with content. This is exactly what I tried to do at the chair of *Problems of Physics and Astrophysics* of the Moscow Institute of Physics and Technology which was set up in 1968. For this purpose, special additional lectures were delivered (there were nearly 70 altogether and they were ended for 'technical reasons'; see reference [2], p 265). For the same purpose, I wrote a paper [1] in 1970, which had almost the same title as the present one. This has been updated many times, the last version opening book [2] published in 2001. In the years that have passed since then, not very many new results have been reported. Such a shortage can be accommodated. Another aspect is worse—over the 30 years my presentation has become morally antiquated. It is difficult to formulate this point clearly but this is the fate of all papers and books of this kind. Incidentally, when I was young, a seminal work for me was O D Khvolson's book *The Physics of Our Days (New Concepts of Contemporary Physics in a Generally Accessible Presentation)* which appeared in 1932 as the fourth 'revised and updated' edition [6]. I think now, that this book was then already somewhat outdated in regard to the latest news (at that time it was quantum mechanics). And O D Khvolson (1852–1934) was, at that time, even a little younger than I am now. All in all, even if I now decided to write the necessary (in my opinion) book anew, I would not be able to do it. But as the well-known proverb says, 'let well alone', and in the hope, perhaps illusory, that my project, if not good, will still useful, I am writing the present article. The 'list of 2001' including the problems which 'seem now to be especially important and interesting' is proposed in the next section. I believe that every physicist should be acquainted with this physical minimum—to know, even if rather superficially, the outlines of each of the enumerated questions.

It need not be emphasized that singling out 'especially important and interesting' questions is not in the least equivalent to a declaration that a great many other physical problems are unimportant or uninteresting. This is obvious but a habit of overcautiousness forces me to make a few more remarks. 'Especially important' problems are distinguished not because others are unimportant but because within the period under discussion they are the focus of attention and follow, to some extent, the main trends of the field. Tomorrow these problems may find themselves in the rear and other problems will come forward in their place. Singling out some problems as 'especially important' is of course subjective and different opinions are needed. But I would like to reject resolutely the reproach that such a distinction is dictated by some personal scientific preferences and personal activity in physics. So, in my scientific activity, the questions associated with radiation by uniformly moving sources [7] (see article 2 in the present volume) were and are most dear to me but I did not and do not include them in the 'list'. Unfortunately, I had to face criticism of the 'list' for the reason that it had not included a subject which was interesting for the critic. I recall in this connection how my senior friend A L Mints (1895–1974) told me after the appearance of paper [1]: "If you had written this paper before you were elected Academician, you would have never been elected." He may have been right but I still believe in the wider outlook of my colleagues.

1.2   The list of 'especially important and interesting problems' in 2001

There is a well-known saying that the proof of the pudding is in the eating. This is why I immediately proceed to the aforementioned 'list'.

(1) controlled nuclear fusion,
(2) high-temperature and room-temperature superconductivity (HTSC and RTSC),
(3) metallic hydrogen and other exotic substances,
(4) two-dimensional electron liquid (anomalous Hall effect and some other effects),

(5) some questions of solid-state physics (heterostructures in semiconductors, metal–dielectric transitions, charge and spin density waves, mesoscopics),

(6) second-order and related phase transitions, some examples of such transitions, cooling (in particular, laser cooling) to superlow temperatures, and Bose–Einstein condensation in gases,

(7) surface physics and clusters,

(8) liquid crystals, ferroelectrics, and ferrotoroics,

(9) fullerenes and nanotubes,

(10) the behavior of matter in superstrong magnetic fields,

(11) nonlinear physics, turbulence, solitons, chaos, and strange attractors,

(12) rasers, grasers, superhigh-power lasers,

(13) superheavy elements and exotic nuclei,

(14) mass spectrum, quarks and gluons, quantum chromodynamics, and quark-gluon plasma,

(15) unified theory of weak and electromagnetic interactions, $W^{\pm}$- and Z^0-bosons and leptons,

(16) The Standard Model, grand unification, superunification, proton decay, neutrino mass, and magnetic monopoles,

(17) fundamental length, particle interaction at high and superhigh energies, and colliders,

(18) non-conservation of CP invariance,

(19) nonlinear phenomena in vacuum and in superstrong magnetic fields and phase transitions in vacuum,

(20) strings and M-theory,

(21) experimental verification of the general theory of relativity,

(22) gravitational waves and their detection,

(23) the cosmological problem, inflation, Λ-term and 'quintessence' (dark energy), and the relationship between cosmology and high-energy physics,

(24) neutron stars and pulsars and supernova stars,

(25) black holes and cosmic strings (?),

(26) quasars and galactic nuclei and the formation of galaxies,

(27) the problem of dark matter (hidden mass) and its detection,

(28) the origin of ultrahigh-energy cosmic rays,

(29) gamma-ray bursts and hypernovae,

(30) neutrino physics and astronomy and neutrino oscillations.

Singling out 30 particular problems (more precisely, items in the 'list') is, of course, absolutely conditional. Moreover, some of them might be divided. In [1], there were 17 problems, in [2] they were already 23. In note [8], 24 problems were listed. In the letters that were published in *Physics Today* in respect of this note, the opinion [9] was expressed that the 'list' should also have included star formation, atomic and molecular physics (true, I am unaware of what exactly was meant), and the question of exceedingly accurate measurements. I had to get acquainted with other suggestions that the 'list' should be extended. Some of them have been taken into consideration but others (for example, those concerning quantum computers, the 'optics' of atomic beams, semiconductor devices, etc) I had to ignore. In preparing the English translation in the middle of 2003, I have not had the feeling that something definitely should be changed in this 'list'.

Any 'list' is undoubtedly not a dogma, something can be discarded and something added depending on the preferences of lecturers and authors of corresponding papers. More interesting is the evolution of the 'list' with time as it reflects the process of the development of physics. In the 1970–71 'list' [1], quarks were given only three lines in the enumeration of the attempts to explain the mass spectrum. This did not testify to my perspicacity, which was admitted in [2]. However, at that time (in 1970), quarks were only five or six years old (I mean the age of the corresponding hypothesis), and the fate of the concept of the quark was indeed vague. Now the situation is, of course, quite different. True, the heaviest t-quark was discovered only in 1994 (its mass, according to the data of 1999, is $m_t = 176 \pm 6$ GeV). The 'list' [1] naturally contains no fullerenes which were discovered in 1985 [10], no gamma-ray bursts (the first report on their discovery was published in 1973; see [2] and later). High-temperature superconductors were synthesized in 1986–87 but in the 'list' [1] this problem was nonetheless considered rather thoroughly for it had been under discussion since 1964. Generally, much has been done in physics in the past 30 years but, I believe, not very much essentially new has appeared with the exception of astrophysics and cosmology. In any case, the 'lists' in [1, 2], as well as that presented here, characterize, to a certain extent, the development and state of physical and astronomical problems from 1970–71 to the present day.

1.3 Some comments (macrophysics)

In [2], the paper with the same title occupies 130 pages. There, each problem of the 'list' is commented on. I cannot do the same here and, therefore, I shall restrict myself to separate, sometimes fragmentary, remarks and comments.[2] The basic goal is to elucidate the development of physics in the years since [2] was published.

---

[2] A large number of references to the literature could be given in connection with practically each item. But this seems to be out of place here. Moreover, the problem of priority would arise and I would not like to touch upon this here. I have tried to make as few references as possible. Sometimes they are of an incidental character and preference was naturally given to papers published in *Uspekhi Fizicheskikh Nauk* (*Physics–Uspekhi*), *Physics Today*, and *Physics World*.

The problem of controlled thermonuclear fusion (number 1 in the 'list') has not yet been solved, although it is already 50 years old. I remember how the work in this direction was started in the USSR in 1950. A D Sakharov and I E Tamm told me about the idea of a magnetic thermonuclear reactor and I was glad to set myself to the solution of this problem because, at that time, I had almost nothing to do in the elaboration of the hydrogen bomb (I have written about all this in article 18). This work was then considered to be supersecret (it was stamped as 'top secret, special file'). Incidentally, I thought at that time and much later that the interest in the thermonuclear problem in the USSR was due to the desire to create an inexhaustible energy source. However, as I have been told by I N Golovin, a thermonuclear reactor was then interesting for 'those who needed it'— largely for quite a different reason—as a source of neutrons (n) for the production of tritium (t) (evidently, with the help of the reaction ^{6}Li + n $\rightarrow$ t + ^{4}He + 4.6 MeV). In any event, the project was treated as so secret and important that, in 1951, I was debarred from participating in it—in the secret department, they simply stopped giving me the working notebooks and my own reports on this work (see comment 11 to article 18). That was the apex of my 'specialized activity'. Fortunately, by the times of Khrushchev, I V Kurchatov and his colleagues had realized that the thermonuclear problem could not be quickly solved and, in 1956, it was declassified and opened to the public. As a reaction to what I had experienced, I published my thermonuclear reports [11] in 1962, although I do not at all claim that I have done anything significant in this field.

Abroad, thermonuclear studies also began (approximately at the same time) mostly as secret, and their declassification in the USSR (which was quite non-trivial for our country at that time) played a great positive role— the discussion of the problem became the subject of international conferences and collaboration. But 50 years have already passed and no operating (energy-producing) thermonuclear reactor has been accomplished and we shall probably have to wait for another 15 years or longer (see [2, section 1]: the latest review on this subject which is known to me and easy to access is paper [12]; and for reference to the Soviet papers see [13]). Work on thermonuclear fusion is being carried out all over the world on a fairly wide front. An especially advanced system—a favorite—is the tokamak. The International Thermonuclear Experimental Reactor (ITER) project has been elaborated for several years. This is a gigantic tokamak which will cost nearly 10 billion dollars. It was supposed to be accomplished by 2005 as a real prototype of the thermonuclear reactor of the future generation. But now that the project is mostly complete, financial difficulties have arisen. Moreover, some physicists find it reasonable to think about alternative smaller-scale constructions first (see [12] and, e. g., [14]). This question is being discussed on the pages of *Physics Today* and other journals but it does not seem pertinent to dwell on it in the present article.[8*]

Generally, the possibility of creating a real thermonuclear reactor is now beyond doubt and the center of gravity of the problem, as I see it, has shifted towards the engineering and economical spheres. However, such a titanic and unique installation as ITER or one of the competing ones remain, of course, interesting for physics.

As for alternative ways of fusing light nuclei for obtaining energy, the hopes for the possibility of 'cold thermonuclear fusion' (e. g. in electrolytic elements) were abandoned [134]; and muon catalysis is very elegant (and should, I think, be elucidated in a general physics course) but seems to be an unrealistic energy source, at least when not combined with uranium fission, etc. There also exist projects with a sophisticated use of accelerators but I am unaware of any success in this field. Finally, inertial nuclear fusion, and specifically 'laser thermonuclear fission', is possible. Gigantic corresponding installations are being constructed but they are not widely known because of secrecy—they are obviously intended to imitate thermonuclear explosions. However, I may simply be ignorant of the situation. In any case, the problem of inertial fusion is important and interesting [218].

The problem of controlled nuclear fusion is now technical rather than physical. In any case, there is no enigma here typical of a number of unsolved physical problems. That is why there exists an opinion that the problem of nuclear fusion may be excluded from our 'list'. This is, however, an exceedingly important and still unsolved problem and, therefore, I would discard it from the 'list' only after the first effective thermonuclear reactor starts operating.

We now proceed to high-temperature and room-temperature superconductivity (abbreviated as HTSC and RTSC, problem 2). To those who are not closely engaged in solid state physics, it may seem that it is time to discard the HTSC problem from the list. In 1970 [1], high-temperature superconductors had not yet been created and to obtain them was a dream which was then mocked here and there. But, in 1986–87, such materials were created and even though they are included in [2] by inertia, maybe it is time to place them among the numerous other substances investigated by physicists and chemists? But this is not the case. Suffice it to say that the mechanism of superconductivity in cuprates (the highest temperature $T_c = 135\,\mathrm{K}$ was reached for $HgBa_2Ca_2Cu_3O_{8+x}$ without pressure, while under a rather high pressure we already have $T_c \approx 164\,\mathrm{K}$ for this cuprate) remains unclear [15–17]. It seems undoubted, to me personally in any case, that a very significant role is played by the electron–phonon interaction with strong coupling but this is not enough. 'Something else' is needed, perhaps an exciton or spin interaction. In any case, the question is open in spite of the great efforts made to investigate HTSC (about 50 000 publications on the subject have appeared in the ten years after HTSC was obtained). But the main question, which is of course intimately related to the preceding one, is the possibility of creating RTSC. Such a possibility does not face any contradiction [15] (see articles 6 and 7 in this volume; many new results are referenced therein) but success is not guaranteed. The situation is here quite similar to that observed before 1986–87 in HTSC.

In the 'list' of [2, section 2], we also find the problem of superdiamagnetism, i.e. the possibility of creating an equilibrium non-superconducting diamagnetic with magnetic susceptibility χ close to $\chi = -1/4\pi$ (it is a well-known fact that, for superconductors, one can formally assume $\chi = -1/4\pi$). From experiment we know that there exist diamagnetics with $\chi = -(10^{-4}-10^{-6})$. Materials with $\chi = -(0.1/4\pi - 0.01/4\pi)$ can be called superdiamagnetics. I do not know why they might not exist but I cannot say anything sensible about this question.

Note, however, that, in this respect, ferrotoroics, which will be discussed later (see problem 8; see also [162, 163]) are under special 'suspicion'.

Metallic hydrogen (problem 3) has not yet been obtained even under a pressure of 3×10^6 atm (at low temperatures). However, the study of molecular hydrogen at high pressure has revealed a whole number of unexpected and interesting features of this substance [18, 142]. Moreover, under compression by shock waves at a temperature of about 3000K, the transition to a metallic (i.e. well-conducting) liquid phase was clearly observed.

Water (more precisely, H_2O) and a number of other substances also exhibited some peculiarities at a high pressure [18]. In addition to metallic hydrogen, fullerenes may also be attributed to 'exotic' substances. Recently, along with the common fullerene C_{60}, the study of fullerene C_{36} began: this substance may have a very high superconducting transition temperature under doping [19]. Examples of exotic substances are numerous.

In 1998, the Nobel Prize in physics was awarded for the discovery and explanation of the fractional quantum Hall effect. Incidentally, the discovery of the integer quantum Hall effect also won the Nobel Prize (1985). I mention the Nobel Prizes here and later not because of some extraordinary respect for them (sometimes one can observe an excessive respect for these prizes). As any deed of the human, awards should not be raised to the rank of the absolute. Even the best of the awards are, in most cases, somewhat conditional, and sometimes errors occur (see, for example, [20, 21] and comment 3). But, on the whole, the Nobel Prizes in physics have gained immense authority and are the landmarks fixing the progress in physics.

The fractional quantum Hall effect was discovered in 1982 (the discovery of the integer quantum Hall effect goes back to 1980). The quantum Hall effect is observed when a current runs in a two-dimensional electron 'gas' (in fact, certainly in a liquid because the interaction between the electrons is substantial, particularly for the fractional effect). The 'system' (a two-dimensional conducting layer on a silicon surface) is, of course, in the magnetic field perpendicular to this current, as under the usual Hall effect. I shall restrict myself here to [22, 23] and the remark that the unexpected and particularly interesting feature of the fractional quantum Hall effect is the existence of quasi-particles with a fractional charge $e^* = (1/3)e$ (e is the electron charge) and other fractional charges (see also [143, 144]). It should be noted that a two-dimensional electron gas (or, generally, a liquid) is interesting not only in respect of the Hall effect but also in other cases and conditions [24, 25, 164, 241].

Problem 5 (some problems of solid state physics) is currently absolutely boundless. In the 'list', I only sketched (in brackets) some possible topics and if I had to deliver a lecture, I would dwell on heterostructures (including 'quantum dots') and mesoscopics just because I am acquainted with these questions better than with some other ones from this area. I shall only mention the whole *Uspekhi Fizicheskikh Nauk* (*Physics–Uspekhi*) issue [24] devoted to this subject (see also [219]) and refer to the recently noticed paper on the metal–dielectric transition [26]. Until recently in electronics (specifically, in superconductors), the motion of charges was mainly considered, while spin variables, so to say, remained in the shadow. Today, much attention is also given to the spins of the carriers: their behavior is studied and provides information. The term 'spintronics' was even proposed. See, for example, [175, 176, 220] about this. It is not at all easy to choose what is most interesting, so the reader and the student should be helped in this respect.

As to problem 6 (phase transitions, etc), I would like to add to [2, section 5] the following. The discovery of low-temperature superfluid ^{3}He phases won the 1996 Nobel Prize in physics [27]. Particular prominence for the past three years has been given to Bose–Einstein condensation (BEC) of gases. These works are undoubtedly of great interest but I am sure that the 'boom' around them was largely due to the lack of historical knowledge. It was as early as in 1925 that Einstein paid attention to BEC [28] and now this question is naturally included in textbooks (see, for example, [29, section 62]). Then, true, BEC had long been ignored and sometimes even called into question. But those are bygone times, especially after 1938 when F London associated BEC with the superfluidity of ^{4}He [30]. Helium II is, of course, a liquid and BEC does not manifest itself, so to say, in a pure form. The desire to observe BEC in a rarefied gas is quite understandable and justified but one should not think of it as a discovery of something unexpected and essentially new in physics (see a similar remark in [31]). The observation of BEC in gases, such as Rb, Na, Li, and finally H, which was made in 1995 and later on, was, in contrast, a great achievement of experimental physics. It only became possible owing to the development of methods of cooling gases to superlow temperatures and keeping them in traps (which, by the way, won the 1997 Nobel Prize in physics [32]).[3*] The realization of BEC in gases initiated a stream of theoretical papers (see reviews [33, 34, 225]; new articles permanently appear, in particular, in *Physical Review Letters*[3]). In a Bose–Einstein condensate, atoms are in a coherent state and interference phenomena can be found, which has led to the appearance of the concept of an 'atomic laser' (see, for example, [35, 36]). BEC in a two-dimensional gas [128, 165] is also very interesting. Problems related to BEC are considered in [223].

The so-called quantum phase transitions also should be mentioned. They occur at $T = 0$, say, as a result of changing pressure (see, for example, [166]; see also [210]).

Problems 7 and 8 touch upon numerous questions which I have not followed and cannot, therefore, distinguish anything new and important. I only wish to point out the acute and justified interest in clusters of various atoms and molecules (i.e. formations containing a small number of particles [135]). The studies on liquid crystals and simultaneously ferroelectrics should also be mentioned. I shall only refer to paper [37] on this subject. The study of thin ferroelectric films [38, 167, 168, 224, 243] is also attractive.

Materials having a spontaneous magnetization M_s (ferromagnetics) and spontaneous electric polarization P_s (pyroelectrics, including ferroelectrics) are well known in physics. Less popular, though known for some years, are ferrotoroics—materials which have spontaneous toroidal moment (more precisely, the toroidal moment density) T_s. It is not the right place here to explain what the toroidal moment is; the interested reader is referred to paper [162]. I restrict myself to referring to paper [169] and the note that ferrotoroics studies have brought no impressive results up to now. However, these studies relate to the hopes of discovering superdiamagnetics mentioned earlier after the comments to problem 2 (HTSC and RTSC).

Fullerenes (problem 9) have already been casually mentioned earlier (see also [10, 19]) and, along with carbon nanotubes, this branch of studies is flourishing [39, 221].

I have not heard anything new either of matter in superstrong magnetic fields (specifically, in the crust of a neutron star) or of the simulation of the corresponding effects in semiconductors (problem 10). Such a remark should not discourage or cause the question of why these problems were introduced into the list. First, in [2, section 8], I tried to elucidate the physical meaning of this problem and to explain why it has, in my opinion, such a charm for a physicist: there are neither particular grounds nor especially spare room to repeat myself here. Second, understanding the importance of a problem is not necessarily related to a sufficient acquaintance with its current state. My whole 'program' is exactly aimed at stimulating interest and prompting specialists to elucidate the state of a

---

[3] This journal has now become the most prestigious in the field of physics. It appears weekly, each issue contains about 60 articles occupying not more than four pages each (with rare exceptions). For example, volume 81 covering the second half of 1998 amounts to nearly 6000 pages.

problem to non-specialists in accessible papers and lectures.

As far as nonlinear physics (problem 11 in the 'list') is concerned, the situation is not as in the previous case. There is a lot of material, *Physical Review Letters* publish papers on this subject in every issue, they even have a special section partly devoted to nonlinear dynamics. Moreover, nonlinear physics and, in particular, the problems listed in item 11 are also presented in other sections of the journal; in total, up to 10–20% of the whole journal is devoted to nonlinear physics (see, e. g., [40]). Generally, it should perhaps be emphasized once again, in addition to [2, section 10], that attention to nonlinear physics is increasingly high. Clearly, this entails the appearance of numerous text-books and monographs (see, for example, [41]). This is largely connected with the fact that the use of modern computer facilities allows the analysis of problems whose investigation was previously no more than a dream.

It is not for nothing that the 20th century was sometimes called not only the atomic age but the laser age as well. The perfection of lasers and the extension of their application are in full swing. But problem 12 concerns not lasers in general but, first of all, superpower lasers. For example, an intensity (power density) $I \sim (10^{20}$–$10^{21})$ W cm^{-2} has already been attained. With such an intensity, the electric field strength is of the order of 10^{12} V cm^{-1}, i. e. this field is two orders of magnitude stronger than the proton field at the ground level of the hydrogen atom. The magnetic field reaches 10^9–10^{10} Oe [42] and very short pulses of duration down to 10^{-15} s (i. e. a femtosecond) are used. Employment of such pulses opens a lot of possibilities, in particular, for obtaining harmonics already lying in the X-ray band and, accordingly, X-ray pulses with a duration of attoseconds $(1 \text{ a} = 10^{-18}$ s$)$ [42, 43, 225]. A related problem is the creation and use of rasers and grasers which are analogs of lasers in X-ray and gamma-ray bands, respectively. Advances in grasers can be expected due to the idea of employing lasing without inversion [145].

Problem 13 is that of nuclear physics. This is, of course, a vast area which is not very familiar to me. For this reason, I distinguished only two points. First, I point out far transuranium elements in connection with the hopes that some isotopes have long lives owing to shell effects (as an example of such an isotope, the nucleus with $Z = 114$ and the number of neutrons $N = 184$, i. e. mass number $A = Z + N = 298$ was pointed out in the literature). The known transuranium elements with $Z < 114$ live only seconds or fractions of a second. The indications of the existence in cosmic rays of long-lived (millions of years) transuranium nuclei, which appeared in the literature (see [2, section 11]), have not yet been confirmed. At the beginning of 1999, a preliminary report [125] appeared on the fact that the 114th element with mass number 289 and a lifetime of nearly 30 s had been synthesized in Dubna (see [170] for further development of this work). Therefore, there are hopes that the element $\binom{114}{298}$ will actually prove to be very long-lived. Second, I mentioned 'exotic' nuclei. These are nuclei of nucleons and antinucleons, some hypothetical nuclei with a heightened density, to say nothing of nuclei having a non-spherical shape and some other specific features. Included here are the problems of quark matter and quark-gluon plasma (see, e. g., [44, 136–138, 146, 171] and references therein).

1.4   Some comments (microphysics)

Problems 14–20 pertain to the field which I refer to as microphysics although it would apparently be more correct to call it elementary particle physics. This name, once seldom used, was considered outdated. The reason was, in particular, that nucleons and mesons at a certain stage were considered to be elementary particles. Now they are known to consist (true, in a conditional sense) of quarks and antiquarks. Quarks, too, were sometimes supposed to consist of other tiny particles (preons, etc). However, such hypotheses are totally ungrounded today and the 'matryoshka' [Russian doll]—the division of matter into successively smaller parts—must one day be exhausted. In any event, we think today that quarks are indivisible and, in this sense, elementary. Without antiquarks, they include six flavors: u (up), d (down), c (charm), s (strange), t (top), and b (bottom); antiquarks are denoted by the same letters but with a bar ($\bar{u}$, etc). Next, leptons are also elementary: the electron and the positron (e^- and e^+), $\mu^{\pm}, \tau^{\pm}$, and the corresponding neutrinos ν_e, ν_μ, ν_τ and antineutrinos $\bar{\nu}_e, \bar{\nu}_\mu, \bar{\nu}_\tau$. Finally, the four vector bosons (the photon γ, the gluon g, Z^0, and $W^{\pm}$) are elementary. I shall not give here a more detailed account of the state of elementary particle physics as a whole because I may refer, besides [2], to the review "The present state of elementary particle physics" published in *Uspekhi Fizicheskikh Nauk* (*Physics–Uspekhi*) in 1998 [45] (see also [147, 160, 171–173]). All that is written there, I attribute to the physical minimum. I shall, however, make some comments and add some points. One of the most topical problems (in [45] it is even called problem number 1) of elementary particle physics is the search for and, as everybody hopes, the discovery of the Higgs—the scalar Higgs-boson with spin zero. According to the estimates, the Higgs mass is below 1000 GeV or rather even below 200 GeV. The Higgs is now being sought and will be sought on the available accelerators and those being reconstructed (in CERN and Fermilab). The main hope of high-energy physics (may also be in the search for Higgs) is the LHC accelerator (Large Hadron Collider) which is now being built in CERN. An energy of 14 TeV (in the center-of-mass of colliding nucleons) will be reached but obviously only in 2006–07. Another very important problem (number 2, according to [45]) is the search for supersymmetric particles (see [159] and later). I cannot but point out the problem of CP violation and, by virtue of CPT invariance (joint spatial inversion P, charge conjugation C, and time reversal T), non-conservation of T-invariance (non-invariance under the time reversal $t \rightarrow -t$). This is, of course, a fundamental question, in particular, from the point of view of the explanation of irreversibility of physical processes (see section 1.6). CP non-conservation was discovered in 1964 in an example of meson decay $K_2^0 \rightarrow \pi^+ + \pi^-$. Incidentally, this discovery won the 1980 Nobel Prize in physics. At the same time, the known processes with CP non-conservation have a small probability (compared to processes that conserve CP invariance). The processes with CP non-conservation are under study; however, their nature is not yet clear. One more process with CP non-conservation, also with a small probability, has recently been discovered [46]. Finally, CP non-conservation is being sought in B-meson decay [47, 172]. Proton decay has not yet been found. According to [126], the mean proton lifetime when determined from the reaction $p \rightarrow e^+ + \pi^0$ is longer than 1.6×10^{33} years. The neutrino mass which is mentioned among the other items of problem 16 will be touched upon later in the discussion of problem 30 (neutrino physics and astronomy).

I shall dwell here on problem 17 or, more concretely, on the fundamental length. 'Elementarists', as those specialized in elementary particle physics are sometimes called, will perhaps scornfully shrug their shoulders wondering what problem is this. If I began compiling the 'list' yesterday (more precisely, as early as in 1999), I would probably not mention such a problem because it was many years ago that it 'sounded at the top of its voice' and was pointed out in [1] and then also in [2]. The point is that, at the end of the 1940s, a technique (the renormalization method, etc; see, e.g., [48]) was developed allowing an unlimited use of quantum electrodynamics. Before this, calculations had sometimes yielded divergent expressions and, to obtain final results, one had to make a cutoff at a certain maximum energy E_{f_0} or at a corresponding length $l_{f_0} = \hbar c / E_{f_0}$ (here $\hbar = 1.055 \times 10^{-27}$ erg s is the quantum constant). The most frequently encountered values were $l_{f_0} \sim 10^{-17}$ cm and $E_{f_0} = \hbar c / l_{f_0} \sim 3$ erg $\sim 10^{12}$ eV $= 1$ TeV. Approximately the same values correspond to the highest energies (in the center-of-mass frame) and the lowest 'impact parameters' reached on modern accelerators. Given this, 'everything is all right', i.e. the conventional physics, for example, quantum electrodynamics, works well. This implies that up to distances $l_{f_0} \sim 10^{-17}$ cm (actually, the length 10^{-16} cm is more often mentioned) and times $t_{f_0} \sim l_{f_0}/c \sim 10^{-27}$ s, the existing space–time concepts are valid. And what is going on at smaller scales? Such a question, along with the difficulties encountered in the theory, led to the hypothesis of the existence of a fundamental length l_f and time $t_f \sim l_f/c$ for which a 'new physics' makes its appearance and, in particular, some unusual spacetime concepts ('granular spacetime' and other things). There are no grounds now to introduce the length $l_f \sim 10^{-17}$ cm (see, however, later). However, another fundamental length, namely the Planck or gravitational length $l_g = \sqrt{G\hbar/c^3} = 1.6 \times 10^{-33}$ cm (here $G = 6.67 \times 10^{-8}$ cm g^{-1} s^{-2} is the gravitational constant) is known and plays an important role in physics: this length corresponds to the time $t_g = l_g/c \sim 10^{-43}$ s and energy $E_g = \hbar c/l_g \sim 10^{19}$ GeV. The Planck mass $m_g = E_g/c^2 \sim \sqrt{\hbar c/G} \sim 10^{-5}$ g is also frequently used. The physical meaning of the length l_g is that, on smaller scales, one cannot already apply the classical relativistic theory of gravity and, in particular, the general relativity (GR) whose construction was accomplished by Einstein in 1915.[4] The point is that, for $l \sim l_g$ and especially on scales $l < l_g$, quantum fluctuations of the metric g_{ik} are already large. Hence, the quantum theory of gravity which should be used here has not yet been completed. So, the length l_g is a certain fundamental length which limits the classical concepts of spacetime. But can one be sure that these classical concepts do not stop 'working' before that, at a fundamental length $l_f > l_g$? As has already been said, we definitely have $l_f < l_{f_0} \sim 10^{-16} - 10^{-17}$ cm but this value of l_{f_0} is 16 orders of magnitude larger than l_g. Physicists have become accustomed to giant-scale extrapolation, for instance, to the assumption that laws obtained on the Earth from various data are identical throughout the whole Universe or at any rate in colossal spacetime regions. An example of such a far-reaching extrapolation is the hypothesis that over the entire interval between $l \sim l_{f_0} \sim 10^{-17}$ cm and $l \sim l_g \sim 10^{-33}$ cm, no other fundamental length l_f exists. Such a hypothesis now seems natural but it has not been proved. The latter should be borne in mind and, for this reason, I included this problem in the 'list' published in 1999 [3]. I did it, so to say, just in case. But already in the second edition of this book (in 2001), I had to make a special remark—that the problem of fundamental length $l_f > l_g$ had become topical again. The point is that for a long time the hypothesis—that, in addition to the usual for us three space dimensions x, y, z and time t, in the real world somehow other 'dimensions' appear—has been discussed (see later). However, until recently, it has been assumed that the fifth and other space dimensions are compactified with a characteristic dimension l_g, i.e. roughly speaking, are 'rolled' into narrow tubes with a radius of the order of l_g. But, in recent years, the possibility has been widely debated that one (and maybe more than one) of the additional dimensions are compacted not with the radius l_g but with another much larger radius l_c. As is clear from the text, this radius, in a certain sense, will play the role of the fundamental length l_f (i.e. $l_c \sim l_f$). I have seen papers in which the additional dimension or several dimensions, by assumption, have a 'radius' $l_c \gg l_g$ and this affects the behavior of the gravitational field. For example, the presence of the

---

[4] In GR, a gravitational field is completely described by the metric tensor g_{ik}. Furthermore, g_{ik} obeys quite definite equations (see, e.g., [49]). Many classical relativistic theories of gravity in which other variables (e.g. a certain scalar field φ) besides g_{ik} appear together with higher-order derivatives exist.

scale l_c could change the dependence of the gravity attraction force on the distance between interacting bodies (particles, etc). Specifically, Newton's law for the potential energy $\varphi \propto 1/r$ at small r becomes steeper (now it is only known that the law $\varphi \propto 1/r$ holds at $r \gtrsim 0.01$ cm). See [148, 226] for more detail. I am convinced that these studies will be the focus of attention in the foreseeable future [174, 211, 227].

As a matter of fact, however, the "length is attacked" on two sides. For comparatively low energies, this is the principle underlying the new accelerators (colliders), primarily the already mentioned LHC (see [45, 50] and chapters 11 and 12 in [51]). This collider, as previously mentioned, is going to reach an energy $E_c = 14$ TeV (in the center-of-mass frame) which corresponds to the length $l = \hbar c / E_c = 1.4 \times 10^{-18}$ cm. In cosmic rays, particles with a maximum energy $E \simeq 3 \times 10^{20}$ eV have been registered (in the laboratory frame, a proton with such an energy, when colliding with a nucleon at rest, in the center-of-mass frame, has an energy $E_c \sim 800$ TeV and $l_c \sim 10^{-20}$ cm). Such particles are, however, very rare, and it is impossible to use them directly in high-energy physics [52, 53]. Lengths comparable with l_g arise in cosmology (and, in principle, inside black-hole event horizons). Energies frequently encountered in elementary particle physics are $E_0 \sim 10^{16}$ GeV. They figure in the yet uncompleted theory of 'Grand Unification'—the unification of electroweak and strong interactions. The corresponding length is equal to $l_0 = \hbar c / E_0 \sim 10^{-30}$ cm and this is still three orders of magnitude larger than l_g. It is obviously very difficult to say what is going on at scales between l_0 and l_g. It may be that a certain fundamental length l_f such that $l_g < l_f < l_0$ is hidden here. Today such an assumption is pure speculation (see, however, earlier and [148, 174]).

As to the terminology, the theory of strong interaction is called quantum chromodynamics (QCD). As has already been said, the scheme uniting the electromagnetic, weak, and strong interactions is referred to as 'Grand Unification'. At the same time, the currently used theory of elementary particles which consists of the theory of electroweak interaction and QCD is called the Standard Model. Finally, the theories with Grand Unification (which is not yet ultimately shaped) which are generalized so as to include gravity are called superunification. No satisfactory superunification has yet been constructed. Superstring theory discussed later claims to be a superunification but this goal has not yet been achieved.

As regards the set of problems 19, one may assert that they are fairly topical but I do not know what is to be added to the material of [2, section 17]. I may have missed some news worthy of note (I shall only point to paper [54] devoted to phase transitions in the early Universe). Incidentally, in [2, section 7], I quoted the remark made by Einstein as far back as 1920 [55]: "the general theory of relativity endows space with physical properties, and so ether does exist in this sense...". Quantum theory 'endowed space' also with virtual pairs of various fermions and zero oscillations of electromagnetic and other Bose fields. This seems to be known to everyone. Nevertheless, *Physics Today*—the organ of the American Physical Society and of another nine analogous societies—begun in 1999 with the article "The persistence of ether" devoted to speculations concerning the physical vacuum named the ether [56]. The 'new ether' now discussed in cosmology under the term 'dark energy' will be touched upon later in relation to problem 23.

Before proceeding to astrophysical and related problems (items 21–30 in the 'list'), I shall dwell on problem 20: strings and M-theory. This is so to say the leading direction (subject) in theoretical physics today. Incidentally, the term 'superstrings' is frequently employed instead of the term 'strings', first, not to confuse them with cosmic strings (see later about problem 25) and, second, to emphasize the use of the concept of supersymmetry (see [159]). In supersymmetric theory, each particle has its partner with other statistics, for example, a photon (a boson with spin unity) corresponds to a photino (a fermion with spin $1/2$), etc. It should be noted at once that supersymmetric partners (particles) have not yet been discovered. Their mass is evidently not less than 100–1000 GeV. The search for these particles is one of the principal problems of experimental high-energy physics both on existing accelerators and those under reconstruction and on LHC.

Theoretical physics cannot yet answer a whole number of questions: for example, how the quantum theory of gravity should be constructed and united with the theory of other interactions, why there exist apparently only six types (flavors) of quarks and six leptons, why the electron neutrino mass is very small, why μ- and τ-leptons differ in their mass from the electron precisely by the factor known from experiment; how the fine structure constant $\alpha = e^2/\hbar c \approx 1/137$ and a number of other constants can be determined from the theory, and so on. In other words, grandiose and impressive as the achievements of physics are, there remain more than enough unsolved fundamental problems. String theory has not yet answered such questions but it promises success in the desired direction. Since I cannot refer to a sufficiently accessible paper on strings, I planned to clarify some essential points. It turned out, however, that I will be unable to do that briefly and at a proper level. I would merely mention the popular reviews [57–60] and [51, chapter 13]; for modern review see [213]; see also [228]. I shall only make some remarks.

In quantum mechanics and quantum field theory, elementary particles are considered point-like. In string theory, elementary particles are oscillations of one-dimensional objects (strings) with characteristic dimensions $l_s \sim l_g \sim 10^{-33}$ cm (or, say, $l_s \sim 100 l_g$). Strings may have a finite length (a 'segment') or may be ring-like. Strings are considered not in the normal four-dimensional space but in multidimensional spaces with, say, ten or eleven dimensions. The theory is supersymmetric. Changing point-like particles for non-point-like ones is not at all a new idea and its main difficulty is the relativistic formulation. As an example, I dare refer to the paper by I E Tamm and myself [61] (see also [62]). No progress had been made in this direction before string theory. The idea of multidimensional spaces, i.e. the introduction of the fifth and higher dimensions, is much older (the Kaluza–Klein theory [63, 64]; see [65, p 296]) but before string theory it had not led to any physical results either. In string theory, however, one can also speak mainly of 'physics hopes', as L D Landau would say, rather than results. But what do we mean by results? The mathematical constructions and the discovery of various symmetry properties are also results. As concerns physics, string theory has not yet given answers to any of the previously listed questions. This did not prevent physicists engaged in the study of strings from speaking not only about the 'first superstring revolution' (1984–85) but also about the 'second superstring revolution' (1994–?) [58].[5] A not very modest terminology has been applied to string theory—it was called 'The Theory of Everything'. It should be noted that string theory is not so young: according to [51, chapter 13], it is already 30 years old and 15 years have passed since the 'first superstring revolution' but no physically clear results have been obtained. In this connection, it is worth noting that the true revolution in physics—the creation of quantum mechanics, mostly by de Broglie, Schrödinger, Heisenberg, Dirac, and Bohr, did not last longer than five to six years (1924–30). It took Einstein eight years (1907–15) to create the general theory of relativity. But I do not set a great deal on these comments. The problems and questions of theoretical physics discussed here are deep and exceedingly involved and nobody knows how much time it will take to answer them. The theory of superstrings seems to be something deep and developing. Its authors themselves only claim the comprehension of some limiting cases and only speak of some hints of a certain more general theory which is called the M-theory. The letter M is chosen because this future theory is called magic or mysterious [57]. Superstring theory would noticeably fortify its position if supersymmetric particles were discovered, although there exist other ways of verifying it [60].

---

[5] In the book *The Structure of Scientific Revolutions* by Kuhn [66], which is widely known and popular in the West, the author writes: 'For me, a revolution is the form of a change including a certain type of reconstruction of the axioms by which the group is guided. But it need not necessarily be a large change or seem revolutionary to those who are outside a separate (closed) community consisting of not more 25 persons' [p 227]. If we adopt such a definition of revolution (I have already had an opportunity to express my opinion of it (see [2, p 201]), then in the majority of fields of physics revolutions break out every few years. Such an approach appears to me to be totally unacceptable.

1.5 Some comments (astrophysics)

Problems 21–30 in our 'list' relate to astrophysics but, in some cases, it is rather conditional. This particularly and even largely concerns the question of experimental verification of GR—the general theory of relativity (problem 21). It would be more logical to discuss the possibility of the analysis of relativistic effects in gravity (see, e. g., [67]). However, in view of the actually existing situation and the history of the corresponding studies, it would be more correct to bear in mind just the verification of GR—the simplest relativistic theory of gravity.[6] The effects of GR in the solar system are rather weak (the strongest effects are of the order of $|\varphi|/c^2$, where φ is the Newton gravitational potential; even on the Sun's surface, $|\varphi|/c^2 = GM_\odot/(r_\odot c^2) = 2.12 \times 10^{-6}$). It is for this reason that the verification which was successfully started in 1919 and has lasted until the present day has not led to accuracies which have become customary in atomic physics. According to such data reported at the 19th Texas "Relativistic Astrophysics and Cosmology" Symposium (December 1998), for the deflection of radio waves by the Sun, the ratio of the observed quantity to the calculated one in GR is $0.999\,97 \pm 0.000\,16$. The same ratio for the rotation of the Mercury perihelion is equal to 1.000 ± 0.001. So, GR has been checked in a weak gravitational field (for $|\varphi|/c^2 \ll 1$) with an error up to a hundredth of a percent; at this level, no deviations from GR were found. A further verification even in a weak field (for example, involving the terms φ^2/c^4) seems to be quite meaningful, although not stimulating because it is hardly probable to observe deviations from GR and the experiments are very complicated. Nevertheless, a whole number of projects of this type exist and will evidently be realized. The verification of the equivalence principle is a special question: the validity of this principle was confirmed up to 10^{-12} but this is not a new result [67].

Within the discussion of light deflection in the field of the Sun, some comments of a historical nature would not be uninteresting. Generally speaking, I do not think that priority questions should take a distinguished place in the lectures and articles whose program is presented here. The point is that such questions are often rather intricate and are decided in the literature in quite an accidental manner. Some statements are adapted by repetition only. To undertake a historical examination in each such case is a troublesome affair and draws attention from the physical essence of the matter. At the same time, some historical excursion might provide insight into a problem and, of course, pay tribute to the pioneers. The deflection of light beams in a gravitational field is a good example of this. A hint

---

[6] The theory in which the gravitational field is described by a certain scalar rather then the metric tensor g_{ik} as in GR is logically the simplest relativistic theory of gravity. But the scalar theory certainly contradicts experiment (for example, light beams are not deflected at all by the Sun in this theory).

of such an effect had already been given by Newton. In the framework of the corpuscular theory of light and in the assumption of equality or even proportionality of a heavy and inert mass, the existence of the deflection is obvious. The deflection of a light ray in the field of the Sun was calculated by Soldner as far back as 1801. The deflection angle turned out to be equal to

$$\alpha' = \frac{2GM_\odot}{c^2 R} = \frac{r_{g\odot}}{R} \tag{1.1}$$

where R is the impact parameter (the shortest distance between the ray and the center of the Sun) and $r_g = 2GM/c^2$ is the gravitational radius ($r_{g\odot} = 3 \times 10^5$ cm because the Sun mass is $M_\odot = 2 \times 10^{33}$ g).

Apparently being unaware of this result, Einstein in his first publication on the way to creating GR (1907), pointed out the deflection of rays and, in 1911, he obtained expression (1.1) on the basis of the then uncompleted GR which allowed only for the variation of the component $g_{00} = 1 + 2\varphi/c^2$. After the creation of GR in 1915, the final result was obtained in the same year:

$$\alpha = \frac{4GM_\odot}{c^2 R} = \frac{2r_{g\odot}}{R} = 1.725'' \frac{r_\odot}{R} \tag{1.2}$$

where $r_\odot = 7 \times 10^{10}$ cm is the Sun's photosphere radius. The distinction between (1.2) and (1.1) is due to the fact that the components of the metric tensor $g_{11} = g_{22} = -(1 - 2\varphi/c^2)$ are important, too. Expressions (1.1) and (1.2) differ exactly by a factor of two but the classical calculation is inconsistent (we mean the application of classical mechanics to a corpuscle moving at the velocity of light) and, therefore, the ratio 2 is accidental. The deflection of a light ray in the field of the Sun was first observed in 1919 and it confirmed the GR expression (1.2) though with a poor accuracy. Later specifications have been discussed previously (references are not given here, they can be found in [67, 68]).

In astrophysics, the deflection of rays in a gravitational field is used more and more frequently in the observation of 'lensing', i. e. focusing electromagnetic waves under the action of a gravitational field in application to galaxies (they lens the light and radio waves emitted by quasars and other galaxies) and stars (microlensing of more remote stars) [68]. This, of course, is not a verification of GR (the accuracy of measurements is rather low) but a use of it. I note that the lensing effect with its characteristic features was, to the best of my knowledge, first considered by Khvolson in 1924 [69] and Einstein in 1936 [70]. The characteristic cone arising upon lensing is called the Einstein cone or the Einstein–Khvolson cone. Only the latter term is correct, of course. Some time ago, the observation of gravitational lenses was believed to be practically impossible (see, for example, [70]). However, the lensing of a quasar was discovered in 1979. At the present time, the observation of lensing and microlensing is a rather widely employed astronomical method (see also [149]). In particular, the data on lensing allow the determination of the Hubble constant H_0. The result is in agreement with other data which are presented later.

The verification of GR in strong fields, i. e. for neutron stars (on their surface $|\varphi|/c^2 \sim 0.1$–0.3) and in the vicinity of black holes and generally for black holes, is very topical. A method [71] was recently proposed to verify GR in a strong field by oscillations of radiation in a binary star, one of whose components is a neutron star (see also [244]). Although black holes might be imagined in pre-relativistic physics, they are essentially a remarkable relativistic object. Black holes will be discussed later on but we can note now that their discovery confirms GR. However, as I understand the situation, one cannot state that what is known about black holes confirms exactly GR but not some relativistic theories of gravity that differ from GR.

A significant verification of GR [up to terms of the order of $(v/c)^5$] is the study of the binary pulsar PSR $1916 + 16$. This has shown that the energy lost by two moving neutron stars in a binary system is in perfect agreement with GR provided allowance is made for the gravitational radiation (whose intensity was calculated by Einstein in 1918). This work won the Nobel Prize in physics in 1993 [72].

The latter work leaves no doubt as to the existence of gravitational waves, though none of the qualified physicists had ever doubted it before (but the quantitative agreement with GR could not be guaranteed in advance). But there exists another problem (number 22 in the 'list')—the detection of gravitational waves coming from space. Technically, the problem is fairly complicated and giant installations are now being built to solve it. For example, the LIGO system (Laser Interferometer Gravitational-wave Observatory, USA) consists of two widely spaced 'antennae' each 4 km long. In this installation, it will be possible to detect a mirror displacement (occurring under the action of an incoming gravitational wave) of 10^{-16} cm and, further on, even smaller displacements. The LIGO and analogous installations now being constructed in Europe and Japan will be put into operation in the near future. This will be the starting point of gravitational wave astronomy (for more detail, see [73]). For orientation, I shall note that radio astronomy was born in 1931 and its intense development began after 1945. Galactic X-ray astronomy appeared in 1962. Gamma-ray astronomy and neutrino astronomy are still younger. The development of gravitational wave astronomy will open up the last known 'channel' through which we can receive astrophysical information.[4*] As in other cases, joint (simultaneous) measurements in the different channels will be of great importance. These may be, for instance, studies of the formation of supermassive black holes simultaneously in neutrino, gravitational-wave and gamma-ray channels [74]. I shall not write here in more detail about the reception of gravitational waves but refer the reader to [2, section 20] and, mainly, to [73] and the references therein.

The set of problems under item 23 in the 'list' represents perhaps the most crucial points in astrophysics. It also includes cosmology (not everybody will agree with such a classification but this does not change the essence of the matter). The cosmological problem is undoubtedly a grand problem. It has always attracted attention, for Ptolemy's and Copernicus' systems are none other than cosmological theories. In the physics of the 20th century, theoretical cosmology was created in the works of Einstein (1917), Friedmann (1922 and 1924), Lemaitre (1927), and many other scientists. But before the late 1940s, all the observations significant from the point of view of cosmology had been made in the optical range. Therefore, only the redshift law had been discovered and, thus, the expansion of the Metagalaxy had been established (the works by Hubble are typically dated 1929, although the redshift had also been observed before and not only by Hubble). The cosmological redshift was justly associated with the relativistic model of the expanding Friedmann Universe but the rapid development of cosmology began only after relic thermal radio emission with a temperature $T_r = 2.7$ K was discovered in 1965. At the present time, it is measurements in the radio wavelength band that play the most prominent role among the observations of cosmological importance. It is impossible to dwell here on the achievements and the current situation in the field of cosmology, the more so as the picture is changing rapidly and can only be discussed by a specialist. I shall restrict myself to the remark that, in 1981, the Friedmann model was developed to the effect that at the earliest stages of evolution (near the singularity existing in the classical models, in particular, those based on GR), the Universe was expanding (inflating) much more rapidly than in the Friedmann models. The inflation proceeds only over the time interval $\Delta t \sim 10^{-35}$ s near the singularity (recall that the Planck time is $t_g \sim 10^{-43}$ s and so the inflation stage can still be considered classically because the quantum effects are obviously strong only for $t \lesssim t_g$). However, the widespread inflation notions and models are criticized [178, 179; see also 204] and I have no clear opinion on this account. But the very existence of inflation is hard to deny and it is important that after the inflation, the Universe develops in agreement with Friedmann's scenario (at any rate, this is the most widespread opinion). The most important parameter of this isotropic and homogeneous model is the matter density ρ or, which is more convenient, the ratio of this density $\Omega = \rho/\rho_c$, where ρ_c is the density corresponding to the limiting model (the Einstein-de Sitter model) in which the space metric is Euclidean. For this model, $\Omega = \Omega_c = 1$. The parameter

$$\rho_c = \frac{3H^2}{8\pi G} \qquad (1.3)$$

where the Hubble constant H appears in the Hubble law

$$v = Hr \qquad (1.4)$$

which relates the velocity of cosmological expansion v (going away from us) with the distance r to a corresponding object, say, a Cepheid in some galaxy. The quantity H varies with time; in our epoch, $H = H_0$. This quantity H_0 has been measured all the time since the Hubble law was established in 1929 (Hubble assumed that $H_0 \approx 500$ km s^{-1} Mpc^{-1}). Now the value $H_0 \approx$ 55–70 km s^{-1} Mpc^{-1} has been reached using various techniques (for example, the value $H_0 = 64 \pm 13$ km s^{-1} Mpc^{-1} has been reported [75], $H_0 = 71 \pm 8$ km s^{-1} Mpc^{-1} has been obtained in [177]). For $H_0 = 64$ km s^{-1} Mpc$^{-1} = 2.07 \times 10^{-18}$ c^{-1}, the critical density is

$$\rho_{c0} = \frac{3H_0^2}{8\pi G} \simeq 8 \times 10^{-30} \ \left(\text{g cm}^{-3}\right). \qquad (1.5)$$

Note that, from considerations of dimensionality, the Planck density is

$$\rho_g \sim \frac{c^3}{\hbar G^2} \sim \frac{\hbar}{cl_g^4} \approx 5 \times 10^{93} \ \left(\text{g cm}^{-3}\right). \qquad (1.6)$$

Probably ρ_g is the maximum density near the singularity in which, according to classical theory, $\rho \to \infty$. Thus, the evolution of the Universe or, more precisely, of its region accessible for us has changed up to the present day (if we now have $\rho \sim \rho_{c0}$) by 123 orders of magnitude (one should not, of course, attach any importance to the last figure).

One of the main, perhaps the principal, goals in cosmology is the determination of the quantity $\Omega = \rho/\rho_c$. If $\Omega > 1$, the expansion of the Universe will stop and contraction will begin (a closed model—we mean the Friedmann models). If $\Omega < 1$, the model is open, i.e. the expansion is unlimited. The simplest model with $\Omega = 1$ is, as previously mentioned, an open one with a Euclidean space metric. To find Ω, it suffices to know ρ_{c0} but the determination of this quantity or the establishment of Ω by other methods is rather a sophisticated task. I refer the reader to the books on cosmology (see [76, 77] and especially [229]; see also [150–152, 214]). An important result which has long been recognized is that it is not only the normal baryon matter (and, of course, electrons) that contribute to Ω (or to ρ, which is the same) but something else which does not contribute to the observed glow of stars and gas. This something is called hidden or dark matter. It is discussed later. An important discovery in recent years has been the compelling proof that $\Omega = 1$, i.e. the Universe is spatially flat [204, 207, 208].[7] In addition, it became clear that some 'vacuum matter',

---

[7] In popular cosmological literature, the cosmological expansion is frequently treated as an explosion of some mass in an empty space with a subsequent expansion of the

frequently termed as 'dark energy' or 'quintessence', also substantially contributes to the value of Ω. This question (which will be discussed later) is closely related to the so-called Λ-term emerging in GR.

This term, which has been considered since 1917, should be discussed first of all. It was in 1917 that Einstein, turning to the cosmological problem in the framework of GR, considered the static cosmological model [78, 153]. He came to the conclusion that a solution existed only if GR equations included a Λ-term and had the form

$$R_{ik} - \frac{1}{2} g_{ik} R - \Lambda g_{ik} = \frac{8\pi G}{c^4} T_{ik}. \tag{1.7}$$

The notation is conventional here and I shall not specify it (see, e.g., [49], section 95]). In his preceding work, Einstein did not introduce the Λ-term (i.e. formally speaking, he assumed $\Lambda = 0$). The physical meaning of the Λ-term (for $\Lambda > 0$) is a repulsion which is absent in Newton's theory of gravity. Since, without the Λ-term, GR in a weak field transforms into the Newton theory, a static solution is clearly impossible without the Λ-term. For this reason, Einstein introduced the Λ-term, which is incidentally the only possible generalization of GR which satisfies the requirements underlying the derivation of equation (1.7). However, after work by Friedmann (1922) and the discovery of the Universe expansion (provisionally in 1929), it became clear that the static model was far from reality and the Λ-term was no longer needed. Moreover, Einstein considered the introduction of the Λ-term to be 'unsatisfactory from the theoretical point of view' [79] and discarded it. Pauli, in the appendix to his well-known book published in English in 1958, totally shared Einstein's opinion (see [65, p 287]). L D Landau hated the idea of the Λ-term but I could not make him give his reasoning. Naturally, I could not put this question to Einstein or Pauli.[8] As has already been mentioned, the introduction of the Λ-term is quite admissible from the logical and mathematical points of view. Why then did the great physicists revolted against it? They must have obviously understood that the introduction of the Λ-term was equivalent to assuming the existence of some 'vacuum matter' with an energy–momentum tensor $T_{ik}^{(v)} = (c^4 \Lambda / 8\pi G) g_{ik}$ [see (1.7) with the momentum tensor T_{ik} of the usual matter]. If we put $g_{00} = 1$, $g_{\alpha\alpha} = -1$, the equation of state of this vacuum

---

products of the explosion. This is, of course, totally incorrect. If the space is Euclidean, homogeneous, and isotropic (i.e. $\Omega = 1$, the Einstein–de Sitter model), its volume is infinite at any time and, for a two-dimensional analog, the expansion is similar to an extending infinite film, which results in the film density (or gas of particles, should they form such a film) decreasing with time.

[8] I automatically wrote the word 'naturally' meaning the impossibility of speaking with Einstein and Pauli. This impossibility is, in fact, not at all natural, it is unnatural! Einstein died in 1955 and Pauli in 1958 when I was already nearly 40. Neither I nor my Soviet colleagues could communicate with them because of the existence of the Iron Curtain. I was first able to go abroad (to Poland) to a scientific conference only in 1962.

matter is as follows.

$$\varepsilon_v = -p_v = \frac{c^4 \Lambda}{8\pi G} \tag{1.8}$$

i. e. for a positive energy density $\varepsilon_v > 0$, the pressure is $p_v < 0$, which corresponds to repulsion. This should be comment on, at least briefly. The point is that, in general relativity, the acting gravitational mass of unit volume is equal to $(\varepsilon + 3p)/c^2$ (see, for example, [77, 229]). This means that in GR the pressure 'has weight'. Hence, for the equation of state (1.8) with $\varepsilon > 0$ (this means that $\Lambda > 0$), the gravitational mass density is $-2\varepsilon_v/c^2$, i. e. it is negative ('antigravity'). In other words, the negative pressure, therefore, acts against the usual gravitational attraction (formally, in GR there are no any 'gravitational masses' and 'forces', so using the classical language, I put these term in quotes). Now this is clear but obviously it was not then widely understood among physicists and cosmologists. In any case, I did not understand it and supported the introduction of the Λ-term only from the previously mentioned formal considerations. As far as I know, Gliner was the first to write about the 'vacuum energy' (1.8) in 1965 [80]. Since *Zh. Eksp. Teor. Fiz.* was then edited by E M Lifshitz, it is clear that he did not consider the work in [80] to be obvious either.

As far as I understand, in GR without any generalizations, the Λ-term (more precisely, the value of Λ) is constant. This value (assuming its constancy and using data for our time) is extremely small, say, $\Lambda_0 \sim 10^{-56}$ cm^{-2} (this value corresponds to the estimate $\Omega_\Lambda \sim 1$ for dimensionless Ω_Λ we introduce later). The question of the Λ-term and its evolution in time has been widely discussed [81, 151–153, 180–182, 204, 229]. Some theoretical evaluations of Λ led to enormous values, while, in reality, the term is presently much less. So attempts have been made to prove or somehow justify that actually $\Lambda = 0$ [81]. It happens that, on observational grounds, one can conclude that $\Lambda \neq 0$ and, moreover, the contribution of Λ dominates in Ω compared to other forms of matter, which causes the Universe to expand with acceleration and not with deceleration at the present time [150–153, 180–183, 204, 239, 240].

In any case, the parameter Ω is currently written in the form

$$\Omega = \Omega_b + \Omega_d + \Omega_\Lambda \tag{1.9}$$

where Ω_b corresponds to the contribution of baryons (and, of course, electrons), Ω_d allows for the dark matter and Ω_Λ for the contribution of the 'vacuum energy' or 'dark energy'. In view of (1.3) and (1.8), we have

$$\Omega_\Lambda = \frac{\rho_v}{\rho_c} = \frac{c^2 \Lambda}{3H^2} \qquad \Lambda = \frac{3\Omega_\Lambda H^2}{c^2}. \tag{1.10}$$

For $\Omega_\Lambda \sim 1$ and $H \sim H_0 \sim 2 \times 10^{-18}$ s^{-1} we have $\Lambda_0 \sim 10^{-56}$ cm^{-2}. The estimates, for our epoch, according to observations, are as follows: $\Omega_b \sim$

0.03 ± 0.015, i. e. there are few baryons. For the dark matter, $\Omega_d \sim 0.3\pm0.1$ and, therefore, if $\Omega = 1$, then $\Omega_\Lambda \sim 0.7 \pm 0.1$ (see comment 9). While the significant role of the 'dark energy' is widely recognized, its description by equation of state (1.8) is questionable. Another equation of state with generally close results is possible:

$$\varepsilon_v = \frac{1}{w}p_v \qquad w < 0. \tag{1.8a}$$

Clearly, the Λ-term corresponds to $w = -1$. Such 'dark energy' is also called 'quintessence' [151, 152, 204]. In paper [152], the quintessence is related to the existence of some scalar field Φ. If both the Λ-term and quintessence are present simultaneously, their contributions into Ω_Λ apparently sum up (see (1.9)). It is possible, in principle, to infer the value of w in (1.8a) from observations [185]. All in all, this problem is, of course, at the early stage of studies but, undoubtedly, here we deal with one of the 'hottest' points of modern cosmology.

The early Universe appeared to be intimately related to elementary particle physics. We mean the region of very high energies which cannot be reached in any other way. I recall that even on the LHC accelerator an energy of 1.4×10^4 GeV will be obtained (hopefully in 2006–07) in the center-of-mass frame, in cosmic rays the energy of up to 3×10^{11} GeV is fixed, and the Planck energy is $m_g c^2 \sim 10^{16}$ erg $\sim 10^{19}$ GeV. In the Grand Unification theory, energies of up to 10^{16} GeV figure (particles of mass $m_{GUT} \sim 10^{-8}$ g). This region is the arena of intense theoretical studies.

Turning to problem 24 (neutron stars and pulsars, supernova stars), I note, first of all, that the hypothesis of the existence of neutron stars was formulated, as far as I know, in 1934. It could hardly appear much earlier because the neutron was discovered experimentally only in 1932. Neutron stars (with a characteristic radius of 10 km and $M \sim M_\odot$) seemed at first to be practically unobservable. But when X-ray astronomy was created in 1962, it seemed that hot neutron stars would be observed in the X-ray range. Now even single neutron stars, to say nothing of binary stars, are actually studied in the X-ray band. But even before this, in 1967–68, the radio emission from neutron stars—pulsars—was discovered. This discovery was rather dramatic and has been described elsewhere, so I shall not write about it here (see, for example, [82]).

Nearly 1000 pulsars are now known with radio pulse periods P (it is also the period of star rotation) from[9] 1.56×10^{-3} to 4.3 s. The magnetic field of millisecond pulsars (on the surface) is of the order of 10^8–10^9 Oe. The majority of pulsars ($P \sim 0.1$–1 s) have a field $H \sim 10^{12}$ Oe. Incidentally, the existence in nature of such strong magnetic fields is also an important discovery. Neutron stars with still stronger fields (magnetars)

---

[9] It is amazing that there exists a star with a mass close to the mass of the Sun and a radius of nearly 10 km which makes 640 revolutions per second!

reaching, according to the estimates, 10^{15}–10^{16} Oe (!) have recently been discovered. These magnetars do not emit radio waves but are observed in soft gamma-rays.

A gamma-ray flare probably from such a magnetar was fixed on 27 August, 1998 (the period of radiation bursts after the flare was 5.16 s; the energy interval of radiation was 25–150 keV [83]). Going back to pulsars, I should note that the creation of the theory of their radiation turned out to be quite a sophisticated task but, on the whole, the theory is constructed [84]. For an up-to-date review of pulsars, see [85].

Neutron stars, both those emitting radio waves (pulsars) and all the other ones (single, stars in binary systems, magnetars), are interesting and unusual physical objects. Their density lies within the limits from 10^{11} g cm^{-3} on the surface up to 10^{15} g cm^{-3} in the center. Meanwhile, in atomic nuclei, $\rho = \rho_n \simeq 3 \times 10^{14}$ g cm^{-3} and there is no such variety of densities. The external crust of a neutron star consists, of course, of atomic nuclei and not of neutrons. The neutronization process with penetration into the depth of the star, the corresponding equation of state, the possibility of pionization (the formation of a pion condensate) and the appearance of quark matter in central regions of the star, superfluidity of the neutron liquid (making up the main component of the star), superconductivity of the proton–electron liquid which is present in the star to the level of several percent (of the number of neutrons)—such are some of the problems of neutron star physics (see also [129]). The possibility of the existence of stars of neutron-star type but consisting of strange quarks, etc has been discussed in the literature. The questions concerning the crust should be specially singled out: the 'cracks' which appear owing to star spin-down caused by the loss due to electromagnetic and corpuscular radiation are appreciable: such cracks are associated with 'starquakes' recorded by variation in pulsar periods. For the physics of pulsars, the structure of the stellar magnetosphere is also important. The question of stellar cooling and, mainly its formation, should be specially emphasized. Apparently, neutron stars are principally formed in supernova explosions. By this, we mean the loss of stability by a normal star and its explosion. A possible but not inevitable product of such an explosion is a neutron star. In a supernova explosion, heavier (compared to helium and some other nuclei) elements are 'boiled', cosmic rays are accelerated in shock waves generated in the interstellar gas and in the envelopes (remnants) of supernovae, electromagnetic radiation of all bands occur. During the explosion itself, neutrinos are also emitted. We were lucky in 1987 for the supernova SN 1987A exploded comparatively close to us (in the Large Magellanic Cloud which is at a distance of 60 kpc from the Earth). I said 'lucky' because the previous supernova observed by the naked eye exploded in the Galaxy in 1604 (the Kepler supernova). The well-known Crab Nebula was formed from a supernova exploded in 1054 and inside it there is a pulsar PSR 0531 radiating even in the gamma-ray band. Neutrino radiation was first registered from supernova SN 1987A. For orientation it is worth noting that the kinetic energy of the remnant of this supernova is $E_K \sim 10^{51}$ erg and the energy output in neutrinos is $E_\nu \sim 3 \times 10^{53}$ erg (recall that $M_\odot c^2 \sim 3 \times 10^{54}$ erg). I hope that what has been said is a clear evidence of how interesting and topical problem 24 is. I believe that a single two-hour lecture or a not very long review will suffice to elucidate this range of questions in the volume necessary for the physical minimum.

Black holes and particularly cosmic strings are much more exotic objects than neutron stars. Cosmic strings (they should not, of course, be confused with superstrings) are some (but not the only possible) topological 'defects' which may result from phase transitions in the early Universe [86, 130]. They are threads which can be closed rings of cosmic scales and may have a characteristic thickness $l_{CS} \sim l_g (m_g/m_{GUT}) \sim 10^{-29}$–$10^{-30}$ cm (here m_{GUT} is the characteristic mass corresponding to Grand Unification, i. e. $m_{GUT} \sim 10^{-8}$ g $\sim 10^{16}$ GeV, whereas $m_g \sim 10^{-5}$ g $\sim 10^{19}$ GeV). Cosmic strings have not yet been observed and I do not even know of any candidates to this role. For this reason, I was on the point of including cosmic strings in the 'list' along with black holes but put the interrogative sign instead. I can repeat once again that it is impossible 'to bound the unbounded' and having thought twice I came to the conclusion that cosmic strings should not be included in the 'list' (see, however, [97, 139]).

As to black holes, the situation is quite different. They are very important astronomical and physical objects. In spite of the fact that it is very difficult to 'seize a black hole's hand', their existence and their great role in the cosmos are now beyond doubt. It is curious that black holes were, in a sense, predicted as far back as the late 18th century by Mitchell and Laplace. They asked themselves the question of whether an object (a star) might exist with such a strong gravitational field that the light from it could not go to infinity. In the framework of Newton's mechanics and the notion of light as corpuscles with a certain mass m, the energy conservation law for the radial corpuscle motion with a velocity v has the form $GMm/r_0 = mv_0^2/2$ (the inert and the heavy masses are assumed to be equal, r_0 is the radius of a star with mass M or, more precisely, the distance from its center from which radiation is emitted and goes to infinity at a velocity v_0). Assuming $v_0 = c$ (the velocity of light), we can see that if $r_0 < r_g$, the light cannot escape from the star and

$$r_g = \frac{2GM}{c^2} = 3\frac{M}{M_\odot} \text{ (km)}. \qquad (1.11)$$

In such a "calculation", the gravitational radius r_g appeared to be exactly coincident with that calculated in GR. The coincidence of even the numerical factor is, of course, accidental (in any case, I do not see any reason for such a coincidence). To the best of my knowledge, the formation of a stationary spherically-symmetric (non-rotating) 'black hole' was first considered within the framework of GR only in 1939 [87] and it was only in the 1960s that black holes entered into astrophysics. Nowadays, black holes and their study is a whole chapter of GR and astrophysics (for a detailed review occupying 770 pages see [88]). Here I can only make a few remarks.

Two types of black holes (with stellar masses $M \lesssim 100 M_\odot$ and giant holes in galaxies and quasars with $M \sim (10^6\text{--}10^9) M_\odot$) are observed[5*] or, to put it more carefully, are most probably observed. Holes with stellar masses are mainly found in observations of binary systems. If one of the stars in such a binary system is invisible (does not radiate) and at the same time its mass is $M \gtrsim 3 M_\odot$, this is most probably a black hole. The point is that another possibility of identifying the invisible component in a binary star is to assume that this is a neutron star. But the mass of neutron stars cannot be greater than approximately $3 M_\odot$ because a star of a larger mass will collapse to become a black hole. Incidentally, one should not think that a black hole, which does not radiate by itself (i. e. does not emit radiation from the region $r < r_g$), cannot be visible—it may emit radiation from the region $r > r_g$ where the matter (the accretion disk) incident on it or rotating around it is located. In the Galaxy, rather many black holes have already been identified in different ways, mainly in binary systems, according to the indicated criterion (the mass of the invisible component is $M > 3 M_\odot$). Giant black holes are located in the nuclei of galaxies and quasars. In the center of a galaxy, there exists a potential well and matter gradually losing its angular momentum flows in to it. Such matter may form star clusters. The fate of the clusters is rather complicated but it is quite natural that in many cases, if not always, a collapse with the formation of a black hole must ultimately occur. However, it is a well-known fact that in the centers of many galaxies bright, sometimes even very bright, nuclei are observed. Such galaxies with very bright nuclei include quasars which were first discovered (or, more precisely, identified as far extragalactic objects) in 1963 with the identification of quasar 3C273. I would not like to go into the history of the problem. Suffice it to say that nuclei bright in optics do not exist in all the galaxies or all the time. Among them, quasars are those which are also bright in the radio band (QSR or QSS—quasi-stellar radio sources). Quasi-stellar objects which are not powerful radio sources are referred to as QSO (quasi-stellar objects). There is apparently some ambiguity in the terminology but it is of no importance for us. Bright galactic nuclei may be compact star clusters or black holes. They can be distinguished by the star motion near the nucleus. If we are dealing with a black hole, then the attracting mass is obviously concentrated within a radius smaller than r_g and, even for $M_{\mathrm{bh}} \sim 10^9 M_\odot$, this radius is $r_g \sim 3 \times 10^{14}$ cm, i. e. negligible on galactic scales (recall that the astronomical unit, i. e. the distance between the Earth and the Sun, makes up 1.5×10^{13} cm). Hence, if it were possible to trace the star motion near the nucleus up to distances comparable with r_g, everything would immediately become clear. But this is impossible even in the case of our Galaxy whose center is at a distance of nearly $8\,\text{kpc} = 2.4 \times 10^{22}\,\text{cm}$ from the Sun. Nevertheless, in this case it has been determined, using a radio interferometer, that the radiation source was of the order of astronomical unit. Optical observations of the velocity field of stars near the galactic center have shown that the motion proceeds around a mass with dimensions smaller than a light week, i.e. smaller than 2×10^{16} cm. As a result, one can say with confidence that it is precisely a black hole of mass $M_{bh} \simeq 2.6 \times 10^6 M_\odot$ (with $r_g \simeq 8 \times 10^{11}$ cm) that is located in the center of the Galaxy [89]. New data [230] fully confirm this. For other galaxies, even nearby ones, the resolution is, of course, worse. Nonetheless, when visible, their nuclei, too, are most likely to be black holes rather than some dense star or gas clusters. Investigations in this field are being intensively carried out.

Besides the previously mentioned black holes, relic miniholes may exist which were formed at early stages of evolution of the Universe. The conclusion, drawn in 1974, that owing to quantum effects black holes must emit all sorts of particles (including photons) [90] is generally significant for miniholes (in this connection see [88] and the recent paper on this subject [91]). The radiation of black holes is thermal (i.e. the same as in the case of a black body) with a temperature

$$T_{bh,r} = \frac{c^3 \hbar}{8\pi G M k_B} \approx 10^{-7} \frac{M_\odot}{M}(\text{K}) = 10^{-7} \frac{2 \times 10^{33}}{M(\text{g})}\,(\text{K}) \qquad (1.12)$$

where $k_B = 1.38 \times 10^{-16}\,\text{erg}\,\text{K}^{-1}$ is the Boltzmann constant. Obviously, even for a black hole of mass $10^{-2} M_\odot$ (no smaller self-luminous stellar object exists), quantum radiation is negligible. But, for miniholes, the situation changes and a minihole of mass smaller than approximately $M_{bh} \sim 10^{15}$ g would not have lived up to our epoch (see [2, section 22]). The radiation of such miniholes can, in principle, be revealed but no indications of the existence of such objects have been reported. One should bear in mind that miniholes can be formed but the efficiency of this process is unknown. It is, therefore, possible that there are either very few or no miniholes in the Universe.

We have, in fact, also touched upon problem 26 or, more precisely, the question of quasars and galactic nuclei. The question of the formation of galaxies, which was somewhat artificially combined with the preceding question, constitutes a special chapter in cosmology. The theoretical part of its contents includes the analysis of the dynamics of density and velocity inhomogeneities of matter in the expanding Universe. At a certain stage, these inhomogeneities increase greatly to form the so-called large-scale inhomogeneities of matter in the Universe. This process ends with the appearance of galaxies and galactic clusters. I repeat again that this is a whole field of cosmology (see, in particular, [127]). The synthesis of chemical elements in the course of Universe expansion is, in a sense, a similar problem. This is also an interesting and important problem which might well have figured in the 'list' but it is already largely inflated and something should be sacrificed (see, however, [184]). The choice is not at all unambiguous.

I shall now dwell on problem 27—the question of dark matter. It has already been briefly discussed. This is essentially quite a prominent and unexpected discovery whose history, as far as I know, goes back to 1933 [92]. The amount of luminous matter is determined from observations, mostly in the visible light. The total amount of gravitating matter has an effect on the dynamics—the motion—of stars in galaxies and galaxies in clusters. The dynamics is manifested in the simplest and most obvious way in the determination of the rotational curves of stars in spiral galaxies and, in particular, in our Galaxy. This method is, in principle, elementary: it was clarified in [2, section 23]. It is, however, convenient to turn to it again since, I am sure, if something can be elucidated at the school level, it will also be useful for specialists in fields of physics far from astronomy. So, we shall consider the motion of a star with mass M in a circular orbit around a spherically symmetric mass cluster. The equality

$$\frac{Mv^2}{r} = \frac{GMM_0(r)}{r^2} \tag{1.13}$$

must obviously hold, where v is the star velocity, r is the radius of its orbit relative to the galactic center and $M_0(r)$ is the mass of the galaxy concentrated inside the region with radius r: Kepler's third law $\tau^2 = (4\pi^2 r^3)/GM_0$, where τ is the star revolution period immediately follows from (1.13). Next, suppose the mass M_0 is concentrated in the region with $r \leq r_0$ and, when $r > r_0$, there are already no masses. Then, obviously, for $r > r_0$, we have

$$v^2(r) = \frac{GM_0(r_0)}{r} . \tag{1.14}$$

Observations testify to the fact that the dependence $v(r)$, which represents the rotation curves, is substantially different from the law $v(r) = \text{constant}/\sqrt{r}$ in the range of values $r > r_0$, where there is already little luminous matter. Briefly speaking, it has been established with confidence that non-luminous matter exists in the Universe which manifests itself owing to its gravitational interaction. Dark matter is distributed not at all uniformly but it is present everywhere—both in galaxies and intergalactic space. Thus, there arose one of the most important, and I would even say the most urgent questions of modern astronomy[10]—what is the nature of dark matter, frequently referred to earlier as hidden mass? It is simplest

_____

[10] The dark matter contribution to Ω was denoted as Ω_d in (1.9).

to assume that this is neutral hydrogen, a strongly ionized (and therefore weakly luminous) gas, planets, weakly luminous stars—brown dwarfs, neutron stars or, finally, black holes. All these assumptions have, however, been disproved by various types of observations. For example, neutral hydrogen is fixed by the radio-astronomy method, hot gas is registered by X-ray emission, neutron stars and black holes are also observed, though with difficulty. It is not easy to observe brown dwarfs which are dwarf stars with such small masses $M \ll M_\odot$ that they glow very weakly. However, such stars have also been discovered [93] and, in all probability, they do not contribute appreciably to the dark matter. The analysis of all these questions is not simple: different opinions concerning the contribution of particular types of baryonic matter to the total matter density exist. We have previously mentioned the estimate $\Omega_b \lesssim 0.05$ (more precisely, $\Omega_b = 0.044 \pm 0.01$). In general, the conventional point of view is now as follows: dark matter is largely of non-baryonic nature. The most natural candidate is the neutrino. But this version is unlikely to hold: the electron neutrino mass ν_e is obviously insufficiently large (the value known to me is $m_{\nu_e} < 3$ eV), while a mass $m_\nu > 10$ eV is needed. The masses ν_μ and ν_τ will be discussed later but they are apparently insufficient as well (the possible role of the neutrino is discussed in [94]). Another very popular hypothesis is the one in which the role of dark matter is played by the hypothetical WIMPs (weakly interacting massive particles) with masses of gigaelectronvolts and higher (the proton mass is $M_p = 0.938$ GeV). WIMPs include hypothetical (I repeat) heavy unstable neutrinos and supersymmetric particles—photinos, neutralinos, etc. There also exist some other candidates for the role of dark matter (for example, pseudoscalar particles or axions) [131]. Cosmic strings and other 'topological defects' should also be mentioned. There are hopes of detecting WIMPs by their radiation of gamma-photons and other particles upon annihilation with corresponding antiparticles. Another way is to observe their collisions, although very rare, with particles of normal matter [95, 132]. The idea of the possibility of WIMPs concentrating into some friable quasi-stars which can, in principle, be detected by microlensing [96] is very elegant. For recent data on microlensing known to me, see [186].

The origin of cosmic rays (CRs) first discovered in 1912 has been enigmatic for many years. But now it is definite that their main sources are supernova stars. In respect of CRs with $E_{CR} < 10^{15}$–10^{16} eV, generally some vague points remain but, on the whole, the picture is clear enough [52] (see article 3 in this volume). It is only the problem of the origin of CRs with ultra- or super-high energies that may be 'particularly important and interesting', according to the terminology adopted here. For example, the origin of the 'break' ('knee') in the energy spectrum of CR at $E_{CR} \sim 10^{15}$–10^{16} eV is not quite clear and especially the situation in the energy range $E_{CR} > 10^{19}$ eV—such CRs are sometimes called ultrahigh-energy CR (UHECR, see [98, 187, 188]) is enigmatic. The highest energy observed in CRs is $E_{CR} \sim 3 \times 10^{20}$ eV as has already been mentioned in another context. It is not easy but it is apparently possible to accelerate particles (say the proton) to such an energy, especially in active galactic nuclei. But then the following difficulty arises: when colliding with cosmic microwave (relic) radiation (with a temperature $T_r = 2.7$ K), particles with ultrahigh energies generate pions and, thus, lose energy and, as a result, cannot reach us from very great distances (the effect of Greizen, Zatsepin, and Kuz'min, 1966). For this reason, a cutoff (steepening) must occur in the CR spectrum; under the simplest assumptions, it occurs at a characteristic energy $E_{BB} = 3 \times 10^{19}$ eV [98]. In fact, however, this cutoff is absent [53, 98]. The question is how the appearance of CRs with $E_{CR} > 3 \times 10^{19}$ and up to 3×10^{20} eV can be explained. Several possibilities are under discussion. The number of active galactic nuclei within distances 20–50 Mpc are apparently insufficient. Moreover, it is unclear whether the known galactic nuclei can provide acceleration up to an energy of 3×10^{20} eV. Particles might be accelerated by cosmic strings and some other 'topological defects' located outside the Galaxy at distances up to 20 Mpc [98]. There are no indications of the existence of such 'defects' especially at comparatively close distances. Another hypothesis suggests that primary UHECR particles are not 'ordinary' particles (protons, photons, nuclei, etc) but some yet unknown particles which, say, have not undergone strong energy losses. Then they may come from a large distance and closer to us, or even in the Earth's atmosphere, transform into ordinary particles and yield an extensive air shower (EAS). Finally, it seems simplest to assume that, in the galactic dark matter which forms the corresponding galactic halo, there exist supermassive particles of mass $M_x > 10^{21}$ eV that have lived longer than the Universe ($t_0 \sim 10^{10}$ years) but which are still unstable. The products of their decay, according to this hypothesis, are observed in the atmosphere and give rise to EAS (for the reader not closely related to this subject, it may not be out of place to explain that UHECR particles, like particles of lower energies, say, $E_{CR} \gtrsim 10^{15}$ eV, are registered in cosmic rays only by EAS). For example, clumps of quark matter ('strangelets') could be such 'primary' particles [238]. On the whole, the problem of very-high-energy CRs is actually enigmatic and, for this reason alone, interesting (see also article 3 for more references).

We now proceed to problem 29, i. e. gamma-ray bursts. A series of Vela satellites were launched in USA in the late 1960s, which were equipped with soft gamma-ray detectors and were intended for the control of the treaty banning nuclear explosions in the atmosphere. No explosions were made but gamma-ray bursts of an unknown origin were registered. Their typical energy was 0.1 1 MeV and the duration amounted to seconds. The received energy flux in the bursts integrated over time was rather large—it reached the values $\Phi_\gamma \sim 10^{-4}$ erg cm^{-2}. If a source located at a distance R radiates isotropically, its total energy output in gamma-ray photons is obviously $W_\gamma = 4\pi R^2 \Phi_\gamma$. This discovery was reported only in 1973 [99]. Gamma-ray bursts have been intensively investigated since then but their nature has long remained unclear. The point is that the angular resolution of gamma-telescopes is not high and observations in other bands (radio wavelength, optical, and X-ray) in the direction of a gamma-ray burst were not carried out immediately. Thus, the source remained absolutely unknown. One of the probable candidates were neutron stars in the Galaxy. In this case, for comparatively close neutron stars at a distance $R \sim 100$ pc $\simeq 3 \times 10^{20}$ cm, the energy output was $W_\gamma \lesssim 10^{38}$ erg. This is already very much if we recall that the total luminosity of the Sun is $L_\odot = 3.83 \times 10^{33}$ erg s^{-1}. However, the distribution of even weak gamma-ray bursts over the sky proved to be isotropic, which means that their sources cannot be located in the galactic disc. If they are located in the giant galactic halo so that $R \sim 100$ kpc (this does not already contradict the data on the angular distribution of sources), then $W_\gamma \lesssim 10^{44}$ erg. Finally, if the bursts are of cosmological origin and, for example, $R \sim 1000$ Mpc then we already have $W_\gamma \lesssim 10^{52}$ erg. This value is so large that many scientists (including me) gave preference to the halo model but, in 1997, a gamma-burst was 'observed' immediately (i.e. its position was located almost instantaneously) and sources with a large redshift were discovered [100, 101]. So, for the burst GRB 971214 (the designation implies that this burst was registered on 14 December, 1997) the redshift parameter[11] was $z = 3.46$ [102]. For the burst GRB 970508, this parameter was $z \geq 0.8$. The sources (already many of them are known) were observed both in the X-ray and optical bands and some of them also in the radio wavelength band. The work is in full swing, and literally a day after the previous sentences were written for the previous edition, on 23 January, 1999, a powerful burst, GRB 990123, was observed over the entire investigated gamma-ray band from 30 keV to 300 MeV, which lasted 100 s. Simultaneously with the gamma-burst, a burst of light was registered whose maximum luminosity reached $L_0 \sim 2 \times 10^{16} L_\odot \sim 10^{50}$ erg s^{-1}. The total energy output in all the electromagnetic bands was $W \sim 3 \times 10^{54}$ erg if the radiation was isotropic (the redshift of the event was $z = 1.61$). More details concerning gamma-bursts are given in the review [103] (see also [245]). GRB 021004 (detected, as clear from its designation, on 4 October, 2002) proved to be very intriguing. It was accompanied by a prolonged optical afterglow. For GRB 000131, the redshift parameter was $z = 4.5$, which corresponds to a distance of 11×10^9 light years [154] from us. It should be emphasized that gamma-ray emission from gamma-ray bursts is most probably anisotropic. Additional strong evidence for this

---

[11] Remember that $z = (\lambda_{obs} - \lambda_{source})/\lambda_{source}$, where λ_{obs} is the observed wavelength of the spectral line and λ_{source} is the wavelength in the source.

has recently been obtained [209]. Then, of course, the energy emitted is smaller than these values. According to [209], some 'mean' energy release in the gamma-ray burst source is 'only' $W \sim 3 \times 10^{51}$ erg, i.e. of the same order as that in supernova explosions. So the term 'hypernovae', used in the title of problem 29 in the 'list', seems to be unsuccessful. Now the problem is to determine the sources of cosmological gamma-ray bursts. These could be the coalescence of two neutron stars, some collision or coalescence of a massive star with a neutron star, explosion of a specific supernova, etc [216]. However, even such sources can marginally radiate 10^{54} erg $\sim M_\odot c^2$. As previously discussed, the actual energy release could be much smaller (this is the case if the emission is highly anisotropic [209]). So a certain, one could even say, fear connected with the huge value of W is now over. In any case, one can hardly doubt that the discovery of the cosmological origin of gamma-bursts (or to be more precise, the detection of X-ray, optical and radio emission associated with gamma-ray bursts) is, side by side with successes in cosmology (I mean first of all revealing the role of dark energy), the most distinguished achievement of astrophysics not only in 1997 but of many years (perhaps since the discovery of pulsars in 1967–68).

It remains to discuss the last problem, number 30, from the 'list': neutrino physics and astronomy. I recall that the hypothesis of the existence of neutrinos was suggested by Pauli in 1930. Neutrinos have long been thought of as practically undetectable because the reaction cross section

$$p + \overline{\nu}_e \to n + e^+ \tag{1.15}$$

(here $\overline{\nu}_e$ is an electron antineutrino) is negligibly small: $\sigma \sim 10^{-43}$ cm^2. However, in 1956, this reaction (1.15) was fixed on an atomic reactor, for which the 1995 Nobel Prize in physics was awarded (more precisely, half of the prize [104], the other half was given for the discovery of the τ-lepton [105]). The question of the neutrino mass probably arose at the very beginning but the mass m_{ν_e} is clearly very small compared to the electron mass. The assumption of zero neutrino mass (only the electron neutrino was discussed at first) did not face any contradictions. After the discovery of the muon and tau neutrinos ν_μ and ν_τ, the same could be said about these neutrinos. However, an idea arose (back in the 1960s) that neutrino oscillations, i.e. mutual transformations of neutrinos of different types (flavors), were possible. This is possible only if the mass of a neutrino of at least one flavor is non-zero. In any event, the question of the neutrino mass arose long ago and remains very topical. There have been attempts to determine the neutrino mass m_{ν_e} by examining the region near the end of the β-spectrum of tritium (the reaction $^3\mathrm{H} \to {}^3\mathrm{He} + e^- + \overline{\nu}_e$: by virtue of the CPT theorem, it is now certain that $m_\nu = m_{\overline{\nu}}$). The maximum decay energy is small in this case—close to 18.6 keV. Measurements are being carried out: as far as I know, it is now believed that $m_{\nu_e} < 3$ eV. More recent measurements yield $m_{\nu_e} < 2.2$ eV [232].[12] The difficulty of making the measurements is connected with the necessity of controlling the energy given to the molecules of the surrounding medium. Incidentally, some of the theoretical estimates (see, e.g., [106]) are as follows:

$$m_{\nu_e} \sim 10^{-5}\,\text{eV} \qquad m_{\nu_\mu} \sim 10^{-3}\,\text{eV} \qquad m_{\nu_\tau} \sim 10\,\text{eV}. \qquad (1.16)$$

I do not know any direct methods of measuring m_{ν_μ} and m_{ν_τ} which are, in principle, possible.[6*] But the study of oscillations opens such possibilities. It is probably pertinent to clarify the very idea of oscillations. This is the assumption that neutrinos of one flavor or another emitted upon decays or born under weak interactions are not eigenstates of the mass operator. That is why, when propagating in spacetime, a neutrino of a certain flavor may gradually become a neutrino of another flavor (for more details, see [106, 107]). Neutrino oscillations have already been sought for 30 years and, in 1998, a definite success was clearly achieved—the transformation of ν_μ into ν_τ was discovered [108, 109]. This is the most prominent discovery in elementary particle physics for many years. It was made on the Japanese–American installation Super Kamiokande whose basic element is a tank (1 km underground) filled with 50 000 tons of perfectly purified water. The tank is surrounded by 13 000 photomultipliers which register the Vavilov–Cherenkov radiation from the muons, electrons, and positrons produced in the water by neutrinos that get into the tank. Here we are speaking of the electron and muon neutrinos formed by cosmic rays in the atmosphere on the opposite side of the Earth. If there are no oscillations then, according to reliable calculations, in the tank there should be twice as many electron neutrinos as muon neutrinos. But, in reality, the numbers of ν_e and ν_μ are the same (their energy is of the order of 1 GeV). The most probable explanation of the observations is that oscillations between ν_μ and ν_τ are observed. Here, the quantity $\Delta m^2 = (m_2^2 - m_1^2)$, where $m_{1,2}$ are neutrino masses, is measured. According to [109], $5 \times 10^{-4} < \Delta m^2 < 6 \times 10^{-3}$ (eV)^2. If one assumes that one mass is much smaller than the other, the heavier mass will be $m_\nu \sim 0.05$ eV. Such a neutrino (this is either ν_μ or ν_τ) is of no interest for cosmology. As has been stated (see [108]), if m_2 and m_1 are very close, then masses are admissible that could be responsible for the dark matter. However, according to [94], the total mass of stable neutrinos of all types does not exceed 2 eV. I cannot judge the significance of the difference of the neutrino mass from zero for elementary particle physics but apparently this is not a very principal fact. However, who knows?

The Sun and the stars are known to radiate at the expense of nuclear reactions proceeding in their interiors and must, therefore, emit neutrinos. These neutrinos, whose energy is $E_\nu \lesssim 10$ MeV, can currently be registered

---

[12] Clearly, it is more correct to measure the mass not in units of eV but in eV/c^2 but this is the question of designations.

only from the Sun. Such observations have already been carried out for more than 30 years using the reaction $^{37}\text{Cl} + \nu_e \rightarrow {}^{37}\text{Ar} + e^-$. Argon atoms in a tank filled with chlorine (more precisely, with a chlorine-containing liquid) are chemically extracted. The observed flux (the first results were presented in 1968) makes up several SNU (solar neutrino units): for a flux of 1 SNU, 10^{36} nuclei of ^{37}Cl or other nuclei capture one neutrino a second on average. According to calculations for different solar models, the flux should be (8-4) SNU (for neutrinos which can be registered by the chlorine detector). I did not try to find more precise values now (at least for me, fluxes are clearer in conventional units, i.e. the number of neutrinos reaching the Earth surface per 1 cm^2 per second). As the Sun emits neutrinos with different energies from different reaction, the flux depends on the observed energy range. For example, the calculations in the framework of the so-called standard solar model yield a flux of neutrinos with energies above 6.75 MeV of 5×10^6 cm^{-2}c^{-1} [184]. Meanwhile, it has been established that the flux registered by the chlorine detector is appreciably smaller than the calculated one, roughly speaking, by a factor of two or three. In view of the complexity of computations for models of the Sun, etc, such a result is not impressive. Therefore, there have been attempts to observe solar neutrinos by other methods. For example, the scattering of neutrinos on electrons $\nu_e + e^- \rightarrow \nu'_e + (e^-)'$ was recorded by the Kamiokande installation (the predecessor of Super Kamiokande). Only neutrinos with energy $E_\nu > 7.5$ MeV, which were emitted by a ^{8}B nucleus, were fixed. The observed flux was again approximately half the calculated one. Finally, two installations were created: the Soviet–American SAGE and the European GALLEX in which the working substance is gallium ^{71}Ga transforming into germanium ^{71}Ge upon capture of neutrinos. Such a detector has a low energy threshold and, as distinct from the chloride one, reacts to the bulk mass of neutrinos emitted by the Sun (these are neutrinos from the reaction $p + p \rightarrow d + e^+ + \nu_e$). And again, the observed flux is smaller than the calculated one [110]. All the available information suggests the conclusion that the neutrino flux from the Sun is, indeed, considerably smaller than calculated one but the calculations disregarded possible neutrino oscillations. This gave rise to an assumption about the existence of such oscillations for ν_e and about their effect upon the observed flux of solar neutrinos (see [107, 111]). Several improved installations for detection of solar neutrinos with different energies are being built or have already been put into operation. Finally, in 2001, success came. The installation SNO (Sudbury Neutrino Observatory) registered the flux of electron neutrinos with energies above 6.75 MeV which proved to be (1.75± 0.14) × 10^6 cm^{-2} c^{-1}. This amounts to only 35% of the flux calculated in the Standard Model (see earlier). However, combining these data with the results obtained by Super Kamiokande, which registers solar neutrinos of all three types or, as one says, flavors, yields the flux $(5.44\pm1)\times10^6$ cm^{-2} c^{-1}, which confirms the standard solar model and, most importantly, the reality of neutrino oscillations [184, 189, 190]. The existing experimental data, as previously mentioned, bound the mass of neutrino by the value (2–3) eV: this is the upper limit, the actual masses can be much smaller. Probably, the neutrino contribution to the dark matter is insignificant (at most percents) but no exact values are unavailable as yet. There is a lot to be done here but the main problem is solved (the earlier data are confirmed [217]). Neutrino oscillations from nuclear reactors have also been observed [232].

Neutrino astronomy is not only solar astronomy. I have already mentioned the detection of neutrinos from the explosion of supernova SN 1987A. Monitoring is now being carried out and if we are lucky and another supernova explodes near the Sun (in the Galaxy or Magellanic Clouds), we shall obtain rich material (supernovae explode in the Galaxy on average approximately once every 30 years but this figure is inaccurate and, what is important, an explosion may occur at any moment). The problem of detecting relic low-energy neutrinos which may contribute to the dark matter is especially noteworthy. Finally, high-energy ($E_\nu \gtrsim 10^{12}$ eV) neutrino astronomy is just opening up. A number of installations for the detection of such neutrinos are under construction [112, 140]. The most probable sources are galactic nuclei, coalescence of neutron stars, and cosmic 'topological' defects. Finally, simultaneous observations in all electromagnetic bands and using gravitational-wave antennae will be carried out. In general, the prospects are most impressive.

My comments on the 'list' are, on the whole, over and there is now every reason to return to the remark made at the beginning of the article. Only slightly more than 70 years have passed since Pauli, with uncharacteristic for him shyness, expressed the idea of the existence of the neutrino in a letter addressed to some physical congress (see, e. g., [104]). And now whole fields of physics and astronomy are devoted to neutrinos. The rate of development is so high that it is difficult to foresee even roughly what physics will look like in a hundred years. But this will be considered in section 7.

1.6   Three more 'great' problems

My whole project—the compilation of the 'list' and the comments on it planned as a pedagogical or educational program and, to some extent, a guide to action—probably is not attractive to everybody. Some will not like the manner and the style of the presentation. This is natural. I can only advocate the right to express my own points of view, which is no obstacle to respecting other opinions. I hope the present article will be beneficial. At the same time, to make the picture complete, I would like to mention three more problems (or ranges of questions) which have not so far been touched upon. Meanwhile, the teaching of physics and discussion of its current state and ways of developing it cannot and should not disregard these three branches, these three 'great' problems. First, I mean the increase in entropy, time irreversibility, and the 'arrow of time'. Second is the problem of interpretation and comprehension of quantum mechanics. And third is the question of the relationship between physics and biology and, specifically, the problem of reductionism.

L D Landau was notable for a clear comprehension of physics, at any rate of something that had already 'settled'. In certain accord with this, he did not like any 'substantiations' (*Neubegrundung*, as he would say using this German word), i. e. obtaining known results in another way or using another method.[13] Of particular value in this connection are the critical remarks made by Landau in respect of the law of entropy increase and the arguments in favor of it. In the *Course* (see [29, section 8]), he definitely said about the ambiguities that remained in this field: "The question of the physical foundations of the law of monotonic increase of entropy thus remains open" ([29, p 33]). The discovery (1964) of CP-parity violation (and, therefore, T-parity violation, i. e. time irreversibility) is clearly related to this subject but all this has not yet been sufficiently investigated and realized. I am ignorant of the present state of the problem and, unfortunately, cannot even suggest an appropriate reference. There is no doubt that the question is still unclear and, in any case, this fact should not be veiled.

The situation with quantum mechanics (I mean non-relativistic theory) is different. The majority of physicists obviously believe that the so-called orthodox or Copenhagen interpretation of quantum mechanics is consistent and satisfactory. This point of view is reflected in the *Course* [113]. Landau often added something like this: "Everything is in general clear but tricky questions are possible which only Bohr is able to answer." In 1939, L I Mandelshtam delivered lectures on the basic principles of quantum mechanics in Moscow State University. These lectures were published posthumously [114]. They were prepared for publication by E L Feinberg and looked through by I E Tamm and V A Fock. As I understand it, L I Mandelshtam completely shared the orthodox interpretation and analysed it thoroughly. Unfortunately, these lectures are not very well known to the scientific community: they were published with great difficulty and in very hard times. Moreover, during that period (in the 1950s), the discussion of the interpretation or, more correctly, the basic principles and understanding of quantum mechanics somewhat faded (in particular, see [2, p 433]). Now this range of problems is given prominence in serious literature. I shall re-

[13] I dare say that I do not agree at all with Landau in this respect and I have already written about this many times (see, e. g., various articles in [2] and in the present volume).

fer to the monographs [115, 116] and the papers [117–119, 156, 158], where a lot of references are given. The current interest in the fundamentals of quantum mechanics is partially due to new experiments, mainly in the field of optics (see [117, 156]). All these experiments testify to the perfect validity and, one could say, the triumph of quantum mechanics. At the same time, they exposed features of the theory which have long and well been known but do not seem obvious. This is not an appropriate place for discussing all these questions. I would only like to note that the discussion of the fundamental principles of non-relativistic quantum mechanics remains topical and should not be ignored.[14] The majority, if not the overwhelming majority, of critics of quantum mechanics are dissatisfied with the probabilistic nature of part of its predictions. Apparently, they would like to return to classical determinism in the analysis of microphenomena and, figuratively speaking, to come ultimately to know exactly where each electron goes in the diffraction experiments. There is no reason to hope for this now.

If we turn to its history, we know that the creation of the theory of relativity and quantum mechanics has led to an understanding of the range of applicability of classical (Newtonian) mechanics. Nevertheless, Newtonian mechanics remained unshakeable. The applicability limits of non-relativistic quantum mechanics associated with relativism are already known. Generalization of the existing relativistic quantum theory (perhaps in the way outlined in string theory) is unlikely to introduce anything new to non-relativistic quantum mechanics and to answer the notorious question of 'where the electron will go'. However, when we speak of the possibilities of the future theory and of its influence on the existing one, we cannot give an *a priori* answer. As has been previously stated, the orthodox (Copenhagen) interpretation seems to be consistent and many scientists are satisfied with it. I can only express my intuitive judgement— non-relativistic quantum mechanics will not undergo substantial changes (we shall not come to know 'where the electron will go') but some deeper understanding (outside the limits of the orthodox interpretation) is still not excluded [156].

I have just used the term 'intuitive judgement'. The notion seems to be clear from the words. But this is, in fact, a deep issue which was

---

[14] The aforesaid is particularly clear if we, for example, take into consideration the fact that, at the end of 1998, an authoritative journal published a paper [120] in which the work of D Bell is called 'the most serious discovery in science' (probably for some period of time). Bell was, in fact, (and remained up to his death in 1990) unsatisfied by the orthodox interpretation of quantum mechanics and tried to replace it by a theory with 'hidden parameters'. However, Bell's analysis and subsequent experiments confirmed quantum mechanics largely against his aspiration. But Bell hoped that a future theory would provide insight into the existing non-relativistic quantum mechanics. However, that was no more than a hope. I failed to find a 'deep scientific discovery' in the work of Bell.

analyzed by E L Feinberg [121].[15] The methodology and philosophy of science are not now respected in Russia. This is a natural reaction to the perversions of the Soviet period when there was no freedom of opinion and dogmatic dialectic materialism was implanted. But the methodology and philosophy of science remain, of course, very important ingredients of scientific *Weltanschauung* (world outlook). Under conditions of freedom of ideology, attention to these problems should be revived in our country. With this goal, a long paper by M B Menskii "Quantum mechanics: new experiments, new applications, and new formulations of old questions" was published in *Physics-Uspekhi* [191]. Then seven responses to this paper were published [192] and the authors had the possibility of presenting all opinions. The free exchange of opinions on the comprehension of quantum theory will be reflected in the proceedings [193] of a conference dedicated to the centenary of quantum theory. Unfortunately, I personally evoked little from these discussions.[7*]

The last 'great' problem to be discussed here concerns the relationship between physics and biology. From the late 19th century until approximately the 1960s or 1970s, physics was, so to say, the prime science, the first and dominant one. All kinds of ranks are conditional in science and we only mean that achievements in physics in the indicated period were particularly bright and, what is important, largely determined the ways and possibilities of the development of the whole of natural science. The structure of the atom and atomic nucleus, i.e. the structure of matter, was established. It is absolutely obvious how important it is, for example, also for biology. The development of physics led in the middle of previous century to a culmination—the mastering of nuclear energy and, unfortunately, atomic and hydrogen bombs. Semiconductors, superconductors, and lasers—all these are also physics which determine the face of modern technology and thus, to a great extent, modern civilization. But the further development of fundamental physics, the basic principles of physics and, concretely, the creation of the quark model of the structure of matter are already purely physical problems which are not essentially significant for biology and other natural sciences. At the same time, using for the most part increasingly perfect physical methods, biology progressed quickly and after the genetic code was deciphered in 1953, its development was particularly rapid. It is biology, especially molecular biology that has now taken the place of the leading science. One may disagree with this terminology and with the essentially unimportant distribution of 'places' in science. I would only like to

---

[15] The term 'intuitive judgement' seems to be particularly suited to judgements that can be neither proved nor disproved. In such cases one customarily applies the word 'belief' (for example, "I believe that ... will be obtained"). But the term 'belief' appeared to be closely related to belief in God and religion. However, belief in God is an intuitive judgement which differs essentially from intuitive judgement in science (see [121, 122] and article 31 in this volume.

emphasize some facts which not all physicists understand, especially in Russia. Physics for us remains the cause of our life, young and beautiful, but for human society and its evolution the place of physics has now been taken by biology. A good illustration of these words is the following detail. The journal *Nature*, whose role and place in science need not be explained, elucidates all the sciences, including physics, astronomy and biology in its weekly issue. At the same time, *Nature* today has sprouted seven satellites— the monthly journals *Nature-Genetics, Nature-Structural Biology, Nature-Medicine, Nature-Biotechnology, Nature-Neuroscience, Nature-Cell Biology,* and *Nature-Immunology*. They are all devoted to biology and medicine. For physics and astronomy, the basic *Nature* issue and certainly the numerous purely physical journals are enough (of course, in biology such journals also exist). The achievements of biology are so widely elucidated even in popular literature that there is no need to mention them here. (Nevertheless, I point out the papers devoted to the 'jubilee' of DNA published in [233].) I am writing about biology for two reasons. First, modern biological and medical studies are impossible without the many-sided use of physical methods and apparatus. Therefore, biological and near-biological subjects must and will occupy more and more space in physics institutes, physics departments and in physics journals. One should understand this well and promote it actively. Second, the question of reductionism is, simultaneously, a great physical and biological problem, and I am convinced that it will be one of the central problems in the science of the 21st century.

We believe now that we know what every living thing consists of, meaning electrons, atoms, and molecules. We are aware of the structure of atoms and molecules and of the laws governing them and radiation. The hypothesis of reduction, i. e. the possibility of explaining all life on the basis of physics, the already known physics, therefore seems natural. The main problems are those of the origin of life and the appearance of consciousness. The formation of complex organic molecules under conditions that reigned on the Earth several billion years ago has already been traced, understood, and simulated. The transition from such molecules and their complexes to protozoa and their reproduction seems to be imaginable. But a certain jump, a phase transition exists here (see [158]). The problem is not yet solved and I think will unreservedly be solved only after 'life in a test-tube' is created. As to the physical explanation of the mechanism of the appearance of consciousness, I am not aware of the situation and can only refer to the discussions of the possibility of creating an 'artificial intellect'. Those who believe in God certainly 'solve' such problems very simply: it was God who breathed life and thinking into inorganic matter. But such an 'explanation' is nothing but a reduction of one unknown to another and lies beyond the scope of the scientific *Weltanschauung* and approach. At the same time, can the possibility of the reduction of biology to (present day) modern physics be taken as certain? The key word here is 'modern'.

And with this word in mind, I think it would be incorrect to answer this question in the affirmative. Until the result is obtained, the possibility that, even at the fundamental level, we do not know something necessary for the reduction cannot be excluded. I make this reservation just to be on the safe side, although my intuitive judgement is as follows: at the fundamental level, no 'new physics' is needed for the reduction—the understanding of all biological processes. No dispute concerning this issue will be fruitful—the future will reveal it.

One cannot but think about this future with jealousy—how many interesting and important things shall we learn even in the next ten years! I shall venture a few remarks on that score in section 1.7.

At the same time, after having singled out three 'great problems', I am afraid that I could somehow distort the present situation in physics and astronomy. Namely, the impression could appear that the aforementioned three 'great problems' are nearly the main problems to be solved. Of course such a conclusion would be quite erroneous. Indeed, as clearly follows from the previous discussion, the main problem in physics is to create a single theory uniting all known interactions, i. e. uniting the 'Standard Model' with gravity. Putting it somewhat differently, one can say the creation of a consistent quantum theory of all fields, including the gravity field, i. e. the generalization of general relativity over the quantum region. Attempts in this direction are related to superstring theory, to revealing the role of supersymmetry and the multidimensionality of space. Undoubtedly, we are at the beginning of a long journey leading to the formulation of some unified theory of all quantum fields and interactions. In relation to this future, a really great and general theory with cosmology is already clear since the cosmological problem cannot be studied in depth outside the quantum gravity frames. The successes in cosmology in recent years are really bright and it is impossible not to see that its face has dramatically changed. Indeed, baryonic substance constitutes only 4% of all matter, the 'dark matter' contributes 23% and the 'dark energy' makes up the other 73%. And we are totally ignorant about what constitutes this 'dark matter' and 'dark energy' or quintessence! It could be dangerous to be too self-sufficient in such a situation and not to admit the possibility that something really fundamental is missed in our understanding of the structure and evolution of the Universe. The problems of the 'birth' of the observed portion of the Universe, of the role and sense of the beginning of expansion (i. e. the Big Bang), of the region near the 'classical singularity', of the possible sense of the state 'before the Big Bang' remain totally unclear.

I did not write in detail about this earlier for I can not say anything essential on this subject. But I write about this here in order not to forget about the wood for the trees.

1.7   An attempt to predict the future

In connection with forecasts for the future, the following phrase may often be heard: forecasting is a thankless occupation. It is meant by this perhaps that life and reality are much richer than our imagination and forecasts often prove to be erroneous. More important is the circumstance that unpredicted and unexpected discoveries are the most interesting. They cannot, of course, be prognosed and, thus, the validity of prognoses seems to be particularly questionable. Nevertheless, attempts to foresee the future seem to be reasonable if one does not attach too much importance to them. This is what I shall do in concluding the present article by a forecast concerning only the problems mentioned earlier (I apologize for some repetition).

The decision to begin the construction of the giant tokamak ITER has been delayed but the research work in the field of thermonuclear fusion continues and alternative systems and projects are being elaborated. There are no doubts about the very possibility of constructing an operating (commercial) reactor. The future of this direction will be determined mainly by economical and ecological considerations. I think that some experimental reactor (naturally with a positive energy output) will, in any case, be constructed in a couple of decades. Laser thermonuclear fusion will also be realized because such an installation is possible and needed for military purposes. Of course, physical experiments will also be carried out on it.

As mentioned in section 1.3, the problem of high-temperature superconductivity (HTSC) has been investigated since 1964 and I had thought of it as quite realistic all the time before the discovery of HTSC in 1986–87. But at that time there was no concrete prediction of HTSC: only that no known fundamental difficulties existed on the way to the creation of HTSC. The present-day situation with room-temperature superconductivity (RTSC) is the same. In 1964, the maximum known critical temperature for superconductors was 23 K and now, for HTSC, we have $T_{c,max} = 164$ K, i.e. the temperature T_c has increased sevenfold. In order to reach room temperature, it now suffices to increase T_c by 'only' a factor of two. Therefore, if we proceed from 'kitchen' considerations, the possibility of obtaining RTSC seems probable. At the same time, there inevitably remain some doubts. If the HTSC mechanism in cuprates, which is still unclear, is basically a phonon or a spin (or a phonon–spin) mechanism, then even a twofold increase in T_c is very difficult. If the exciton (electron) mechanism is decisive, then the creation of RTSC is, on the contrary, quite plausible. I can only express here an intuitive judgement. Namely, I believe that RTSC will be obtained in the not very remote future (maybe tomorrow or maybe in several decades) (see also articles 6 and 7 in the present volume).

I remember the times when the creation of metallic hydrogen seemed to be 'a matter of technique'. One can, of course, say the same thing today but the static pressures of nearly 3×10^6 atm now attained to obtain the metallic phase turned out to be insufficient, the pressure should be increased by one and a half times [157]. It is unknown (at least to me) how the pressure can be heightened appreciably if new materials stronger than diamond are not discovered. Dynamical compression leads to heating and it is unclear how to avoid it. I am of the intuitive opinion that these difficulties may rather soon be overcome. At the same time, the hopes (which existed) to obtain a 'piece' of metallic hydrogen and to use it do not seem to be realistic.

In respect of all the other problems (4–13) of section 1.3, it is clear that they will be intensively investigated and many interesting things will be clarified. But being perhaps insufficiently informed, I cannot point to any vivid expectations. A surprise may, however, be expected from fullerene C_{36} or K_3C_{36} type compounds if they show HTSC properties. The study and application of nanotubes is promising. Long-lived transuranium nuclei may obviously be obtained.

Macrophysics should also include the fireball (ball-lightning) problem which I did not include in the 'list'. The existence of ball lightning is beyond doubt. The problem of its origin has long been discussed. Many models and hypotheses have been proposed but a notorious consensus has not been attained. The origin of the fireball, I believe, will clearly and unambiguously be established only after these objects are created in the laboratory where all the conditions and parameters can be controlled. Incidentally, such attempts were repeatedly made and claims expressed that fireballs were born. But apparently all such statements have not been confirmed. Successes in laser physics are impressive, especially the use and studies of attosecond pulses ($1\,\mathrm{as} = 10^{-18}\,\mathrm{s}$) [197, 242] and the creation of the 'atomic nanoscope' [198] allowing study of individual atoms by light.

In the field of microphysics (elementary particle physics), an obvious recession (in the number of discoveries, etc) has been observed within the last two decades compared to the previous period. This is perhaps largely due to the want of a new generation of accelerators. But hopefully the LHC will go on line in 2007 and some other existing accelerators that are now under reconstruction will become operational even before that date. Therefore, one can expect the discovery of the scalar Higgs-boson or even of several 'Higgses'. If such a particle is not discovered (which is difficult to believe), the theory will face great difficulty. However, if new particles or, more specifically, supersymmetric partners of already known particles are not found even on the LHC, this may only signify that the masses of these particles exceed $14\,\mathrm{TeV} = 1.4 \times 10^{13}$ eV. As I understand it, this will not mean anything special. Among the anticipated results, we can point out a further investigation of neutrino oscillations and the determination of the masses of the neutrinos ν_e, ν_μ, and ν_τ. New results concerning the non-conservation of, in particular, CP invariance, at higher energies will also be obtained. It may appear to be important in the analysis of the 'arrow of time' problem. Magnetic monopoles have been sought for many years and the hope for their discovery is now practically gone. But who knows? On new installations (especially on Super Kamiokande), attempts are continuing to discover proton decay. In collisions of relativistic heavy nuclei, progress can be expected in the question of quark-gluon plasma and, generally, quark matter.

In spite of the fact that the forefront of physics—elementary particle physics—is no longer the 'queen of sciences', studies in this field have scaled up and diversified. The future will undoubtedly bring us many new results in this field, too, but it is senseless to enumerate scrupulously here the projects, tasks and separate questions. What is, however, necessary to distinguish is the 'question of questions'—quantum gravity and its unification (superunification) with other (strong and electroweak) interactions. Something of the kind is claimed by string (superstring) theory. To think that string theory is already nearly 30 years old would be an overestimation but even the notorious 'first superstring revolution' took place already 15 years ago (see section 1.4). Nevertheless, an accomplished theory, the 'theory of everything' is out of the question. And the theory of superstrings may not be the way at all in which the future theory will evolve. But can such remarks be treated as a reproach to or an underestimation of string theory? I ask the reader not to think so. This is an exceedingly deep and difficult problem. What are 15 or even 30 years on this way? We have got so used to the rapid development of physics and its successes that we seemingly lose perspective. The same as in economics and population, an exponential growth, in this case the gain of our physical knowledge, cannot last very long. I do not dare make forecasts in the field of quantum cosmology and generally a new and really fundamental theory. Large-scale work on fundamental problems is carried out in all the civilized world and there are a lot of suggestions and hopes (some references were already made in this connection; see [199–202, 205] for more details).

I shall now proceed to what was attributed in the 'list', sometimes conditionally, to astrophysics.

An experimental verification of GR in weak and strong fields is under way and will continue. The most interesting thing would certainly be at least the slightest deviation from GR in the non-quantum region. I am of the intuitive opinion that GR does not need any modification in the non-quantum region (some changes in superstrong gravitational fields may however be necessary but these changes are most likely to be of a quantum nature, i.e. they will disappear as $\hbar \to 0$). Such an assumption is not at all the absolutization of GR. I only mean to say that the applicability range of GR is exclusively classical. Logically, some other restrictions are possible. To make this clear, I shall give an example from Newtonian (classical) mechanics. We know that this system of mechanics is restricted, so to speak, from two sides—relativistic and quantum. Some other restrictions, for instance, in the case of very small accelerations (see [123] and [2, section 23]), are also logically imaginable. The generalization of GR associated with quantum theory is already a great problem which was discussed earlier.

From the very beginning of the 21st century, gravitational waves will be detected by a number of installations now being constructed, first of all LIGO in USA. The first pulses to be received will apparently be those generated by the coalescence of two neutron stars. Correlations with gamma-bursts and with high-energy neutrino radiation are possible and even quite probable. So, gravitational-wave astronomy will be born (its possibilities are described in [73, 203]).

The whole of extra-galactic astronomy, which is now rapidly developing, is connected with cosmology to this or that extent. New wide-aperture telescopes are already operating. For example, two Keck telescopes (on the Hawaiian Islands) have mirrors 10 m in diameter (they were put into service in 1992 and 1996, respectively), while the famous Palomar telescope which has been in operation since 1950 has a mirror 5 m in diameter; the Russian telescope in Zelenchuk (operating since 1976) has a mirror 6 m in diameter. The Hubble Space Telescope launched in 1990 (mirror diameter 2.4 m) is very efficient. New telescopes for various bands (from X-ray to radio wavelength) are continually being built. Worthy of special note are satellites—gamma-ray observatories and installations for the detection of cosmic neutrinos (they can, of course, be called neutrino telescopes). As a result of titanic work on all these telescopes, the value of the Hubble constant will finally be specified and the parameters Ω_b, Ω_d, and Ω_Λ (see section 1.5) evaluated. Thus, the cosmological model, at least at the stage after the formation of cosmic microwave background (i. e. for the redshift parameter $z \lesssim 10^3$) will eventually be selected. Now (at the beginning of 2003) new outstanding observations of cosmic microwave background were obtained by the WMAP satellite [234] (see also [237]). The role of the Λ-term and the contribution of dark matter not only on the average (the parameter Ω_d) but for various objects (the Galaxy, galactic clusters, superclusters, etc) will be determined. I have somehow got to the enumeration of various astronomical problems and objects, which is beyond the scope of this article. New material will be obtained for practically all the problems and questions but disputable, unclear and, to some extent, problematic issues are particularly noteworthy. Such issues include the discovery of black miniholes and cosmic strings (they may be of different types) and some other 'topological defects'.

Since the nature of dark matter and dark energy is absolutely unclear, the solution of this problem may now be thought of as the most important in astronomy if we do not touch upon the principal question of cosmology (the region near the classical singularity, i. e. the quantum region—our Universe as part of a branched and apparently infinite system). The possible means of dark matter studies were already discussed in section 1.5. This is a truly enigmatic problem and success can only be hoped for. But I shall not be surprised if it is solved soon.

In respect of problem 28—the origin of ultrahigh-energy cosmic rays—there is an essential vagueness, as was mentioned in section 1.5. The situation resembles that associated with the origin of dark matter and it is not excluded that these questions are interrelated. The directions of further studies are obvious and they are under way (see comments 10 and 11 to article 3). The same can be said about gamma-ray bursts and neutrino astronomy. Incidentally, the most significant achievements in physics and astrophysics for the past five years has been the proof of the cosmological origin of gamma-ray bursts (more precisely, of their considerable part) and the discovery of neutrino oscillations and, thus, the proof of the fact that at least one sort of neutrinos has a non-zero mass. This should be added by the progress in understanding neutrino emission from the Sun, establishing accelerated expansion of the Universe and revealing the (preliminary) role of the 'dark energy'. Neutrino 'telescopes' for detecting high-energy neutrinos are soon to be put into operation. As has already been mentioned, their simultaneous operation with gravitational antennae and gamma-ray telescopes will undoubtedly be beneficial. As to the detection of relic neutrinos and relic gravitational waves, I am not aware of the situation (in respect of gravitational waves see [141]).

As has already been emphasized, the distinction of some problems among others is rather conditional and is connected with some awkwardness—quite a lot of significant and interesting ones appear to have fallen overboard! I felt this especially keenly when I singled out gamma-ray bursts and did not mention the development of other branches of gamma-ray astronomy (see, e. g., [124]).

Summarizing I can state that almost all the directions discussed earlier are fairly promising. I think that within the coming twenty to thirty years, we shall answer to all the previously mentioned questions except perhaps for the fundamental problems of elementary particle physics (superstrings, etc) and quantum cosmology near classical singularities. I simply dare not foretell anything in these two directions.

Concluding, I would like to return to the three 'great' problems mentioned in section 1.6. As far as the 'arrow of time' is concerned, I do not see any new experiments which might provide an insight into this problem. My intuition suggests that CP and, thus, T invariance violation is of importance. But what can be contributed by new experiments? As to the basic principles of non-relativistic quantum mechanics, the question of interpretation is largely of gnoseological nature. The new refined experiments which are now being carried out to verify the uncertainty relations, the notorious teleportation, etc in no respect go beyond the limits of the known theory. My intuitive judgement is that we shall never be able to predict 'where the electron goes' in diffraction experiments. Future theory (conditionally, superstring theory and its development) may provide some new insights but I cannot imagine what particular results they could be (the concept of time is under suspicion in quantum mechanics). As concerns the third of the 'great problems'—reductionism—I acknowledge my incompetence. Perhaps it is for this reason that I would not be surprised if 'life in a test-tube' were created in the 21st century. But if at all, this may only be achieved by biochemical methods, while physics may play an auxiliary role. One way or another, I cannot make predictions in this field. This relates to the fundamental problems mentioned at the end of section 1.6.

Having finished the article, I clearly see its shortcomings. The large scope of the article accounts for the sketchy manner of the presentation and perhaps for some superficiality. Everything has its price. But the reader will judge whether the price is too high. However, the very idea of the article cannot be discredited by some shortfalls. I call on those who disagree with it for constructive criticism—maybe someone will succeed better where I failed.

Finally I shall make one further remark.

From the previous presentation, it is clear that very many new, important, and interesting things may be anticipated in the coming years and the more so in the first half of the 21st century. The rather pessimistic foresight encountered in the literature concerning the development of physics and astrophysics in the foreseeable future seem to be a result of a lack of information, incompetence, or simply misunderstanding. Another thing is that the exponential law of the development of science in respect of some 'indices' (the number of research workers, the number of publications, etc) is limited in time and a certain saturation sets in (for more details see [2, section 27] and [121]). However, this circumstance does not, on the whole, contradict what has been said earlier, for we have discussed the near future. I think that in about ten years, it will be quite pertinent to write a new article with the same title as the present one. It will be interesting to see what will be realized and how my 'list' will have to be updated by discarding the outdated and adding new items. I hope that there will be a physicist who will do this work and that *Uspekhi Fizicheskikh Nauk* (*Physics–Uspekhi*) will offer some space for it.

In conclusion, I take the opportunity to thank all those whom I consulted on this or that question and who made critical remarks on the manuscript (I do not mention the names because I do not want anybody to be responsible, even indirectly, for the shortfalls of the article.)

Comments

1*. In 1971, I first published the paper "What problems of physics and astrophysics seem now to be especially important and interesting?" [1]. The title speaks for itself and, indeed, the paper contains the list of problems which were distinguished for some reasons. However, in fact, the aim of the paper is more general—some educational program is addressed. The present article explains this in more detail. By essence, its content should change with time. This is what has been done. For example, paper in [2] includes the version 1995 and finally [3] corresponding to 1999 was published in 1999. Then, in 2001, the updated English translation of this paper from [2] and paper [3] were included in my book *The Physics of a Lifetime* (Springer, 2001). The present article is a version as of the middle of 2002 based on the text of [3]. The changes are not very large in volume, which is obvious: only three years (by the time of the final proof of this volume) have passed since paper [3] was published. Nonetheless, several notes and references have been added. Several more small comments have been added during the English translation.

2*. On 21 November, 2001, the 1700th meeting took place where I announced the end of these seminars. I was already 85 years old and everything comes to its end. My talk at the 1700th seminar has been published as FIAN preprint no 48 (2001) and the work of the seminar as a whole will probably be described some day in a book, which is going to be prepared by my collaborators from the Theoretical Department of the FIAN.

3*. Comments on BEC in the main text are absolutely identical to those in the previous edition. I write this remark in connection with the Nobel Prize in physics of 2001. This prize was awarded to three physicists 'for the achievement of Bose–Einstein condensation in dilute gases of alkali atoms, and for early fundamental studies of the properties of the condensates' [165]. It is undoubtedly a bright success in experimental physics but the race for achieving the set objective resembles a sports competition in a swimming pool or on a running-track. As in such competitions, medals are awarded to three people, while tens can participate. Even paper [165] mentions, in addition to the names of the laureates, the names of 21 more physicists (nine of them being direct collaborators of the laureates) who participated in the 'competitions'. Interestingly, the experiments with hydrogen began much earlier than those with alkali metals but the objective (i. e. BEC) was achieved three years later (in 1998) compared to the work by the 2001 Prize laureates. Apparently, only the choice of gas is 'responsible' for this.

A similar situation happened with the 1997 Prize [32]. What I have said has no intention of casting aspersions on the laureates. I have no doubt that the Nobel Committee sought to select, however difficult this could be, the three most distinguished researchers in the chosen field. But the very character of these prizes contrasts too dramatically with the vast majority of prizes awarded for individual and mostly original discoveries and achievements. The bright example can be the award of the first Nobel Prize in physics in 1901 to W Roentgen for the discovery of X-rays. At the same time, as mentioned in the main text of this article, Nobel Prizes in physics as a whole have gained authority, so I think that, in the future, most really great achievements in physics will sooner or later win such a prize (see also [21, 206] and articles 21 and 22 of this volume).

4*. Here a stipulation is in order, considering the possibility of the existence in space of some so far non-detected particles contribution to the 'dark matter' (see later). Naturally, should such particles be discovered, their study will provide one more method (channel) of obtaining astronomical information (see also comment 9 to article 3 from this book).

5*. Recently, a black hole with, so to say, intermediate mass $M \sim (10^3 - 10^4)M_\odot$ has been reported. See [236] on the search for black holes.

6*. See, however, [155] where some possibility of measuring m_{ν_μ} and m_{ν_τ} is suggested.

7*. I took the opportunity and in paper [194], which entered the collection [193], I tried to provide, though modest, but a contribution to the history of the creation and development of quantum mechanics. Let me explain the point here. As is well known, the uncertainty relation, which is connected to the name of W Heisenberg, lies at the focus of quantum mechanics. So, some 30 years ago in England and in the USA, I had many communications with the well-known physicist Felix Bloch (1905–83). Bloch is only four years younger than Heisenberg (1901 76), he finished Leipzig University in 1928 and then for some time (I do not know dates) was Heisenberg's assistant, at that time a professor in the same Leipzig University. Naturally, Bloch was closely connected with Heisenberg and literally told me the following. In his (Bloch's) opinion, the uncertainty relation should be credited not to Heisenberg but to Bohr. When Bloch said this to Heisenberg, he, in Bloch's words, agreed with this and remarked something along the following lines: Bohr expresses himself vaguely, so I put all this in a clearer way. The role of Bohr in understanding quantum mechanics clearly follows from the corresponding paper by Heisenberg [195][16] and some other well-known publications but Bloch's statement goes much farther.

Undoubtedly, Heisenberg was one of the outstanding theoreticians in the physics of the 20th century but he was a very contradictory person with

---

[16] This paper appeared in a Russian translation in the recently published selected works of Heisenberg [196]. Clearly, the translators have worked hard to publish this book. However, for absolutely unclear reasons, they have not provided this book with notes or comments and have also included in this book both the papers by Heisenberg (written alone or with co-authors) and two papers by other authors (true, their authorship is indicated).

different aspects: this will be touched upon in article 30 from this book.

Incidentally, I note that, in 1959, Heisenberg visited the USSR for a physics conference (in Kiev). Unfortunately, I could not attend this conference and, in particular, meet Heisenberg. I was just refused permission to attend the conference under the pretence that my work was secret, though I had not been involved in any classified research at that time and, in general, I have never been at the center of top secret studies. Many more 'well-informed' physicists participated in the conference. Nor had Landau initially been admitted to this meeting and I remember he told me that he would go there anyway, creating quite a scandal. Because the eyes of the world were on this international conference, they simply had to let him attend. I did not even try to do anything. I have no ability to demand anything and, most probably, would not achieve any result despite my best intentions. What could I do—I have never been beloved by the KGB or the Central Committee of the CPSU.

8*. As of July 2002 (see 2002 *Physics World* **15**(7) 44), a somewhat reduced variant of the ITER installation costing four billion dollars is planned. It should be completed about 10 years after the installation site is chosen. For orientation, I would like to chose 2016 as the operational starting date, i.e. one hundred years after my date of birth. I also refer to the review on the thermonuclear fusion problem [212, 213].

9*. In paper [215], the value $\Omega_m = \Omega_d + \Omega_b = 0.30^{+0.04}_{-0.03}$ is quoted. According to [184], $\Omega_b = 0.044 \pm 0.01\Omega_c$. According to [222], $\Omega_b = 4\%$, $\Omega_d = 23\%$, $\Omega_\Lambda = 73\%$, and the 'age' of the Universe $t_U = 13.7 \times 10^9$ years (this is time counted from the beginning of expansion). Paper [234] gives $\Omega_m = 0.27 \pm 0.04$, $\Omega_\Lambda = 0.73 \pm 0.04$, and $\Omega = 1 \pm 0.02$.

10*. In my presentation of cosmological problems, I share the commonly accepted point of view: I think it is the correct one. But it should be noted that, in the astronomical community, as well as in others, there are 'heretics' who object and criticize the 'orthodox' views. Among such objections, I would like to note only the point of view of G Burbidge, who is a qualified astronomer and the editor of the *Annual Review of Astronomy and Astrophysics*. In essence, based on observational data, Burbidge argues that the redshift in the spectra of some cosmological objects arises not only from the expansion of the Universe but also in part from some other reasons [235]. Astronomers whose opinion I could canvass have a negative opinion about Burbidge's statement. However, here we deal with observational data which can be analyzed to know who is actually right. Of course, it is inadmissible to reject non-orthodox opinions on 'general grounds' without testing them. Hopefully, the corresponding analysis will be carried out.

References[17]

[1] Ginzburg V L 1971 *Usp. Fiz. Nauk* **103** 87 [1971 *Sov. Phys. Usp.* **14** 21]; 1985 *Physics and Astrophysics. A Selection of Key Problems* (New York: Pergamon Press)

[2] Ginzburg V L 2001 *The Physics of a Lifetime* (Berlin: Springer)[18]

[3] Ginzburg V L 1999 *Usp. Fiz. Nauk* **169** 419 [1999 *Phys. Usp.* **42** 353]

[4] Ginzburg V L 1999 *Vestn. Ross. Akad. Nauk* **69** 546 [1999 *Herald Russ. Acad. Sci.* **69**(4) 27]—article 31 in this volume

[5] 1996 *Priroda* no 9 84

[6] Khvolson O D 1932 *Fizika Nashikh Dnei* [*Physics of Our Days*] 4th edn (Leningrad-Moscow: GTTI)

[7] Ginzburg V L 1996 *Usp. Fiz. Nauk* **166** 1033 [1996 *Phys. Usp.* **39** 973]

[8] Ginzburg V L 1990 *Physics Today* **43**(5) 9

[9] 1991 *Physics Today* **44**(3) 13

[10] Smalley R E 1997 *Rev. Mod. Phys.* **69** 723 [Translated into Russian 1998 *Usp. Fiz. Nauk* **168** 323]

Curl R F 1997 *Rev. Mod. Phys.* **69** 691 [Translated into Russian 1998 *Usp. Fiz. Nauk* **168** 331]

Kroto H 1997 *Rev. Mod. Phys.* **69** 703 [Translated into Russian 1998 *Usp. Fiz. Nauk* **168** 343] (Nobel Lectures in Chemistry)

[11] Ginzburg V L 1962 *Tr. Fiz. Inst. Akad. Nauk SSSR* **18** 55

[12] Todd T N and Windsor C G 1998 *Contemp. Phys.* **39** 255

Kulcinski G L and Santarius J F 1998 *Nature* **396** 724

Smirnov V P 2003 *Vestn. Ross. Akad. Nauk* **73** 299 [2003 *Herald Russ. Acad. Sci.* **73** 136]

[13] 2001 *Usp. Fiz. Nauk* **171** 877 [2001 *Phys. Usp.* **44** 835]

[14] Hoffman A 1998 *Phys. World* **11**(12) 25; 2000 *Nature* **408** 302

[15] Ginzburg V L 1997 *Usp. Fiz. Nauk* **167** 429; 1998 **168** 363 [1997 *Phys. Usp.* **40** 407; 1998 **41** 307]

[16] Ruvalds J 1996 *Supercond. Sci. Technol.* **9** 905

[17] Ford P J and Saunders G A 1997 *Contemp. Phys.* **38** 63

[18] Hemley R J and Ashcroft N W 1998 *Physics Today* **51**(8) 26

Kitamura H *et al* 2000 *Nature* **404** 259

[19] Côté M *et al* 1998 *Phys. Rev. Lett.* **81** 697

Collins P G *et al* 1999 *Phys. Rev. Lett.* **82** 165

[20] Crawford E, Sime R L and Walker M 1997 *Physics Today* **50**(9) 26

[21] Ginzburg V L 1998 *Vestn. Ross. Akad. Nauk* **68** 51 [1998 *Herald Russ. Acad. Sci.* **68**(1) 56]—article 21 in this volume. See also article 22

[22] 1998 *Physics Today* **51**(12) 17

[23] Dorozhkin S I *et al* 1998 *Usp. Fiz. Nauk* **168** 135 [1998 *Phys. Usp.* **41** 127]

---

[17] The literature on the problems touched upon in this article is inconceivably extensive. I have tried to make only a minimum number of references allowing the reader to somehow catch hold of the corresponding material. Preference was given to references to the most easily accessible journals (*Physics-Uspekhi, Physics Today, Physics World*, etc) and the most recent publications known to me and those containing many references.

[18] The English version was a revised and updated translation of the Russian edition 1995 *O Fizike i Astrofizike* [*On Physics and Astrophysics*] (Moscow: Byuro Kvantum).

[24] 1998 *Usp. Fiz. Nauk* **168**(2) [1998 *Phys. Usp.* **41**(2)]

[25] 1998 *Physics Today* **51**(12) 22

[26] Altshuler B L and Maslov D V 1999 *Phys. Rev. Lett.* **82** 145

[27] Lee D M 1997 *Rev. Mod. Phys.* **69** 645 [Translated into Russian 1997 *Usp. Fiz. Nauk* **167** 1307]

Osheroff D D 1997 *Rev. Mod. Phys.* **69** 667 [Translated into Russian 1997 *Usp. Fiz. Nauk* **167** 1327]

Richardson R C 1997 *Rev. Mod. Phys.* **69** 683 [Translated into Russian 1997 *Usp. Fiz. Nauk* **167** 1340] (Nobel Lectures in Physics)

[28] Einstein A 1925 *Sitzungsber. Preuss. Akad. Wiss., Phys.-Math. Kl.*, 3 [Translated into Russian 1965 *Sobranie Nauchnykh Trudov* [*Collection of Scientific Papers*] vol 1 (Moscow: Nauka) p 489; Einstein A 1925 *Berl. Ber.* (1/2) 3

[29] Landau L D and Lifshitz E M 1988 *Statisticheskya Fizika* [*Statistical Physics*] (Moscow: Fizmatlit) [Translated into English 1994 (Oxford: Pergamon Press)]

[30] London F 1938 *Nature* **141** 643

[31] Kleppner D 1996 *Physics Today* **49**(8, Pt 1) 11

[32] Chu S 1998 *Rev. Mod. Phys.* **70** 685 [Translated into Russian 1999 *Usp. Fiz. Nauk* **169** 274]

Cohen-Tannoudji C N 1998 *Rev. Mod. Phys.* **70** 707 [Translated into Russian 1999 *Usp. Fiz. Nauk* **169** 292]

Phillips W D 1998 *Rev. Mod. Phys.* **70** 721 [Translated into Russian 1999 *Usp. Fiz. Nauk* **169** 305] (Nobel Lectures in Physics)

[33] Kadomtsev B B and Kadomtsev M B 1997 *Usp. Fiz. Nauk* **167** 649 [1997 *Phys. Usp.* **40** 623]

[34] Pitaevskii L P 1998 *Usp. Fiz. Nauk* **168** 641 [1998 *Phys. Usp.* **41** 569]

[35] Kleppner D 1997 *Physics Today* **50**(8, Pt 1) 11

[36] Hutchinson D A W 1999 *Phys. Rev. Lett.* **82** 6

[37] Lehmann W *et al* 2001 *Nature* **410** 447

[38] Auciello O, Scott J F, and Ramesh R 1998 *Physics Today* **51**(7) 22

Bune A V *et al* 1998 *Nature* **391** 874

[39] Chesnokov S A *et al* 1999 *Phys. Rev. Lett.* **82** 343; 2000 *Physics World* **13**(6) 29

Critical politics of carbon sinks 2000 *Nature* **408** 501

Ball P 2001 *Nature* **414** 142

Zhang P and Crespi V H 2002 *Phys. Rev. Lett.* **89** 056403

Eletskii A V 2002 *Usp. Fiz. Nauk* **172** 401 [2002 *Phys. Usp.* **45** 369]

[40] 1999 *Phys. Rev. Lett.* **82**(3)

[41] Ryskin N M and Trubetskov D I 2000 *Nelinejnye Volny* [*Nonlinear Waves*] (Moscow: Fizmatlit)

[42] Mourou G A, Barty C P J, and Perry M D 1998 *Physics Today* **51**(1) 22

Niikura H *et al* 2002 *Nature* **417** 917

[43] Kapteyn H and Murnane M 1999 *Phys. World* **12**(1) 33

[44] Wilczek F 1998 *Nature* **395** 220

[45] Okun L B 1998 *Usp. Fiz. Nauk* **168** 625 [1998 *Phys. Usp.* **41** 553]

[46] Mavromatos N 1998 *Phys. World* **11**(12) 21; 1999 *Physics Today* **52**(2) 19

[47] 1999 *Phys. World* **12**(1) 5

1999 *Physics Today* **52**(1) 22

[48] Heitler W 1954 *The Quantum Theory of Radiation* 3rd edn (Oxford: Clarendon Press) [Translated into Russian 1956 (Moscow: IL)]

[49] Landau L D and Lifshitz E M 1988 *Teoriya Polya* [*The Classical Theory of Fields*] (Moscow: Fizmatlit) [Translated into English 1983 (Oxford: Pergamon Press)]

[50] Sessler A M 1998 *Physics Today* **51**(3) 48

[51] 1997 *Critical Problems in Physics* ed V L Fitch, D R Marlow and M A E Dementi (Princeton, NJ: Princeton University Press)

[52] Ginzburg V L 1996 *Usp. Fiz. Nauk* **166** 169 [1996 *Phys. Usp.* **39** 155]

[53] O'Halloran T, Sokolsky P and Yoshida S 1998 *Physics Today* **51**(1) 31

[54] Gleiser M 1998 *Contemp. Phys.* **39** 239

[55] Einstein A 1920 *Äther und Relativitätstheorie* (Berlin: von Julius Springer) [Translated into Russian 1965 *Sobranie Nauchnykh Trudov* (*Collection of Scientific Papers*) vol 1 (Moscow: Nauka) p 682]

[56] Wilczek F 1999 *Physics Today* **52**(1) 11

[57] Witten E 1997 *Physics Today* **50**(5) 28

[58] Schwarz J H 1998 *Proc. Natl Acad. Sci. USA* **95** 2750

[59] Gauntlett J P 1998 *Contemp. Phys.* **39** 317

[60] Kane G 1997 *Physics Today* **50**(2) 40

[61] Ginzburg V L and Tamm I E 1947 *Zh. Eksp. Teor. Fiz.* **17** 227

[62] Ginzburg V L and Man'ko V I 1976 *Fiz. Elem. Chastits At. Yadra* **7** 3 [1976 *Sov. J. Part. Nucl.* **7** 1]

[63] Kaluza Th 1921 *Berl. Ber.* 966

[64] Klein O 1926 *Nature* **118** 516; 1927 *Z. Phys.* **46** 188

[65] Pauli W 1958 *Theory of Relativity* (New York: Pergamon Press) [Translated into Russian 1991 (Moscow: Fizmatlit)]

[66] Kuhn T S 1970 *The Structure of Scientific Revolutions* (Chicago: The University of Chicago Press) [Translated into Russian 1975 (Moscow: Progress)]

[67] Will C M 1993 *Theory and Experiment in Gravitational Physics* (Cambridge: Cambridge University Press) [Translated into Russian 1985 (Moscow: Energoatomizdat)]; 1999 *Physics Today* **52**(10) 38

[68] Zakharov A F 1997 *Gravitatsionnye Linzy i Mikrolinzy* [*Gravitation and Microlenses*] (Moscow: Yanus-K)
Zakharov A F and Sazhin M V 1998 *Usp. Fiz. Nauk* **168** 1041 [1998 *Phys. Usp.* **41** 945]

[69] Khvolson O 1924 *Astron. Nachrichten* **221** 329

[70] Einstein A 1936 *Science* **84** 506 [Translated into Russian 1966 *Sobranie Nauchnykh Trudov* (*Collection of Scientific Papers*) vol 2 (Moscow: Nauka) p 436]

[71] Stella L and Vietri M 1999 *Phys. Rev. Lett.* **82** 17

[72] Hulse R A 1994 *Rev. Mod. Phys.* **66** 699 [Translated into Russian 1994 *Usp. Fiz. Nauk* **164** 743]
Taylor J H (Jr) 1994 *Rev. Mod. Phys.* **66** 711 [Translated into Russian 1994 *Usp. Fiz. Nauk* **164** 757] (Nobel Lectures in Physics)

[73] Barish B C and Weiss R 1999 *Physics Today* **52**(1) 44
Braginskii V B 2000 *Usp. Fiz. Nauk* **170** 743 [2000 *Phys. Usp.* **43** 691]

58   *References*

[74] Shi X, Fuller G M and Halzen F 1998 *Phys. Rev. Lett.* **81** 5722

[75] Kundic T *et al* 1997 *Astrophys. J.* **482** 75

[76] Novikov I D 1983 *Evolyutsiya Vselennoi* [*Evolution of the Universe*] (Moscow: Nauka) [Translated into English 1983 (Cambridge: Cambridge University Press)]

[77] Peebles P J E 1993 *Principles of Physical Cosmology* (Princeton: Princeton University Press) [Translation of the previous edition: Peebles P J 1975 *Fizicheskaya Kosmologiya* [*Physical Cosmology*] (Moscow: Mir)]

[78] Einstein A 1917 *Berl. Ber.* (1) 142 [Translated into Russian 1965 *Sobranie Nauchnykh Trudov* (*Collection of Scientific Papers*) vol 1 (Moscow: Nauka) p 601]

[79] Einstein A 1931 *Berl. Ber.* (12) 235 [Translated into Russian 1966 *Sobranie Nauchnykh Trudov* (*Collection of Scientific Papers*) vol 2 (Moscow: Nauka) p 349]

[80] Gliner E B 1965 *Zh. Eksp. Teor. Fiz.* **49** 542 [1966 *Sov. Phys. JETP* **22** 378]

[81] Weinberg S 1989 *Usp. Fiz. Nauk* **158** 639 [Translated from 1989 *Rev. Mod. Phys.* **61** 1]

[82] Ginzburg V L 1970 *Pul'sary* [*Pulsars*] (*New Phenomena in Life, Science, Technology.* Ser. Fiz., Astron., issue 2) (Moscow: Znanie)

[83] Hurley K *et al* 1999 *Nature* **397** 41
Sancetta C 1999 *Nature* **398** 27

[84] Beskin V S, Gurevich A V and Istomin Ya N 1986 *Usp. Fiz. Nauk* **150** 257 [1986 *Sov. Phys. Usp.* **29** 946]; 1993 *Physics of the Pulsar Magnetosphere* (Cambridge: Cambridge University Press)

[85] Beskin V S 1999 *Usp. Fiz. Nauk* **169** 1169 [1999 *Phys. Usp.* **42** 1071]

[86] Vilenkin A and Shellard E P S 1994 *Cosmic Strings and Other Topological Defects* (Cambridge: Cambridge University Press)

[87] Oppenheimer J R and Snyder H 1939 *Phys. Rev.* **56** 455

[88] Frolov V P and Novikov I D 1998 *Black Hole Physics* (*Fundamental Theories of Physics*, vol. 96) (Dordrecht: Kluwer) [The first edition of this book was also published in Russian Novikov I D and Frolov V P 1986 *Fizika Chernykh Dyr* [*Physics of Black Holes*] (Moscow: Nauka)]

[89] 1998 *Physics Today* **51**(3) 21
Ghez A M *et al* 2000 *Nature* **407** 349

[90] Hawking S 1974 *Nature* **248** 30

[91] Bekenstein J D and Schiffer M 1998 *Phys. Rev. D* **58** 064014

[92] Zwicky F 1933 *Helv. Phys. Acta* **6** 110
Oort J H 1940 *Astrophys. J.* **91** 273
Blitz L, Fich M and Kulkarni S 1983 *Science* **220** 1233
Rubin V C 1983 *Science* **220** 1339

[93] Tinney C G 1999 *Nature* **397** 37

[94] Elgarøy Ø *et al* 2002 *Phys. Rev. Lett.* **89** 061301

[95] Pretzl K P 1993 *Europhys. News* **24** 167

[96] Gurevich A V, Zybin K P, and Sirota V A 1997 *Usp. Fiz. Nauk* **167** 913 [1997 *Phys. Usp.* **40** 869]

[97] Gill A J 1998 *Contemp. Phys.* **39** 13

[98] Berezinsky V 1999 *Nucl. Phys. B* (Proc. Suppl.) **70** 419

Schwarzschild B 1998 *Physics Today* **51**(10) 19

[99] Klebesadel R W, Strong I B, and Olson R A 1973 *Astrophys. J. Lett.* **182** L85

[100] Schwarzschild B 1997 *Physics Today* **50**(6) 17; **50**(7) 17

[101] McNamara B J and Harrison T E 1998 *Nature* **396** 233

[102] Kulkarni S R *et al* 1998 *Nature* **393** 35

[103] Postnov K A 1999 *Usp. Fiz. Nauk* **169** 545 [1999 *Phys. Usp.* **42** 469]

[104] Reines F 1996 *Rev. Mod. Phys.* **68** 317 (Nobel Lectures in Physics) [Translated into Russian 1996 *Usp. Fiz. Nauk* **166** 1352]

[105] Perl M L 1996 *Usp. Fiz. Nauk* **166** 1340; 1997 *Physics Today* **50**(10) 34

[106] Perkins D H 1997 in *Critical Problems in Physics* ed V L Fitch, D R Marlow and M A E Dementi (Princeton: Princeton University Press) p 201

[107] Wolfenstein L 1996 *Contemp. Phys.* **37** 175

[108] Schwarzschild B 1998 *Physics Today* **51**(8) 17

[109] Fukuda Y *et al* 1998 *Phys. Rev. Lett.* **81** 1562

[110] Fukuda Y *et al* 1998 *Phys. Rev. Lett.* **81** 1158
Fukuda S *et al* 2000 *Phys. Rev. Lett.* **85** 3999

[111] Baltz A J, Goldhaber A S, and Goldhaber M 1998 *Phys. Rev. Lett.* **81** 5730

[112] Bahcall J N *et al* 1995 *Nature* **375** 29

[113] Landau L D and Lifshitz E M 1989 *Kvantovaya Mekhanika. Nerelyativistskaya Teoriya* [*Quantum Mechanics. Non-Relativistic Theory*] (Moscow: Fizmatlit) [Translated into English 1977 (Oxford: Pergamon Press)]

[114] Mandelshtam L I 1950 *Polnoe Sobranie Trudov* [*Complete Collection of Papers*] ed M A Leontovich vol 5 (Leningrad: Izd. Akad. Nauk SSSR) p 347

[115] Kadomtsev B B 1997 *Dinamika i Informatsiya* [*Dynamics and Information*] (Moscow: Red. Zhurn. Usp. Fiz. Nauk); 2nd edn 1999 (Moscow: Red. Zhurn. Usp. Fiz. Nauk)

[116] Bub J 1997 *Interpreting the Quantum World* (Cambridge: Cambridge University Press)

[117] Klyshko D N 1998 *Usp. Fiz. Nauk* **168** 975 [1998 *Phys. Usp.* **41** 885]

[118] Haroche S 1998 *Physics Today* **51**(7) 36

[119] Goldstein S 1998 *Physics Today* **51**(3) 42; **51**(4) 38; see also the discussion of this subject: 1999 *Physics Today* **52**(2) 11

[120] Whitaker A 1998 *Phys. World* **11**(12) 29

[121] Feinberg E L 1992 *Dve Kul'tury: Intuitsiya i Logika v Iskusstve i Nauke* [*Two Cultures: Intuition and Logic in Arts and Sciences*] (Moscow: Nauka); revised edition Feinberg E L 1998 *Zwei Kulturen* (Berlin: Springer); 1997 *Voprosy Filosofii* no 7 54; [expanded 3rd Russian edn 2004 (Fryazino: Vek 2)]

[122] Ginzburg V L 1998 *Poisk* nos 29–30

[123] Milgrom M 1993 *Astrophys. J.* **270** 363; 1994 *Ann. Phys.* (N. Y.) **229** 384

[124] Gehrels N and Paul J 1998 *Physics Today* **51**(2) 26

[125] 1999 *Nature* **397** 289; 1999 *Phys. World* **12**(2) 7; 1999 **12**(3) 19

[126] Shiozawa M *et al* 1998 *Phys. Rev. Lett.* **81** 3319

[127] Collins C A 1999 *Contemp. Phys.* **40** 1

[128] Hijmans T 1999 *Physics Today* **52**(2) 17

[129] Bildsten L and Strohmayer T 1999 *Physics Today* **52**(2) 40

60   *References*

[130] Williams G A 1999 *Phys. Rev. Lett.* **82** 1201

[131] Ellis J 1998 *Proc. Natl Acad. Sci. USA* **95** 53

[132] Rosenberg L J 1998 *Proc. Natl Acad. Sci. USA* **95** 59

[133] Peebles P J E 1999 *Nature* **398** 25

[134] 1999 *Phys. World* **12**(3) 12

[135] Smirnov B M 1993 *Usp. Fiz. Nauk* **163** 29; 1994 **164** 1165; 1997 **167** 1169 [1993 *Phys. Usp.* **36** 933; 1994 **37** 1079; 1997 **40** 1117]

[136] Rafelski J and Müller B 1999 *Phys. World* **12**(3) 23

[137] Sorge H 1999 *Phys. Rev. Lett.* **82** 2048

[138] Heiselberg H 1999 *Phys. Rev. Lett.* **82** 2052

[139] Contaldi C, Hindmarsh M, and Magueijo J 1999 *Phys. Rev. Lett.* **82** 2034

[140] 1999 *Physics Today* **52**(3)

[141] Grishchuk L P *et al* 2001 *Usp. Fiz. Nauk* **171** 3 [2001 *Phys. Usp.* **44** 1]

[142] Maksimov E G and Shilov Yu I 1999 *Usp. Fiz. Nauk* **169** 1223 [1999 *Phys. Usp.* **42** 1121]

[143] Laughlin R B 1999 *Rev. Mod. Phys.* **71** 863 [Translated into Russian 2000 *Usp. Fiz. Nauk* **170** 292]

      Störmer H L 1999 *Rev. Mod. Phys.* **71** 875 [Translated into Russian 2000 *Usp. Fiz. Nauk* **170** 304]

      Tsui D C 1999 *Rev. Mod. Phys.* **71** 891 [Translated into Russian 2000 *Usp. Fiz. Nauk* **170** 320] (Nobel Lectures in Physics)

[144] Scarola V M, Park K and Jain J K 2000 *Nature* **406** 863

[145] Kocharovskaya O, Kolesov R and Rostovtsev Yu 1999 *Laser Phys.* **9** 745

      Kocharovskaya O 1992 *Phys. Rep.* **219** 175

[146] 2000 *Phys. World* **13**(10) 37

[147] Kalmus P 2000 *Contemp. Phys.* **41** 129

[148] Abel S and March-Russell J 2000 *Phys. World* **13**(11) 39

[149] 2000 *Phys. World* **13**(6) 3

[150] 2000 *Gen. Relat. Gravit.* **32**(6)

[151] Caldwell R and Steinhardt R 2000 *Phys. World* **13**(11) 31

[152] Armendariz-Picon C, Mukhanov V and Steinhardt P J 2000 *Phys. Rev. Lett.* **85** 4438

[153] Harvey A and Schuking E 2000 *Am. J. Phys.* **68** 723

[154] 2000 *Phys. World* **13**(11) 3

[155] Beacom J F, Boyd R N and Mezzacappa A 2000 *Phys. Rev. Lett.* **85** 3568

[156] Wilczek F 2000 *Physics Today* **53**(6) 11; 2002 **55**(8) 10

[157] Loubeyre P, Occelli F and Le Toullec R 2002 *Nature* **416** 613

[158] Chernavskii D S 2000 *Usp. Fiz. Nauk* **170** 157 [2000 *Phys. Usp.* **43** 151]

[159] Kane G L 2000 *Contemp. Phys.* **41** 359

[160] Tigher M 2001 *Physics Today* **54**(1) 36

[161] Nagamatsu J *et al* 2001 *Nature* **410** 63

      Bud'ko S L *et al* 2001 *Phys. Rev. Lett.* **86** 1877

[162] Ginzburg V L 2001 *Usp. Fiz. Nauk* **171** 1091 [2001 *Phys. Usp.* **44** 1037]

[163] Ginzburg V L 2001 *Nedodumannoe, nedodelannoe...* [*Left Unthought, Unfinished...*] Preprint 34 (Moscow: FIAN)

[164] Feng X G *et al* 2001 *Phys. Rev. Lett.* **86** 2625

[165] 2001 *Physics Today* **54**(12) 14

[166] Si Q *et al* 2001 *Nature* **413** 804

[167] Blinov L M *et al* 2000 *Usp. Fiz. Nauk* **170** 247 [2000 *Phys. Usp.* **43** 243]

[168] Bratkovsky A M and Levanyuk A P 2001 *Phys. Rev. Lett.* **86** 3642

[169] Sannikov D G 2001 *Pis'ma Zh. Eksp. Teor. Fiz.* **73** 447 [2001 *JETP Lett.* **73** 401]

[170] Oganessian Y 2001 *Nature* **413** 122; 2001 *Vestn. Ross. Acad. Nauk* **71** 509 [2001 *Herald Russ. Acad. Sci.* **71** 379]

[171] Hands S 2001 *Contemp. Phys.* **42** 209

[172] Green M G and McMahon T R 2001 *Contemp. Phys.* **42** 339
       Abe K *et al* 2002 *Phys. Rev. Lett.* **89** 071801

[173] Mukhin K N and Tikhonov V N 2001 *Usp. Fiz. Nauk* **171** 1201 [2001 *Phys. Usp.* **44** 1141]

[174] Rubakov V A 2001 *Usp. Fiz. Nauk* **171** 913 [2001 *Phys. Usp.* **44** 871]

[175] Weiss D 2000 *Phys. World* **13**(3) 23

[176] Salis G *et al* 2001 *Nature* **414** 619
       Ganichev S D *et al* 2002 *Nature* **417** 153; 2002 *J. Supercond.* **15**(1)

[177] Hogan C J 2000 *Nature* **408** 47

[178] Dymnikova I G 1986 *ZhETF* **90** 1900 [1986 *Sov. Phys. JETP* **63** 1111]

[179] Gliner E B 2002 *Usp. Fiz. Nauk* **172** 221 [2002 *Phys. Usp.* **45** 213]

[180] Taylor A and Peacock J 2001 *Physics World* **14**(3) 37
       Schwartschild B 2001 *Physics Today* **54**(6) 17

[181] Novikov I D 2001 *Vestn. Ross. Acad. Nauk* **71** 886 [2001 *Herald Russ. Acad. Sci.* **71** 493]

[182] Volovik G E 2001 *J. Low Temp. Phys.* **124** 25

[183] Polarski D 2001 *J. Low Temp. Phys.* **124** 17

[184] Turner M S 2001 *Physics Today* **54**(12) 10

[185] Sarbu N, Rusin D, and Ma C 2001 *Astrophys. J. Lett.* **561** L147

[186] Alcock C *et al* 2001 *Nature* **414** 617

[187] Cronin J W 1999 *Rev. Mod. Phys.* **71** S165

[188] Nagano M and Watson A A 2000 *Rev. Mod. Phys.* **72** 689

[189] 2001 *Physics Today* **54**(8) 13

[190] Pennicott K 2001 *Phys. World* **14**(7) 5

[191] Menskii M B 2000 *Usp. Fiz. Nauk* **170** 631 [2000 *Phys. Usp.* **43** 585]

[192] 2001 Responses to M B Menskii's paper *Usp. Fiz. Nauk* **171** 437; 2002 *Usp. Fiz. Nauk* **172** 843 [2001 *Phys. Usp.* **44** 417; 2002 **45** 783]

[193] 2002 *100 Let Kvantovoi Teorii (Istorija, Fizika, Filosofija)* [*100 Years of Quantum Theory (History, Physics, Philosophy)*] editor-in-chief E A Mamchur (Moscow: NIA–Priroda)

[194] Ginzburg V L 2002 K stoletnemu jubileju kvantovoi teorii [To quantum theory centenary] paper in [193]

[195] Heisenberg W 1927 *Z. Phys.* **43** 172

[196] Heisenberg W 2001 *Izbrannye Trudy* [*Selected Works*] (Moscow: URSS) paper [195] is on p 209

[197] Hentschel M *et al* 2001 *Nature* **414** 509; see also Silberberg Y 2001 *Nature* **414** 494; 2002 *Phys. World* **15**(1) 25

[198] Guthöhrlein G R *et al* 2001 *Nature* **414** 49; see also Steane A 2001 *Nature* **414** 24

[199] Rivers D J 2001 *J. Low Temp. Phys.* **124** 41

[200] Smolin L 2001 *Phys. World* **14**(11) 19

[201] Wilczek F 2001 *Physics Today* **54**(6) 12; **54**(11) 12

[202] 2001 *Quantization, Gauge Theory, and Strings.* vols I, II. (*Proc. Int. Conf.
       Dedicated to the Memory of Professor Efim Fradkin, Moscow, June 5–10,
       2000*) ed A Semikhatov, M Vasiliev and V Zaikin (Moscow: Scientific
       World)
       Voronov B L 2001 *Usp. Fiz. Nauk* **171** 869 [2001 *Phys. Usp.* **44** 827]

[203] Andersson N and Comer G L 2001 *Phys. Rev. Lett.* **87** 241101

[204] Chernin A D 2001 *Usp. Fiz. Nauk* **171** 1153 [2001 *Phys. Usp.* **44** 1099]

[205] Rodgers P 2001 *Phys. World* **14**(12) 5

[206] Ginzburg V L 2002 *Usp. Fiz. Nauk* **172** 213 [2002 *Phys. Usp.* **45** 205]

[207] de Bernardis P *et al* 2000 *Nature* **404** 955

[208] Silk J 2000 *Phys. World* **13**(6) 23

[209] Frail D *et al* 2001 *Astrophys. J. Lett.* **563** L55
       Hattori M and Taylor T D 2001 *Nature* **414** 854

[210] Greiner M *et al* 2002 *Nature* **415** 39
       Osterloh *et al* 2002 *Nature* **416** 608

[211] Dienes K R 2002 *Phys. Rev. Lett.* **88** 011601

[212] Hazaltine R D and Prager S C 2002 *Physics Today* **55**(7) 30

[213] Marshakov A V 2002 *Usp. Fiz. Nauk* **172** 977 [2002 *Phys. Usp.* **45** 915]

[214] Peebles P J E 2000 in *New Cosmological Data and the Value of the Funda-
       mental Cosmological Parameters: Proc. IAU Symp. 201* ed A Lasenby,
       A Wilkinson and A W Jones (San Francisco: Pacific Astronomical Soci-
       ety) p 1

[215] Allen S W, Schmidt R W, and Fabian A C 2002 *Mon. Not. R. Astron. Soc.*
       334 L11

[216] Blau S K 2002 *Physics Today* **55**(7) 18
       Bloom J S *et al* 2002 *Astrophys. J. Lett.* **572** L45
       Price P A *et al* 2002 *Astrophys. J. Lett.* **572** L51
       Lazzati D, Ramirez-Ruiz E, and Rees M J 2002 *Astrophys. J. Lett.* **572**
       L57
       Mazzali P A *et al* 2002 *Astrophys. J. Lett.* **572** L61

[217] Ahmad Q R *et al* 2002 *Phys. Rev. Lett.* **89** 011301 011302; 2002 *Physics
       Today* **55**(7) 13

[218] Krokhin O N 2002 *Usp. Fiz. Nauk* **172** 1466 [2002 *Phys. Usp.* **45** 1305]

[219] Alferov Zh I 2001 *Rev. Mod. Phys.* **73** 767 [Translated into Russian 2002
       Usp. Phys. Nauk **172** 1068] (Nobel Lecture in Physics)

[220] Vedyaev A V 2002 *Usp. Fiz. Nauk* **172** 1458 [2002 *Phys. Usp.* **45** 1296]

[221] Goldberg J *et al* 2003 *Nature* **422** 599

[222] Fukugita M 2003 *Nature* **422** 489

[223] Pitaevskii L and Stringari S 2003 *Bose–Einstein Condensation* (Oxford:
       Clarendon Press)

[224] Junquera J and Ghosez P 2003 *Nature* **422** 506

[225] Tisch J and Marangos J 2003 *Phys. World* **16**(4) 19

[226] Long J C *et al* 2003 *Nature* **421** 922

[227] Rubakov V A 2003 *Usp. Fiz. Nauk* **173** 219 [2003 *Phys. Usp.* **46** 211]

[228] 2001 *Current Science* **81**(12) [This journal is published by the Indian
       Academy of Sciences]

[229] Peacock J A 2002 *Cosmological Physics* (Cambridge: Cambridge University

Press)

[230] 2003 *Physics Today* **56**(2) 19

[231] Fox D W *et al* 2003 *Nature* **422** 284

[232] 2003 *Physics Today* **56**(3) 14

[233] Ahn M H *et al* 2003 *Phys. Rev. Lett.* **90** 041801; 2003 *Phys. World* **16**(3)

[234] Ferreira P G 2003 *Phys. World* **16**(4) 27
       Schwarzschild B 2003 *Physics Today* **56**(4) 21

[235] Burbidge G R 2001 *Publ. Astron. Soc. Pacific* **113** 899; 2003 *Astrophys. J.*
       585 112

[236] Cherepashchuk A M 2003 *Usp. Fiz. Nauk* **173** 345 [2003 *Phys. Usp.* **46**
       335]

[237] Linder E V 2003 *Phys. Rev. Lett.* **90** 091301

[238] Madsen J and Larsen J M 2003 *Phys. Rev. Lett.* **90** 121102

[239] Perlmutter S 2003 *Phys. Today* **56**(4) 53

[240] Turner M S 2003 *Phys. Today* **56**(4) 10

[241] 2003 *Phys. Today* **56**(4) 24

[242] 2003 *Phys. Today* **56**(4) 27

[243] Junquera J and Ghosez P 2003 *Nature* **422** 506

[244] DeDeo S and Psaltis D 2003 *Phys. Rev. Lett.* **90** 141101

[245] Coburn W and Boggs S E 2003 *Nature* **423** 415
       Waxman E 2003 *Nature* **423** 388

Article 2

Radiation by uniformly moving sources (Vavilov–Cherenkov effect, Doppler effect in a medium, transition radiation, and other phenomena)[1*]

2.1 Introduction

In beginning this lecture, I wish to thank the Presidium of the Russian Academy of Sciences for awarding me the M V Lomonosov Great Golden Medal, the highest award of the Academy. Certainly it is a great honor for everybody to be distinguished in such a way but I have more reasons to be proud. The fact is that I have been with the Academy (at the P N Lebedev Physical Institute [FIAN]) since 1940, i.e. almost as long as my scientific career. Equally important for me is that I E Tamm, my teacher, was also awarded the M V Lomonosov Medal in 1968. Finally, it is a pleasure to share the honor with such an eminent physicist as A Abraham.

Now, a few words about this choice of lecture. I have been interested in many problems during my long life as is the case with many who work in theoretical physics. As the subject of the present talk, I could choose among the theory of superconductivity, cosmic-ray astrophysics, and radiation by uniformly moving sources. I have chosen the last option for two reasons. To begin with, I really love this problem. The word 'love' is not commonly used in scientific literature but this is just a matter of the historically established style. Indeed, each of us likes one thing in science and dislikes another, exactly as she does in everyday life. Perhaps I love radiation by uniformly moving sources because my early studies were devoted to this problem and I was young at that time. The second reason for my choice is that radiation by uniformly moving sources appears to be a traditional field of special interest for Russian physicists and, besides, a purely academic problem. Indeed, a most conspicuous achievement in this area, the Vavilov–Cherenkov (VC) effect, was discovered by S I Vavilov and P A Cherenkov in 1934 [1, 2] and explained by I E Tamm and I M Frank in 1937 [3]. Transition radiation was first investigated by I M Frank and myself in 1945 [4]. All these authors worked at the Physical Institute and all were members of the Academy. It is worthwhile to note that I E Tamm, I M Frank, and P A Cherenkov won the 1958 Nobel Prize in physics for the discovery and explanation of the VC effect (S I Vavilov died in 1951, before he was 60 and Nobel Prizes are not awarded posthumously).

2.2   Vavilov–Cherenkov effect

The VC effect in the true, somewhat narrow, sense of the term is essentially emission of electromagnetic waves (light) with a continuous spectrum and a specific angular distribution by an electric charge (e. g. an electron) moving in a medium at a constant velocity v. Radiation with a cyclic frequency ω occurs only if the charge speed v exceeds the phase velocity of light in a given transparent medium $v_{\mathrm{ph}} = c/n(\omega)$, i. e.

$$v > \frac{c}{n(\omega)} \tag{2.1}$$

where $n(\omega)$ is the refractive index (at a frequency ω) in the medium (c is the velocity of light in vacuum). The said specificity of radiation angular distribution is reflected in the angle θ_0 between the wavevector of emitted waves k and the speed v, with

$$\cos\theta_0 = \frac{c}{n(\omega)\,v}. \tag{2.2}$$

The results (2.1), (2.2) can be obtained using the Huygens principle according to which each point on the path of a charge uniformly and rectilinearly moving with speed v is a source of a spherical wave emitted as the charge passes through the point (figure 2.1).[2*] If condition (2.1) is fulfilled, these spheres have a common envelope, i. e. a cone whose apex coincides with the instantaneous charge position, the angle θ_0 being defined by expression (2.2).

Disregarding the dispersion (i. e. ω-dependence of n), the angle θ_0 is the same for all frequencies ω and radiation has a clear-cut front which forms a cone with the angle of opening $\pi - 2\theta_0$ and the charge (source) in its apex (figure 2.1). This cone is totally analogous to the Mach cone that characterizes a shock wave generated by the motion of a supersonic source (bullet, shell, aircraft, missile) in the air or other media, the velocity of the shock wave or sound u playing the role of the phase velocity of light

$v_{\text{ph}} = c/n$ in expressions (2.1) and (2.2). The hydrodynamic (acoustic) front at the Mach cone is very sharp and easy to observe (for example, as a supersonic plane flies by) because the dispersion of sound, i.e. the dependence of its velocity u on frequency, is normally very small.

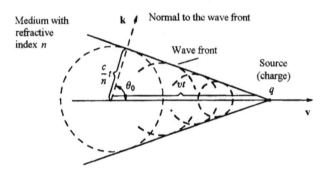

Figure 2.1. Formation of Vavilov–Cherenkov radiation: ct/n is the path of light within a time t; $vt = ct/(n\cos\theta_0)$ is the path of the charge (source) within the same time.

Hence, VC radiation is the electrodynamic (optical) analogue of the well-known (since the last century) acoustic phenomenon. Why then was it first discovered and explained only about 60 years ago? Doubtless, it might have been observed much earlier but, on the whole, the delay is readily explicable. First, to observe the VC effect in a more or less pure form, one must have a beam of relativistic or near-relativistic charged particles. However, such beams were first available only in the 1930s (suffice it to say that there had been no accelerators until that time). Second, the motion of sources (charges) in electrodynamics (in sharp contrast to both hydrodynamics and acoustics) is, in the first place and most frequently, considered in vacuum. The VC effect in vacuum is impossible since the velocity v of particles is always lower than the velocity of light $c = 3 \times 10^{10}$ cm s^{-1} (here, we do not consider tachyons, hypothetical and, in all probability, non-existent particles faster than light). True, this statement should be made with some reservation (see, for instance, [5, chapter 9] and [6, 7]) but, on the whole, the old assertion: "a uniformly moving charge does not radiate" is quite applicable.

This dogma appears to have prejudiced the prediction of the VC effect in the past. Nevertheless, it was actually foretold by O Heaviside, an outstanding English physicist, as early as 1888 [8]. At that time, however, even the electron was unknown and any talk about fast particles travelling in a dielectric was out of the question. For this reason, apparently, Heaviside's prediction was put to rest and surfaced again only in 1974 [9, 10]. Calculations performed by a well-known German physicist A Sommerfeld [11] made another portent of the theory of Tamm and Frank of which they became aware only after they had reported their own findings [3]. In 1904, Sommerfeld considered the uniform motion of a charge in vacuum and came to the conclusion that it radiates at a velocity exceeding that of light ($v > c$). However, the special theory of relativity that appeared only a year later (in 1905) argued that a charge cannot propagate at a speed higher than c and Sommerfeld's paper was forgotten. It ensues from the theory of relativity (disregarding tachyons) that c is the maximum velocity for an individual charge (as $v \to c$, the particle mass $m_0/\sqrt{1 - v^2/c^2}$ tends to infinity), whereas a radiation source (consisting, say, of many particles) may have any speed (see [5–7]). This is an interesting problem but I do not touch upon it here for lack of space. Neither Sommerfeld nor physicists that worked for 30 years after him came to consider the motion of a charge in a medium instead of a vacuum.

This was done by Tamm and Frank [3] who calculated radiation of a charge q moving with a constant velocity v in a medium with the refractive index $n(\omega)$. This automatically led to expression (2.2) and yielded radiation intensity (power) per unit time (i.e. at a path v)

$$\frac{\mathrm{d}W}{\mathrm{d}t} = \frac{q^2 v}{c} \int_{c/[n(\omega)v] \leq 1} \left(1 - \frac{c^2}{n^2(\omega)v^2}\right) \omega \, \mathrm{d}\omega. \qquad (2.3)$$

Clearly, the integration here is performed over all frequencies satisfying the condition (2.1).

Tamm and Frank forwarded a reprint of their paper [3] to Sommerfeld who sent his reply[1] on 8 May 1937 via Austria (direct postal communication with addressees in the USSR was difficult to maintain because the Nazis were already in power). Sommerfeld wrote: "I never thought that my calculations made in 1903 could ever have any physical implication. This confirms that the mathematical aspect of a theory outlasts changing physical concepts."

The history of the discovery of the VC effect is described at greater length in a book by Frank [14]. The theory of this effect has already been mentioned in this lecture. As regards its experimental aspects, VC radiation was observed by P Curie and M Curie in bottles with radium salt solutions. Nowadays, blue luminescence of water which is largely due to VC radiation can be seen by excursion parties visiting a nuclear reactor in a water tank. In 1926–29, Malle (France) carried out a series of special studies on radiation in fluids subjected to gamma-rays. However, nobody

---

[1] This letter was reprinted in full in *Memoirs about I E Tamm* [12, p 120]. It is worth noting that Sommerfeld mentions obtaining some publications of the USSR Academy of Sciences as its foreign member. He probably received *Doklady Akademii Nauk SSSR*. It is a great pity that foreign members of the Russian Academy now have no such opportunity [13].

before Vavilov and Cherenkov understood that the observed effect was a new phenomenon rather than simple luminescence induced by gamma-rays.

Cherenkov's experiments prompted by Vavilov were first intended to study the luminescence of uranyl salt solutions caused by gamma-irradiation using the original measuring technique previously developed by Vavilov and co-workers who employed the ability of the human eye to detect such luminescence after an adaptation to complete darkness [14, 15]. Cherenkov happened to observe that the fluid (sulfuric acid) was luminous even in the absence of a salt solution which induced him to believe that further work on his dissertation should be given up as a bad job [15]. It was S I Vavilov who understood that the observed effect was radiation other than luminescence. This suggestion required new measurements which, in the end, provided unequivocal evidence of the discovery of a previously unknown phenomenon [1, 2]. According to Vavilov [2], the radiation was not due to the direct effect of gamma-rays but was associated with Compton electrons released into the fluid under the action of gamma-rays. Subsequent findings by Cherenkov [16] obtained and interpreted in collaboration with Vavilov and Frank [14, 15] revealed a number of properties of this radiation which laid the framework for deeper insights into its nature by Tamm and Frank [3].

It is clear from the foregoing that the role of S I Vavilov as the co-author of this discovery can hardly be questioned, and only the term 'VC effect' is really true. I wish to emphasize this because the opposite view can be encountered in our (Russian) literature which all physicists aware of the facts regard as groundless (see [14, 15, 17, 18]). S I Vavilov himself gave the reason for the term 'Cherenkov effect' to be widely used both in this country and abroad because he only published a short note on this VC effect [2] and submitted a paper [19] to the *Physical Review* signed by Cherenkov alone.[2] I cannot say why he chose to do this: perhaps it was due to his unwillingness to outshine his pupil—a noble action which would be characteristic of him. Unfortunately, S I Vavilov was frequently attacked as a man, a physicist, an organizer of research, and President of the USSR Academy of Sciences. I consider all this 'criticism' to be absolutely unfounded as I have had the opportunity to state elsewhere ([20, pp 395, 397], see also [21]). See also article 16 of the present volume.

The VC effect has been extensively employed in physics (to say nothing of its value for the understanding of electrodynamics of continuous media and physics in general). First, the VC effect may be used to determine the particle velocity v by measuring the angle θ_0 (see (2.2)) or (proceeding from (2.1)) to demonstrate directly, in the absence of effect, that $v < c/n(\omega)$ (certainly, the refractive index $n(\omega)$ in the transparent medium can

---

[2] Curiously, paper [19] was first submitted to *Nature* but was declined there. This shows that the VC effect was apprehended as something non-trivial at that time.[3*]

and must be thought of as known). Second, as the radiation intensity is proportional to the squared particle charge q (see (2.3)), it is easy to distinguish between particles with elementary charge e (electrons, protons, etc) and nuclei with charge Ze (Z is the atomic number of the element). Indeed, even for the nucleus of helium ($Z = 2$), the radiation intensity is four times that of hydrogen isotopes ($Z = 1$); and it is 676 times higher for the iron nucleus ($Z = 26$) than for protons of the same velocity. Naturally, 'Cherenkov counters' (as such instruments are commonly known) are widely used in accelerators and in high-energy physics at large [22, 23]. Special emphasis should be placed on the use of the VC effect in cosmic-ray studies (VC radiation from atmospheric showers) and in facilities for detecting high-energy neutrino.

The theory of VC radiation as a whole is of course beyond the scope of the present lecture (see [5–7, 14, 22, 24] for details) and I only wish to dwell on a few problems that I was engaged in myself.

In 1940, L I Mandelstam, acting as an opponent to P A Cherenkov's thesis for the degree of Doctor of Sciences, suggested that the VC effect is equally likely to occur when a charge (source) propagates not in a continuous medium but inside a narrow empty channel made in such matter. This inference is physically relevant because VC radiation is formed not only on the charge path itself but also near it, at as large a distance as the wavelength of emitted light $\lambda = 2\pi c/[n(\omega)\omega]$. I M Frank and myself calculated the corresponding radiation intensity [25] which naturally decreases with increasing radius r of the empty channel in which a charge propagates axially. When $\sqrt{1 - v^2/c^2} \sim 1$, the radiation is almost as high as it is in the absence of the channel provided $r/\lambda \lesssim 0.01$ (in optics, this means that $r \lesssim 5 \times 10^{-7}$ cm). A qualitatively similar picture is obtained if a channel is replaced by a slit or when a charge moves near the medium (dielectric). This is important because when a charge moves in a medium, its energy loss due to VC radiation is relatively small and ionization loss in the immediate proximity to the trajectory prevails. Therefore, ionization losses are excluded when a charge propagates in channels, slits or near the medium, while VC radiation persists. This is important, but not crucial, for charges. However, when the Doppler effect occurs in a medium, i.e. in the case of the motion of excited atoms, it is possible to observe the phenomenon only if channels or gaps are available; otherwise, the atom is destroyed. The Doppler effect can be and is actually observed also in very rarefied media, e.g. in plasmas.

It may be appropriate to recall here that I used the analysis of radiation associated with the charge propagation near a medium to discuss various modes of microwave generation [26–28].

We shall now proceed to the methods for the calculation of VC radiation intensity. Tamm and Frank [3] obtained expression (2.3) from the solution of equations of electrodynamics in a medium. They estimated radiation intensity as a flux of the Poynting vector through the cylindrical surface surrounding the charge trajectory. Another approach consists in determination (using, of course, the same equations) of the force which slows down the moving charge. The work of this force in a transparent medium is equal to the radiation energy (2.3). Such calculations were made, e. g., by Fermi [29] and are included in the *Course* developed by Landau and Lifshitz [30, section 115]. Finally, there is one more way to obtain the same intensity (power) (2.3) which consists in computing the electromagnetic field energy produced by a charge per unit time [31].

An explicit (the so-called Hamiltonian) method is very convenient for this purpose. For a homogeneous isotropic stationary medium, it consists in the expansion of the field vector potential A in the series (see, for instance, [5] for more details):

$$A(r,\ t) = \sum_{\lambda,\ i=1,2} q_{\lambda i}(t)\, A_{\lambda i}(r)$$

$$A_{\lambda 1} = e_\lambda \sqrt{8\pi}\, \frac{c}{n}\, \cos(k_\lambda r) \tag{2.4}$$

$$A_{\lambda 2} = e_\lambda \sqrt{8\pi}\, \frac{c}{n}\, \sin(k_\lambda r)$$

where e_λ is the polarization vector ($e_\lambda = 1$) and the refractive index $n = \sqrt{\varepsilon}$ (ε is the dielectric permittivity of the medium assumed, for simplicity, to be non-magnetic). The transverse electromagnetic field being examined is

$$E_{\text{tr}} = -\frac{1}{c}\frac{\partial A}{\partial t} \qquad\qquad H = \operatorname{rot} A$$

and its energy

$$\mathcal{H}_{\text{tr}} = \int \frac{\varepsilon E_{\text{tr}}^2 + H^2}{8\pi}\, \mathrm{d}V = \frac{1}{2} \sum_{\lambda,\ i=1,\ 2} \left(p_{\lambda i}^2 + \omega_\lambda^2 q_{\lambda i}^2 \right) \tag{2.5}$$

where

$$p_{\lambda i} = \frac{\mathrm{d}q_{\lambda i}}{\mathrm{d}t} \qquad\qquad \omega_\lambda^2 = \frac{c^2}{\varepsilon} k_\lambda^2 \equiv \frac{c^2}{n^2} k_\lambda^2. \tag{2.6}$$

The field equation has the form

$$\Delta A - \frac{\varepsilon}{c^2}\frac{\partial^2 A}{\partial t^2} = -\frac{4\pi}{c}\, j.$$

For a point charge q which moves with a velocity v, the current density is $j = qv\delta[r - r_q(t)]$, where $r_q(t)$ is the radius-vector of the charge and δ is the delta-function. Following the substitution of the expansion (2.4), the field equation looks like

$$\frac{\mathrm{d}^2 q_{\lambda 1}}{\mathrm{d}t^2} + \omega_\lambda^2 q_{\lambda 1} = \sqrt{8\pi}\, \frac{c}{n}\, (e_\lambda v)\, \cos(k_\lambda r_q)$$

$$\frac{\mathrm{d}^2 q_{\lambda 2}}{\mathrm{d}t^2} + \omega_\lambda^2 q_{\lambda 2} = \sqrt{8\pi}\, \frac{c}{n}\, (e_\lambda v)\, \sin(k_\lambda r_q). \tag{2.7}$$

Therefore, the field equations are reduced to equations (2.7) for the 'field oscillators' $q_{\lambda i}(t)$. Integrating these equations and substituting the solution into (2.5), one obtains the field energy as summarized over the energies of all oscillators. For a uniformly and rectilinearly moving charge, we have $\boldsymbol{r}_q(t) = \boldsymbol{v}t$ and equations (2.7) are readily integrable since they are equations for an oscillator moving under the effect of a harmonic force proportional to $\cos(\boldsymbol{k}_\lambda \boldsymbol{v}t)$ or $\sin(\boldsymbol{k}_\lambda \boldsymbol{v}t)$, i. e. with a frequency

$$\omega = \boldsymbol{k}_\lambda \boldsymbol{v} = k_\lambda v \cos\theta = \frac{\omega_\lambda n v}{c} \cos\theta. \tag{2.8}$$

For $\omega = \omega_\lambda$, there is resonance and amplitudes $q_{\lambda i}$ grow with time, i. e. radiation occurs. Obviously, in a vacuum where $n = 1$, the frequency ω is always lower than ω_λ provided, of course, that $v < c$. This means precisely that a charge uniformly moving in a vacuum does not radiate while both resonance and radiation are possible in a medium. According to (2.8), the condition for this is the condition $(nv/c)\cos\theta = 1$, i. e. the condition for VC radiation (2.2). The substitution of the solution for $q_{\lambda i}(t)$ into (2.5) leads to the expression $\mathcal{H}_{\text{tr}} = (dW/dt)t$, where dW/dt is defined by formula (2.3).

So, in this case, the calculation by the Hamiltonian method is demonstratively and technically very simple. The character of the present lecture allows me to note here that it is exactly this simplicity which stimulated my interest in theoretical physics (I graduated from Moscow State University in 1938 and trained to experiment in optics believing that I should not become a theoretician for want of faculties for mathematics). Certainly, this remark is not the main excuse for my having included the Hamiltonian method in this lecture. It is of importance that this mode of computing the radiation energy, unlike the two others mentioned earlier, can almost trivially be extended to the case of an anisotropic medium, i. e. non-cubical crystals and plasma in a magnetic field. In this case, the field must be expanded in normal waves capable of propagating in a proper medium (in an isotropic medium, as in a vacuum, normal waves are reduced to waves $\boldsymbol{A}_{\lambda i}$ present in (2.4), due to degeneration). Therefore, the VC effect is easy to examine in an anisotropic medium, especially in uniaxial crystals [32]. Under these conditions, VC radiation gives rise to two cones which are generally non-circular and have different polarizations (direction of the electric field in the waves). The VC effect in crystals was experimentally investigated by V P Zrelov [22].

Different aspects of the theory of VC radiation have been considered in many other papers besides those cited here. They are concerned with the extension to magnetic media, detailed analysis of radiation in crystals, the role of boundaries, etc (see [5–7, 14, 22, 33, 34] and references therein). Absorption studies [29, 35] and investigations into VC radiation for various dipoles and higher multipoles (as opposed to that for charges) are especially noteworthy (see [5–7, 14] containing references to the original works). Studies on VC radiation of multipoles are far from being completed [5–7], probably owing to the fact that the known particles have very small magnetic moments (to say nothing about other multipoles), while the associated radiation is equally weak and of no practical value. Radiation of a magnetic charge (monopole) might be much stronger but such monopoles have never been observed and are unlikely to exist in nature.

Detailed analysis of all these problems and experiments using VC radiation is beyond the scope of this lecture (see [22, 23]) but I cannot help considering the quantum interpretation of the VC effect.

2.3   Quantum theory of the Vavilov–Cherenkov effect

The previously discussed classical theory of the VC effect is sufficiently correct in the optical part of the spectrum. Nevertheless, methodological considerations alone are enough to make us dwell on the quantum theory of the effect [36] as well (see also [5–7, 14]).

What is the way to explain, in quantum language, the absence of radiation of a uniformly moving charge (or another source with zero eigenfrequency) in vacuum? To this effect, it suffices to use the laws of the conservation of energy and momentum for the case of photon emission by a particle:

$$E_0 = E_1 + \hbar\omega \qquad E_{0,1} = \sqrt{m^2c^4 + c^2p_{0,1}^2}$$

$$p_0 = p_1 + \hbar k \qquad k = \frac{\omega}{c} \qquad p_{0,1} = \frac{mv_{0,1}}{\sqrt{1 - v_{0,1}^2/c^2}} \qquad (2.9)$$

where $E_{0,1}$ and $p_{0,1}$ are, respectively, the energy and momentum of a charge with the resting mass m before (subscript 0) and after (subscript 1) the emission of a photon with energy $\hbar\omega$ and momentum $\hbar k = (\hbar\omega/c)(k/k)$. It is easy to see that equations (2.9) have no solution (with $\omega > 0$) for $v < c$, i.e. radiation is impossible (see formula (2.11) with $n = 1$).

To examine the radiation of a source in a medium, only one thing should be known, namely the amount of radiation energy and momentum, because the particle energy $E = \sqrt{m^2c^4 + c^2p^2}$ does not change in the medium. This question is not so trivial (see [5, chapter 13]) but it is easy to obtain the correct answer by intuition. Indeed, a stationary immutable medium does not affect the frequency ω, while the wavelength $\lambda = \lambda_0/n$, where $\lambda_0 = 2\pi c/\omega$ is the wavelength in vacuum. Further, the wavenumber is $k = 2\pi/\lambda = \hbar\omega n/c$. With this in mind, equations (2.9) should be replaced by

$$E_0 = E_1 + \hbar\omega \qquad E_{0,1} = \sqrt{m^2c^4 + c^2p_{0,1}^2}$$

$$\boldsymbol{p}_0 = \boldsymbol{p}_1 + \hbar\boldsymbol{k} \qquad k = \frac{\omega n(\omega)}{c} \qquad \boldsymbol{p}_{0,1} = \frac{m\boldsymbol{v}_{0,1}}{\sqrt{1 - v_{0,1}^2/c^2}}. \qquad (2.10)$$

The solution of these equations for ω and θ_0, where θ_0 is the angle between $\boldsymbol{v}_0$ and $\boldsymbol{k}$, leads to[7*]

$$\cos\theta_0 = \frac{c}{n(\omega)\,v_0}\left[1 + \frac{\hbar\omega(n^2 - 1)}{2mc^2}\sqrt{1 - \frac{v_0^2}{c^2}}\right] \qquad (2.11)$$

$$\hbar\omega = \frac{2(mc/n)(v_0\cos\theta_0 - c/n)}{(1 - 1/n^2)\sqrt{1 - v_0^2/c^2}}. \qquad (2.12)$$

Under the condition

$$\frac{\hbar\omega}{mc^2} \ll 1 \qquad (2.13)$$

(or a somewhat more accurate inequality following from (2.11)), expression (2.11) turns into the classical expression (2.2). It cannot be otherwise because the condition (2.13) is explicitly the classicality condition (it is invariably fulfilled if the quantum constant $\hbar \to 0$). The classical limit corresponds to the neglect of recoil (a change in the particle momentum $\boldsymbol{p}_0$) due to emission of a 'photon in a medium' with momentum $\hbar\boldsymbol{k}$. It has already been mentioned that, according to (2.12), radiation for $\omega > 0$ is only possible if $v_0 > c/n$ (the relation $\cos\theta_0 \leq 1$ is always true). In the classical limit when the result (expression (2.2)) does not depend on $\hbar$, the quantum computation is of purely methodological value: it may prove convenient but is optional. This is really so and the conservation laws may be formulated in the classical domain as well provided that the relationship between the emitted electromagnetic energy $\mathcal{H}_{tr}$ and the radiation momentum is taken into account. The pertinent simple calculations are reported in [5–7]. Certainly, quantum computation of the radiation intensity is equally possible [36] by generalizing (2.3).

In the optical region, the only one where applications of the VC effect are normally feasible, the ratio $\hbar\omega/mc^2 \sim 10^{-5}$ even for electrons, i.e. quantum corrections are immaterial. When told about my work [36] in 1940, L D Landau stated that it was of no interest (see [20, p 382]). See also article 10 of the present volume. As is clear from the previous discussion, that conclusion of his was fully justified and his comment hit the mark as was usual with his criticism. However, another approach to a problem or another way of obtaining a result is sometimes useful to apply to the solution of other problems. An example has been given earlier to illustrate the various methods for calculating VC radiation power (2.3).

The same appeared to refer to the application of conservation laws to the analysis of radiation in a medium, which turned out to be very insightful in studies of the Doppler effect in a medium.

2.4   Doppler effect in the medium

The sources examined earlier (i. e. charges) have no natural frequency. Another important case is a source without a charge or any multipole moment constant in time but having a natural frequency ω_0. A classical example is an oscillator and a quantum one is an atom emitting frequency ω_{00} on a certain transition (this is a frequency in the reference frame in which the source is at rest).

If such a source travels in a vacuum with constant velocity v (in the laboratory reference frame), the frequency of the emitted waves in this frame is

$$\omega(\theta) = \frac{\omega_{00}\sqrt{1 - v^2/c^2}}{1 - (v/c)\cos\theta} = \frac{\omega_0}{1 - (v/c)\cos\theta} \tag{2.14}$$

where θ is the angle between the wavevector $\boldsymbol{k}$ (the direction of observation) and $\boldsymbol{v}$: the frequency ω_0 in (2.14) is the oscillation frequency in the laboratory reference frame. A change in the frequency of waves emitted by a moving source is known as the Doppler effect. Certainly, this effect also occurs in acoustics and holds for all waves regardless of their nature.

Let an oscillator or an atom (molecule) move in a transparent medium with a refractive index $n(\omega)$, which rests in the same laboratory reference frame. Do not be confused by the fact that a source travelling in continuous matter is at risk of being destroyed, for a channel or a gap can be used in such a medium (see earlier).

In the presence of a medium, formula (2.14) is replaced by [37, 14]

$$\omega(\theta) = \frac{\omega_{00}\sqrt{1 - v^2/c^2}}{|1 - (v/c)n(\omega)\cos\theta|} = \frac{\omega_0}{|1 - (v/c)\,n(\omega)\cos\theta|}. \tag{2.15}$$

This expression can be derived following the general rule, i. e. by changing the velocity of light in vacuum c by the phase velocity in the medium $c/n(\omega)$ (of course, there is no need to substitute c/n for c under the root $\sqrt{1 - v^2/c^2}$ because this root is related to the slowing down time for a moving source and has nothing in common with radiation). Certainly, expression (2.15) can also be obtained automatically from the solution of the field equation for a moving emitter. Non-trivial in (2.15) is the appearance of the modulus which is necessary to ensure the positiveness of the frequency. If the motion occurs at a velocity lower than that of light (i. e. $v < c/n$) or higher than that of light but outside the cone (2.2), i. e. under the condition

$$\frac{v}{c} n(\omega) \cos \theta < 1 \qquad (2.16)$$

we are dealing with the ordinary, normal Doppler effect. True, the so-called complex Doppler effect is equally feasible in this situation due to dispersion, i.e. the dependence of n on ω [37, 14].

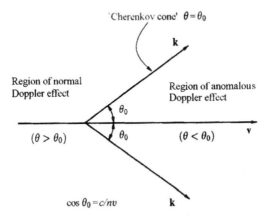

Figure 2.2. Regions in which the normal and anomalous Doppler effect occur.

In the case of motion with a speed higher than the velocity of light (when the condition (2.1) is fulfilled), formula (2.15) without modulus would lead to negative frequency values in the angular region where

$$\frac{v}{c} n(\omega) \cos \theta > 1. \qquad (2.17)$$

Radiation in the region (2.17), i.e. inside the cone (2.2) frequently referred to as the Cherenkov cone (Figure 2.2), is known as the anomalous Doppler effect. The whole picture is rather complicated if dispersion is allowed for (there is an individual cone for each frequency or several cones if n shows a non-monotonic dependence on ω). But here we shall confine ourselves to the case where dispersion is not involved, i.e. $n(\omega) = n = $ constant. Then, according to (2.15), the frequency $\omega \to \infty$ at the Cherenkov cone itself, where $(v/c)n \cos \theta = (v/c)n \cos \theta_0 = 1$ and this is true of either side of the cone (as $\theta \to \theta_0$). There is nothing more to add based on (2.15) and the difference between the normal and anomalous Doppler effects is not very significant.

It turned out that the quantum approach (to be more precise, the use of the laws of the conservation of energy and momentum) allows a very important feature of the anomalous Doppler effect to be revealed [38, 5–7, 14]. Let us assume that an emitter represents 'a system' (atom) with two levels—the lower level 0 and the upper one 1 (figure 2.3). Then, using type (2.10) conservation laws, one needs only to change the expression for the emitter energy with allowance for the internal degrees of freedom (levels). This means that the energy

$$E_{0,1} = \sqrt{(m + m_{0,1})^2 c^4 + c^2 p_{0,1}^2} \qquad (2.18)$$

where $(m + m_0)c^2 = mc^2 + W_0$ is the total energy of the system (atom) in the lower state 0 and $(m + m_1)c^2 = mc^2 + W_1$ is the same energy in the upper state 1. The energy $W_1 > W_0$ and the resting atom emits the frequency $\omega_{00} = (W_1 - W_0)/\hbar$ during the transition $1 \to 0$.

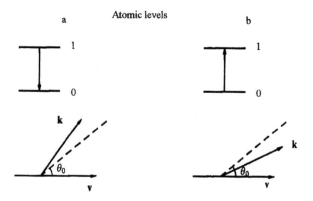

Figure 2.3. Transitions between levels 0 and 1 in the case of the normal and anomalous Doppler effect.

The use of the conservation laws in the classical limit (2.13) leads to formula (2.14) or (in an accurate computation [38]) to a somewhat more complicated expression containing terms of the order of $\hbar\omega/mc^2$. However, the quantum corrections are not so important as the following unexpected fact. Examining the signs (this is elementary algebra), it is easy to notice that in the region of the normal Doppler effect as well as in a vacuum, the atom passes over from the upper level 1 to the lower one 0 (the direction of this transition is determined by the requirement that the emitted photon energy $\hbar\omega$ be positive, i.e. from the condition that $\omega > 0$). In the region of the anomalous Doppler effect, in contrast, photon emission entails excitation of the atom, i.e. it passes from level 0 to level 1 (see figure 2.3), the energy being drawn from the kinetic energy of the translational motion.

Thus, at the supraluminal speed $v > c/n$ (the only condition for the anomalous Doppler effect), an emitting atom, which is initially even unexcited (i.e. lies at the lowest level 0), will get excited (i.e. pass over to level 1) and will simultaneously emit a photon inside the Cherenkov cone. The initially excited atom emits outside the Cherenkov cone, i. e. at angles $\theta > \theta_0$, and passes from level 1 to level 0. As a result, an atom moving faster than light undergoes continuous excitation and radiates. In the context of the classical oscillator model, this means that the oscillator is permanently excited. The anomalous Doppler effect has important implications for plasma physics. The crucial role of the VC effect in plasma and the related concepts and analogies were emphasized by I E Tamm in his Nobel lecture [17]. He also suggested that the acoustic analog of the anomalous Doppler effect in optics is of importance for the analysis of vibrations arising in the flight of a supersonic aircraft (the so-called flutter).

I believe it would have been difficult to conjecture specific features of the anomalous Doppler effect unless the quantum approach had been applied [38] (to put it precisely, it follows from the previous discussion that the quantum approach itself is not as important as the use of the conservation laws). It is certainly possible to confirm this finding and go further by means of the classical or quantum computation of radiation reaction associated with the motion of an emitter in a medium. Specifically, the influence of the radiation force on oscillator vibrations can be determined for an oscillator travelling in a medium [39; 5, section 7]. The emission of waves outside the Cherenkov cone (i. e. under the normal Doppler effect) turns out to suppress the oscillations. In contrast, radiation going inside the Cherenkov cone and corresponding to the anomalous Doppler effect swings vibrations of the oscillator and, thus, causes its excitation. Excellent agreement between this finding and the previous reasoning in the quantum language is quite obvious.

Note that several papers developing [39] and some other studies in the same field were published by B E Nemtsov [40], formerly a gifted theoretician and presently the noted governor of the Nizhegorodskaya province.[4*]

The foregoing discussion facilitates the understanding of the excitation mechanism of uniformly accelerated 'detectors' [41, 7]. This problem is known to be extensively considered in the literature (see references in [41]) in connection with the investigation of radiation of black holes and uniformly accelerated systems (acceleration radiation).

2.5   Transition radiation at the boundary between two media

When a source with no frequency (charge, multipole) of its own is in uniform rectilinear motion in a medium, it emits radiation (VC radiation) only at velocities that exceed the velocity of light (2.1). The medium is, however, assumed to be homogeneous throughout and constant in time.

If the medium is inhomogeneous or (and) variable in time, some radiation is also possible for a source uniformly travelling slower than light. Such radiation, first described in 1945 [4], is now called transition radiation.

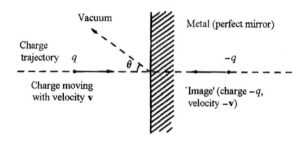

Figure 2.4. Transition radiation of a charge q crossing the vacuum–metal interface.

The simplest case of transition radiation is exemplified by a charge that crosses the interface when moving rectilinearly and uniformly at any speed. Then, the charge–interface intersection point becomes a source of transition radiation. Such a conclusion most readily offers itself when the charge comes from a vacuum and falls on a good (high-conductivity) metal that may serve as an ideal mirror (figure 2.4). It is known from electrodynamics that, under these conditions, the field of a charge in the vacuum is the sum of fields of a charge q moving in the vacuum in the absence of a mirror and a charge $-q$ which moves in a mirror to run into q (i.e. with velocity $-v$); the charge $-q$ is referred to as 'the image' of charge q. As soon as the charge q crosses the interface, it finds itself in a high-conductivity medium and produces practically no field in the vacuum; concurrently, the image $-q$ is lost. Therefore, from the 'point of view' of an observer in the vacuum, a pair of charges q and $-q$ annihilate. Also, it is known from electrodynamics that annihilation, as any charge acceleration (both 'charges', q and $-q$, stop abruptly at the interface), results in radiation which, in the present case, is transition radiation.

For an ideal mirror, the energy emitted into the vacuum is

$$W_1(\omega, \theta) = \frac{q^2 v^2 \sin^2 \theta}{\pi^2 c^3 \left[1 - (v^2/c^2)\cos^2 \theta\right]^2}$$

$$W_1(\omega) = 2\pi \int W_1(\omega, \theta) \sin \theta \, d\theta = \frac{q^2}{\pi c}\left(\frac{1 + v^2/c^2}{2v/c}\ln\frac{1 + v/c}{1 - v/c} - 1\right).$$

$$(2.19)$$

In the ultrarelativistic limit (as $v \to c$),

$$W_1(\omega) = \frac{q^2}{\pi c} \ln \frac{2}{1 - v/c} = \frac{2q^2}{\pi c} \ln \frac{2E}{mc^2} \qquad\qquad E = \frac{mc^2}{\sqrt{1 - v^2/c^2}} \gg mc^2.$$

$$(2.20)$$

Formulae (2.19) and (2.20) are easy to obtain [5–7, 42]. However, in the general case of two media with complex permittivities ε_1 and ε_2, calculations are cumbersome [4, 5, 42], and I shall not present even their results. It is only worth noting that the previously mentioned 'backward' transition radiation (figure 2.4) is of no practical value. True, it apparently accounts for the observed anticathode luminescence in X-ray tubes. In principle, such transition radiation can also be used to measure the particle energy E because it is included in expression (2.20) for emitted energy. However, equation (2.20) contains only logarithmic E-dependence while the absolute energy value W_1 is small. However, it turned out in 1959 [43, 44] that, for relativistic particles, it is more reasonable to consider 'forward' transition radiation spreading in the direction of particle velocity, e. g. when it leaves matter for vacuum. In this case, very high frequencies are also emitted and the total radiation energy of a particle with charge q and mass m

$$W_2 = \int W_2(\omega)\, d\omega = \frac{q^2 \omega_p}{3c} \frac{E}{mc^2}. \qquad (2.21)$$

Here ω_p is the plasma frequency of matter (for high frequencies all substances are equivalent to a plasma with dielectric permittivity

$$\varepsilon = n^2 = 1 - \frac{\omega_p^2}{\omega^2} \qquad\qquad \omega_p^2 = \frac{4\pi e^2 N}{m_e}$$

where N is the electron concentration in the substance and e and m_e are, respectively, the electron charge and mass).

The radiation energy W_2 is proportional to the particle energy E. Hence, the energy E can be found by measuring W_2, which is of paramount importance in high-energy particle physics. Also, it is essential that the use of the VC effect for energy measurements is of not efficient when the energies are high. The point is that in the ultra-relativistic region with $v \to c$, both the Cherenkov angle θ_0 (see (2.2)) and the radiation intensity (2.3) exhibit a very low sensitivity to the particle energy $E = mc^2/\sqrt{1 - v^2/c^2}$. The so-called transition radiation counters extensively employed in high-energy particle physics [45, 46] are based just on measurements of 'forward' transition radiation energy W_2. To avoid misunderstanding, we shall emphasize that these counters should contain a 'sandwich' of many sheets (plates) separated by, say, air-gaps because the energy W_2 (see (2.21)) for a single interface is very small. But the presence of many interfaces imposes limitations on the construction of a counter that, in turn, give rise to very interesting physics (radiation formation zones) which we shall not discuss here to save room (see [5–7, 14, 42]).

2.6   Transition radiation (general case), transition scattering, and transition bremsstrahlung

Transition radiation emitted when a source crosses a clear-cut interface is the simplest case. Generally, transition radiation is generated whenever a source (charge) moves uniformly in an inhomogeneous or/and a non-stationary medium or near it. Apart from the previously discussed 'annihilation' of a source and its image, transition radiation may be interpreted in a different very general fashion. This can be illustrated using an isotropic transparent medium with the refractive index n. Then, in the general case, the phase velocity of light in a medium is $v_{\rm ph} = c/n(\omega, r, t)$, where r are coordinates and t is time (of course, $n(\omega, r, t) = n(\omega)$ in a homogeneous and stationary medium). Light emission by a charge with velocity v is specified by the ratio $v/v_{\rm ph} = vn/c$. In the vacuum, $n = 1$ and there is no emission at $v = $ constant (assuming that $v < c$): it is only possible upon charge acceleration when $v = v(t)$ and, therefore, the acceleration is $w = dv/dt \neq 0$. For a uniform rectilinear motion in a medium, when $v = $ constant and $w = 0$, the vn/c ratio can change owing to the dependence of n on r or (and) t. This represents transition radiation, with $n(\omega, r, t)$ undergoing alteration at the point of charge location or near it (within the zone of radiation formation).

If a source crosses the interface of two media, the index n undergoes variation at this interface. A somewhat different variant is an inhomogeneous medium (emulsion, plasma in an inhomogeneous magnetic field, etc). One more situation is also interesting, even though not important in practice, when a charge uniformly moves in a homogeneous medium and the index n changes throughout it at an instant $t = t_0$ (or within a certain time interval near the moment t_0), e. g. owing to compression. Then, a point on the trajectory occupied by the charge at an instant t_0 plays (although not literally) the same role as the boundary between the two media [47, 42]. An important case of an inhomogeneous medium is a periodically inhomogeneous medium, e. g. a block of plates used in transition radiation counters [48, 42]. Under such conditions, transition radiation is sometimes referred to as resonance transition radiation or transition scattering. Indeed, when a charge moves in a periodically inhomogeneous medium (e. g. a sinusoidal one, see (2.22), or a medium consisting of a set of clear-cut interfaces, etc.), one may argue (from 'the point of view' of a charge) that a wave of dielectric permittivity (refractive index) is incident on this charge. Scattering of this wave by the charge gives rise to transition radiation. Nevertheless, the term 'transition scattering' is hardly relevant unless the effect persists for a resting charge as well. Then the term 'transition radiation' is unnatural, while 'transition scattering' appears to be adequate. In fact, this effect is apparent, for instance, when a permittivity wave is incident on a stationary (fixed) charge q giving rise to an electromagnetic wave coming from (scattered by) the charge (figure 2.5).

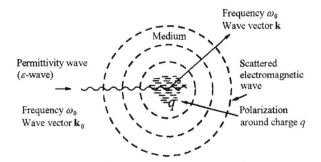

Figure 2.5. Schematic diagram of the formation of transition scattering of a permittivity wave by a motionless (fixed) charge q.

This inference is easy to understand even beyond the scope of the general transition radiation theory. For example, let us consider a transparent isotropic medium with dielectric permittivity $\varepsilon = n^2$. If an acoustic wave propagates in such a medium, the density of the medium is $\rho = \rho^{(0)} + \rho^{(1)} \sin(k_0 r - \omega_0 t)$, where k_0 and ω_0 are, respectively, the wave vector and the acoustic wave frequency. But a change in the density of the medium ρ entails a change in ε, which causes the permittivity wave to propagate through the medium:

$$\varepsilon(r, t) = \varepsilon^{(0)} + \varepsilon^{(1)} \sin(k_0 r - \omega_0 t) \tag{2.22}$$

where $\varepsilon^{(0)}$ is the permittivity in the absence of an acoustic wave and $\varepsilon^{(1)}$ is the change in ε due to an alteration in the density. Certainly, the permittivity wave is not necessarily induced by an acoustic wave; e. g. it may be associated with a longitudinal plasma wave.

Let us now place a fixed or infinitely heavy charge q in a medium. An electric field E and induction $D = \varepsilon E$ occur around the charge. In the absence of the wave, E is a Coulomb field equal to

$$E^{(0)} = \frac{q r}{\varepsilon^{(0)} r^3} \qquad D^{(0)} = \varepsilon^{(0)} E = \frac{q r}{r^3}. \tag{2.23}$$

In the presence of the wave (2.22), an additional polarization

$$\delta P = \frac{\delta D}{4\pi} = \frac{\varepsilon^{(1)}}{4\pi} E^{(0)} \sin(k_0 r - \omega_0 t) \tag{2.24}$$

is observed to a first approximation (provided $|\varepsilon^{(1)}| \ll \varepsilon^{(0)}$).

Such a polarization, which possesses no spherical symmetry (for $k_0 \neq 0$), is responsible for an electromagnetic wave with frequency ω_0 propagating from the charge (figure 2.5). The wavenumber of this wave is

$k = 2\pi/\lambda = (\omega_0/c)\sqrt{\varepsilon^{(0)}}$. If, as we presumed, the permittivity wave is caused by an acoustic one, then $k \ll k_0 = \omega_0/u$, where u is the velocity of sound (assuming, of course, that $u \ll c/\sqrt{\varepsilon^{(0)}}$).

An arising electromagnetic wave may be regarded as scattered in the same sense as other types of scattering, e. g. Thomson scattering of an electromagnetic wave by a resting electron (we certainly mean 'rest' here disregarding the effect of the incident wave). If the medium is an isotropic plasma and the incident wave is a longitudinal (plasma) wave, the discussed transition scattering is actually the transformation of the longitudinal to the electromagnetic (transverse) wave. This accounts for the important role of transition scattering in plasma physics, which has been confirmed in [5–7, 42] and can be illustrated by the following example. There is an electric field in a plasma longitudinal wave with a frequency close to $\omega_p = \sqrt{4\pi e^2 N/m_e}$ and ε changes. Therefore, with the longitudinal wave travelling in the plasma, its particles (electrons and ions) are affected by both the electric field wave and the permittivity wave. Plasma electrons undergo oscillations in the electric field and, thus, give rise to scattered electromagnetic waves (the so-called Thomson scattering) whose intensity is inversely proportional to the squared mass m of the scattering particle. Due to this, the Thomson scattering by ions is $(m_i/m_e)^2$ less intensive than that by electrons (m_e and m_i are the electron and ion masses, respectively). Hence, the intensity of Thomson scattering even by the lightest ions (protons with mass $m_p = 1836 m_e$) is $(1836)^2 \approx 3.4 \times 10^6$ times lower than that by electrons. In contrast, transition scattering in the first approximation is totally independent of the scattering particle mass m and occurs even as $m \to \infty$. Therefore, all longitudinal wave scattering by ions in plasma is practically transition scattering with an intensity of the same order as the intensity of longitudinal wave scattering by electrons. Generally, it is impracticable to analyze plasma processes taking no account of transition scattering.

Another phenomenon related to transition scattering is transition bremsstrahlung [42, 50]. Ordinary bremsstrahlung is known to result from particle collisions which cause their acceleration (deceleration) with the emission of electromagnetic waves. Since light particles (electrons) are accelerated faster than heavy ones propagating with the same speed, the bremsstrahlung of electrons is much more intense under similar conditions than that of heavy particles (protons, etc). This is true, however, only of particle collisions and the resulting bremsstrahlung in a vacuum. The situation is quite different in a medium. It has been shown earlier that radiation (transition radiation) may occur in the absence of particle acceleration. Therefore, if in a medium (plasma), a charge q flies close to a charge q' even without an appreciable acceleration of either of them, radiation is produced which is natural to call transition bremsstrahlung. The physical nature of transition bremsstrahlung is readily understood if the field E and polarization $P = [(\varepsilon - 1)/4\pi]\,E$ of a uniformly moving charge q are expanded in waves with the wavevector k_0 and the frequency $\omega_0 = k_0 v$, where v is the charge velocity. In a medium, such waves are associated with permittivity waves having the same ω_0 and k_0 values. Such permittivity waves are scattered by other charges q' giving rise to transition bremsstrahlung.

Transition radiation and closely related transition scattering and transition bremsstrahlung have been objects of many studies, including the monograph [42].

This communication presents merely an overview of all these phenomena. Nevertheless, the previous consideration made it clear, I hope, that this range of questions is fairly interesting for physics (of special importance are transition radiation counters and applications in plasma physics).

2.7   Concluding remarks

Analogy, i. e. the transfer of concepts from one field of science to another, is an important tool in physics (and, doubtless, in other sciences). That is why to work fruitfully a scientist must be a broad-minded person rather than a specialist in a narrow field. This trivial thought is advocated in my book [20] and I dare say this has guided the bulk of my scientific activities. The scope of problems dealt with in the present report appears to be an illustration of this view. For example, the VC effect is an analogue of the Mach supersonic radiation (cone), the excitation of mechanical oscillations in supersonic flows is analogous to the anomalous Doppler effect and various forms and types of transition radiation are also interrelated via common notions. On the whole, it appears that the analysis of various problems and effects pertaining to the investigation of radiation produced by uniformly moving sources helps to develop a certain 'ideology' and create a specific 'language'. This is easy to deduce from the examples provided earlier in this report and from those to follow (see also [5–7, 14, 42]; a popular account of this problem can be found in [51]).

In 1946, L D Landau found that some attenuation of longitudinal (plasma) waves in an isotropic plasma occurs even in the absence of collisions [52]. This effect, referred to as Landau (or collisionless) damping, is of paramount importance in the physics of plasma and plasma-like media (specifically, in physics of metals and semiconductors where conduction electrons form a sort of plasma). Landau arrived at his result without any relation to VC radiation and, in fact, the mechanism of Landau damping can undoubtedly be understood without any reference to VC radiation. At the same time, the Landau damping condition is precisely the VC condition (2.8) for the emission of a longitudinal wave by an electron (n in (2.8) is certainly the refractive index for the longitudinal wave). For this reason, those who understand the mechanism of VC radiation can easily perceive the nature of Landau damping.

It has been emphasized earlier that the VC and Doppler effects can be observed not only when a source travels in a medium but also inside an empty channel in the medium or near it. The same is true of transition radiation and transition scattering. Let, e.g. a charge move uniformly and rectilinearly above the flat surface of a medium composed of two different materials. Then, the charge crossing the boundary between the two phases induces transition radiation. Generally, such radiation is always emitted if inhomogeneities appear near the charge path, e.g. the edges of a metallic waveguide which a charge enters or leaves; another example is a charge flying over a diffraction grating [53, 54], etc. This type of transition radiation is sometimes called diffraction radiation. Its physical nature is best understood on the basis of the previously mentioned concept of charge 'images' propagating in the medium ('mirror') around the trajectory. The 'images' emit radiation during their non-uniform motion (another demonstrable explanation of this effect can be given, see, for instance [51]).

As early as 60 years ago, at the inception of quantum electrodynamics, it became clear that, quantum effects (in the first place, creation of electron – positron pairs e^+e^-) being taken into consideration, the vacuum in a sufficiently strong electromagnetic field is no longer the 'absolute emptiness' of classical physics, in which electromagnetic waves of any frequency can freely propagate (without interaction with one another). However, when virtual pair creation is feasible, the vacuum in a strong field behaves like a nonlinear anisotropic medium. The field (e.g. magnetic field H) will be thought of as strong if it is comparable with a characteristic field

$$H_c = \frac{m_e^2 c^3}{e\hbar} = 4.4 \times 10^{13} (\text{Oe}). \tag{2.25}$$

The characteristic electric field E_c is defined by the same expression (2.25) and its meaning is quite obvious: on the Compton electron wavelength $\hbar/(m_e c) = 3 \times 10^{-11}$ cm, the field $2E_c$ does the work $2\hbar e E_c/(m_e c) = 2m_e c^2$ over the electron charge e, necessary to create a e^+e^- pair with the rest mass $2m_e c^2 \sim 10^{-6}$ erg $\sim 10^6$ eV. The field (2.25) is so strong that nonlinear vacuum polarization for a long time seemed to be pure abstraction. However, in 1967–68, magnetized neutron stars (pulsars) were discovered and in these $10^{12} - 10^{13}$ Oe fields are typical. Also, it became clear that the situation characteristic of strong fields (2.25) in the vacuum can be simulated in semiconductors. These findings made strong fields an object accessible for astrophysical and physical studies. For our purpose, all these facts are interesting because the VC effect, transition radiation, and transition scattering may occur in strong fields (see [42] and references herein). A vacuum also behaves as a certain medium in a gravitation field, which also makes it possible to consider, say, transition radiation associated with the transformation of gravitational waves to electromagnetic ones [42].

Apart from the aforementioned acoustic analogue of the VC effect, acoustic analogies of electromagnetic transition radiation and transition scattering also exist [55]. It was somewhat surprising for me to learn that transition radiation of elastic waves plays an important role in elastic systems, e. g. for an inhomogeneous track interacting with the wheels of a uniformly moving railway car [56].

Generally, analogies of the VC and Doppler effects, transition radiation, and transition scattering are obviously possible for wave fields of any type, hence (bearing in mind the quantum theory) for particles of any type with the transformation (emission) of fields (particles) of another type. An example is transition radiation (creation) of electron–positron pairs produced by a charge crossing an interface, e. g. the boundary of atomic nucleus. In a word, radiation during uniform motion of various sources is a universal phenomenon rather than an eccentricity. Unsurprisingly, more and more theoretical and experimental studies of this problem are being reported. To my knowledge, the papers published in 1995 are concerned with transition (diffraction) radiation of relativistic electrons travelling over a diffraction grating [54], transition radiation in elastic systems [56], transition radiation of a neutrino with magnetic moment [57], further development of the theory of transition radiation [58, 59], transition bremsstrahlung polarization in plasma [61], and an in-depth allowance for transition scattering in the analysis of bremsstrahlung in plasma [61] with an important application to the solar neutrino problem [62].[5*,6*]

To summarize, there can be little doubt that the scope of physical problems first raised in the P N Lebedev Physical Institute over 50 years ago [1–4] and discussed in the present lecture has become an inalienable self-sufficient branch of modern physics.

Comments

1*. The Presidium of the Russian Academy of Sciences awarded me the 1995 Lomonosov Great Golden Medal of the Russian Academy of Sciences. According to the statute, a laureate should give a talk at the general meeting of Russian Academy of Sciences, which I did on 1 November, 1996. I prepared the corresponding text for publication (my talk was certainly much shorter). The communication was published in 1996 *Usp. Fiz. Nauk* **166** 1033 [1996 *Phys. Usp.* **39** 973]; a brief summary was placed in 1996 *Vestn. Ross. Akad. Nauk* (*Herald Russ. Acad. Sci.*) **66**(4) 381. Note that, in the present issue, the title of the paper additionally contains the words

"The Doppler effect in a medium", which were not included earlier.

2*. In the past, when the concept of a light-carrying ether was used, the Huygens principle was mostly employed in the study of the emission and propagation of electromagnetic waves in a vacuum. And at the present time, it seems to be pertinent to mention and apply this principle in the course of electrodynamics and, concretely, optics. The same can be said in respect of any type of waves, e. g. sound waves. In the case of electrodynamics in a medium, the Huygens principle was already known not to have a conditional character. So, the VC radiation is indeed a superposition of waves coming from particles of the medium under the influence of the source field (in particular, a moving charge). As to the VC effect, see also article 8 of the present book.

3*. The same is evidenced by the following fact. After the appearance of paper [19], Collins and Reiling in the USA observed VC radiation using a beam of relativistic electrons [63]. They confirmed the validity of formula (2.2) but believed that it described bremsstrahlung radiation.

4*. B E Nemtsov became later Vice-Prime Minister of the Russian Government and now (2001) he is leader of the Duma's fraction the "Union of Right Forces". Having become a statesman, B E Nemtsov is no longer engaged in physics, about which he wrote to me. It is, of course, natural with his full-time job of a politician.

5*. I do not follow the literature devoted to the problems discussed in this paper. But this field of physics is naturally 'alive'. As an example, I shall point to papers [64, 65] mentioned in the note to the 1997 Russian edition. The first of these papers discussed VC and transition radiation in the case of a thin crystalline plate. The VC effect for a non-point charge is considered in paper [65]. Recently, I have noticed papers [66, 67] devoted, respectively, to VC radiation during the motion of vortex lines in superconductors and to transition radiation during the motion of the same vortices. I shall also point out review [68] devoted to diffraction radiation, i. e. one of the types of transition radiation, and paper [74] about X-ray transition radiation.

6*. In a letter to *Uspekhi Fizicheskikh Nauk* [69], in view of the appearance of several new publications (the corresponding references are given in [69]; see also [70]), I have recently dwelt on some issues of the theory of radiation of uniformly moving sources.

7*. It is clear from (2.10) that solution (2.11) also holds for a "medium" (material) with $n < 0$, i. e. with a negative refractive index (see [71, 72]).

The use of the conservation laws, i. e. the "quantum" approach, is of course also valid for an anisotropic medium (this is explained at length in [5, chs 6 and 7]). There exists a curious possibility that the group velocity $d\omega/d\boldsymbol{k}$ of waves is aligned with $-\boldsymbol{k}$ or that its component $d\omega/dk_{\mathrm{r}}$, which is perpendicular to the particle velocity $\boldsymbol{v}$, is negative. A similar situation, enlightened in [5, ch 7] (see figure 7.1), is analogous to that observed for a medium with $n < 0$. See also [73, section 6.4.3] and [75].

References

[1] Cherenkov P A 1934 *Dokl. Akad. Nauk SSSR* **2** 451 [1934 *C. R. Acad. Sci. URSS* **2** 451]

[2] Vavilov S I 1934 *Dokl. Akad. Nauk SSSR* **2** 457 [1934 *C. R. Acad. Sci. URSS* **2** 457]

[3] Tamm I E and Frank I M *Dokl. Akad. Nauk SSSR* **14** 107 [1937 *C. R. Acad. Sci. URSS* **14** 107]

[4] Ginzburg V L and Frank I M 1946 *Zh. Eksp. Teor. Fiz.* **16** 15; 1945 *J. Phys. USSR* **9** 353 (brief version).

[5] Ginzburg V L 1979 *Theoretical Physics and Astrophysics* (Oxford: Pergamon Press) [Russian original, later edn, 1987 (Moscow: Nauka)]; Ginzburg V L 1989 *Application of Electrodynamics in Theoretical Physics and Astrophysics* (New York: Gordon and Breach)

[6] Ginzburg V L 1986 *Trudy FIAN* **176** 3

[7] Ginzburg V L 1993 in *Progress in Optics* vol 32 ed E Wolf (Amsterdam: North Holland) p 267

[8] Heaviside O 1889 *Electrician* (23 November) 83; 1889 *Phil. Mag.* **27** 324

[9] Tyapkin A A 1974 *Usp. Fiz. Nauk* **112** 731 [1974 *Sov. Phys. Usp.* **17** 288]

[10] Kaiser T R 1974 *Nature* **247** 400

[11] Sommerfeld A 1904 *Göttingen Nachr. Math. Phys. Kl.* **2** 99; **5** 363; 1905 **3** 201

[12] *Vospominaniya o I E Tamme* [*Memoirs about I E Tamm*] ed E L Feinberg 3rd edn (Moscow: Nauka 1995)

[13] Ginzburg V L 1995 *Vestn. Ross. Akad. Nauk* **65** 848

[14] Frank I M 1988 *Izluchenie Vavilova–Cherenkova* (*Voprosy Teorii*) [*VC Radiation: Theoretical Aspects*] (Moscow: Nauka)

[15] Dobrotin N A, Feinberg E L and Fock M V 1991 *Priroda* 11 58

[16] Cherenkov P A 1937 *Dokl. Akad. Nauk SSSR* **14** 99, 103; 1938 **21** 117 [1937 *C. R. Acad. Sci. URSS* **14** 99, 103; 1938 **21** 117]

[17] Tamm I E 1959 The Nobel lecture *Usp. Fiz. Nauk* **68** 387; 1960 *Science* **131** 206; 1975 *Sobranie Nauchnykh Trudov* T 1 [*Collected Scientific Papers* vol 1] (Moscow: Nauka) p 121

[18] Bolotowskii B M and Vavilov Yu N 1995 *Phys. Today* **48**(12) 11; 1996 *Phys. Today* **49**(9) 120

[19] Cherenkov P A 1937 *Phys. Rev.* **52** 378

[20] Ginzburg V L 2001 *The Physics of a Lifetime: Reflections on the Problems and Personalities of 20th Century Physics* (Berlin: Springer)

[21] Feinberg E L 1990 *Nauka i Zhizn'* 8 34

[22] Zrelov V P 1968 *Izluchenie Vavilova–Cherenkova i ego Primenenie v Fizike Vysokikh Energiy* [*VC Radiation and Its Applications in High-Energy Physics*] vols 1, 2 (Moscow: Atomizdat) [Translated into English 1970 *Cherenkov Radiation in High-Energy Physics* (Jerusalem: Israel Program for Scientific Translations)]

[23] 1990 *Cherenkovskie Detektory i Ikh Primenenie v Nauke i Tekhnike*

[Cherenkov Detectors and Their Applications in Science and Technology] ed A M Baldin (Moscow: Nauka)

1994 CERN Courier **34**(1) 22

[24] Tamm I E 1975 Sobranie Nauchnykh Trudov T 1 [Collected Scientific Papers vol 1] (Moscow: Nauka) p 77; 1939 J. Phys. USSR **1** 439

[25] Ginzburg V L and Frank I M 1947 Dokl. Akad. Nauk SSSR **56** 699 [1947 C. R. Acad. Sci. URSS **56** 699]

[26] Ginzburg V L 1947 Dokl. Akad. Nauk SSSR **56** 145 [1947 C. R. Acad. Sci. URSS **56** 145]

[27] Ginzburg V L 1947 Dokl. Akad. Nauk SSSR **56** 253 [1947 C. R. Acad. Sci. URSS **56** 253]

[28] Ginzburg V L 1947 Izv. Akad. Nauk SSSR, Ser. Fiz. **11** 165

[29] Fermi E 1940 Phys. Rev. **57** 485

[30] Landau L D and Lifshitz E M 1992 Elektrodinamika Sploshnykh Sred (Moscow: Nauka) [Translated into English 1984 Electrodynamics of Continuous Media (Oxford: Pergamon)]

[31] Ginzburg V L 1939 Dokl. Akad. Nauk SSSR **24** 130 [1939 C. R. Acad. Sci. URSS **24** 130]

[32] Ginzburg V L 1940 Zh. Eksp. Teor. Fiz. **10** 608; 1940 J. Phys. USSR **3** 101

[33] Pafomov V E 1961 Trudy FIAN **16** 94

[34] Bolotovskii B M 1957 Usp. Fiz. Nauk **62** 201; 1961 **75** 295 [1962 Sov. Phys. Usp. **4** 781]

[35] Kirzhnits D A 1989 in Nekotorye Problemy Sovremennoi Yadernoi Fiziki: k 80-Letiyu Akademika I M Franka [Some Problems of Modern Nuclear Physics: On the Occasion of the 80th Birthday of I M Frank] ed I S Shapiro, compiler L B Pikelner (Moscow: Nauka) p 144

[36] Ginzburg V L 1940 Zh. Eksp. Teor. Fiz. **10** 589; 1940 J. Phys. USSR **2** 441

[37] Frank I M 1942 Izv. Akad. Nauk SSSR, Ser. Fiz. **6** 3

[38] Ginzburg V L and Frank I M 1947 Dokl. Akad. Nauk SSSR **56** 583 [1947 C. R. Acad. Sci. URSS **56** 583]

[39] Ginzburg V L and Eidman V Ya 1959 Zh. Eksp. Teor. Fiz. **36** 1823 [1959 Sov. Phys. JETP **36** 1300]

[40] Nemtsov B E 1986 Zh. Eksp. Teor. Fiz. **91** 44 [1986 Sov. Phys. JETP **64** 25]; 1985 Izv. Vyssh. Uchebn. Zaved. Radiofizika **28** 1549; 1987 **30** 968 [1985 Radiophys. Quantum Electron. **28** 1076; 1987 **30** 718]

Nemtsov B E and Eidman V Ya 1984 Zh. Eksp. Teor. Fiz. **87** 1192 [1984 Sov. Phys. JETP **60** 682]; 1987 Izv. Vyssh. Uchebn. Zaved. Radiofizika **30** 226 [1987 Radiophys. Quantum Electron. **30** 171]

[41] Ginzburg V L and Frolov V P 1986 Pis'ma Zh. Eksp. Teor. Fiz. **43** 265; [1986 JETP Lett. **43** 339]; 1989 Trudy FIAN **197** 8;

Frolov V P and Ginzburg V L 1986 Phys. Lett. A **116** 423

[42] Ginzburg V L and Tsytovich V N 1984 Perekhodnoe Izluchenie i Perekhodnoe Rasseyanie (Moscow: Nauka) [Complete translation into English 1990 Transition Radiation and Transition Scattering (Bristol, New York: A Hilger)]

[43] Garibyan G M 1959 Zh. Eksp. Teor. Fiz. **37** 527 [1960 Sov. Phys. JETP **37** 372]

Garibyan G M and Yan Shi 1983 Rentgenovskoe Perekhodnoe Izluchenie

[*X-ray Transition Radiation*] (Erevan: Izd. Akad. Nauk Arm. SSR)

[44] Barsukov K A 1959 *Zh. Eksp. Teor. Fiz.* **37** 1106 [1960 *Sov. Phys. JETP* **37** 787]

[45] Fabjan C W and Fischer H G 1980 *Rep. Prog. Phys.* **43** 1003

[46] Kleinkhecht K 1982 *Phys. Rep.* **84** 85

[47] Ginzburg V L 1973 *Izv. Vyssh. Uchebn. Zaved. Radiofizika* **16** 512 [1973 *Radiophys. Quantum Electron.* **16** 386]

[48] Ter-Mikaelyan M L 1972 *High Energy Electromagnetic Processes in Condensed Media* (New York: Wiley-Interscience)

[49] Ginzburg V L and Tsytovich V N 1973 *Zh. Eksp. Teor. Fiz.* **65** 1818 [1974 *Sov. Phys. JETP* **38** 909]

[50] Tsytovich V N 1973 *Trudy FIAN* **66** 173

[51] Bolotovskii B M and Davydov V A 1989 *Zaryad, Sreda, Izluchenie* [*Charge, Medium, Radiation*] (Moscow: Znanie)

[52] Landau L D 1946 *Zh. Eksp. Teor. Fiz.* **16** 574 [1946 *J. Phys. USSR* **10** 25]

[53] Smith S J and Purcell E M 1953 *Phys. Rev.* **92** 1069

[54] Woods K J *et al.* 1995 *Phys. Rev. Lett.* **74** 3808

[55] Pavlov V I and Sukhorukov A I 1985 *Usp. Fiz. Nauk* **147** 83 [1985 *Sov. Phys. Usp.* **28** 784]

[56] Vesnitskii A I, Kononov A V, and Metrikin A V 1995 *Prikl. Mekhan. Tekh. Fiz.* **36**(3) 170 [1995 *J. Appl. Mech. Tech. Phys.* **36** 468]

Vesnitskii A I and Metrikin A V 1996 *Usp. Fiz. Nauk* **166** 1043 [1996 *Phys. Usp.* **39** 983]

[57] Sakuda M and Kurihara Y 1995 *Phys. Rev. Lett.* **74** 1284

[58] Krechetov V V 1995 *Izv. Vyssh. Uchebn. Zaved. Radiofizika* **38** 639 [1995 *Radiophys. Quantum Electron.* **38** 423]

[59] Kalikinsky I I 1995 *Zh. Tekh. Fiz.* **65**(10) 131 [1995 *Tech. Phys.* **40** 1047]

[60] Korsakov V B and Fleishman G D 1995 *Izv. Vyssh. Uchebn. Zaved. Radiofizika* **38** 887 [1995 *Radiophys. Quantum Electron.* **38** 577]

[61] Tsytovich V N 1995 *Usp. Fiz. Nauk* **165** 89 [1995 *Phys. Usp.* **38** 87]

[62] Tsytovich V N *et al. Collective Plasma Processes and the Solar Neutrino Problem.* Technical Report RAL-TR-95-066 [*Collected Works Published in Different Journals* (1995, 1996)]; 1996 *Usp. Fiz. Nauk* **166** 113 [1996 *Phys. Usp.* **39** 103]

[63] Collins G B and Reiling V G 1938 *Phys. Rev.* **54** 499

[64] Yamamoto N, Sugiyama H and Toda A 1996 *Proc. Roy. Soc.* A **452** 2279

[65] Villavicencio M V, Roa-Neri J and Jimenez J L 1996 *Nuovo Cim.* B **111** 1041

[66] Goldobin E, Wallraffa A and Ustinov A V 2000 *J. Low Temp. Phys.* **119** 589; 2000 *Phys. Rev.* B **62** 1414

[67] Dolgov O V and Schopohl N 2000 *Phys. Rev.* B **161** 12389

[68] Bolotovskii B M and Galstjan E A 2000 *Usp. Fiz. Nauk* **170** 809 [2000 *Phys. Usp.* **43** 755]

[69] Ginzburg V L 2002 *Usp. Fiz. Nauk* **172** 373 [2002 *Phys. Usp.* **45** 341]

[70] Wahlstrand J K and Merlin R 2003 *Phys. Rev.* B **68** 054301

[71] Veselago V G 1967 *Usp. Fiz. Nauk* **92** 517 [1968 *Sov. Phys. Usp.* **10** 509]; 2002 **172** 1215 [2002 *Phys. Usp.* **45** 1097]

[72] Parazzoli C G *et al.* 2003 *Phys. Rev. Lett.* **90** 107401

[73] Agranovich V M and Ginzburg V L 1984 *Crystal Optics with Spatial Dispersion and Exitons* (Berlin: Springer)
[74] Grichine V M 2002 *Phys. Lett.* B **525** 225
[75] Agranovich V M *et al* 2004 *Phys. Rev.* B **69** 165112

Article 3

On the birth and development of cosmic-ray astrophysics[1*]

3.1

At the present time and indeed ever since the 1950s, the astronomical aspect (including investigations of different variations) has played a dominant role in the study of cosmic rays. This is explained, firstly, by the creation of accelerators with which cosmic rays cannot compete for the purposes of high-energy physics. (I mean, of course, at energies reached by accelerators.) Secondly, the development of both astronomy itself and equipment for the investigation of primary cosmic rays has increased the possibilities of new discoveries in cosmic-ray astrophysics and stimulated interest in the subject. There is hardly any doubt that the situation will remain the same in the foreseeable future and this will be touched upon again at the end of this article.

This book[1*] should presumably reflect the history of the birth and development of cosmic-ray astrophysics or, as it is now often called, high-energy astrophysics, although the term 'cosmic-ray origin' is even more often used by tradition. To argue about terminology is hardly fruitful and not very important either but it is still necessary to come to an agreement in this respect. Later I shall use the term cosmic-ray astrophysics when speaking of the field of investigations devoted to the determination of the characteristics and the role in astronomy of relativistic charged particles (cosmic rays) in space. High-energy astrophysics is wider, it includes uncharged high-energy particles, i.e. gamma rays and neutrinos with an energy, say, higher than 10–20 MeV. Sometimes X-rays are included here, too. As to the 'origin of cosmic rays', this term, as I think, is only reasonable to apply to a narrower problem, i.e. to clarifying the origin of cosmic rays observed near the Earth.

In spite of the fact that, as has been said, the terminology is rather conditional, it seems more correct to give essentially different answers to the questions of when cosmic-ray astrophysics was born and when the problem of the origin of cosmic-rays appeared. Indeed, the very discovery of cosmic rays about 70 years ago already raised the question of their origin. But how can we speak of astrophysics, if for almost 15 years after the discovery of V Hess[1] (i. e. until about 1927), there had still been doubts as to the extraterrestrial origin of cosmic rays. Over the same period, cosmic rays (irrespective of their origin) had been believed to be gamma rays. It was only later that the discovery of the geomagnetic effect and the use of equipment on balloons made it possible to establish the nature of primary cosmic rays: in about 1939–41, it became clear that these are mainly protons and, in 1948, the nuclei of a number of elements were discovered in the composition of primary cosmic rays [1, 2].

Nevertheless, cosmic rays remained something of secondary importance for astronomy and this is quite natural: cosmic rays were only observed near the Earth, and because of the high degree of their isotropy no evidence existed as to the character and location of their sources. This situation is analogous to the one which would exist in optical astronomy if one did not observe separate stars and nebulae but analyzed only the spectrum of all the sources taken together. True, before 1950 there did appear some papers anticipating future developments. Among them one should mention the supposition by Baade and Zwicky (1934) [3] of the connection of supernova flares with the formation of neutron stars and with cosmic-ray generation; and we should also mention the paper by Fermi (1949) [4] devoted to the acceleration of cosmic rays when they propagate in interstellar magnetic fields.

However, the situation changed quickly and radically in 1950–53 when the connection was established between cosmic rays (or, more precisely, their electron component) and cosmic radio emission of synchrotron origin. As a result, there appeared information on cosmic rays far from the Earth both inside our Galaxy and outside its limits. Thus, cosmic rays turned out to be a universal phenomenon and a source of important astronomical information since they radiate radio waves and also electromagnetic waves of other wavelengths (true, apart from radio emission only optical synchrotron radiation was observed in the 1950s). Another, not less important, circumstance also became clear: there are many cosmic rays in the Universe in the sense that their energy and pressure in some regions are very substantial in the general energy balance and for the dynamics of interstellar medium, supernova remnants, radio-emitting 'clouds' in radio galaxies, etc. In other words, it turned out that cosmic rays play an important role in astronomy and that 'elements' of the Universe are not only

[1] The question of the discovery of cosmic rays will clearly be touched upon in other articles of the present collection.[1*] Proceeding from the data known to me [1, 2], I think it is well established (and, probably, commonly accepted) that cosmic rays were discovered by V Hess in 1912.

stars, planets, electromagnetic radiation, and non-relativistic interstellar and intergalactic plasma but also cosmic rays as an inalienable ingredient. This was, naturally, just how cosmic-ray astrophysics was born as one of the branches of astronomy.

3.2

The aim of this book[1*] is 'to take the evidence' of those who have participated in the process of cosmic-ray studies. As for me, this means that I should first of all dwell on the birth of cosmic-ray astrophysics in 1950–53. As is well known, participants in the events often fail to notice many things or see them in the wrong light. Therefore, it is, as a rule, only a historian of science basing his studies on all available material (and first of all publications) who can reconstitute an objective picture of events. The birth and the development of cosmic-ray astrophysics is no exception in this respect but, on the contrary, particularly needs the judgement of a detached onlooker. The point is that, firstly, there were many participants in the events. Secondly, as far as I can judge, some of those participants are of different opinions as to their own and other people's role. Thirdly, quite erroneous statements on this subject can be met in the literature. In such a situation, I only wish to make a few remarks and to give the relevant references. I hope all this will help historians of science.

Cosmic radio emission was discovered in 1931 by K Jansky whose publications appeared between 1932 and 1935. The second radio astronomer was G Reber, his first results were published in 1940 (for some interesting historical details and references see [5]). Solar radio emission was registered in 1942–43 in several places and, in particular, by Reber. Finally, immediately after the Second World War, radio astronomy began rapidly developing in different countries. Note that the first (after the Sun) identified source of cosmic radio emission was the Crab Nebula (J Bolton, 1948; see [5]).

There were two reasons for the progress in radio astronomy. The first was the improvement of radio equipment whose sensitivity had reached quite a fantastic level. The second one was the existence of a much more intensive cosmic radio emission than had been supposed. For example, even in the 19th century there was no doubt as to the presence of some solar radio emission corresponding, say, to the radiation of a black body with a photosphere temperature $T \approx 6000\,°C$. Actually, however, in the metric waveband, the intensity of radio emission from the quiet Sun is two orders of magnitude greater and the intensity of sporadic solar emission is a few more orders of magnitude higher than that expected from an equilibrium photosphere. The situation is the same, as is now known, with X-ray astronomy.

3.3

Here is evidently just the moment for an autobiographical remark. I graduated from the physics faculty of Moscow State University in 1938 as a specialist in optics but immediately changed to theoretical physics—quantum electrodynamics, the theory of higher spin particles, etc. But when in 1941 the war broke out, I, like many of my colleagues, began looking for an application of my efforts in a more practical field and so, rather occasionally, I devoted myself to radio-wave propagation in the ionosphere. These studies, together with other topics, occupied me over many years and the results are summarized in [6]. It was the work on radio-wave propagation in the ionosphere that brought me to radio astronomy.

The well-known Soviet radiophysicists, L I Mandelstam and N D Papaleksi, thought of radiolocation of the Moon and, then, probably, of radiolocation of other celestial bodies as far back as the 1930s. In any case, at the end of 1945 or at the beginning of 1946, N D Papaleksi asked me to clarify the conditions for radio-wave reflection from the Sun. While solving this problem, I saw at once that metric waves, to say nothing of longer ones, must be reflected above the photosphere—in the corona. Simultaneously it became clear that radio waves are noticeably absorbed in the corona and, consequently, must be radiated by it. The effective temperature of such an equilibrium radiation may reach the temperature of the coronal plasma which is estimated as $T \sim 10^6$ K [7]. Non-equilibrium (sporadic) radio emission may, of course, be still stronger just as is observed.

In March and November of 1947, I published the reviews [8] devoted to solar and galactic radio emission. In particular, I indicated in that review that galactic radio emission at a waveband of 4.7 m and shorter can be associated with thermal radiation of the interstellar gas at $T \approx 10\,000$ K. Although cited in [8], Jansky's data (the effective temperature $T_{\mathrm{eff}} \approx 1.5 \times 10^5$ K at a wavelength $\lambda = 14.6$ m) could not be fully relied upon and needed confirmation and, moreover, they could hardly be reconciled with thermal radiation of interstellar gas. But Jansky was right and, by 1948 or thereabouts, it had become quite clear that non-solar cosmic radio emission included a non-thermal component. By that time, it had already been established that an effective temperature of sporadic radio emission of the Sun may reach colossal values of the order of 10^{13} K. Therefore, it was easy to suppose the existence of stars still more active in the radio-frequency band. So, quite naturally, the 'radio-star hypothesis' arose and was developed which connected non-thermal cosmic radiation with the presence of a rather large number of 'radio stars' in the Galaxy [9–12].

Gradually it became clear, however, that to explain the observations one needed to postulate a huge number of 'radio stars' possessing quite unusual properties and spatial distribution. The corresponding assumptions were not confirmed and, what is even more important, an alternative hypothesis appeared which associated non-thermal radio emission with the synchrotron mechanism.

3.4

Electromagnetic radiation due to relativistic particle motion in a magnetic field, i.e. synchrotron radiation, was analyzed in detail by Schott [13] as early as 1912. However, the practical importance of this radiation became apparent only in the 1940s, in connection with the creation of cyclic electron accelerators (particularly the synchrotron), and a large number of articles appeared which repeated and developed the results of Schott. As an example, we shall mention papers [14–17].

It seems today that, in such a situation, the idea of applying the synchrotron mechanism to cosmic conditions should have appeared in 1946–47, the more so as Pomeranchuk considered the radiation of ultra-relativistic electrons as they moved in the Earth's magnetic field as early as 1939 [18]. *Post factum*, however, many ideas and hypotheses often seem obvious. In reality, however, to apply the synchrotron mechanism in astronomy, one should have known the theory of synchrotron radiation and should have realized the possible conditions for its emergence in certain concrete situations far from the Earth. In any event, it was not until 1950 that Alfvén and Herlofson used the synchrotron mechanism to explain radiation from radio stars [19] (this article is reproduced in [1]). In [19], for example, a more or less ordinary star surrounded by a magnetosphere (or a 'trapping field') is considered in which relativistic electrons move. Such an assumption corresponds to the models of the solar origin of cosmic rays and, as we now know, has nothing in common with either the solar system or the discrete sources of cosmic radio emission—supernova remnants, etc. So the value of paper [19] was not in the choice of the model but in the fact that attention was paid (as far as I know, for the first time) to a possible connection between cosmic radio emission and cosmic rays. This line was continued in the paper by Kiepenheuer [20], where there is a reference to [19] and the intensity of synchrotron radiation is estimated which should appear in interstellar space in the field $H \sim 10^{-6}$ G for the relativistic electron concentration $N_e \sim 3 \times 10^{-11}$ cm^{-3}.

Both papers [19, 20] appeared in *Physical Review* in the form of rather short letters and did not attract much attention. Apart from brevity, a still more important role here was apparently played by other factors: most of the astronomers were unacquainted with the synchrotron mechanism, the notes were not published in an astronomical journal, and the radio-star hypothesis, as has already been mentioned, was attractive (or, in any case, popular). I, on the contrary, noticed papers [19, 20], was acquainted with the synchrotron mechanism and did not see particular grounds for supposing the existence of a huge number of radio stars. In short, using the formulae from paper [16], I immediately repeated in more detail all the estimations of [19, 20]. As far as I know, my article [21] which was submitted for publication in October, 1950, was the first and for a rather long time the only response to papers [19, 20]. As has been mentioned, [21] suggested no new ideas (I do not touch upon the discussion contained in [21] referring to cosmic-ray radio emission in the Earth's magnetic field—that is another question) but, of course, promoted the 'introduction' of the synchrotron mechanism into astronomy. And this process turned out to be very long and not simple even in the USSR, where I made reports and generally popularized the synchrotron theory in every possible way.[2]

In such cases, however, talks do not suffice—one should have demonstrated the defects of the alternative approach (the radio-star hypothesis) and found some bright arguments in favor of the synchrotron mechanism. I may permit myself to express a supposition that I did not achieve a success on this way not so much for the lack of imagination as for almost a complete lack of knowledge of astronomy—it so happened that I was then unacquainted not only with the university but even with the school course in astronomy. The only astronomical object I had been acquainted with before was the Sun. Therefore, it was only to the Sun that G G Getmant-sev[3], my post graduate student of those days, and I applied the synchrotron theory [22]. But, in the case of the Sun, there were other possibilities as well and it was not easy to separate the synchrotron component from the total radiation. I would like to mention also the work by Getmantsev [23], performed at my suggestion, wherein he obtained an important result concerning synchrotron radiation for electrons with a power-law spectrum. The papers by Getmantsev [23, 24] and the analysis of the possibility of using radio-wave diffraction on the Moon to determine the dimensions of the sources [8, 25] promoted the development of radio astronomy but did not bring any change even into the minds of astronomers acquainted with our work. So, as late as 1952, I S Shklovsky not only continued advocating the radio-star hypothesis but also considered the synchrotron hypothesis inadmissible [11].[4] As an argument confirming such a conclusion, Shklovsky

[2] Quite recently (in 1996), I learnt about the paper by G W Hutchinson (1952 *Phil. Mag.* **80** 847) which was also devoted to the relation between galactic radio emission and cosmic rays. The author knew and cited papers [19, 21] but, for some reason, paid no attention to paper [20]. The new suggestion from Hutchinson, compared to the previous papers, was a possible relation between radio emission and bremsstrahlung gamma radiation from the same relativistic electrons that enter in the composition of cosmic rays. Some interesting things concerning Hutchinson's work and generally the history of the synchrotron theory of cosmic radio emission can be found in book [79].

[3] Dr G G Getmantsev died on 30 April, 1980 in his 55th year.

[4] In [11, p 445], we find: "Apart from the notion that the sources of galactic radio emission are the interstellar ionized gas and 'radio stars' (in the previously mentioned sense), there also exists another concept developed most extensively by V L Ginzburg [35] [this is paper [21] from the literature cited here, VLG]. According to [35], the source

presents the following: a magnetic field $H \sim 10^{-5}$–10^{-6} G exists only in ionized gas clouds, whereas outside the clouds "the magnetic field strength will be in any case several orders of magnitude lower" [11]. At that time, before the appearance of the work by Pikelner [26] mentioned later, such an erroneous idea had probably been spread. I dwell here upon the paper by Shklovsky [11] in more detail only because it is this particular paper that has repeatedly been mentioned in the West as the main one to express and develop the hypothesis of the synchrotron nature of cosmic radio emission. To some extent, such a misunderstanding is surely connected with a poor knowledge of Russian.

The next very important step in this direction was made in 1952 by I M Gordon (due to some circumstances his paper [27] was published only in 1954 but he made reports in 1952 and the main content of [27] has been known in the USSR at least since the end of 1952). Gordon applied the synchrotron mechanism to optical radiation in solar flares and even discussed the possibility of synchrotron X-ray radiation. The main thing is that he paid attention to the importance and the feasibility of polarization measurements—the presence of polarization is rather characteristic of synchrotron radiation and, consequently, the detection of polarization could provide decisive evidence. To the particularly important results of this period one should refer the work by S B Pikelner [26] who, as a matter of fact, introduced the concept of the gas galactic halo and 'cosmic-ray halo'. He also stressed that the magnetic field with $H \gtrsim 3 \times 10^{-6}$ G exists in the entire Galaxy. Note, by the way, that, in my opinion, S B Pikelner, who died in 1975 in the 55th year of his life, was the most prominent Soviet astronomer–theoretician and he left a lasting trace in astrophysics (see the obituary [28]). The contribution made by Pikelner is not limited to his publications, for he disinterestedly helped everybody who needed his advice and there was no end to those who came to this noble and kind man.

From the work of Pikelner [26], it became particularly clear that the previously mentioned objection by Shklovsky to the application of the synchrotron mechanism to the Galaxy was quite groundless. And it was precisely with reference to paper [26] that Shklovsky changed his mind drastically and in the article [29] stated that the attempt to explain non-thermal galactic radio emission "by a summary effect of galactic stars turns out quite groundless". This is how the synchrotron hypothesis won in the USSR.

But in the West the radio-star hypothesis was used for a few more years. Moreover, my paper submitted to the Manchester Symposium on Radio Astronomy (1955) was not even included in the Proceedings of the

of galactic radio emission may be bremsstrahlung of relativistic electrons in interstellar magnetic fields. We are not in a position to make here a detailed analysis of this hypothesis which for some reason seems to us to be inadmissible."

symposium [30]. Meanwhile, the paper by Unsöld defending the radio-star hypothesis was published there (see also [12]). And only at the next, 1958, Paris Symposium on Radio Astronomy was the synchrotron mechanism, so to say, officially accepted and my report published [31]. By the way, I had previously published two papers in English on the synchrotron mechanism and on the origin of cosmic rays [32].

3.5

The study of the Crab Nebula turned out to be very important for the development of high-energy astrophysics. This object was the first identified discrete source of cosmic radio emission [5] and since the angular dimension of the Crab makes up several minutes of the arc, its very powerful radio emission could not certainly be regarded as thermal (see, for example, [24]). On the contrary, it was quite natural to associate radio emission from the Crab with the presence in it of relativistic electrons. However, even Gordon, who had already turned to considering synchrotron radiation over the entire spectrum, including radiation at optical wavelengths, did not apply this approach to the Crab Nebula and this (the supposition of the synchrotron origin of optical radiation) was done by Shklovsky [33]. Paper [33] was undoubtedly of considerable importance but it should be mentioned that the only source of information about the synchrotron mechanism cited therein was the article by Shklovsky himself [29]. Neither is there (in [33]) any suggestion of measuring the radiation polarization, although in the literature this suggestion is often ascribed to Shklovsky and even called a 'brilliant prediction'.[2*] This suggestion was, in fact, made by Gordon [27]. Moreover, Shklovsky objected to the possibility of measuring polarization. The discussion of this topic took place at the meeting, devoted to the origins of cosmic rays, held in Moscow in May 1953 [34]. I shall quote here some remarks made at this meeting [34].

> **I M Gordon**: I would like to put such a question: in December in Leningrad I suggested an analogous scheme in respect of the continuum of flares, where for a possible nature of the continuum and a verification of the hypothesis, according to which such a spectrum is due to relativistic electron radiation I planned a rather simple experiment—the measurement of the polarization of light. There is no doubt that if the continuum of the Crab Nebula is indeed caused by radiation of relativistic electrons in a magnetic field, then the radiation will be partially polarized owing to the magnetic field anisotropy. Unfortunately we failed to perform such an experiment with flares, since we are short of adequate equipment, whereas the Crab Nebula is a very suitable object for this purpose. Maybe such an attempt has been made? [34, p 253]

I S Shklovsky: It is much more difficult to undertake such a verification for the Crab Nebula than for flares. Taking into account the fact that the magnetic field in the region of supernova remnants is extremely inhomogeneous and some regions of homogeneity in the field have dimension of the order of 10^{16} cm, the appearance of such an effect can hardly be expected at all, since polarization will be statistically averaged. [34, p 254]

V L Ginzburg: And, finally, the last point—the polarization of radiation. I do not agree with what I S Shklovsky answered to I M Gordon and believe that it is interesting to investigate the polarization of optical radiation from supernova remnants. Subject to the magnetic field orientation, polarization will be different and the radiation intensity will also vary. If the dimensions of the quasi-homogeneous field regions are very small, the effect will be small too, but it is of course worth measuring. I think that such attempts may also be undertaken in the radio-frequency band, although there are much less possibilities here. Maybe one would succeed in using the fact that the source in the Taurus is covered by the Moon, and then by the diffraction on the edge of the Moon one can improve the angular resolution of radio astronomical apparatus by about two orders of magnitude. [34, p 260]

I M Gordon: The latter remark refers to radiation polarization. Polarization of optical light (here I cannot agree with I S Shklovsky) must undoubtedly exist because the spherical symmetry which most probably took place during the matter ejection was unlikely to vanish completely because of the turbulence. [34, p 268]

I S Shklovsky: First of all about polarization of the visible radiation of the Crab Nebula. I cannot agree that the polarization can be found. The polarization of several percent even for such a powerful source as the Sun is practically very difficult to register. Here we are dealing with a 9th magnitude object. Knowing that the dimension of the region with a homogeneous magnetic field hardly exceeds 0.01 of the whole system, I think it is difficult to expect a somewhat noticeable polarization. [34, p 276]

I hope that the inclusion of this information is justified in view of some erroneous statements encountered in the literature and, so to say, 'adopted by repetition only'. The decision to present the facts[5] here was also stimulated by the circumstance that, in paper [36], Shklovsky himself stated (pp 385, 386):

[5] These facts have been in brief and partially implicitly presented in the introduction to [35]. Note that the expediency of polarization measurements is also mentioned in my paper [44] submitted for publication before the meeting [34].

It is of particular importance that on the basis of this theory [the synchrotron theory, VLG] we succeeded in predicting several essentially new and important phenomena, the existence of which was then unknown and which, soon after the predictions, were discovered in specially undertaken observations (polarization of optical and radio emission of the Crab Nebula, ...).[3*]

Not to be misunderstood, I should add that, in spite of what has been said, I consider the contribution made by I S Shklovsky to high-energy astrophysics to be substantial. Moreover, as for me personally, I have never claimed much in this respect and always referred to the earlier works I knew, in particular to papers [19, 20]. Thus, the previously mentioned remarks are not at all aimed at supporting my own priority.[4*]

To make the picture complete, we should add that the measurements of polarization of the optical radiation from the Crab [37–39] appeared to be quite successful and also very important from the point of view of the formation of 'astronomical public opinion'. But as I have always believed, the extraordinary attention paid to the Crab was due not so much to the essence of the matter as to the bright history of the discovery and investigation of the envelope of the supernova which flared in 1054 of our era. In other words, if the Crab Nebula did not exist at all, the development of high-energy astrophysics would, in practice, change very little since the Crab, although 'the first among equal', is not a unique object—a supernova remnant, a pulsar, etc.

3.6

We have already mentioned at the beginning of this article that the turning point in the assessment of the astrophysical role of cosmic rays was reached when their connection with cosmic radio emission was established. To fully describe the modern state of high-energy astrophysics and especially to follow its development over 30 years would require a whole encyclopedia. In 1963, when there was much less material, S I Syrovatskii and I tried to solve this problem in the book [40]. We have probably managed to do it to some extent since this book is rather widely known and, somewhat surprisingly for me, I come across references to it even in current literature. The preparation of the second edition, which we repeatedly thought of, was hampered by several circumstances. One of them was the abundance of new material.[6] Another is the impossibility to answer several 'damned questions' which arose from the outset (I mean round about, say, 1953). The latter remark should be explained.

---

[6] The book [40] contains more than 500 references. If one tried today to cover the material with the same degree of completeness, the number of references would increase many times.[5*]

Once we had obtained information about cosmic rays far from the Earth, the situation with regard to the whole problem of the origin of cosmic rays observed near the Earth changed radically. It became clear, for example, that one cannot use models of the solar origin of the main part of cosmic rays [41]. However, there appeared a basis for choosing the halo galactic model in which cosmic rays originate in the Galaxy, occupying a rather extensive halo. Supernova stars were put forward as the main sources of cosmic rays with still more grounds than before [3, 41, 42].[7] Such a 'galactic halo model' was rather thoroughly developed [34, 40, 41, 43–47] and my paper [46] written in 1953 ended like this:

> ... the main thing is done here, and the picture outlined above will not undergo drastic changes, like it had to suffer up to recently, until radio astronomical methods were used to clarify this range of questions. With the development of radio astronomy, and also cosmic electrodynamics, the question of cosmic ray origin became a truly astrophysical problem and overstepped the limits of mainly hypothetical constructions inaccessible for control by observation. For this reason and also bearing in mind the progress made in cosmic ray physics one could be sure that the further development of the theory of cosmic ray origin will be a rapid one.

Such an optimistic prognosis, which I repeated many times later on, was, however, confirmed only in general. In some particular branches, uncertainty remained for decades! So, objections were raised to the existence of the radio halo [48], metagalactic models [49] were still discussed, and the hypothesis of the role of supernovae as the dominating sources of cosmic rays was debated. The latter question is, to a certain extent, quantitative (a certain role in cosmic-ray generation may be played by stars and various types of processes) and is still being discussed [50]. However, I believe as before that supernovae are dominating galactic sources but if the role of other stars turned out to be substantial, this would only enrich the picture. The questions of the halo and metagalactic models (in the latter ones all cosmic rays observed near the Earth come from the Metagalaxy) are, in contrast, qualitative, they determine the very type of the model. For physical reasons (we speak of energetic estimates and dynamical considerations) presented in detail, for example, in [40, 48] (see also [49, p 463]), I never doubted for a single moment either the existence of the halo or the validity of the galactic model. But the standpoint of the opponents of these models was also justifiable, they required observations not words.

---

[7] A new argument here was provided by the radio data testifying to the presence in supernova remnants of large numbers of relativistic electrons [43–45]. I should stress that the appearance of my paper [44] was stimulated by the paper by Shklovsky [43]. I cannot say whether or not Shklovsky knew about papers [3, 42]; in [44] the review [42] is cited.

However, it turned out to be extremely difficult to determine the energy density of cosmic rays outside the Galaxy (in the intergalactic space) and to find the 'trapping region' of cosmic rays in the Galaxy (and, thus, to establish the existence and the dimensions of the cosmic-ray halo or, at least, the radio halo—the halo of relativistic electrons). *Post factum*, this is more or less clear: radio data indicate only the average number of radiating electrons along the line of sight and one can only relate this number to the spatial dimension of the radiating region surrounding the solar system by way of recalculations which were insufficiently convincing bearing in mind the accuracy and thoroughness of the sky survey is of that time. As to the main, proton-nuclear component of cosmic rays, there were not direct ways to estimate its intensity far from the Earth. Therefore, my optimism was probably based, to a great extent, on an underestimation of the difficulties with observations, which is typical of theoreticians. Of course, success came from time to time. In 1961, relativistic electrons were finally discovered directly in the composition of primary cosmic rays near the Earth (only the upper limit—about 1% of protons, for their intensity was known earlier). In 1965, a thermal relict radio emission with a temperature of about 2.7 K was discovered, from which a conclusion was soon drawn that the electron component of cosmic rays must originate in the Galaxy (owing to inverse Compton losses due to relic radiation, electrons with an energy $E \gtrsim 10^{10}$ eV cannot reach the Earth even from the nearest radio galaxies).

But all this was not enough for impatient theoreticians (including myself). I remember that, in 1967 (if I am not mistaken), I gave a talk on the origin of cosmic rays in the California Institute of Technology. Since the audience was wide, I spoke of supernovae, radio halos, etc. Then R Feynman asked me somewhat impatiently about as follows: 'Well, but we know all this, is there anything new?' I felt ashamed and after that I did not love to speak about cosmic rays for nearly 10 years.[6*] Since then, S I Syrovatskii and I worked much less in this field. I have always had many other interesting things to do in various fields;[8] S I Syrovatskii, apart from his interests in cosmic-ray astrophysics, also worked in the field of magnetic hydrodynamics and later took great interest in solar physics. In his last years, when he was already seriously ill, S I Syrovatskii went on developing very successfully and actively the theory of current sheets and a number of other problems. The life of this talented physicist theoretician ceased in September 1979 in his 55th year [52].

---

[8] The list of practically all my publications up to 1977 can be found in [51].

3.7

At the symposium on radio astronomy in 1966, the following exchange of remarks took place (see [48, p 436]).

> **H Alfvén**: Ginzburg has said that it is absolutely clear that cosmic radiation plays a decisive role in the Galaxy. I am not at all sure about this, because what we observe and what we conclude from observation are so different... It may very well be that 99% of the cosmic radiation is a local phenomenon confined to our environment in the same way as the Van Allen radiation belts are confined to the Earth's magnetic field.
>
> **Ginzburg answers**: The arguments against the solar or local origin of cosmic rays are numerous... The radio-astronomical evidence is quite strong. According to this, relativistic electrons are present in a gigantic region outside the solar system; although the halo is open to discussion, there is no question about the presence of cosmic-ray electrons in the disk, in densities comparable to those near the Earth... However, I agree that is extraordinarily difficult to disprove anything.
>
> **Alfvén**: To disprove anything is very difficult but also to prove it.
>
> **Ginzburg**: Fortunately it is possible to do something. I have worked in the field for some years, and I can say in the course of time the argument slowly improves. So I hope during my lifetime, I shall see the full victory of these things.
>
> **Alfvén**: I hope you will live very long.[10*]

Now that I am writing this article in March, 1980, I am in my 64th year and by no means do I think that it is 'too long'. But the proof of the validity of the galactic halo model already exists and I have lived to see this. The limits of this article do not allow me to discuss this point in more detail. I shall restrict myself to a brief explanation and some references to the literature.

The most important and essential success in high-energy astrophysics for the past decade is the appearance of observational gamma astronomy (we do not touch upon X-ray astronomy, for it is 'not typical' of high-energy astrophysics and may be related to it only conditionally). One of the most important sources of gamma-rays in space is the decay of π^0-mesons produced in collisions of cosmic rays (protons and nuclei) with interstellar gas particles. It is clear that the intensity of such gamma-rays is proportional to the gas concentration n and the intensity of cosmic rays I and, after some recalculation, also to their energy density w_{cr}. Thus, gamma astronomy offers the opportunity to determine the density w_{cr} far from the Earth (the gas concentration n being known), exactly as the recording of synchrotron radiation makes it possible to find the energy density of the electron component $w_{cr,e}$ of cosmic rays, the magnetic field strength H being known. In metagalactic models of the origin of cosmic rays, the density everywhere in the Galaxy and around it is approximately the same. Therefore, in the framework of such models, one can predict a gamma-ray flux from the decay of π^0-mesons coming from the Magellanic Clouds [53, 49] or, say, in the direction of the galactic anticenter [54]. Unfortunately, Magellanic Clouds have not yet been observed in gamma-rays but measurements in the direction of the anticenter have been carried out and provide evidence of a decrease of the density w_{cr} with distance from the solar system to the galactic boundaries [54]. More accurate measurements must, of course, specify the quantitative results but, as far as I know, no arguments exist that would cast doubt on the previously mentioned decrease of the density w_{cr}. That is why I consider metagalactic models to be disproved.[7*]

The proof of the existence of the radio halo and, thus, of a not smaller (and rather a more strongly pronounced) halo of cosmic rays is based on the observation of 'edge-on' galaxies NGC 4631 and NGC 891 [55, 56]. It is worth mentioning that there were some earlier attempts to observe the galaxy NGC 4631 [57] but the radio halo was not then observed (I remember I was somewhat discouraged by this, the more so as I had promoted such observations: the cause of the failure with the discovery of the halo in the measurements [57] is not quite clear to me). Processing the radio data for our Galaxy is less obvious but also now leads to the conclusion that a radio halo does exist [58, 59]. Looking back I think that the considerable delay in solving the problem of the existence of radio halo is, first, the result of misunderstandings (including in particular the confusion in the definitions of radio disk and radio halo; see [60, 61]) and, second, is due to the difficulty of observing the radio halo at centimetric and even decimetric waves: at metric waves, observations with a high angular resolution are difficult and have not, in fact, been carried out for galaxies seen edge-on (for some additional remarks in this respect, see [48, p 365; 50, vol 2, p 148; 60, 61]).

Thus, by the late 1970s the largest blank spots in the general picture of the origin of cosmic rays origin had vanished at last, thus concluding an important stage in the development of cosmic-ray astrophysics. In most cases, such conclusions are of course somewhat conditional and, besides, we shall be quite sure of their validity only in about a decade. However, I have never advocated caution just for the sole fear of making a mistake and this also refers to predictions. The present article could, of course, be entirely limited to the past but I would not like to do that. The history of science, like general history, is of course interesting by itself. At the same time, one of the strongest stimuli for our attempts to fall back on the history is still the desire to look into the future by extrapolation. There is a grain of truth in the remark imbued with certain bitterness: "History teaches us only that history teaches us nothing". But this refers rather to social life and politics than to science. In any case, while writing this article, I was thinking about the present and the future not less than about the past. I would like to hear the talks to be made on 7 August, 2012, when, I hope, the centenary of the discovery of cosmic rays will be celebrated (on 7 August, 1912, V Hess made his most successful flight and this date is best of all suited for the 'birthday' of cosmic rays). But, alas... people of my generation can hardly hope even for a smaller thing—to come to know the state of cosmic-ray studies by the beginning of the next century—by 1 January, 2001. At the same time, as we know from our past experience, it is quite possible to have the general outlook of the development of high-energy astrophysics for 20–30 years ahead. Unexpected events will happen, of course, and this is one of the most attractive features of science. However, under conditions, when the creation of some grandiose installations (e. g. satellite HEAO type observatories, the deep-water neutrino station DUMAND, etc) takes not less than a decade, a prognosis for 20–30 years does not seem Utopian. So, what can be expected in the foreseeable future? I shall mention very briefly several key problems and tendencies.[9]

(a) The new installations which are being created for satellites and high-altitude balloons and partially also the ones which now exist will allow us already in the next decade and, in any case, before 2000 to gain much new information on the chemical and isotope composition of cosmic rays (including radio-active ^{10}Be and other nuclei) and on the spectra of different components (including electrons, positrons, and antiprotons).

(b) The use of all these data for establishing the composition of cosmic rays in the sources and the character of their propagation in interstellar space, etc requires the development of theory and some calculations beyond the scope of the widely used and, I am sure, quite unrealistic leaky box model [60, 64, 65].

(c) The theory of particle acceleration in space including various plasma effects is now being developed although it must be admitted that this has been in progress for a long time [4, 40, 41]. Here one should clarify the role of acceleration in interstellar medium by shock waves generated in supernova explosions, to analyze some effects in young supernova remnants during their explosions and near pulsars. This is connected with the entire problem of cosmic-ray sources in the Galaxy (the contribution from super-

---

[9] Immediately before the present article I have written another one intended for the collection of papers [62] in honor of the 80th birthday of Jan Oort, one of the most prominent astronomers of our century. Simultaneously, the paper 'The origin of cosmic rays (introductory remarks)' is being written, which is to be submitted for the IAU IUPAP symposium 'The Origin of Cosmic Rays' (Italy, June, 1980). The Proceedings of this symposium [63] may be expected to reflect the modern state of the problem rather completely (see also [50]). By virtue of what has been said, some overlap of the present article and the previous ones [62, 63] turned out to be inevitable.

novae and other stars, the role of acceleration near the galactic center). Also related is the problem of cosmic-ray acceleration and propagation in radio galaxies and quasars.

(d) In spite of the remarkable achievements of radio astronomy as a whole, the progress in some branches has been slow. This refers particularly to the metric wave band. Meanwhile, remote regions of the radio halo of galaxies and galactic clusters must be most clearly discernible precisely at relatively long wavelengths. An analysis of the frequency dependence of brightness distribution in the radio halo over a wide wave band from centimetres to 3–10 m will apparently allow valuable information to be obtained on the halo and the character of cosmic-ray propagation in the halo (diffusion, galactic wind, and convection, etc [50]).

(e) The cosmic-ray spectrum is extended at least up to energies of about 10^{20} eV. The superhigh-energy range ($E > 10^{16}$–10^{17} eV) is accessible to investigation only by way of observing air showers. The chemical composition of cosmic rays in this region is unknown, the origin of particles is unclear. The most probable now seems the model in which particles with $E < 10^{19}$ eV originate mostly in the Galaxy and those with an energy $E > 10^{19}$ eV—in the Local Supercluster. But this is unclear and the question is, generally, open. To solve this problem, one has to make further cumbersome measurements of anisotropy (this is, however, of interest also at lower energies) and to study the structure of showers, etc.[8*]

Apart from their role in astrophysics, superhigh-energy cosmic rays are (and will probably be for many decades) also of interest for physics. It is a known fact that from 1927–29 and to the early 1950s cosmic rays played an exclusive role in high-energy physics they helped to discover: positron, $\mu^{\pm}$-leptons, $\pi^{\pm}$-mesons, $K^{0,+}$-mesons, and a few hyperons [1, 2]. Later on, physical investigations passed over to accelerators but for energies inaccessible to accelerators, cosmic rays naturally remain the only source of particles. In the 1980s, one can, as far as we know, reckon at most on the use of colliding beams of protons with an energy $E_c = 10^{12}$ eV in each beam. In a recalculation from the center-of-mass system to a laboratory frame of reference, this is equivalent to the use of protons with an energy $E = 2E_c^2/Mc^2 \approx 2 \times 10^{15}$ eV. Consequently, at energies $E > 2 \times 10^{15}$ eV, cosmic rays will for a long time remain the only source of particles.[11*] True, there are very few particles with such energies (for example, the intensity of particles with energy $E \geq 10^{16}$ eV makes no more than 10^2 particles km^{-2} sr^{-1} h^{-1}) but the difficulties are, to a certain extent, compensated by a present-day unprecedented increase of technical possibilities.

(f) The role of gamma astronomy has already been mentioned. The current decade will lead, as may be expected, to progress in this field analogous to that in X-ray astronomy in the 1970s. A new generation of gamma-telescopes will make it possible not only to specify the results obtained on the satellites SAS-2, COS-B, and others but also to investigate a large number of discrete sources, including Magellanic Clouds. The already established fact that the gamma-luminosity of quasar 3C 273 and of some galactic sources is very high (for 3C 273, the luminosity is $L_\gamma [50 < E_\gamma < 500 \text{ MeV}] \approx 2 \times 10^{46}$ erg s^{-1}, for pulsar PSR0532 in the Crab, $L_\gamma [E_\gamma > 100 \text{ MeV}] \approx 3.5 \times 10^{34}$ erg s^{-1}) seems rather significant. Very interesting results may also be expected from the study of gamma-lines, gamma-rays with an energy $E_\gamma \gtrsim 10^{11} - 10^{12}$ eV, detected by the flashes of Cherenkov radiation in the atmosphere, etc.[12*]

(g) The study of high-energy neutrinos should also be referred to the key problems of high-energy astrophysics and astrophysics as a whole. This field of study, if we mean experiment, is only emerging. However, an underground detection of neutrinos from supernova flares in the Galaxy is already realistic. The creation of deep-water optical and (or) acoustic systems (the DUMAND project) is expected to allow us to register with a rather high angular resolution of nearly 1°, neutrinos of $E_\nu \gtrsim 10^{12}$ eV from remote extragalactic sources. Since such neutrinos can only be generated by cosmic rays (protons and nuclei) with a sufficiently high energy, the potential importance of these and related gamma-ray measurements is obvious. We shall mention here that combined measurements of a high-energy neutrino flux and a gamma-ray flux are a promising method to investigate the cores of quasars and active galactic nuclei [66]. In general, I am sure that neutrino astronomy (both at high energies and at energies $E_\nu \lesssim 10$ 20 MeV), along with gravitational wave astronomy, is the main and yet unused reserves of astronomy.

Apart from the previously mentioned problems (a)–(g), one could point out some other ones but we believe that what has been said testifies already to the richness of the current problems and possibilities. So, there is every reason to suppose that, within the next 20–30 years, the relative weight of high-energy astrophysics in astronomy will only increase or at least will not be diminished.[8*]

And what will happen afterwards? Of course, many unclear details will remain and, what is even more important, quite new problems and questions are likely to appear of which we have no notion today. At the same time, I think it quite possible and even probable that there will be practically no other sources of astronomical information except those already known (electromagnetic waves, cosmic rays, neutrinos, and gravitational waves).[9*] In this case, some saturation and a qualitative change in the character of the development of astronomy is inevitable. How interesting it would be to go back to this question, say, in 30 years!

In conclusion, I would like to apologise to the reader for the frequent use in this article of personal pronouns (I, me, etc). In English, it is true, it does not seem to sound so sharply egocentric as in Russian. In any case, if it is possible to avoid the use of personal pronouns in scientific papers, I was unable to do it here.

Comments

1*. The article was written in 1980 for the collection which was eventually published in 1985: Ginzburg V L 1985 "On the birth and development of cosmic ray astrophysics" in *Early History of Cosmic Ray Studies. Some Personal Reminiscences* ed Y Sekido and H Elliot (Dordrecht, Holland: D. Reidel).

2*. See Morrison P 1961 *Handbuch der Physik* **46**(1) S 1.

3*. In the letter by Ya B Zeldovich and myself, mentioned in article 17 of the present book (see item 9 and comment 21), I S Shklovsky's attention was drawn, in particular, to what has been said here. It might be hoped that, in his paper [36], I S Shklovsky was so carried away that he erred inadvertently and ungroundedly ascribed to himself the suggestion of observing the polarization of the radiation from the Crab. Unfortunately, it was not so. The recent collection of papers [67] published the answers given by I S Shklovsky as late as 1981 to the questions of B Oliver. He said again (see [67, p 68]) that he 'had predicted the polarization of optical radiation from Crab Nebula'. How can Shklovsky's position be explained? All I know in this respect is presented in articles 3, 4, and 17 of the present collection and in the yet unpublished materials mentioned in article 17 and comment 21 therein. The facts are established and a public analysis of some biographical and psychological aspects seems to be out of place here.

4*. I think that, after such a remark, it is simply unfair to state (and I have unfortunately encountered such statements) that my criticism of some of the priority claims by I S Shklovsky was prompted by the desire to establish my own priority.

This was how comment 4 to this article concluded in the 1997 Russian edition of the present book. I would like to take the opportunity, perhaps the last one, and say in addition a few words about my relations with I S Shklovsky. What is written in this respect in the present article and in articles 4 and 17 is the absolute truth and was not dictated by indecent priority claims or the desire to settle some score. At the same time, I am very sorry that we never reconciled, although in the last years of his life we felt no animosity towards each other. As to myself, I am aware of my feelings. I was told that the same was true about ISh. Moreover, in 1976 (when ISh and I were already 60), N S Kardashev called me up to suggest, obviously by ISh's consent, that we should resume our relations. I refused but said that I was not feeling malevolence and would keep loyalty towards ISh. I kept the latter promise and my refusal was explained by a suspicion that I was wanted as a person possessing some authority. I am not sure it was actually so but, all the same, I think it was a mistake of mine to have refused to shake a stretched hand and I now sincerely regret it. I Shklovsky was a talented man and did much in astronomy. His attitude towards science and the rotten Soviet regime was progressive. As concerns the points mentioned in my articles 3, 4, and 17, I am inclined to believe that it was something morbid, an idée fixe. I cannot find another excuse of why, regardless of the well-known facts, ISh persistently asserted that it had been his suggestion that the polarization of optical radiation of Crab Nebula be measured (in this connection, see also article 17 in the present collection and, in particular, the notes therein).

What a pity that a human life is still so short (I hope the situation will change with the development of biology) and one cannot reverse time.

5*. For this and some other reasons, S I Syrovatskii and I never managed to prepare a new edition of book [40]. An attempt to somehow replace monograph [40] was then made by a group of five authors. I hope the corresponding book [68] is of certain value. See also the review [83].

6*. For some more details in this respect, see the book Ginzburg V L 2001 *The Physics of a Lifetime: Reflections on the Problems and Personalities of 20th Century Physics*. (Berlin: Springer) p 447.

7*. The results of measurements of gamma-ray fluxes from Magellanic Clouds were published only in 1993. These measurements were conducted in the Compton gamma-ray observatory (CGRO) and confirmed that the cosmic-ray flux in Metagalaxy is much lower than that in the Galaxy [69]. This was further evidence to support the opinion about the inapplicability of metagalactic models.

8*. The text of the present article, written in 1980, has not been changed. I should state, with a certain satisfaction, that I have had no particular reason for serious revision. New things have mostly appeared in gamma astronomy (see [69–71]). High-energy neutrino studies had given no new results by the mid-2000 but they may be expected to appear in the near future on installations that are now under construction [72]. As to cosmic-ray astrophysics, it became particularly clear that essentially new results may only be anticipated in studies of cosmic rays with the highest energy or, more concretely, energies $E > 3 \times 10^{18}$ eV. Precisely this fact is stressed in paper [71] to which I refer the reader. See also papers [73, 74] and article 2, problem 28 in this book.

9*. The possibility of the presence of still unobserved particles (neutralinos, photinos, axions, etc) in cosmic space is now widely discussed. Some of these particles may appear to be the main component of the dark matter that fills the Universe. Naturally, it is quite urgent to discover all such hypothetical particles if they do exist. In that case, it will become possible to speak of the appearance of new channels of astronomical information.

10*. I am taking the opportunity and recall a story associated with this exchange of remarks. In late 1967 or early 1968 in Cambridge (England), I delivered the so-called "Scott lectures" (if I am not mistaken, Scott was someone, not a physicist, who left money to the Cavendish Laboratory for the annual delivery of two lectures). My lectures were devoted to cosmic-ray astrophysics. The lectures were attended by many people among whom was P Dirac sitting in the front row. When I read the remarks by Alfvén and myself (which are presented in the text), Dirac roared with laughter and, as far as I remember, began to applaud. Why does this episode seem interesting to me? First, after Einstein and Bohr, Dirac was, I think, the most outstanding physicist of the 20th century (I would not like to say "great" instead of "outstanding", although in this case it may be well justified). That is why his reaction to different things is interesting. Second, Dirac was known for his taciturnity and self-restraint and it was thorough that he was interested in only a narrow range of problems, often rather unpopular in scientific circles. I was not close to Dirac but saw him and heard his talks several times. And never did I witness anything resembling even remotely that behavior of his at the Scott lecture. It might perhaps be caused by Dirac's opinion that a long life would be necessary to clarify the fate of some of his ideas.

11*. On the Large Hadron Collider (LHC) which is to come into operation in CERN, evidently, in 2006 or 2007, the proton energy in each of two counter-propagating beams will make up 7 TeV. This is the highest energy that will be reached on accelerators in the foreseeable future (the total energy in the center-of-mass system is, of course, equal to 14 TeV). A proton of energy 3×10^{20} eV $= 3 \times 10^8$ TeV which collides with a resting nucleon has an energy of nearly 400 TeV in the center-of-mass system. So, cosmic rays will obviously long be of certain interest for high-energy physics (for superhigh-energy cosmic rays, see reviews [75, 76, 81]).

12*. The study of gamma-ray bursts is of particular interest today (see article 1, problem 29). We shall point also to the comprehensive review in gamma astronomy [77] and, for instance, to the discovery of hard gamma-ray emission by the Cas A supernova remnant [78]. Generally, the detection of X-ray and gamma-ray photons from the remnants of different supernovae is continued. The sources of such photons are obviously relativistic electrons accelerated as a result of bursts. I am taking the opportunity to refer to paper [80] in which the estimates $L_{cr} \simeq 5 \times 10^{40}$ erg s^{-1} and $L_e \sim 10^{39}$ erg s^{-1} of "luminosity" (power) for cosmic rays (L_{cr}) and their electron component (L_e) in the Galaxy (see [68]) are confirmed. Several new references to papers on gamma bursts are presented in article 1. Paper [81] elucidates the state of the problem of cosmic rays of highest energies (see also [84]). Paper [82] is devoted to the discussion of several interesting possibilities in neutrino physics which will appear as a result of the observation of superhigh-energy neutrinos coming from space.

References

[1] Hillas A M 1972 *Cosmic Rays* (Oxford: Pergamon Press)

[2] Dorman I V 1981 *Kosmicheskie Luchi (Istoricheskii Ocherk)* [*Cosmic Rays:*

Historical Outline] (Moscow: Nauka) [Translated into English 1985 *Early History of Cosmic Ray Studies* eds Y Sekido and H Elliot (Dordrecht: D. Reidel)]

[3] Baade W and Zwicky F 1934 *Natl Acad. Sci. USA* **20** 259; 1934 *Phys. Rev.* **46** 76

[4] Fermi E 1949 *Phys. Rev.* **75** 1169

[5] Kraus J D 1967 *Radio Astronomy* (New York: McGraw-Hill)

[6] Ginzburg V L 1970 *The Propagation of Electromagnetic Waves in Plasmas* (Oxford: Pergamon Press)

[7] Ginzburg V L 1946 *Dokl. Acad. Nauk SSSR* **52** 487

[8] Ginzburg V L 1947 *Usp. Fiz. Nauk* **32** 26; 1947 **34** 13

[9] Unsöld A 1949 *Z. Phys.* **126** 176; 1951 *Phys. Rev.* **82** 857

[10] Ryle M 1949 *Proc. Phys. Soc. London* **62A** 491; 1950 *Rep. Prog. Phys.* **13** 184

[11] Shklovsky I S 1952 *Astron. Zh.* **29** 418

[12] Unsöld A 1955 *Z. Phys.* **141** 70

[13] Schott G A 1912 *Electromagnetic Radiation* (Cambridge: Cambridge University Press)

[14] Artsimovich L A and Pomeranchuk I Ya 1946 *Zh. Eksp. Teor. Fiz.* **16** 379

[15] Schiff L 1946 *Rev. Sci. Instrum.* **17** 6

[16] Vladimirskii V V 1948 *Zh. Eksp. Teor. Fiz.* **18** 392

[17] Schwinger J 1949 *Phys. Rev.* **75** 1912

[18] Pomeranchuk I Ya 1939 *Zh. Eksp. Teor. Fiz.* **9** 915

[19] Alfvén H and Herlofson N 1950 *Phys. Rev.* **78** 616

[20] Kiepenheuer K O 1950 *Phys. Rev.* **79** 738

[21] Ginzburg V L 1951 *Dokl. Akad. Nauk SSSR* **76** 377

[22] Getmantsev G G and Ginzburg V L 1952 *Dokl. Akad. Nauk SSSR* **87** 187

[23] Getmantsev G G 1952 *Dokl. Akad. Nauk SSSR* **83** 557

[24] Getmantsev G G 1951 *Usp. Fiz. Nauk* **44** 527

[25] Getmantsev G G and Ginzburg V L 1950 *Zh. Eksp. Teor. Fiz.* **20** 347

[26] Pikelner S B 1953 *Dokl. Akad. Nauk SSSR* **88** 229

[27] Gordon I M 1954 *Dokl. Akad. Nauk SSSR* **94** 813

[28] Pikelner S B (Obituary) 1976 *Usp. Fiz. Nauk* **119** 377 [1976 *Sov. Phys. Usp.* **19** 540]

[29] Shklovsky I S 1953 *Astron. Zh.* **30** 15

[30] Van de Hulst H C (ed) 1957 *Radio Astronomy: Symposium No. 4 held at the Jodrell Bank Experimental Station near Manchester, August 1955* (Cambridge: University Press)

[31] Bracewell R N (ed) 1959 *Paris Symposium on Radio Astronomy: IAU Symposium No. 9 and URSI Symposium No. 1, 30 July–6 August 1958* (Stanford, CA: Stanford University Press)

[32] Ginzburg V L 1956 *Nuovo Cimento Suppl. Ser. 10* **3** 38 (this paper also appeared in [1]); 1958 *Prog. Elem. Part. Cosmic Ray Phys.* (Amsterdam) **4** 339

[33] Shklovsky I S 1953 *Dokl. Akad. Nauk SSSR* **90** 983

[34] 1954 *Proceedings of 3rd Conference on Questions Cosmogony (Origin of Cosmic Rays)* (Moscow: Acad. Sci. USSR)

[35] Ginzburg V L and Syrovatskii S I 1965 *Annu. Rev. Astron. Astrophys.* **3** 297

[36] Shklovsky I S 1962 *Usp. Fiz. Nauk* **77** 3 [1962 *Sov. Phys. Usp.* **5** 365]

[37] Vashakidze M A 1954 *Astron. Circular* 147 11

[38] Dombrovskii V A 1954 *Dokl. Akad. Nauk SSSR* **94** 1021

[39] Oort J H and Walraven Th 1956 *Bull. Astron. Inst. Netherlands* **12** 285

[40] Ginzburg V L and Syrovatskii S I 1964 *The Origin of Cosmic Rays* (Oxford: Pergamon Press)

[41] Rosen S (ed) 1969 *Selected Papers on Cosmic Ray Origin Theories* (New York: Dover)

[42] ter Haar D 1950 *Rev. Mod. Phys.* **22** 119

[43] Shklovsky I S 1953 *Dokl. Akad. Nauk SSSR* **91** 475

[44] Ginzburg V L 1953 *Dokl. Akad. Nauk SSSR* **92** 1133

[45] Hayakawa S 1956 *Prog. Theor. Phys.* **15** 111

[46] Ginzburg V L 1953 *Usp. Fiz. Nauk* **51** 343; 1954 *Fortschr. Phys.* **1** 659

[47] Shklovsky I S 1960 *Cosmic Radio Waves* (Cambridge, MA: Harvard University Press)

[48] 1967 *Radio Astronomy and the Galactic System. IAU Symposium No. 31, Noordwijk, The Netherlands, 25 August–1 September 1966* ed H van Woerden (New York: Academic Press)

[49] 1975 *Phil. Trans. R. Soc. London* **277A** 317

[50] Miyake S (ed) 1979–1980 *16th Int. Cosmic Ray Conf., Kyoto, Japan, August 6–18, 1979: Conf. Papers V 1–14* (Tokyo: Institute of Cosmic Ray Research, University of Tokyo Press)

[51] Kuz'menko R I and Makhrova I A (Compiler) 1978 *Vitalii Lazarevich Ginzburg* (Materialy k Bibliographii Uchenykh SSSR. Ser. Fiziki, Vyp. 21 (Bibliography of Acad. Sci. USSR Scientists. Ser. Phys., Issue 21)) (Moscow: Nauka)

[52] Sergei Ivanovich Syrovatskii (Obituary) 1980 *Usp. Fiz. Nauk* **131** 73 [1980 *Sov. Phys. Usp.* **23** 274]

[53] Ginzburg V L 1972 *Nature* (Phys. Sci.) **239** 8

[54] Dodds B, Strong A M and Wolfendale A V 1975 *Mon. Not. R. Astron. Soc.* **171** 569; 1979 *Pramana* **12** 631

[55] Ekers R D and Sancisi R 1977 *Astron. Astrophys.* **54** 973

[56] Allen R J, Baldwin J E and Sancisi R 1978 *Astron. Astrophys.* **62** 397

[57] Pooley G G 1969 *Mon. Not. R. Astron. Soc.* **144** 143

[58] Bulanov S V, Dogiel V A and Syrovatskii S I 1976 *Astrophys. Space Sci.* **44** 267

[59] Webster A 1978 *Mon. Not. R. Astron. Soc.* **185** 507

[60] Ginzburg V L 1978 *Usp. Fiz. Nauk* **124** 307 [1978 *Sov. Phys. Usp.* **21** 155]; Ginzburg V L and Ptuskin V S 1976 *Rev. Mod. Phys.* **48** 161

[61] Ginzburg V L 1979 *The Large-Scale Characteristics of the Galaxy: Symp. No. 84, College Park, Maryland, USA, 12–17 June, 1978* (ed W B Burton) (Dordrecht, Holland: D. Reidel) p 485

[62] Ginzburg V L 2001 *The Physics of a Lifetime: Reflections on the Problems and Personalities of 20th Century Physics* (Berlin: Springer) p 457

[63] Setti G, Spada G and Wolfendale A W (eds) 1981 *Origin of Cosmic Rays: Symp. No. 94, Bologna, Italy, June 11–14, 1980* (Dordrecht, Holland: D. Reidel)

[64] Ginzburg V L, Khazan Ya M, and Ptuskin V S 1980 *Astrophys. Space Sci.*

68 295

[65] Wallace J M 1980 *Astrophys. Space Sci.* **68** 27

[66] Berezinsky V S and Ginzburg V L 1981 *Mon. Not. R. Astron. Soc.* **194** 3

[67] Shklovsky I S 1996 *Razum, Zhizn', Vselennaya [Mind, Life, Universe]* (Moscow: Yanus)

[68] Berezinskii V S, Bulanov S V, Ginzburg V L, Dogiel V A and Ptuskin V S 1990 *Astrophysics of Cosmic Rays* ed V L Ginzburg (Amsterdam: North-Holland) [Russian original 1984 *Astrofizika Kosmicheskikh Luchei* ed V L Ginzburg (Moscow: Nauka)]

[69] Ginzburg V L 1993 *Usp. Fiz. Nauk* **163**(7) 45 [1993 *Phys. Usp.* **36** 587]

[70] Ginzburg V L and Dogel V A 1989 *Usp. Fiz. Nauk* **158**(1) 3 [1989 *Sov. Phys. Usp.* **32** 385]

[71] Ginzburg V L 1996 *Usp. Fiz. Nauk* **166** 169 [1996 *Phys. Usp.* **39** 155]

[72] Bahcall J N *et al* 1995 *Nature* **375** 29

[73] Ginzburg V L 1999 *Usp. Fiz. Nauk* **169** 419 [1999 *Phys. Usp.* **42** 353]

[74] Biermann P L *et al* 2001 *Current Topics in Astrofundamental Physics: the Cosmic Microwave Background. 7th Course, Erice Lectures 1999* (NATO Science Series, ser. C, vol. 562) ed N G Sánchez (Dordrecht: Kluwer)

[75] Nagano M and Watson A A 2000 *Rev. Mod. Phys.* **72** 689

[76] Cronin J W 1999 *Rev. Mod. Phys.* **71** S165

[77] Hoffman C M *et al* 1999 *Rev. Mod. Phys.* **71** 897

[78] Aharonian F A *et al* 2001 *Astron. Astrophys.* **370** 112

[79] Breus T K 2002 *Institute* (Moscow: Izd. "KoNT")

[80] Dogiel V A, Schönfelder V and Strong A W 2002 *Astrophys. J. Lett.* **572** L157

[81] Watson A A 2002 *Contemp. Phys.* **43** 181

[82] Fodor Z, Katz S D, and Ringwald A 2002 *Phys. Rev. Lett.* **88** 171101

[83] Macdonald F B and Ptuskin V S 2001 *The Century of Space Science* eds J A M Bleeker, J Geiss and M C E Huber (Dordrecht: Kluwer) p 677 Ptuskin V S and Zirakashvili V N 2003 *Astron. Astrophys.* **403** 1

[84] Isola C and Sigl G 2002 *Phys. Rev. D* **66** 083002

Article 4

Remarks on my work in radio astronomy[1*]

4.1

An attempt to elucidate the history of the development of radio astronomy in its different aspects and using different sources seems to be quite relevant and useful. But unfortunately, in spite of sincere aspirations, I have not been able to write a paper which would correspond to the requirements of the editor of the present book.[1*] The main reason is evidently that I am not an astronomer but a physicist both by education and by the experience of many years. Astronomy became for me a 'part-time job'. While in my personal life it was a rather accidental occurrence, from a more general point of view it happened on the wave of the truly great process of the transformation of astronomy from being optical to 'all-wave'. In the course of this process, many physicists, radio engineers, and other professionals took an interest in astronomical methods and problems and 'came' to astronomy (often irritating professional astronomers by their poor knowledge of classical astronomy and even of the proper terminology). Some of the neophytes became real astronomers, while others did not give up their previous specialities and devoted only part of their time to astronomy. It is actually not even a question of time but rather of the style of the work. A physicist, say, may devote all his/her efforts to astrophysics (which is now rather difficult to separate from physics proper) but if s(he) insufficiently commands purely astronomical material, s(he) nevertheless does not become a real astronomer, who must know well the astronomical classics and literature, observational methods and results, etc. In a word, I am not a professional astronomer and, therefore, my astronomical works are somewhat fragmentary and episodic except, possibly, those connected with cosmic-ray astrophysics.

In such a situation, I can here only outline and briefly comment on my radio-astronomical papers in the hope that it will be useful for the history of radio astronomy.

4.2

When the war with Germany broke out on 22 June, 1941, I made up my mind to devote myself to something which would be at least potentially useful for defense; before, I had been mainly engaged in the theory of elementary particles or, as it would now be called, high-energy physics. Taking rather accidental advice, I turned to radio-wave propagation in ionosphere. I shall not dwell here on this activity, which lasted many years and was eventually reflected in monograph [1]. But it was, in fact, the ionosphere that was the starting point for my passing over to, first, solar and then extra-solar radio astronomy.

Well-known Soviet physicists and radio specialists, L I Mandelshtam and N D Papaleksi, contemplated the problem of Moon radiolocation long before the war. In 1944, stimulated by the progress in radiolocation (for more details, see [2, 3]), Papaleksi returned to this idea, also considering radiolocation of the planets and the Sun. In this connection, at the end of 1945 or the beginning of 1946, he asked me to clarify the conditions of radio-wave reflection from the Sun. As a matter of fact, it was a typical ionospheric problem and all the formulae were at hand. The results of the calculations did not seem very optimistic since for a large set of parameters such as electron concentration and temperature in the corona and chromosphere, which then remained unknown in many respects, radio waves would be strongly absorbed in the corona or chromosphere and not even reach the level of reflection. (The question of reflection due to inhomogeneities was not considered; the 'point' $n(\omega) = \sqrt{1 - \omega_p^2/\omega^2} = 0$ was, roughly speaking, taken for the level of reflection.) But this immediately suggested a more interesting conclusion: the source of solar radio emission must not be the photosphere but rather the chromosphere and, for longer waves, the corona. At that time the corona was already assumed to be heated to hundreds of thousands or even a million degrees. Thus, even under equilibrium conditions, i. e. in the absence of perturbations and sporadic processes, the temperature of emission from the solar corona at waves longer than about a metre must reach approximately a million degrees at the photosphere temperature of $T_{ph} \approx 6000$ K. This is what was presented in my first astronomical paper [4]. In that same year, D F Martyn [5] and I S Shklovsky [6] reported similar conclusions. It is beyond doubt that these papers were written independently, I can only say that my paper [4] was submitted for publication on 27 March, 1946, while the dates of submission of papers [5, 6] are not indicated; they appeared respectively in *Nature* of 2 November, 1946, and in *Astronomicheskii Zhurnal* of November–December, 1946.[2*] In my calculations, I used the absolutely clear and reliable formulae known from the theory of ionospheric propagation [1, 4, 7]. Martyn [5] did not present formulae but he apparently acted in the same way. Shklovsky [6] believed that one should take separate account of, and then sum up the absorption due to 'free–free' transitions and the absorption due to electron–proton collisions. As a matter of fact, this is one and the same mechanism [1] and, therefore, Shklovsky's quantitative results [6] were incorrect. But this circumstance was then of no importance since the parameters of the corona were not exactly known.

The paper by Pawsey [8], published immediately after Martyn's paper [5], confirmed the existence of thermal radio emission from the corona with $T \sim 10^6$ K. It was shown that such radiation plays the role of the lower limit, reached when the sporadic component of solar radio emission becomes sufficiently weak.

4.3

A weak point in radio astronomy during that period was its low angular resolution, preventing investigation of the Sun even for regions with sizes of minutes of arc (it is difficult to believe this today when radio interferometers are far ahead of the best optical telescopes in angular resolution power). In this connection, N D Papaleksi suggested measurement of solar radio emission during the total eclipse of 20 May, 1947, with the help of a 1.5-m wavelength antenna, which was installed on board a ship and had a wide (several degrees) directivity pattern. These measurements were successful [2, 3, 9, 10] and turned out to be the first of their kind. Although the intensity of optical emission from the Sun during the total eclipse diminishes by several orders of magnitude, at 1.5-m wavelength the intensity during the eclipse diminished no more than by 60%. Thus, metre-wave radio emission was proved to come from the corona, which remains uncovered by the Moon even under a total optical eclipse. Using these measurements, one also succeeded in observing some details concerning the distribution of active radio-emitting regions in the solar disk.

I took part in the Brazil expedition of the USSR Academy of Sciences, on board the ship "Griboyedov", which conducted the previously mentioned radio observations of the Sun. It seems that I was included on the staff of the expedition as recognition of my work in the development of the young Soviet radio astronomy. I did not participate, however, in the solar radio emission measurements carried out on board the ship. I was on the main part of the expedition which made its way inside Brazil to perform optical measurements that unfortunately failed because of bad weather. This main part of the expedition also included a small ionospheric group, headed by Ya L Alpert, of which I was a part: weather could not, of course, prevent ionospheric measurements.

As a result of the expedition and related activities, I became for a while almost a professional radio astronomer (see earlier). I tried to get acquainted with all the available materials, methods of measurements, etc. As a result I wrote two reviews of radio astronomy [10, 11], perhaps the first in the world literature (I cannot, however, guarantee this).[1] Now that 35 years have passed, it is difficult to judge the value of these papers and I do not want to analyze them in detail. I shall only mention the proposal made in [10] to use radio-wave diffraction on the Moon edge in order to improve the angular resolution of details on the Sun during eclipses. This question was further considered more thoroughly in my paper with G G Getmant-sev [12] where we were already thinking more of discrete sources of cosmic radio emission than the Sun. The method of radio-wave diffraction on the Moon edge has since been widely used and, therefore, I have nothing more to add here except that in [10] and, more extensively, in [12], we also discussed the possibility of enhancing the angular resolution still more if the source under observation is located on the line connecting the Moon center with the point of observation (here we are evidently dealing with the Arago–Poisson light spot).[2] Such observations are, of course, strongly hampered by the non-spherical shape of the Moon and the necessity to have a source on or very near to the previously mentioned line and I am unaware of any attempts to exploit this method. But perhaps one should nevertheless analyze such a possibility in more detail? This could work not only for the Moon but also planets and their satellites, as well as artificial screens (both flat and spherical).

4.4

If I dwelt on the entirety of my subsequent papers in so much detail, it would take too much space. As has already been mentioned, I worked in the field of astrophysics in a rather sporadic and chaotic way: that portion closer to radio astronomy can perhaps be divided into three main trends:

(a) Twinkling of radio sources in and beyond the ionosphere, oscillations of the intensity of solar radio emission, the use of polarization mea-

---

[1] How far I remained from astronomy as a whole, in spite of what has been said, is seen from the incident I mentioned in note [29] dedicated to J Oort. On the way back from Brazil, the participants of the expedition were lucky to visit Leiden, although quite accidentally. And there, instead of meeting J Oort and generally taking part in the discussion of astronomical problems, I rushed to the Kamerlingh Onnes Cryogenic Laboratory, since I was then most of all interested in low-temperature physics.

[2] As an objection to the wave theory of light, Poisson pointed out a consequence of this theory which seemed to him quite absurd: on the axis of the geometrical shadow from a round opaque screen a light spot must be observed (the source is considered to be point-like and is located behind the screen on the axis perpendicular to the screen plane). The presence of a light spot was confirmed by an experiment conducted immediately after this by Arago. For an opaque sphere as screen, the conditions of observation of a central peak are facilitated [12].

surements and earth satellite measurements [13–20]. An approximate idea of the contents of these papers is suggested already by their titles.

(b) The theory of sporadic radio emission from the Sun. V V Zheleznyakov and I engaged ourselves in this range of questions beginning in 1958 [21]. A number of other papers followed but it would be out of place to refer to them here since the corresponding results (with references) are presented in the book by Zheleznyakov [22].

(c) The theory of cosmic synchrotron radio emission and its connection with the problem of the origin of cosmic rays and with high-energy astrophysics as a whole. This was the subject of my main astrophysical interest and I am still interested in it, although less than before. From the point of view of historical information which I can offer, these problems also play the most important role. For this reason, item 5 of the present note will be devoted to this subject.

Now I shall dare to mention a few more of my radio-astronomical works. The origin of radio galaxies and, specifically, the question of the energy source of their radio emission was not immediately clear. For instance, one can mention the hypothesis of colliding galaxies, which clearly proved incapable of explaining the mechanism of energy output in radio galaxies. Another idea suggested was a sharp increase in the number of supernova flares in radio galaxies, an idea which has always been rather groundless.

Therefore, I believe that my note [23] was not without value since it stated that the required energy output and cosmic-ray acceleration in radio galaxies is, in principle, easily provided by the gravitational energy release, in particular in the process of star formation. On the whole, it was pointed out that "it seems more attractive to associate the galactic flares not with supernova flares, but rather with another large-scale mechanism, for example, the gravitational instability of a galaxy or of its central part, as schematically outlined above". This trend of thought continued to a certain extent in later papers devoted to quasars (see [24] and the literature cited therein, as well as [25]).

The discovery of pulsars, naturally, gave rise to the temptation to clarify the mechanism of their radio emission. Zheleznyakov and I, and partially V V Zaytsev, published several articles on this subject (the last of them was the review [26]). But the problem proved to be much more complicated than it seemed at first (it happens sometimes that a problem turns out to be particularly sophisticated, which is not suspected in advance). For this reason, I decided long ago to leave this field and, from a recent review [27], I see that I was right. I think that only representatives of the younger generation (or even generations) will be able to gain real insight into this very interesting but essentially complicated and many-sided range of questions—the mechanisms of pulsar radiation and the whole theory of pulsar magnetospheres.[3*]

4.5

Now I shall finally turn to the synchrotron theory of cosmic radiation, as well as cosmic-ray astrophysics (traditionally often called the problem of the origin of cosmic rays). Fortunately (at least for me) I can restrict myself mainly to brief remarks and refer the reader to my paper [28], where I have already touched upon the history of the problem.

Approximately in 1947–49 (as far as I remember) it became quite clear that comparatively long-wave, non-solar cosmic radio emission (including, in particular, the very first radio astronomical results, namely, the measurements by K Jansky carried out on a wavelength of about 15 m) possesses an effective temperature T_{eff} exceeding 10^4 K. Therefore, it was impossible to interpret such radio emission as thermal radiation from interstellar gas, for this gas generally has a temperature $T \lesssim 10^4$ K, and in any case radiation with $T_{\text{eff}} \gg 10^4$ K cannot be explained by thermal radiation of a gas. Thus, one had to assume the existence of a non-thermal source analogous, for example, to the sporadic sources of non-thermal solar radio emission. This was how the 'radio-star hypothesis' appeared in a quite natural way. According to this idea, some stars were anomalously powerful radio sources responsible for the non-thermal, cosmic radio emission with its continuum and diffuse directional distribution [30–32]. The radio-star hypothesis, however, faced a lot of difficulties, mainly involving assumptions, sometimes arbitrary and unrealistic, concerning the hypothetical radio stars. Furthermore, an alternative soon appeared and with time became stronger—the synchrotron hypothesis of the origin of non-thermal radio emission, which in the end proved to be valid.

Competition or struggle between these two hypotheses took several years. From the physical point of view, synchrotron radiation had been known and quite clear for many years [33] and, in the 1940s, it was especially widely discussed in the physical literature in connection with the analysis of synchrotrons (see the references cited, e. g., in [28]). But it was only in 1950 that the first papers [34, 35] appeared in which the synchrotron mechanism was considered as related to cosmic radio emission: radiation from radio stars was discussed in [34] and that from interstellar space was considered in [35]. It seems rather strange to me that these papers appeared in a journal of physics and, besides, in the form of short letters. In any case, as far as I know, they did not attract astronomers' attention. But I, on the contrary, believed at once that the synchrotron mechanism was responsible for non-thermal cosmic radio emission. I ascribe this not to my keen insight but to the previously mentioned fact that I was closer to physics and rather far from classical astronomy. In this situation, the synchrotron mechanism seemed clear and realistic, whereas the hypothetical strange 'radio stars' remained purely speculative. I immediately verified the calculations of [34, 35] and, if I am not mistaken, added little essential. (I have not now compared all the expressions and estimates from [34, 35] and with those of my paper [36], submitted for publication on 31 October, 1950, because it does not seem important here, the more so as I have never claimed priority.) But the fact is that my paper [36] was the first, and remained for some time the only one, that responded to the proposals [34, 35] to use the synchrotron mechanism in astronomy. The point is probably that the reaction of astronomers was quite the opposite to mine, i. e. the synchrotron mechanism seemed mysterious and speculative, whereas 'radio stars', although posing riddles, were more acceptable—for what kinds of stars cannot exist? In this respect, I S Shklovsky was not an exception. He not only developed the radio-star hypothesis [32, 37] but also positively denied the synchrotron hypothesis ("which for a number of reasons does not seem to us to be acceptable" [37][3]). I dare make this remark here only because it is just the paper [37] by Shklovsky that has been repeatedly cited in the world literature as almost the first and the principal one discussing the application of the synchrotron hypothesis. A similar error is spread concerning the question of exploiting synchrotron radiation polarization as a criterion for establishing the validity of a synchrotron origin of cosmic radiation. Polarization measurements were proposed by I M Gordon [38; 39, pp 253, 268] and supported by me [39, p 260; 40], whereas I S Shklovsky denied their efficiency (see [39, pp 254, 276]; the English translation of the corresponding discussion [39] is given in [28]). Concerning the papers by I S Shklovsky, I shall make only two more remarks. After the appearance of a very important work by S B Pikelner [41], who emphasized that the interstellar magnetic field exists in the entire galactic volume, Shklovsky realized [42] that his earlier objection [37] to the efficiency of the synchrotron mechanism was groundless. In this same paper [42], he came to the conclusion that the radio-star hypothesis, which he had supported not long before [32, 37], was 'a complete failure'. Furthermore, the 'interpretation of "radio stars" as a very numerous category of galactic objects' was described in this paper as "quite groundless". Moreover, paper [42] included several remarks of concrete astronomical character of which I shall not judge. But in view of what is said later, it is relevant to note that paper [42] did not, in any respect, develop the theory of synchrotron radiation (it contains not a single formula concerning this theory). In a subsequent paper [43], Shklovsky proposed the synchrotron interpretation of optical radiation from the Crab Nebula, which was undoubtedly an achievement and played an important role. Polariza-

---

[3] This quotation from [37] is presented at greater length in [28], i. e. in article 3 of the present book, where one can also find more details concerning the discussion presented here and reflected in [39]. In article 3, I also mentioned the paper by G Hutchinson which, unfortunately, came to my knowledge with great delay

tion measurements were not, however, mentioned in [43] (see also earlier) and, in any case, the possible efficiency of the synchrotron mechanism in astrophysics not only in the radio but also in the optical and even the X-ray bands, had been emphasized before by Gordon [38]. Finally, I would like to note that the only source of information about the synchrotron mechanism cited in Shklovsky [43] was his previous paper [42] characterized earlier.

Here and in [28], I have made some remarks concerning several papers by I S Shklovsky which were devoted to the origin of cosmic radio emission. I had to emphasize an obvious discrepancy between their actual content and what is often ascribed to them in the literature. I have already tried to do this in a brief and somewhat implicit form at the beginning of paper [44] but without any success. That is why I came to the conclusion that I must either not write about the history of radio astronomy at all or tell the truth as I see it. It is the concern of others, especially historians of science, to verify the facts, the more impartially and thoroughly, the better.

Judging from the proceedings of the 1955 Manchester Symposium on Radio Astronomy [45], which included a paper on the 'radio-star' hypothesis while my paper sent to the symposium was not even published, 'astronomical public opinion' in 1955 was still on the side of the 'radio-star' model. But already at the next, the 1958 Paris Radio Symposium, the synchrotron mechanism was unconditionally accepted as dominating in the production of non-thermal cosmic radio emission (we do, of course, not mean the radiation coming from the solar atmosphere and generally from relatively dense regions). This time my report "Radio astronomy and the origin of cosmic rays" was included in the proceedings of the Symposium [46], although I myself was not able to participate.[4]

The establishment of a relationship between radio astronomy and cosmic rays has led, as a matter of fact, to the appearance of a new field of research in astronomy, namely cosmic-ray astrophysics and then high-energy astrophysics.[5] This has been just the main road I have taken (since 1950-53) in my own astronomical activities, which is characterized in paper [28] (see also [47-49, 29] in the historical context and with a number of references. It is hardly relevant to repeat it here.[4*]

---

[4] In the proceedings [46], the section "Mechanisms of Solar and Cosmic Radio Emission" opens with a long report, including even a 'historical review', by G R Burbidge (paper 98). This review, as well as other sections of the same paper, however, contains statements quite opposite to those presented here (and in [28]) on the basis of the available published materials. It would be of interest to know which sources were used by G Burbidge.

[5] Cosmic rays is the term used at the present time only in application to charged particles. Therefore, gamma-ray astronomy, X-ray astronomy, high-energy neutrino astronomy, and cosmic-ray astrophysics are altogether more and more often referred to as high-energy astrophysics. This term is used here in this sense.

4.6

As is clear from what has been said at the beginning of the present article, I am able to compare the situation in physics and in astronomy, at least in some respects and aspects. It seems to me that it would not be out of place to touch here upon such a comparison. The point is that a person, if compared with a measuring device, reacts mainly not to a function (i. e. to the actual value of the quantity or to the 'state') but to its derivative. As a result, some astronomers do not seem to appreciate sufficiently the situation which is very favorable for them as a 'scientific community'. The principal circumstance is the very existence of such a community united by the International Astronomical Union (IAU). The IAU and its services (half a hundred commissions, continuously held symposia and colloquia, exchange of information, a well-organized system of publication of journals, various proceedings, books, etc) contribute much to the development of astronomy. Physics as a whole does not experience anything of the kind at the present time. This can, of course, be explained by a much larger number of physicists compared to astronomers, by the multiplicity of scientific fields of research, by the ties with industry and applied fields, etc. But the fact is, I think, that at the present time one cannot speak of a united scientific community of physicists. The existence of separate narrower communities, unions, and commissions with their congresses and symposia is, on the one hand, useful but, on the other hand, leads to a still deeper separation of physicists working in different fields of research. All this very much hampers work both from the psychological and organizational points of view. I, for example, am 75–85% a physicist, judging by the number of my papers and results. But by the amount of literature received (reprints, preprints, journals), by the number of invitations to symposia, etc., and by the number and 'strength' of contacts with colleagues, I am rather an astronomer (though I include here in astronomy the range of questions particularly close to physicists, namely high-energy and cosmic-ray astrophysics). Specifically, considering that the most important part of my physics research lies in the field of superconductivity and superfluidity, I feel myself in bad need of contacts (even indirect ones, to say nothing of personal contacts) with corresponding specialists outside the USSR. My work definitely suffers much from this circumstance.[5*] Were these astronomical problems, the situation would be quite different.

It would be out of place to dwell here longer on these comments, which are aimed only at emphasizing how valuable the activity of the IAU is. Therefore, in spite of all the difficulties, for progress in the further development of astronomy it seems necessary to maintain a strong, united IAU as long as possible.

Comments

1*. This paper was written in 1982 and published in the collection 1984 *The Early Years of Radio Astronomy* ed W Sullivan (Cambridge: Cambridge Univ. Press).

2*. In the booklet [50], part of which was reprinted in [51] (see p 293), I S Shklovsky said that the idea underlying his paper [6] had struck him in the early summer of 1946 during his visit to FIAN (P N Lebedev Physical Institute), where I worked at that time. I do not remember any contact with Shklovsky then.

3*. See Beskin V S, Gurevich A V and Istomin Ya N 1993 *Physics of the Pulsar Magnetosphere* (Cambridge: Cambridge University Press).

4*. What has been done after 1982 is reflected in book [52] and in papers [53–55].

5*. It should not be forgotten that it was written in 1982, i. e. in the period when I could go abroad only with great difficulties. Moreover, censorship existed, which hampered correspondence with other countries, to say nothing of the inadmissible work of the post in combination with censorship (all correspondence that came to me was opened and inspected and I certainly got it with some delay, if at all). I am glad to say that the current situation is quite different—there is practically no censorship and one can go abroad freely. Unfortunately, another, financial, difficulty appeared. As to me personally, the main difficulty is my age.

References

[1] Ginzburg V L 1970 *The Propagation of Electromagnetic Waves in Plasmas* 2nd edn (Oxford: Pergamon Press) [Russian original 1967 (Moscow: Nauka)]

[2] Ginzburg V L 1948 N D Papaleksi and radio astronomy *Izv. Akad. Nauk SSSR Ser. Fiz.* **12** 34; 2001 *The Physics of a Lifetime: Reflections on the Problems and Personalities of 20th Century Physics* (Berlin: Springer)

[3] To the centenary of N D Papaleksi 1981 *Usp. Fiz. Nauk* **134** 3 [1981 *Sov. Phys. Usp.* **24** 341]

[4] Ginzburg V L On solar radiation in the radio spectrum 1946 *Dokl. Akad. Nauk SSSR* **52** 491 [1946 *C. R. Acad. Sci. URSS* **52** 487]

[5] Martyn D F 1946 Temperature radiation from the quiet Sun in the radio spectrum *Nature* **158** 632

[6] Shklovsky I S 1946 On the radiation of radio waves by the Galaxy and by upper layers of the Solar atmosphere *Astron. Zh.* **23** 333

[7] Ginzburg V L 1949 On the absorption of radio waves in the solar corona *Astron. Zh.* **26** 84

[8] Pawsey J L 1946 Observation of million degree thermal radiation from the Sun at a wave-length of 1.5 metres *Nature* **158** 633

[9] Khaykin S E and Chikhachev B M 1947 Investigation of solar radio emission

during the solar eclipse on May 20, 1947 by the Brazil expedition of the USSR Academy of Sciences *Dokl. Akad. Nauk SSSR* **58** 1923; 1948 *Izv. Akad. Nauk SSSR, Ser. Fiz.* **12** 38

[10] Ginzburg V L 1948 New data on solar and galactic radio emission *Usp. Fiz. Nauk* **34** 13

[11] Ginzburg V L 1947 Solar and galactic radio emission *Usp. Fiz. Nauk* **32** 26

[12] Getmantsev G G and Ginzburg V L 1950 On solar and cosmic radio emission diffraction on the Moon *Zh. Eksp. Teor. Fiz.* **20** 347

[13] Ginzburg V L 1952 Interstellar matter and ionospheric perturbations leading to radio-star twinkling *Dokl. Akad. Nauk SSSR* **84** 245

[14] Gershman B N and Ginzburg V L 1955 On the mechanism of the appearance of ionospheric inhomogeneities *Dokl. Akad. Nauk SSSR* **100** 647; 1959 *Izv. Vyssh. Ucheb. Zaved. Radiofiz.* **2**(1) 8

[15] Ginzburg V L and Pisareva V V 1956 On the nature of solar radio emission intensity oscillations and inhomogeneities in the solar corona in *Proc. 5th Conf. on Cosmogony: Radioastronomy* (Moscow: Akad. Nauk SSSR) p 229

[16] Ginzburg V L 1956 On the mechanisms of formation of ionospheric inhomogeneities leading to 'radio star' twinkling in *Proc. 5th Conf. on Cosmogony: Radio Astronomy* (Moscow: Akad. Nauk SSSR) p 512

[17] Ginzburg V L 1956 On non-ionospheric intensity oscillations of radio emission from nebulae *Dokl. Akad. Nauk SSSR* **109** 61

[18] Ginzburg V L 1960 On the possibility of determining the magnetic field strength in the external solar corona by transmission through it of polarized radio emission from discrete sources *Izv. Vyssh. Ucheb. Zaved. Radiofiz.* **3** 341

[19] Benediktov E A, Getmantsev G G and Ginzburg V L 1962 Radioastronomical investigations employing artificial satellites and space rockets *Planet. Space Sci.* **9** 109

[20] Ginzburg V L and Pisareva V V 1963 Polarization of radio emission from discrete sources and the study of metagalactic, galactic and near-solar space *Izv. Vyssh. Ucheb. Zaved. Radiofiz.* **6** 877

[21] Ginzburg V L and Zheleznyakov V V 1958 On possible mechanisms of sporadic solar radio emission (radiation in an isotropic plasma) *Astron. Zh.* **35** 694 [1958 *Sov. Astron.* **2** 653]

[22] Zheleznyakov V V 1970 *Radio Emission of the Sun and Planets* (Oxford: Pergamon Press) [Russian original 1964 (Moscow: Nauka)]

[23] Ginzburg V L 1961 On the nature of radio galaxies *Astron. Zh.* **38** 380 [1961 *Sov. Astron.* **5** 282]

[24] Ginzburg V L and Ozernoi L M 1977 On the nature of quasars and active galactic nuclei *Astrophys. Space Sci.* **50** 23

[25] Ginzburg V L 1964 On magnetic fields of collapsing masses and the origin of superstars *Dokl. Akad. Nauk SSSR* **156** 43 [1964 *Sov. Phys. Dokl.* **9** 329]

[26] Ginzburg V L and Zheleznyakov V V 1975 On the pulsar emission mechanisms *Annu. Rev. Astron. Astrophys.* **13** 511

[27] Michel F C 1982 Theory of pulsar magnetospheres *Rev. Mod. Phys.* **54** 1

[28] Ginzburg V L 1985 On the birth and development of cosmic ray astrophysics in *Early History of Cosmic Ray Studies: Some Personal Reminiscences*

eds Y Sekido and H Elliot (Dordrecht, Holland: D. Reidel) [This is the preceding (article 3) of the present collection]

[29] Ginzburg V L 2001 *The Physics of a Lifetime: Reflections on the Problems and Personalities of 20th Century Physics* (Berlin: Springer) p 457

[30] Ryle M 1949 Evidence for the stellar origin of cosmic rays *Proc. Phys. Soc. London* **62A** 491; 1950 Radio astronomy *Rep. Prog. Phys.* **13** 184

[31] Unsöld A 1949 Uber den Ursprung der Radiofrequenzstrahlung und der Ultrastrahlung in der Milchstrasse [On the origin of radio frequency and ultrashort-wave radiation from the Milky Way] *Z. Astrophys.* **26** 176; 1951 Cosmic radiation and cosmic magnetic fields. I. Origin and propagation of cosmic rays in our Galaxy *Phys. Rev.* **82** 857; 1955 Astrophysikalische Bemerkungen zur Enstehlung der Kosmischen Ultrastrahlung [Astrophysical comments on the origin of cosmic radiation] *Z. Phys.* **141** 70

[32] Shklovsky I S 1951 Radio stars *Dokl. Akad. Nauk SSSR* **79** 423

[33] Schott G A 1912 *Electromagnetic Radiation and the Mechanical Reactions Arising from It* (Cambridge: Cambridge University Press)

[34] Alfvén H and Herlofson N 1950 Cosmic radiation and radio stars *Phys. Rev.* **78** 616

[35] Kiepenheuer K O 1950 Cosmic rays as the source of general galactic radio emission *Phys. Rev.* **79** 738

[36] Ginzburg V L 1951 Cosmic rays as a source of galactic radio emission *Dokl. Akad. Nauk SSSR* **76** 377

[37] Shklovsky I S 1952 On the nature of galactic radio emission *Astron. Zh.* **29** 418

[38] Gordon I M 1954 On the problem of the physical nature of chromospheric eruptions *Dokl. Akad. Nauk SSSR* **94** 813 [Publication of this paper was delayed for technical reasons. Its essential points were reported and were known in the USSR at least from the late 1952]

[39] Ginzburg V L 1954 in *Proc. Third Conf. on Cosmogony: The Origin of Cosmic Rays* (Moscow: Akad. Nauk SSSR) pp 253–268

[40] Ginzburg V L 1953 Supernova and nova stars as sources of cosmic and radio emission *Dokl. Akad. Nauk SSSR* **92** 1133

[41] Pikelner S B 1953 Kinematic properties of interstellar gas in connection with cosmic ray isotropy *Dokl. Akad. Nauk SSSR* **88** 229

[42] Shklovsky I S 1953 The problem of cosmic radio emission *Astron. Zh.* **30** 15

[43] Shklovsky I S 1953 On the nature of the Crab Nebula's optical emission *Dokl. Akad. Nauk SSSR* **90** 983

[44] Ginzburg V L and Syrovatskii S I 1965 Cosmic magnetobremsstrahlung [synchrotron radiation] *Annu. Rev. Astron. Astrophys.* **3** 297

[45] Van de Hulst H C (ed) 1957 *Radio Astronomy Symp. No. 4 held at the Jodrell Bank Experimental Station near Manchester, August 1955* (Cambridge: Cambridge University Press)

[46] Bracewell R N (ed) 1959 *Paris Symp. on Radio Astronomy: IAU Symp. No. 9 and URSI Symp. No. 1. 30 July – 6 August 1958* (Stanford, CA: Stanford University Press)

[47] Ginzburg V L 1956 The nature of cosmic radio emission and the origin of cosmic rays *Nuovo Cim.* Ser. 10 Suppl **3** 38; 1958 *Prog. Elem. Part. Cosmic Ray Phys.* **4** 339

[48] Ginzburg V L and Syrovatsky S I 1964 *The Origin of Cosmic Rays* (Oxford: Pergamon Press) [The Russian edition did not contain some supplementary material and was published in 1963 (Moscow: Akad. Nauk SSSR)]

[49] Ginzburg V L and Ptuskin V S 1976 On the origin of cosmic rays: some problems in high-energy astrophysics *Rev. Mod. Phys.* **48** 161

[50] Shklovsky I S 1982 *From the History of the Development of Radioastronomy in the USSR* [*News on Life, Science, and Technology* Ser. Cosmonautics, Astronomy, No. 11] (Moscow: Znanie)

[51] Shklovsky I S 1996 *Razum, Zhizn', Vselennaya* [*Mind, Life, Universe*] (Moscow: Yanus)

[52] Berezinskii V S, Bulanov S V, Ginzburg V L, Dogiel V A and Ptuskin V S 1990 *Astrophysics of Cosmic Rays* ed V L Ginzburg (Amsterdam: North-Holland) [Russian original 1984 *Astrofizika Kosmicheskikh Luchei* ed V L Ginzburg (Moscow: Nauka)]

[53] Ginzburg V L and Dogel' V A Some aspects of gamma-ray astronomy 1989 *Usp. Fiz. Nauk* **158** 3 [1989 *Sov. Phys. Usp.* **32** 385]

[54] Ginzburg V L 1993 The origin of cosmic rays (Forty years later) *Usp. Fiz. Nauk* **163** 45 [1993 *Phys. Usp.* **36** 587]

[55] Ginzburg V L 1996 Cosmic ray astrophysics (history and general review) *Usp. Fiz. Nauk* **166** 169 [1996 *Phys. Usp.* **39** 155]

Article 5

Some remarks on ferroelectricity, soft modes, and related problems[1]*

5.1 Introduction

The history of science can, as a rule, be most objectively described not by participants in the events but by specialists in a respective field or professional historians of science, who are ready to consider thoroughly and to compare all the available material. As for me, first of all I myself participated in the theoretical studies of ferroelectric phenomena and soft modes and, second, I have no taste for historic researches and for analysis of priority questions. Therefore, I have made up my mind to make here some remarks of a historical character only because of the existing rather peculiar situation. Indeed, in all Russian books that deal with the relevant problems, I am referred to as the author or one of the few authors of the phenomenological (thermodynamic) theory of ferroelectric phenomena and of the concept of a soft mode. Almost with the same persistence, my work on the theory of ferroelectricity and soft modes is never mentioned in books published in the West. Those translated into Russian are always supplied with notes by translators or editors of translation informing the reader that the theory of the $BaTiO_3$-type ferroelectrics and, in general, the thermo-dynamic theory of ferroelectrics has been developed also by Ginzburg and that the concept of a soft mode also belongs to Ginzburg. The suspicion may arise in this connection that such a situation is a result of either a misplaced patriotism of the Soviet authors or of some priority claims on my part (such claims, if any, could have influenced the Soviet authors). I believe neither is true. I have never put forward priority claims and do not intend ever to do this. The Soviet authors often refer to my papers cited here simply because they are known to them and easily available. As far as L D Landau[1] is concerned, my first work on the theory of ferroelectrics [2]

---

[1] Here I mean the book by L D Landau and E M Lifshitz [1], in section 19 of which it is stated that "the quantitative theory of ferroelectricity can be developed in terms of the

was done, in fact, before his eyes and in concluding paper [2] I acknowledge his 'discussion of the problem'.

It can be added that I have not participated in any of the previous international conferences on ferroelectricity, I cannot attend the present Meeting and most probably will not be able to in future. Neither do I have any personal contacts with physicists working on the problems of ferroelectricity outside the USSR. In this connection, I hope, the remarks to follow will not seem to be immodest or irrelevant.

5.2   On the phenomenological theory of ferroelectricity

Long ago (see, for example, [3]), it was realized that in Rochelle salt, between two Curie points there exists spontaneous polarization P_0. Then, in 1937, Jaffe [4] suggested, as far as I know for the first time, that phase transitions should occur at the Curie points. In the ferroelectric range (255 K $< T <$ 297 K), the crystal possesses a monoclinic symmetry and outside this range, an orthorhombic symmetry. This hypothesis was confirmed by Müller [5] who also developed essentially the phenomenological (thermodynamic) theory of the properties of Rochelle salt. Rochelle salt is known to be an especially complicated example of a ferroelectric since its spontaneous polarization is non-zero only in a comparatively narrow temperature range and the piezoelectric effect takes place both inside and outside the ferroelectric region. It is, therefore, natural that the thermo-dynamic theory of the behavior of Rochelle salt is less clear and more cumbersome than that of barium titanate. Moreover, the general approach to the phase transition theory, known as the Landau theory of phase transitions [7], was not used in the respective studies (see [5] and [6] where the earlier papers are cited too). It should be emphasized here that, in simple cases (say, for one order parameter), the Landau theory practically coincides with the earlier treatments of van der Waals and other authors, which were used in the liquid–gas critical point theory and in some other cases. To a great extent, this also concerns the previously mentioned theories of Rochelle salt [5, 6]. The main feature of the Landau theory of phase transitions is the generality of approach and the consistency of inclusion of symmetry requirements. Furthermore, in the simplest cases, the Landau theory 'works', so to say, almost automatically. But as far as I know, before the appearance of my paper [2] the Landau theory had not been used in an explicit form for the description of ferroelectrics.

This was the situation in 1944 when Wul and Goldman [8] discovered that polycrystals (ceramics) of barium titanate possessed a very high dielectric permittivity ε which depended strongly on the temperature T and

general theory of second-order phase transitions; this has been done by V L Ginzburg (1945)".

had a rather sharp maximum at $T \approx 400$ K. Scarcity of the available data and polycrystallinity of specimens made it impossible to realize at once that this was a new ferroelectric (paper [8] concludes with the remark that "neither these facts nor the composition and structure of barium titanate permit it to be clashed with the group of ferroelectrics").

I worked and am still working at the same institute and naturally took interest in the results [8]. I also knew the Landau theory [7, 9] and could discuss the new questions with Landau himself. This is how paper [2] appeared: it was submitted for publication on 31 July, 1945, i. e. almost exactly 40 years before this meeting. In this paper, after a brief review of the properties of ferroelectrics (with references to the previously mentioned book [3] and papers [4, 5]), a thermodynamic theory for the transition from a non-pyroelectric modification to a pyroelectric one was formulated in the now quite standard way. The polarization P was taken as the order parameter, as a result of which near the second-order transition point the thermodynamic potential was written in the form

$$\Phi = \Phi_0 + \alpha P^2 + \frac{\beta}{2} P^4 - EP. \tag{5.1}$$

The field E in (5.1) 'is assumed to have the polarization direction and anisotropy is disregarded'. Supposing that at the transition point $T = \theta$, the coefficient is $\alpha(\theta) = 0$ ($\alpha < 0$ at $T < \theta$ and, $\alpha > 0$ at $T > \theta$) and, in the temperature range under consideration $\beta > 0$, we obtained for the spontaneous polarization

$$P_0^2 = -\alpha/\beta \qquad T < \theta. \tag{5.2}$$

Then the entropy and the jump in the specific heat were found:

$$\Delta C_\theta = \frac{T}{\beta_\theta} \left(\frac{\partial \alpha}{\partial T} \right)_\theta^2. \tag{5.3}$$

Taking into account that $E = \partial F/\partial P$, $F = \Phi + EP$, we obtained $E = 2\alpha P + 2\beta P^3$. In a weak field, $P = P_0 + [(\varepsilon - 1)/4\pi]E$ and, accordingly,

$$\alpha = \frac{2\pi}{\varepsilon - 1} \ (T > \theta) \qquad \alpha = -\frac{\pi}{\varepsilon - 1} \ (T < \theta). \tag{5.4}$$

It was shown further that 1 had to be replaced here with ε_0 and that (for $\varepsilon \gg \varepsilon_0$) ε_0 could be neglected. So, assuming for a second-order transition that $\alpha = (\partial \alpha/\partial T)_\theta (T - \theta)$, we obtained, for the permittivity, the Curie–Weiss law

$$\varepsilon = \frac{2\pi}{\left(\frac{\partial \alpha}{\partial T} \right)_\theta (T - \theta)} \qquad (T > \theta, \ (T - \theta) \ll \theta). \tag{5.5}$$

In [2] we also dealt with the case of first-order transitions close to second-order transitions, i. e. close to the tricritical point (at that time the tricritical point was called the critical Curie point). In this case, the term $(\gamma/6)P^6$

was added to expansion (5.1). Further, we discussed polarization in strong electric fields and pointed out, in particular, that for a field stronger than

$$E_k = \frac{4}{3^{3/2}}|\alpha||P_0| = 4\beta \left(\frac{|\alpha|}{3\beta}\right)^{3/2} \tag{5.6}$$

the curve $\Phi(P)$ has only one minimum. Finally, some experimental data concerning the ferroelectrics KH_2PO_4, KD_2PO_4, and KH_2AsO_4 were analyzed and compared with the theory. As concerns $BaTiO_3$, we emphasized, with reference to [8] and some then yet unpublished results by the same authors (probably these results were included in [10, 11]), that its behavior is analogous to that of KH_2PO_4-type ferroelectrics and the following conclusion was drawn: "It seems almost doubtless that the dielectric properties of barium titanate are due to its transition from a non-pyroelectric crystal class (for $T > \theta$) into a pyroelectric one (for $T < \theta$)." Wul and Goldman agreed with this opinion and cited it in their paper [10] with reference to [2] which had not yet been published by that time.

Paper [2] concludes with some remarks concerning the properties of $BaTiO_3$. Note that the structure of the pyroelectric (ferroelectric) phase of $BaTiO_3$, which had not then been known from experiment, was assumed in [2] to be either tetragonal or rhombohedral. For each of these cases, the table of piezoelectric coefficients was presented and it was emphasized that, for $BaTiO_3$, the Curie point is, at the same time, a point below which there occurs not only pyroelectricity but also piezoelectricity.[2] However, "cooling of barium titanate in the absence of field and deformations should result (on the average) in a non-piezoelectric state because of the occurrence of orientational twinning." From this it is clear that cooling in a field in the presence of deformations may help to avoid orientational twinning. In paper [2] we proposed to apply this method to the $\alpha \rightleftarrows \beta$ transition in quartz.

Thus, in paper [2], it was realized for the first time that $BaTiO_3$ is a ferroelectric and the thermodynamic theory of ferroelectric transition was formulated in terms of the Landau theory without taking into account anisotropy. To tell the truth, I do not at all remember why I did not start writing more general expressions involving anisotropy and elastic stresses [12, 13]. I think the explanation is that I am a theoretician who worked and is still working in various fields. So, there were plenty of problems to be solved, while the theory of crystal lattice and the solid-state theory as a whole were of relatively small interest for me. As a result, it

---

[2] It is a curious fact that somewhere in the 1950s I gave evidence in court concerning the piezoelectric effect in $BaTiO_3$. Someone in the USA, who had a patent, brought a suit against the US government in connection with the use of $BaTiO_3$ piezoelectric elements. The American government used my testimony (and, in fact, my paper [2]) to reject the claim. I should note that although I am the author of several hundred papers and a number of innovations, I have never applied for any patents.

was only the new experimental data on BaTiO$_3$ [14–16] that stimulated me to develop the theory allowing for the possibility of different orientations of the polarization vector and also the piezoelectric effect [12]. The starting point was the expression for the thermodynamic potential

$$\Phi = \Phi_0 + \alpha(P_x^2 + P_y^2 + P_z^2)$$
$$+ \frac{1}{2}\beta_1(P_x^4 + P_y^4 + P_z^4) + \beta_2(P_x^2 P_y^2 + P_x^2 P_z^2 + P_y^2 P_z^2)$$
$$+ \frac{1}{2}s_{11}(\sigma_{xx}^2 + \sigma_{yy}^2 + \sigma_{zz}^2) + s_{12}(\sigma_{xx}\sigma_{yy} + \sigma_{xx}\sigma_{zz} + \sigma_{yy}\sigma_{zz})$$
$$+ \frac{1}{4}s_{44}(\sigma_{xy}^2 + \sigma_{xz}^2 + \sigma_{yz}^2) - \gamma_1(\sigma_{xx}P_x^2 + \sigma_{yy}P_y^2 + \sigma_{zz}P_z^2)$$
$$- \gamma_2[\sigma_{xx}(P_y^2 + P_z^2) + \sigma_{yy}(P_x^2 + P_z^2) + \sigma_{zz}(P_x^2 + P_y^2)]$$
$$- 2\gamma_3(\sigma_{xy}P_x P_y + \sigma_{xz}P_x P_z + \sigma_{yz}P_y P_z)$$
$$- (E_x P_x + E_y P_y + E_z P_z), \tag{5.7}$$

where $P = \{P_i\} \equiv \{P_x, P_y, P_z\}$ is the polarization and σ_{ik} is the stress tensor. In equilibrium we have

$$E_i = [\partial(\Phi + E_i P_i)]/\partial P_i$$

and from (5.7) one can obtain the relation between E_i, P_i, and σ_{ik}. At $T < \theta$, when $\alpha < 0$, solutions are possible which correspond to tetragonal and rhombohedral symmetry. Further in [12], the coefficients β_1 and β_2 in (5.7) were so chosen that the absolute minimum of Φ correspond to a tetragonal structure because precisely this fact had been established in experiment [14]. I will not present here the corresponding formulae, for they basically coincide, of course, with the expressions given in numerous books on ferroelectrics. The expansion (5.7) is, however, limited owing to a disregard of sixth-order terms with respect to P_i. That is why no solutions exist that would correspond to the orthorhombic phase and one cannot consider first-order transitions close to the tricritical point. As has been pointed out earlier, the term $(\gamma/6)P^6$ was included in [2] but only one ferroelectric direction was considered. Devonshire [17, 18] wrote an expansion that differs from (5.7) only by the addition of the term $1/6\zeta'(P_x^6 + P_y^6 + P_z^6)$. Meanwhile, to take into account of sixth-order terms, one should write down all the invariants admitted by the symmetry. In the case of BaTiO$_3$, this means that the general sixth-order expression has the form

$$\Phi_6(P) = \frac{1}{6}\delta_1 P^6 + \delta_2[P_x^4(P_y^2 + P_z^2) + P_y^4(P_x^2 + P_z^2)$$
$$+ P_z^4(P_x^2 + P_y^2)] + \delta_3 P_x^2 P_y^2 P_z^2. \tag{5.8}$$

This is just what Kholodenko and Shirobokov did [19]. Papers [19] were submitted for publication in late 1950 and were, in fact, a continuation of paper [12] (in [19], paper [17] is also cited). I think, that the absence in [17] of all the sixth-order terms is due to the fact that Devonshire did not follow the Landau theory with its general approach (in any case, in [17, 18] there are no references to the Landau theory). At the same time, in his paper [17], Devonshire used the experimental data that were new at that time (see, in particular, [20]) and, undoubtedly, the properties of $BaTiO_3$ were described there in more detail than in [12, 13]. To say nothing of the previously mentioned inconsistency in respect of terms of the order of P_i^6, note that in [17], which was submitted for publication on 26 July, 1949, there is a reference to [2], while [12] was submitted on 7 July, 1948, i.e. a year before [17]. What has been said was not, of course, aimed at belittling the value of Devonshire's papers [17, 18, 21]. I only wanted to elucidate the subject and the proper place of papers [2, 12, 13, 19] cited in the Russian literature (see, for example, [1, 22, 23]).

5.3   On the concept of a soft mode

Under structural second-order phase transitions or first-order transitions close to the tricritical point, the frequency of one or several normal modes of a crystal lattice tends to zero or greatly decreases. True, the vanishing of only one mode frequency only occurs in the simplest cases since a 'soft mode' usually interacts with other modes and with other degrees of freedom in the system. Generally, for real crystals, the picture may be rather complicated, particularly because crystal behavior near the transition point is also affected by long-range forces and defects. Nevertheless, these complications, on the whole, do not diminish the significance of the 'soft modes' as a basis for understanding the dynamics of the phase transition processes. The ideas connected with relation between phase transitions and the decrease in frequency for a number of modes is somewhat vaguely referred to as a 'soft mode concept'.

The concept of a soft mode has crystallized gradually as a result of a large number of theoretical and experimental studies. It would be out of place to dwell here on the history of this range of questions and I shall restrict myself to only a few remarks of a 'personal' character.

In experiment, the soft mode was encountered for the first time, as far as I know, by Landsberg and Mandelstam in the course of investigating the combinational (Raman) scattering of light near the $\alpha \rightleftarrows \beta$ transition in quartz [24]. They revealed that the Raman line with a frequency of 207 cm^{-1} (at room temperature) broadens and spreads out as temperature increases and in β-quartz (i.e. at $T > \theta = 846$ K) vanishes altogether. In this connection, it was noted in [24] that such a behavior of the line is an indication of "a substantial change or disappearance of bonds responsible for corresponding oscillation". It was further established [25–27] that the frequency of the previously mentioned line decreases particularly strongly as temperature increases. These facts were taken into account and discussed in the paper by Levanyuk and myself [28] devoted to the spectral composition of light scattered near second-order phase transition points. In paper [28], published in 1960, the concept of a 'soft mode' is presented as something known and obvious in application both to quartz and to ferroelectrics (in this connection, reference is given to papers [12, 13] and also to [32], which is cited later and was in press at that time). Specifically, in [28], we read:

> As the parameter η [we mean the order parameter, VLG] in quartz we can choose the displacement of the mean position of the silicon atom in α-quartz compared with its position in β-quartz. Since the non-equilibrium value of η corresponds to a mutual displacement of various sublattices, this displacement will vary with the frequency of the optical vibration of the crystal $\Omega_i(q)$. At the transition point we have $\eta_0 = 0$, that is, there is no elastic force and the frequency of the vibration $\Omega_i(0)$ becomes zero; here $\Omega_i(0) \approx \Omega_i(q)$ since $q \ll q_{\max} = \pi/d \sim 10^8$ ($d \sim 3 \times 10^{-8}$ is the lattice constant).

We shall not dwell here in more detail on paper [28] and a number of subsequent papers devoted to light scattering near phase transition points. First, we are short of place and, second, it is not really necessary for one can refer to the reviews [29–31].[3] I would like to draw attention to soft modes in ferroelectrics which were considered not only in [28] and in more detail in [32] in 1960 but as far back as 1949 in the already cited papers [12, 13]. Meanwhile the widespread opinion (so to say, adopted by repetition) is that soft modes 'appeared' in 1960 in the papers by Cochran and Anderson. For example, in the recent paper by Cowley [33], we read: 'The concept of a "soft mode" has been a useful concept in the understanding of structural phase transition ever since it was introduced by Cochran and Anderson in 1960' (in [33], reference is given to papers [34, 35]).[4] The first paper by Cochran appeared, in fact, in 1959 [36], while the report by Anderson [35] was made in 1958. Incidentally, in [32], there are references to paper [36] and report [35] which had not yet been published by that time. But in these papers I saw (and today I am of the same opinion) only a confirmation and a certain development of the remarks made ten years before in [12, 13].

Indeed, in [12, 13], the analysis of the equilibrium properties of BaTiO$_3$ is followed with the words:

[3] I am not the one to judge of the quality of reviews [29, 31]. As far as the review by Krishnan [30] is concerned, I would like to note that it differs advantageously from some others in its completeness and objectiveness.

[4] More precisely, in [33] there is a reference to W Cochran *Adv. Phys.* **29** 219. This is evidently a misprint because in *Adv. Phys.* as in 1960 we find only paper [34] by Cochran.

Let us also consider the dispersion of the dielectric constant of barium titanate. This problem is much more complicated than the one considered above because it does not admit a purely thermodynamical treatment. It can, in fact, be solved only through the lattice mode investigation. Some statements and dispersion estimates can however be made without such consideration.

To this end, the equation for polarization in equilibrium

$$2\alpha P + 2\beta_1 P^3 = E$$

is generalized as follows

$$\mu\ddot{P} + \nu\dot{P} + \alpha P + \beta_1 P^3 = \frac{1}{2}E_0 e^{i\omega t} \tag{5.9}$$

i. e. only one mode is considered. We quote further:

In fact, polarization is of course determined by different lattice modes and formula (5.9)[5] is, generally, invalid. But, for small frequencies, this formula can be regarded as a result of expansion of P in a frequency series ($\ddot{P} = -\omega^2 P$, $\dot{P} = i\omega P$, etc.). In this sense, formula (5.9) is certainly valid if the term $\mu\ddot{P} + \nu\dot{P}$ is comparatively small. As is readily seen, it follows from (5.9) that[6]

$$\varepsilon = \frac{2\pi}{\alpha + i\omega\nu - \omega^2\mu} \qquad T > \theta$$

$$\varepsilon = \frac{2\pi}{-2\alpha + i\omega\nu - \omega^2\mu} \qquad T < \theta. \tag{5.10}$$

Note that at $T < \theta$ we consider here only the dispersion which is not connected with reorientation of regions and is due exclusively to variations in polarizability of the material itself. Formula (5.9) was derived not directly by expansion in a frequency series but proceeding from the rough oscillator model in order one could estimate the μ value.

We present such a long quotation here because papers [12, 13] were published only in Russian.

Expressions (5.10) are obviously the usual dispersion formulae and for second-order transitions, which alone have been considered, the self(nor-

---

[5] In [12] this is formula (16) and in [13]—formula (49). As before, we shall not pay attention to the numeration of formulae in different papers.

[6] In [12, 13] in the formula for the case $T < \theta$ the factors 2 are absent because of an obvious misprint or a slip which was corrected in [28, 32, 37].

mal)-frequencies are given by

$$\omega_i^2 = \frac{\alpha}{\mu} = \frac{(\partial\alpha/\partial T)_\theta}{\mu}(T - \theta) \qquad T > \theta$$

$$\omega_i^2 = \frac{2|\alpha|}{\mu} = \frac{2(\partial\alpha/\partial T)_\theta|T - \theta|}{\mu} \qquad T < \theta.$$

(5.11)

In [28], the frequency ω_i was denoted by Ω_i. In [32], formulae (5.10) and (5.11) were extended to the anisotropic case and, for second-order transitions, the self-frequencies (denoted by $\omega_t, \omega_{z,t}$ and $\omega_{x,y,t}$) tend to zero as $\sqrt{|T - \theta|}$, when $T \to \theta$. First-order transitions are discussed too and, specifically, for BaTiO$_3$ at the upper transition point $\theta \approx 373$ K, the following estimates are presented (the value $\mu \sim 10^{-27}$ is used):

$$\omega_t(\theta) = \frac{\sqrt{\alpha(\theta)}}{\mu} \sim 6.3 \times 10^{11} \qquad \omega_{z,t}(\theta) = 2\omega_t(\theta)$$

$$\lambda_t(\theta) = \frac{2\pi c}{\omega_t}(\theta) \sim 0.3 \text{ (mm)}.$$

(5.12)

But this was already 1960 (paper [32]). The estimate $\alpha \sim \omega^2\mu$ for $\lambda \approx 3$ cm is presented in [13] on the basis of experimental data and formulae (5.11) are not written down.

It was possibly just this last circumstance that gave grounds to Cochran [38] to characterize the corresponding place in paper [13] as "obscurely worded even with benefit of hindsight". In accordance with what has been said at the beginning of this paper, the participants in the studies (in this case, Cochran and I) cannot lay claim to an objective estimation of their role. But they have every right to express their opinions. Inasmuch as Cochran has exercised this right and, it seems to me, mentioned [13] in a rather slighting tone (see the earlier quotation from his paper [38]) I shall also use my right.[2*]

I regard the results of [12, 13] that have been cited here to be quite valid and correct even today, after 36 years, and "even with benefit of hindsight". True, some expressions of the type of (5.11) should have been added, as was done later in [28, 32, 37]. Such a conclusion is, however, a consequence of understanding from experience, that even the dispersion formulae (5.10) are not well enough known to everybody. But it is simply not serious to suspect me of such ignorance in 1949 (suffice it to say that in the same year my monograph [39] devoted to radio-wave propagation was published). The experimenters [40], too, have understood correctly the essence of papers [12, 13] and, in particular, the dispersion character of formulae (5.10) and the ensuing consequences concerning the nature of the soft mode (i. e. oscillation with a frequency $\omega_i \to 0$ as $T \to \theta$).

I will not continue this dispute provoked by paper [38] and will try to explain the reason for such differing opinions and assessments of the role of

different approaches. I believe the whole point here is that the place of and
the relation between the phenomenological and the microscopical theories
were understood differently. For me and for all those who use, first of all,
the phenomenological theory, the whole information on electromagnetic
wave propagation in a medium is contained in the permittivity $\varepsilon(\omega)$ or,
in a more complicated case, in the permittivity tensor $\varepsilon_{ij}(\omega, k)$, where k
is the wavevector (denoted in [28] by q). For transverse waves, we have
$k^2 = (\omega^2/c^2)\varepsilon(\omega)$ and for longitudinal waves $\varepsilon(\omega) = 0$ (for an appropriate
generalization in the case of anisotropy of spatial dispersion see, e.g., [41]).
Self-frequencies of the medium are poles of the function $\varepsilon(\omega)$, etc. In this
respect, the identity of expressions (5.10) and (5.11) is obvious, and the
microscopic theory is only needed to calculate the function $\varepsilon(\omega)$ or, say,
the frequencies ω_i and other parameters in the dispersion formula

$$\varepsilon = \varepsilon_\infty + \frac{2\pi/\mu}{\omega_i^2 - \omega^2 + (i\nu/\mu)\omega}.$$

For an isotropic and homogeneous medium, the phenomenological theory
reduces to what has been said earlier and is fairly trivial. Therefore, the
centre of gravity lies with the microscopic theory. But for an anisotropic
and (or) inhomogeneous medium, the solution of a number of problems in
terms of the microscopic theory is usually quite inefficient and is a sign
of poor skill. Generally speaking, both approaches—phenomenological and
microscopic—are quite legitimate and they should be combined. In the general form, such a conclusion is evident. But in practice, misunderstandings
often occur. One of the recent examples is, in my opinion, the introduction
of the concept of polaritons, which are simply 'normal' electromagnetic
waves in a medium and were considered in the last century. True, the spatial dispersion was, as a rule, disregarded earlier but its inclusion does not
at all affect the essence of the matter and the new terminology does not
change anything either (in this connection, see [41]).

By virtue of what has been said, I believe that the approaches to soft
modes in ferroelectrics using the phenomenological theory [12, 13] and the
lattice theory [34, 36, 42] have equal rights or, more precisely, are comple-
mentary. By the way, paper [42] ends with the phrase: "Broadly similar
conclusions on the order of magnitude and temperature dependence of vi-
brational modes in ferroelectrics and quartz have been reached by Ginzburg
(1960; our reference [32]) and by Ginzburg and Levanyuk (1960; our reference [28]), whose treatment is essentially a phenomenological one." It
remains only to guess why quite the same phenomenological approach,
used in [12, 13], 20 years after the appearance of his own paper [42] seemed
to Cochran [38] an 'obscurely worded' one.

To make the picture complete, it is necessary to mention that the
microscopic approach is in certain respects, of course, richer and wider
than the phenomenological one. For instance, for those vibrations which

are inactive in electromagnetic radiation, i. e. do not at all or weakly affect the tensor $\varepsilon_{ij}(\omega, \boldsymbol{k})$ and also in the presence of several essential modes and large wavenumbers k (i. e. for short waves comparable with the lattice parameter) the phenomenological treatment is either less clear and familiar or inefficient and insufficient [43]. In short, I am not going to criticize the use of crystal lattice theory, where its application is justified by the character of the problem. In particular, papers [34, 42] have undoubtedly played an important role in the development of the soft-mode concept. I only claim that it was just this concept, not some vague idea, that had been developed in [12, 13] and in a whole number of other papers cited here and in [38]. The widespread citation of the paper by Anderson [35] in connection with the soft-mode concept seems to be another example of statement 'adapted by repetition'. This was also noticed by Cochran [38], who wrote that the paper by Anderson published only in Russian and in a not very easily available book [35] "had been little read although frequently cited". The remarks in [35] concerning the microscopic theory of $BaTiO_3$-type ferroelectrics are qualitative in character and suggest ideas that do not overstep the limits of what was known from the literature. This also concerns the role of the correction to the effective or internal field (Lorentz correction) particularly stressed by Cochran [38]. For instance, in [35] we read: "When the Lorentz correction nearly cancels the repulsive term, we have a high dielectric constant material; when it is greater for $k = 0$, we get a ferroelectric." Exactly the same can be found, for example, on page 517 of paper [13], where the Lorentz correction is denoted by fP (an effective field $E_{\text{ef}} = E + fP$). But I am not at all claiming any priority here, the more so as the role of the Lorentz correction in the theory of ferroelectrics for models had been clarified long before (see, for example, [3]). Only four lines in [35] are devoted to a soft mode itself and even those are written in an indirect form:

> Another interesting physical effect which has not yet been tested is also implied: near but somewhat above T_c the stiffness of transverse optical modes, and thus the frequency of optical rays, must decrease to very low values. Rough estimates suggest $\lambda \sim 1$–3 mm.

Thus, as has been justly noted by Cochran [38], the paper by Anderson [35], "contrary to what many who have cited the paper believe, is not primarily about soft modes". I should add that whatever is the attitude to the parts of the papers [12, 13] devoted to the dispersion of permittivity near the phase transition point $\theta \equiv T_c$, the soft mode is, in any case, discussed more thoroughly in [12, 13] than in [35] published 10 years later. In 1958 I listened to the report made by Anderson [35] in Moscow and, as far as I remember, discussed with him the theory of barium titanate and handed over to him the relevant reprints. Irrespective of whether it was so, or whether, which is more probable, Anderson has not read these reprints (only the reprint of paper [2] was in English) I have no grudge against him. As far as I know and as follows from [38], Anderson himself does not lay claim to the creation of the soft-mode concept, while the opposite was stated by those who are, so to say, 'more royalists than the king himself' or 'more faithful Catholics than the Pope'.

It would be not out of place to emphasize here that I do not claim some unique priority in the creation of the soft-mode concept—a number of authors have contributed to its formulation and development (see earlier and [38]). My contribution to this field is reflected in papers [12, 13, 28, 29, 31, 32, 37, 44]. The value of this contribution can only be judged by other people. I will only express the opinion that references to papers [12, 13] and others in Russian books (see, in particular, [22, 23]) have a real basis.

The question of the 'laws' of citation and the establishment of priority is of some interest. In this connection, I will suggest the hypothesis concerning the reason for which my papers on ferroelectrics and soft modes are practically ignored in the West. The main reason is apparently the fact that papers [12, 13] were published only in Russian (at that time neither *Zhurnal Eksperimental'noi i Teoreticheskoi Fiziki* nor *Uspekhi Fizicheskikh Nauk* were translated into English). The second reason is that I have never attended international conferences on ferroelectrics and since 1960 have practically not been engaged in the study of ferroelectricity with the exception of light scattering near phase transition points [29, 31] and an episodic paper [44]. The third reason is that old papers are seldom read, whereas many think it necessary to refer to them from the words of others (just such a tendency has been characterized here as false statements 'adopted by repetition': how strong this tendency is seems to be clear from comparison of papers [33] and [38]). Finally, the fourth reason is simply the matter of chance and individual attitudes. For example, the paper by L D Landau and myself [45] devoted to the theory of superconductivity also appeared only in Russian in 1950, i.e. immediately after papers [12, 13]. Nonetheless, paper [45] has become widely known and often cited (either with a reference to [45] or, more often now, as the 'Ginzburg–Landau theory'). Apparently, some role was played here by the name of Landau. An essential part was probably the fact that paper [45] was translated into English by D Shoenberg (Cambridge, England) on his own initiative and this unpublished translation somehow circulated. I think, however, that paper [45] could quite easily have also been rediscovered only after many years in some review.

I would like to conclude this historical excursion by saying that I have no illusions that this report can change the situation. The scientific front has moved far ahead and very few people are now interested in facts that took place 40 or 30 years ago. And just for those few I would like to make their acquaintance with the history of the development of the theory of ferroelectricity and the soft mode concept easier.

5.4 On the range of applicability of the Landau theory in the case of ferroelectric phase transitions

In its main approximation (where mean values are calculated), the Landau theory of phase transitions is equivalent to the theory of a self-consistent (mean molecular) field and disregards order parameter fluctuations. Using the Landau theory, one can, however, calculate the contribution of fluctuations but only provided that this contribution is small. The latter does not generally take place near second-order phase transition points and, under certain conditions, near first-order phase transition points. In the cases where fluctuations are large, one often uses nowadays the fluctuation theory of phase transitions and, more concretely, scale invariance theory and renormalization group theory [9].

The formulation of the fluctuation theory of phase transitions is undoubtedly a great achievement. At the same time, the degree of generality and the applicability limits of the results of the fluctuation theory are often overestimated and the possibility of using the Landau theory is denied even where it is, in fact, valid. In such a situation, it seems reasonable to discuss briefly the applicability limits of the Landau theory in the case of ferroelectrics. This question has recently been given a wider and a more detailed consideration in paper [46].

In the case of one order parameter η, we write the thermodynamic potential as

$$\Phi[\eta(r)] = \Phi_0 + \alpha\eta^2 + \frac{\beta}{2}\eta^4 + \frac{\gamma}{6}\eta^6 + \delta(\nabla\eta)^2. \tag{5.13}$$

This expression differs from (5.1) with $\eta = P$ by an account of the sixth-order term, the absence of the field E, and, what is most important, by the introduction of the correlation energy $\delta(\nabla\eta)^2 \equiv \delta(\nabla\eta)^2$. For a second-order transition, assuming $\gamma = 0$, $\beta = \beta_0 = $ constant below the phase transition point $\theta \equiv T_c$, we have (see (5.2))

$$\eta_e^2 = -\frac{\alpha}{\beta} = \frac{\alpha_\theta'(\theta - T)}{\beta_\theta} \qquad \left(\frac{\partial^2\Phi}{\partial\eta^2}\right)_{\eta=\eta_e} \equiv \Phi_e'' = -4\alpha. \tag{5.14}$$

To calculate the fluctuations, we expand the deviation

$$\Delta\eta(r) = \eta(r) - \eta_e = \sum_k \eta_k e^{ikr}$$

in a power series and, in a lower (Gaussian) approximation, obtain (V is the volume of the body, k_B is the Boltzmann constant)

$$\int [\Phi(\eta_e + \Delta\eta) - \Phi(\eta_e)]dV = \sum_k \left(\frac{1}{2}\Phi_e'' + \delta k^2\right)|\eta_k|^2 V$$

$$\overline{|\eta_k|^2} = \frac{k_B T}{(\Phi_e'' + 2\delta k^2)V} \tag{5.15}$$

$$\overline{(\nabla\eta)^2} = \frac{1}{V}\int (\Delta\eta(r))^2 dV = \sum_k \overline{|\eta_k|^2} = \frac{k_B T}{2\pi^2}\int \frac{k^2 dk}{(\Phi_e'' + 2\delta k^2)}.$$

For convenience, we repeat briefly the calculation made in [32], where we further estimated the part $\overline{(\Delta\eta)^2}$ which strongly depends on temperature near θ and is equal to

$$\overline{(\Delta\eta)_T^2} = \frac{k_B\theta}{8\pi\delta}\sqrt{\frac{2|\alpha|}{\delta}} = \frac{k_B\theta^{3/2}(\alpha_\theta')^{1/2}\tau^{1/2}}{4\sqrt{2\pi}\delta^{3/2}} \qquad \tau = \frac{\theta - T}{\theta}.$$

The value of τ_L' determined from the condition

$$\overline{(\Delta\eta)_T^2} = \eta_e^2 = \frac{\alpha_\theta'\theta\tau_L'}{\beta_\theta} \tag{5.16}$$

can be conditionally thought of as the limit of applicability of the Landau theory in an ordered phase. From (5.16), we have

$$\tau_L' = \frac{(k_B)^2\theta\beta_\theta^2}{32\pi^2\alpha_\theta'\delta^3} = \frac{1}{32\pi^2}\left(\frac{k_B}{\Delta Ca^3}\right)^2\left(\frac{a}{\sqrt{2}r_{co}'}\right)^6 \tag{5.17}$$

where $\Delta C = (\alpha_\theta')^2\theta/\beta_\theta$ is the jump in the specific heat (see (5.3)), a is the interatomic distance or the lattice constant, and

$$r_{co}' = (\delta/2\alpha_\theta'\theta)^{1/2} \tag{5.18}$$

is the correlation radius for spatial variations of the order parameter conditionally extrapolated to $T = 0$ (more precisely, the correlation radius near the transition point θ for $T < \theta$ is determined as $r_c' = r_{co}'\tau^{-1/2}$). Conditions of the type (5.17) can be obtained not only by comparison of $\overline{(\Delta\eta)_T^2}$ with η_e^2 but also by comparison of the fluctuation correction to specific heat with the specific heat (more precisely, with ΔC) and in other ways [9, 31, 46, 47]. It is convenient to introduce a lattice constant a into (5.17) because for many transitions, including structural ones, $k_B/\Delta Ca^3 \sim 1$ and it becomes immediately clear that the value of τ_L' is determined by the ratio $(a/r_{co})^6$ (note that for $T > \theta$, i.e. for a disordered phase, the formula obtained for τ_L is similar to (5.17); in this phase, $r_{co} = \sqrt{2}r_{co}'$, where the prime denotes, as usual, the values which belong to the ordered phase;

see [46]). Note, finally, the obvious fact that corrections to the Landau theory are small under the conditions

$$\tau \gg \tau_{\rm L}' \ (T < \theta) \qquad |\tau| \gg |\tau_{\rm L}| \ (T > \theta). \tag{5.19}$$

The range of $\tau \lesssim \tau_{\rm L}' \sim |\tau_{\rm L}|$ values is called a critical region. Unfortunately, the existing fluctuation theory does not, generally, give the answer to the question of the temperature dependence of the quantities in the whole critical region with a gradual transition to the Landau theory. Indeed, the fluctuation theory usually leads only to establishing the temperature dependence for sufficiently small τ, for example, the dependence $\eta_{\rm e} = \eta_{\rm e0}\tau^\beta$, where the value of the critical index β (it should not be confused with the coefficient β in (5.13)!) is, in some cases, calculated and, say, $\beta = 1/3$. In the Landau theory, we have $\beta = 1/2$ but this only refers to a second-order transition which is far from the tricritical point. In the other limiting case where the tricritical point is quite near, $\eta_{\rm e} \sim \tau^{1/4}$ (i.e. one can say that $\beta = 1/4$: we mean the approach to the tricritical point along the line $\beta(p, T) = 0$; see [31, 46]). Between these extremes, intermediate cases exist and, as a result, if the character of the transition is unknown (specifically, if it is not clear that we deal with a second-order transition far from the tricritical point), the deviation from the dependence $\eta_{\rm e} = \eta_{\rm e0}\tau^{1/2}$ and even closeness to the law $\eta_{\rm e} = \eta_{\rm e0}\tau^{1/3}$ is not at all evidence for the inapplicability of the Landau theory. It should be emphasized here that many structural (in particular, ferroelectric) transitions are just close to the tricritical point. At this point, $T = \theta_{\rm cr}$, as is known, $\beta(\theta_{\rm cr}) \equiv \beta_{\theta_{\rm cr}} = 0$ and, as is obvious from (5.17), the nearer the tricritical point is, the narrower the critical region is (according to (5.17), as $\beta_\theta \to 0$, the value $\tau_{\rm L}' \to 0$). It should be added that the substance ('specimen') inhomogeneity and the presence of internal stresses and defects can also essentially distort the picture and, so to speak, imitate the fluctuation-type behavior (for the references concerning this and other questions discussed here, see [46]).

From the aforesaid, it has become clear that the determination of the critical region, where the Landau theory is inapplicable, is a very complicated and delicate problem for phase transitions in a solid state (investigation of light scattering and X-rays near transition points [46] may appear to be very helpful in the solution of this problem). As a result, as far as I know, there is not a single reliable example of structural transition in a solid state for which a critical (fluctuational) region is established.[7]

There is an impression, therefore, that for structural phase transitions the critical region is especially very narrow. Here and later we mean even 'pure' second-order transitions (i.e. transitions far from the tricritical point) for which the critical region is larger than for transitions close to the tricritical point.

---

[7] The same, generally, refers to transitions in liquid crystals.[3*]

The reason why the critical region is narrow for ferroelectric displacement-type transitions has already been discussed in paper [32]. Specifically, even if a comparatively long-range dipole interaction is predominant in structural transitions, the coefficient is $\delta \sim a^2$, i.e. the correlation energy, is not anomalously large (as is the case, e.g., with superconductors). Hence, as is clear from (5.17) and (5.18), if we disregard the smallness of the numerical coefficient in (5.17), the smallness of the critical region $\tau'_{\rm L}$ can only be due to the smallness of the derivative $\alpha'_\theta = (d\alpha/dT)_\theta$. Such a smallness actually exists according to experimental data for a number of transitions. Further, if we use a reasonable estimate $\alpha(0) \sim \alpha'_\theta \theta$ or evaluate the coefficient α in a somewhat different way [46], we come to the conclusion that in ferroelectrics this coefficient is relatively small compared to its 'normal' value $\alpha \sim 1$ (here we assume $\eta = P$ and, thus, if only the second term is involved in (5.13), the thermodynamic potential density is $\Phi = \alpha P^2$). This 'normal' value corresponds only to the dipole interaction energy or to the elastic energy. The real value of α in a ferroelectric is obtained by involving both the elastic energy and the term of the type $-(f/2)P^2$ which results from the difference between the effective electric field and the mean one (see section 5.2 and [13]). In other words, the term αP^2 for a ferroelectric is the difference between two contributions and is much smaller than either of them. The latter circumstance is reflected in the fact that spontaneous polarization P_0 in many ferroelectrics (even far from the transition temperature θ) is small compared with polarization of 'strong' pyroelectrics, which corresponds to the appearance of a dipole moment $p \sim ea \sim 10^{-18}$ in a crystal cell. In brief, there is every reason to believe that at least for displacement-type ferroelectrics $r'_{\rm co} \gg a$ and the width of the critical region is small [32], and amounts to only negligible fractions of a degree. This question is discussed in some more detail in [46] but it cannot be considered to be completely solved because for this purpose the radius $r'_{\rm co}$ or, in fact, the coefficient δ should be measured for a known α'_θ (see (5.18)).

Thus, for known ferroelectrics (we mean three-dimensional crystals), the critical region is small and, apparently, has not yet been observed. Thus, the use of the Landau theory, which is typically applied in this case, is quite grounded. At the same time, it would be very interesting to further specify the size of the critical region both from measurements that determine the values of $\tau'_{\rm L}$ and $\tau_{\rm L}$ and from direct measurements of the temperature dependence of certain quantities (first of all, the polarization P_0), as well as finding corresponding critical indices. It may appear quite reasonable to investigate quasi-two-dimensional specimens (for which the critical region is wider than for three-dimensional systems). Most important here is a thorough control over the specimen quality, and if the transition is close to a tricritical point, one should process the measurements using (in the case of Landau theory) the formulae in which the coefficient γ in expansion (5.13) is not assumed to be zero. Finally, small fluctuation corrections can be involved already on the basis of the Landau theory. Only such a complex approach, a many-sided analysis of the problem will completely clarify the situation. Anticipating the final results, I shall express the opinion that the application of the Landau theory for typical three-dimensional ferroelectrics does not practically entail any noticeable limitations.

5.5 Ferroelectricity, soft modes, and high-temperature superconductivity[4*]

The problem of high-temperature superconductivity consists in finding the possibility of creating superconductors or superconducting 'systems' (for instance, 'sandwiches' consisting of metal and dielectric layers) for which the critical temperature T_c greatly exceeds the presently known values and reaches, say, the liquid nitrogen boiling point at an atmospheric pressure $T_{b,N_2} = 77.4$ K.[8] The problem of high-temperature superconductivity can now be considered as open. The theoretical considerations give no evidence against the possibility for the critical temperature T_c to reach the values $T_c \sim 100$–300 K even for not so exotic materials as, for example, metallic hydrogen. However, as a result of a rather long study [48–51], it has become clear that it is not at all easy to solve the problems positively (i. e. to create a substance with $T_c \gtrsim 70$ K) because for this purpose very rigorous conditions are needed which are not yet sufficiently clear. Moreover, it is not excluded that to create high-temperature superconductors under more or less usual conditions (say, without very high pressures) is altogether impossible.[5*] The theoretical analysis of the corresponding questions encountered considerable difficulties but they have been realized and are being studied, naturally with the hope to clarify the situation in the near future.

This is, roughly speaking, my opinion at the present time and it is reflected in [48–50]. In such a situation, I would not have touched here upon the problem of high-temperature superconductivity at all if it had not been for the following reasons. First, when inviting me to give a talk at the conference, the organizers pointed to the "connection between ferroelectricity and superconductivity" as a possible topic. Second, such a presumed connection, as well as the connection between soft modes and high-temperature superconductivity, is mentioned in the literature [52, 53]. Third, and this is the main point, an interesting aspect of the connection between ferroelectricity and superconductivity has not yet been elucidated in the literature with the exception of [49, 50]). I mean an account of the

---

[8] The term 'high-temperature superconductors' is often used in the literature for compounds of the type Nb_3Sn (with $T_c = 18.1$ K) or Nb_3Ge (in this case $T_c \approx 23.2$ K, which is the highest reliably established critical temperature before 1986).[4*]

difference between the effective (acting) electric field E_{ef} and the mean macroscopical field E. The following remarks will mostly concern this aspect.

How the question of the effective field is connected with the theory of ferroelectricity is well known and was, in particular, mentioned in section 5.2. We shall restrict ourselves to the isotropic case and the simplest chain of relations which do not need any comments:

$$E_{\text{ef}} = E + fP \qquad P = \frac{\varepsilon - 1}{4\pi} E = \alpha N E_{\text{ef}} \qquad (5.20)$$

$$\varepsilon = 1 + \frac{4\pi\alpha N}{1 - f\alpha N}. \qquad (5.21)$$

The model is obviously used here involving soft dipoles with polarizability α; their concentration is N. However, expression (5.21) is, in fact, wider than such a model. It follows from (5.21) that

$$\frac{\varepsilon - 1}{1 + (\varepsilon - 1)f/4\pi} = 4\pi\alpha N$$

and, for $f = 4\pi/3$, one obtains the Clausius–Mossotti or Lorenz–Lorentz formula

$$\frac{\varepsilon - 1}{\varepsilon + 2} = \frac{4\pi}{3}\alpha N.$$

For a low-density plasma, $f = 0$ (see [39]) and already from this example it is clear that the f values depend strongly on the character of the system. If $f\alpha N \to 1$, then $\varepsilon \to \infty$ and, for $f\alpha N > 1$, we would have $\varepsilon < 0$. However, the permittivity ε, considered earlier and denoted by $\varepsilon_l(0, k \to 0)$ in what follows, cannot be negative (for $\varepsilon < 0$, the system loses stability). Thus, as $f\alpha N \to 1$, the system must be reconstructed and, if we bear in mind a crystal lattice, we must expect a structural phase transition. This may turn out to be a transition into a pyroelectric state, i.e. a ferroelectric transition.

We shall now proceed to superconductivity, the appearance of which in a metal upon a weak effective interaction of conduction electrons and, under certain simplifying assumptions (isotropy, etc), is determined by the effective interelectron interaction

$$V_{\text{ef}}(\omega, k) = \frac{4\pi e^2}{k^2 \varepsilon_l(\omega, k)}. \qquad (5.22)$$

Here $\varepsilon_l(\omega, k)$ is the longitudinal permittivity of the system (metal) at a frequency ω and for a wavevector k. The tensor of the total permittivity for an isotropic system is (see, for instance, [41])

$$\varepsilon_{ij}(\omega, k) = \varepsilon_{\text{tr}}(\omega, k)\left(\delta_{ij} - \frac{k_i k_j}{k^2}\right) + \varepsilon_l(\omega, k)\frac{k_i k_j}{k^2}. \qquad (5.23)$$

The dispersion equation for longitudinal waves is $\varepsilon_l(\omega, k) = 0$ and, for transverse waves, it is

$$k^2 = \frac{\omega^2}{c^2} \varepsilon_{tr}(\omega, k).$$

In the approximation in question, the critical temperature for super-conductivity (for more details see [48–51]) is

$$T_c = \frac{\overline{\omega}}{1.45} \exp\left(-\frac{1}{\lambda - \mu^*}\right) \qquad \mu^* = \frac{\mu}{1 + \mu \ln(E_F/\overline{\omega})} \qquad (5.24)$$

where E_F is the electron energy on the Fermi surface and λ, μ, and $\overline{\omega}$ are some constants which are expressed in an integral way through $\varepsilon_l(\omega, k)$. For what follows, it is only of importance that

$$\mu - \lambda = N(0) \left\langle \frac{4\pi e^2}{k^2 \varepsilon_l(0, k)} \right\rangle \qquad (5.25)$$

where $N(0)$ is the electron density of state on the Fermi surface and the brackets $\langle \ \rangle$ stand for some averaging over k.

Not so long ago, it was customarily considered that for a system to be stable, the requirement must hold that the static permittivity should be non-negative,

$$\varepsilon_l(0, k) \geq 0. \qquad (5.26)$$

If we accept this condition, then $\lambda \leq \mu$, in the approximation in question $T_c^{max} = (E_F/1.45)e^{-3/\lambda}$ and, even under some generalizations, $T_c^{max} \sim 10$ K. Thus, if the inequality (5.26) were really necessary, the problem of high-temperature superconductivity could actually be considered as 'closed' [54].

Fortunately, as has been found by Kirzhnits *et al* [48–51, 55], the requirement (5.26) refers only to the limiting case of long waves

$$\varepsilon_l(0, k \to 0) \geq 1. \qquad (5.27)$$

If $k \neq 0$ (and in the theory of superconductivity large k are significant), then the stability is provided by the requirement

$$1/\varepsilon_l(0, k) \leq 1 \qquad k \neq 0 \qquad (5.28)$$

i.e.

$$\varepsilon_l(0, k) \geq 1 \quad \text{or} \quad \varepsilon_l(0, k) < 0 \qquad k \neq 0. \qquad (5.29)$$

Thus, negative values of $\varepsilon_l(0, k)$ are admissible and, moreover, in many metals are, indeed, $\varepsilon_l(0, k) < 0$ [50, 51].

It is, therefore, possible that $\lambda > \mu$ and that T_c is large. In ordinary (low-temperature) superconductors, we probably have $\lambda > \mu$ too but, in this case, the values $\overline{\omega}$ in (5.24) are not large; for the phonon mechanism of superconductivity $\overline{\omega} \to \omega_{\mathrm{D}}$, where $T_{\mathrm{D}} = \hbar\omega_{\mathrm{D}}/k_{\mathrm{B}}$ is the Debye temperature. For the exciton (non-phonon) mechanism of superconductivity, the frequency $\overline{\omega} \to \omega_l$ can be large ($\hbar\omega_l$ is the exciton energy which can easily reach 0.1–1 eV $\sim 10^3$–10^4 K). But for a non-phonon mechanism, the question of how negative $\varepsilon_1(0, k)$ values can be reached without loss of stability is not yet clear enough, for corresponding cases formulae of the type (5.24), (5.25) are invalid (the coupling is always strong), and the possible T_{c} values have not yet been reliably estimated [50]. We meant just this fact when characterizing the present-day state of the problem of high-temperature superconductivity.

The only important thing for us now is that for achieving high T_{c} values, we must have a substance with $\varepsilon_1(0, k) < 0$. To this end it is, in turn, necessary that the effective field should be significantly different from the mean field, i. e. the factor f in (5.20), (5.21) should be sufficiently large. True, in (5.20), (5.21), the expressions were written for $k = 0$ but a formula of the type (5.21) also remains valid for $k \neq 0$, where $f = f(k)$ (see [50]).

Now, the connection between ferroelectricity and superconductivity, which we would like to point out here, is already clear. Namely, for the appearance of ferroelectricity, the 'correction' due to the effective field in the static case and as $k \to 0$ (i. e. for a homogeneous field) must be large enough as to provide the system (lattice) reconstruction and the occurrence of spontaneous polarization. For the appearance of superconductivity with a large critical temperature T_{c}, the 'correction' due to the effective field must also be sufficiently large in order that it could also guarantee a large negative value of the permittivity $\varepsilon_1(0, k)$ for a static field but for large k (we practically mean the values $k \sim k_{\mathrm{F}}$, where k_{F} is the wavevector on the Fermi surface).

Such a connection between ferroelectricity and superconductivity expressed in the language of the permittivity $\varepsilon_1(0, k)$ is not an immediate but an indirect one. The same can be said, however, about the connection between ferroelectricity and superconductivity reflected in terms of soft modes, closeness to structural transitions, etc (see [52, 53] and [48, chapter 5]). Both the approaches, both the languages are legitimate and can perhaps provide insight into the problem of high-temperature superconductivity. I can only state that the 'language of permittivity' is closer and more clear to me. The aim of section 5.4 of the present report is, obviously, to acquaint the reader, at least superficially, with this language.[6*]

Comments

1*. This is the report prepared for the Sixth International Meeting on Ferroelectricity (August 1985, Kobe, Japan). I was unable to participate in this and all preceding meetings like this. The report was published in 1987

Ferroelectrics **76** 3 and in the Russian language in 1987 *Trudy P N Lebedev Physical Institute* **180** 3. The editor of the Russian version discarded the phrase "I cannot attend the present meeting and most probably will not be able to in future". In the English version, this phrase did appear, of course, but the editor reduced two phrases in section 2 of the paper (see the text).

In some of my papers, in particular, in those included in the present book (see, e. g., article 17), I have emphasized that I am against all debates aimed at establishing priority and corresponding claims. The present paper may seem to refute such assertions. I hope, however, that the reader will understand that this is not so. This report, which was invited by the Organizing Committee of the Meeting on Ferroelectricity, refers to the 50-year-old story (1945–49). I have no claims or grudges against anybody except the Soviet regime which deprived me of the possibility of attending conferences and, for the most of the period discussed here, I might not even publish my papers in English.

True, the introduction to the paper seems to sound somewhat plaintive and I myself do not like to read it. Because of this and for the reason mentioned earlier, the paper was not included in the first edition of the collection. But now I have come to the conclusion that some other aspects (for example, the desire to make the material more complete) are more important. It should be noted that in 2001, on the invitation of the Organizing Committee of 10th International Meeting on Ferroelectricity (Madrid, Spain, September 2001), I prepared a new report devoted to the history of the study of ferroelectricity. This report [56] mostly repeats the present one but involves some new material as well.

2*. In the English version of the paper, the editor discarded, without my consent, part of the quotation from paper [38] ("even with the benefit of hindsight"), as well as the words "and, it seems to me, mentioned [13] in a rather slighting tone" (see the quotation from paper [38]).

3*. One should not forget that I have long been unaware of the literature in this field and, thus, it is quite possible that some assertions made in the text have become outdated.

4*. The report was written in 1985 and high-temperature superconductivity was discovered in 1986–87 (see article 6 in this book). That is why, the text of section 5.4 in this paper is clearly partially outdated. Nevertheless, this item was neither omitted nor revised because, I believe, from the point of view of physics, it seems interesting even today.

5*. Recall once again that the text had been written before the creation of high-temperature superconductors.

6*. In 2001 I was invited to give a talk at the 10th International Conference on Ferroelectricity held in Madrid (Spain) and this time (in fact, for the first time) I came to a conference on ferroelectricity, although I was already 85. The corresponding paper was published [56]. It largely repeats the present article 5 (as has already been mentioned, it is the talk prepared for the sixth analogous conference held in 1985). In paper [56], however, in its concluding section 7, I also dwelt briefly on a number of works, apart from my own ones, carried out in the USSR and Russia (of course, I mean works on ferroelectricity and related phenomena). Special emphasis should be laid on ferrotorics, i.e. substances with toroidal momentum (in this respect, see also references [158] and [160] to article 7). In this book, it would perhaps be pertinent to replace the present article 5 by a more compact and somewhat more informative paper [56]. I did not do this because I wanted the second edition to be changed as little as possible.

References

[1] Landau L D and Lifshitz E M 1960 *Electrodynamics of Continuous Media* (Oxford: Pergamon Press); 2nd edn 1984 [The corresponding Russian editions are published in 1957 and in 1982. The quotation in the text was taken from the 1st edn (the 2nd edn was not essentially changed in this respect)]

[2] Ginzburg V L 1945 *Zh. Eksp. Teor. Fiz.* **15** 739; 1946 *J. Phys. USSR* **10** 107

[3] Kurchatov I V 1933 *Ferroelectrics* (Moscow: Gostekhteorizdat) [French translation 1936 *Le champ moléculaire dans les diélectriques* (Paris: Hermann)

[4] Jaffe H 1937 *Phys. Rev.* **51** 43; 1938 **53** 917

[5] Mueller H 1940 *Phys. Rev.* **57** 829; **58** 805

[6] Cady W G 1946 *Piezoelectricity: An Introduction to the Theory and Applications of Electromechanical Phenomena in Crystals* (New York: McGraw-Hill) [The Russian translation was published in 1949 (Moscow: IL)]

[7] Landau L D 1937 *Zh. Eksp. Teor. Fiz.* **7** 19 627; *Phys. Z. Sovjetunion* **11** 26 545 [In English Landau L D 1965 *Collected Papers* ed D ter Haar (Oxford: Pergamon Press) p 193, 209]

[8] Wul B M and Goldman I M 1945 *C. R. Acad. Sci. URSS* **46** 139

[9] Landau L D and Lifshitz E M 1980 *Statistical Physics* part 1 (Oxford: Pergamon Press) [In paper [2] we cite the 2nd Russian edn of this book published in 1940]

[10] Wul B M and Goldman I M 1945 *C. R. Acad. Sci. URSS* **49** 177

[11] Wul B M and Vereshchagin L F 1945 *C. R. Akad. Sci. URSS* **48** 634

[12] Ginzburg V L 1949 *Zh. Eksp. Teor. Fiz.* **19** 36

[13] Ginzburg V L 1949 *Usp. Fiz. Nauk* **38** 490 [Unfortunately, at that period JETP and UFN were not translated into English]

[14] Megaw H D 1946 *Proc. Phys. Soc. London* **58** 133

[15] Roberts S 1947 *Phys. Rev.* **71** 890

[16] Rzhanov A V 1949 *Usp. Fiz. Nauk* **38** 461

[17] Devonshire A 1949 *Phil. Mag.* **40** 1040

[18] Devonshire A 1951 *Phil. Mag.* **42** 1065

[19] Shirobokov M Ya and Kholodenko L P 1951 *Zh. Eksp. Teor. Fiz.* **21** 1237; **21** 1250

[20] Kay H F and Vousden P 1949 *Phil. Mag.* **40** 1019

[21] Devonshire A 1954 *Adv. Phys.* **3** 85

[22] Smolenskii G A *et al* 1971 *Segnetoelektriki i Antisegnetoelektriki [Ferro-electrics and Antiferroelectrics]* (Leningrad: Nauka)

[23] Vaks V G 1973 *Vvedenie v Mikroskopicheskuyu Teoriyu Segnetoelektrikov [Introduction to Microscopic Theory of Ferroelectricity]* (Moscow: Nauka)

[24] Landsberg G S and Mandelstam L I 1929 *Z. Phys.* **58** 250

[25] Ney M J 1931 *Z. Phys.* **68** 554

[26] Nedungadi T 1940 *Proc. Indian Acad. Sci.* **A11** 85 [The abstract of this paper is contained in: Raman V C and Nedungadi T 1940 *Nature* **145** 147]

[27] Narayanaswamy P K 1947 *Proc. Indian Acad. Sci.* **A26** 521

[28] Ginzburg V L and Levanyuk A P 1960 *Zh. Eksp. Teor. Fiz.* **39** 192 [1961 *Sov. Phys. JETP* **12** 138]

[29] Ginzburg V L, Levanyuk A P, and Sobyanin A A 1980 *Phys. Rep.* **57** 151 [The short version in Russian: 1980 *Usp. Fiz. Nauk* **130** 615]

[30] Krishnan R S 1981 *Ferroelectrics* **35** 9

[31] Ginzburg V L, Levanyuk A P, and Sobyanin A A 1983 in *Light Scattering Near Phase Transitions (Modern Problems in Condensed Matter Sciences)* vol 5, eds H Z Cummins and A P Levanyuk (Amsterdam: North-Holland) p 3

[32] Ginzburg V L 1961 *Fiz. Tverd. Tela* **2** 2031 [1960 *Sov. Phys. Solid State* **2** 1824]

[33] Cowley R A 1984 *Ferroelectrics* **53** 27

[34] Cochran W 1960 *Adv. Phys.* **9** 387

[35] Anderson P W 1960 in *Fizika Dielektrikov* ed G I Skanavi (Moscow: Akad. Nauk SSSR) p 290

[36] Cochran W 1959 *Phys. Rev. Lett.* **3** 412

[37] Ginzburg V L 1962 *Usp. Fiz. Nauk* **77** 621 [1963 *Sov. Phys. Uspekhi* **5** 649]

[38] Cochran W 1981 *Ferroelectrics* **35** 3

[39] Ginzburg V L 1949 *Theory of the Radio Wave Propagation in the Ionosphere* (Moscow–Leningrad: Gostekhteorizdat) [This book was developed into the monograph: Ginzburg V L 1970 *The Propagation of Electromagnetic Waves in Plasmas* (Oxford: Pergamon Press)]

[40] Murzin V N, Pasinkov R E, and Solov'ev S P 1967 *Usp. Fiz. Nauk* **92** 427 [1968 *Sov. Phys. Usp.* **10** 453]

[41] Agranovich V M and Ginzburg V L 1984 *Crystal Optics with Spatial Dispersion and Excitons* 2nd edn (Berlin: Springer) [This book is a revised translation of the book by V M Agranovich and V L Ginzburg 1979 *Kristallooptika s Uchetom Prostranstvennoi Dispersii i Teoriya Eksitonov [Crystal Optics with an Account of Spatial Dispersion and the Theory of Excitons]* (Moscow: Nauka)]

[42] Cochran W 1961 *Adv. Phys.* **10** 401

[43] Cochran W and Zia A 1968 *Phys. Status Solidi* **25** 273

[44] Ginzburg V L and Sobyanin A A 1983 *Fiz. Tverd. Tela* **25** 2017 [1983 *Sov. Phys. Solid State* **25** 1163]

[45] Ginzburg V L and Landau L D 1950 *Zh. Eksp. Teor. Fiz.* **20** 1064 (For the English translation see [7])

[46] Ginzburg V L, Levanyuk A P and Sobyanin A A 1986 *Ferroelectrics* **73** 171

[47] Levanyuk A P 1959 *Zh. Eksp. Teor. Fiz.* **36** 810 [1959 *Sov. Phys. JETP* **9** 571]

[48] 1982 *High-Temperature Superconductivity* ed V L Ginzburg and D A Kirzhnits (New York: Consultants Bureau) [The Russian original 1977 (Moscow: Nauka)]

[49] Ginzburg V L 1983 *Waynflete Lectures on Physics: Selected Topics in Contemporary Physics and Astrophysics* (Oxford: Pergamon Press) ch 2

[50] Dolgov O V and Maksimov E G 1982 *Usp. Fiz. Nauk* **138** 95 [1982 *Sov. Phys. Usp.* **25** 688]

[51] Dolgov O V, Kirzhnits D A, and Maksimov E G 1981 *Rev. Mod. Phys.* **53** 81

[52] 1977 *Proc. Int. Conf. on Low Modes Ferroelectrics and Superconduction Ferroelectrics* **16** 17 nos 1/2

[53] Lefkowitz I and Bloomfields P E 1984 *Ferroelectrics* **51** 173

[54] Cohen M L and Anderson P W 1972 in *Superconductivity in d- and f-band Metals (American Institute of Physics Conf. Proc. 4)* ed D H Douglass (New York: American Institute of Physics) p 17

[55] Dolgov O V, Kirzhnits D A and Losyakov V V 1982 *Zh. Eksp. Teor. Fiz.* **83** 1894 [1982 *Sov. Phys. JETP* **56** 1095]

[56] Ginzburg V L 2001 *Usp. Fiz. Nauk* **171** 1091 [2001 *Phys. Usp.* **44** 1037]

Article 6

Superconductivity: the day before yesterday, yesterday, today, and tomorrow[1*]

6.1   Introduction

Actively working physicists usually take little interest in the past and I myself am not an exception—I began studying the theory of superconductivity in 1943 but only in 1979 did I find time to look through the classical papers of Kamerlingh Onnes (1853–1926). And I found them fairly interesting.

Briefly presenting the history of the study of superconductivity, I shall divide it, although rather conditionally, into three periods: the day before yesterday (1911–41), yesterday (1942–86), and today (1987–?).

6.2   The day before yesterday (1911–41)

Helium was first liquefied in 1908 and, which is rather significant, up to 1923 liquid helium had been obtained and used only in Kamerlingh Onnes's laboratory in Leiden, where superconductivity was discovered in 1911. This happened in the course of systematic measurements of the low-temperature electric resistance of metals. This point was quite obscure at that time. True, Drude hypothesized in 1900 that metal contained an electron gas responsible for the electrical conductivity. Drude also proposed the well-known formula for electrical conductivity $\sigma = e^2 n / m\nu$, where n is the electron concentration and ν is the frequency of electron collisions with the lattice. However, the temperature dependencies $n(T)$ and $\nu(T)$ remained quite unknown and the electron model itself was contradictory (in classical theory the electron gas had to make a large contribution to the specific heat, which is not observed). Kamerlingh Onnes, and probably not he alone, at first believed that with decreasing temperature T the electrical conductivity σ should decrease, i.e. according to modern terminology, metals were considered to be semiconductors. This hypothesis was not confirmed and, as T decreased, a fall in the resistance $R(T) = \rho l/S$, $\rho(T) = 1/\sigma(T)$ (l is the length of the wire and S its cross-section), was observed. Furthermore, Kamerlingh Onnes was inclined to think that for a pure metal (platinum) the resistance $R = 0$ at $T > 0$ ('the conclusion seems quite grounded that the resistance of pure platinum within the experimental errors due to the achieved degree of purity is already equal to zero at helium temperatures' [1]). Exceedingly pure samples were to be examined to confirm this hypothesis. But, for platinum and gold, the purification from impurities was a very complicated task, especially at the beginning of the 20th century. It was for this reason that Kamerlingh Onnes passed over to the examination of mercury which is comparatively easy to purify and distill. This was an especially lucky choice because, in addition, the helium boiling temperature at atmospheric pressure $T_{b,\,He} = 4.2$ K appeared to be close to the critical superconducting transition temperature $T_c(Hg) = 4.15$ K for mercury. The latter fact (i.e. the closeness of $T_{b,\,He}$ and T_c) was a certain detail which facilitated the discovery of a superconducting jump [2].[1] The main point is that mercury becomes superconducting at temperatures which were attainable in those times. If Kamerlingh Onnes had continued measurements on platinum, gold, silver, or copper, he would obviously have never discovered superconductivity unless by chance he tried to measure the resistance of some superconductor. This circumstance might have put off the discovery of superconductivity for years. With mercury, success came with the first experiments [3] (since the original papers [1, 3–5] are hardly available, I shall refer to the fact that some of their results are presented in paper [6], where paper [4] is placed as an appendix). Here I shall only note that in [3], which became known in April and May of 1911, the resistance of mercury at $T = 3$ K was shown to be immeasurably small. But the crucial point, i.e. the detection of a sharp superconducting transition (which may be considered as the discovery of superconductivity), was made in paper [4] reported on 25 November, 1911.

Of course, the studies were continued and, in 1913, the discovery of superconductivity in white tin ($T_c = 3.69$ K) and lead ($T_c = 7.26$ K) was reported [5] and the disappearance of superconductivity upon the passage of a rather strong current was also stated. H Kamerlingh Onnes was awarded the 1913 Nobel Prize in physics for "his investigations into the properties

---

[1] As is clear from [2], the superconducting jump and superconductivity, in general was first observed in an explicit form by Gilles Holst who conducted measurements of the resistance of mercury. Holst was a highly qualified physicist (and later, the first director of the Philips Research Laboratories and Professor of Leiden University). Kamerlingh Onnes, however, did not even mention his name in the corresponding publication. As mentioned in [2], Holst himself had not apparently thought of such behavior by Kamerlingh Onnes as unjust or unusual. The situation is not clear to me and, for our generation, it is quite unnatural; perhaps 90 years ago morals and manners in the scientific community were totally different.

of matter at low temperatures which led, *inter alia*, to the production of liquid helium". As we see, the prize was awarded not for the discovery of superconductivity but this issue was also touched upon in Kamerlingh Onnes's Nobel lecture [7]. In particular, he noticed:

> Mercury has passed into a new state, which on account of its extraordinary electrical properties may be called the superconducting state. There is left little doubt, that, if gold and platinum could be obtained absolutely pure, they would also pass into the superconducting state at helium temperatures.

Thus, Kamerlingh Onnes still supported the wrong hypothesis that all metals were superconducting at helium temperatures. I am unaware of his arguments but they undoubtedly could not have been serious because the theory of metals did not exist at that time. In this connection, the discovery of superconductivity was not obviously so amazing. Moreover, there was nobody to repeat the Leiden experiments, for as has already been mentioned, no other laboratory obtained liquid helium until 1923. And it so happened that the discovery of superconductivity, which was an event of paramount importance, had quite a moderate resonance, at least by today's measure.[2]

Let us mention some consequent Leiden works. The existence of a critical magnetic field $H_{c\,m}(T)$ was revealed in 1914: for mercury, $H_{c\,m}(0) = 411$ Oe and, for lead, $H_{c\,m}(0) = 803$ Oe. The first magnet with a superconducting winding was constructed in 1914. Particularly noteworthy is the fact that as far back as 1922 an attempt was made [9] to observe the isotope effect, namely the dependence of T_c on the isotope mass in a metallic lattice in lead samples with different isotopic compositions (common lead with atomic weight $A = 207.30$ and uranium lead with $A = 206.06$ were employed). Unfortunately, the isotope effect for such samples makes up only about 10^{-2} K and it was not observed. As is well known, the isotope effect in superconductors was discovered only in 1950 (see references, e.g., in book [10]) and it played an outstanding role in showing the importance of the electron–phonon interaction for the appearance of superconductivity.

It is no less interesting that, in 1924, Kamerlingh Onnes was also close [11] to the discovery of the Meissner–Ochsenfeld effect (this effect, which is most significant for the study of superconductivity, was discovered only in 1933 [12]). Namely, Kamerlingh Onnes was investigating the behavior of a lead ball in an external magnetic field and failed to register the field pushing out the ball upon its transition to the superconducting state probably

---

[2] For instance, the bibliography placed at the end of monograph [8] devoted to superconductivity contains 450 references embracing the period of 1911–44 (some other data are also presented in [6]). Of them, only 34 refer to the interval of 1911–25, and the author or a co-author of 19 of these references is Kamerlingh Onnes. For comparison, we can say that within ten years of the discovery of high-temperature superconductivity (HTSC) in 1986–87, nearly 50 000 publications were devoted to this subject.

only because he used an empty ball to spare liquid helium which was scarce at that time. In the case of an empty ball, a closed superconducting ring analogous to a doubly connected torus may form. Under such conditions the Meissner effect is masked.

In spite of this failure, the contribution of the Leiden laboratory and certainly of Kamerlingh Onnes himself can hardly be overestimated. To what has already been said, I shall add that not a single superconductor had been discovered outside Leiden before 1928. In addition, Kamerlingh Onnes began liquid helium studies that led to the discovery of its superfluidity in 1938. The first step in this direction was made in 1911, the year of the discovery of superconductivity. Namely, the curve of the temperature dependence of the helium density $\rho(T)$ was found [13] to exhibit a kink corresponding to the λ-point. After that, liquid helium studies were conducted over many years [14, 15], which in the 1930s resulted in the discovery (for the most part by W H Keesom, Kamerlingh Onnes's successor in the Leiden laboratory, and his colleagues) of clearly pronounced anomalies—the λ singularity in the specific heat and super-thermal conductivity of He II [14, 15]. Finally, in 1938, these studies ended in the discovery of superfluidity of helium II by Kapitza [16] and Allen and Missner [17]. Undoubtedly, the long path (27 years!) to the discovery of superfluidity compared to the expeditious discovery of superconductivity (see earlier) is first of all explained by methodological reasons: to measure electrical resistance is easy, while to observe helium flow through a narrow gap or a capillary is difficult and, besides, one must hit upon the idea of carrying out such experiments.

The discovery and further study [18] of superfluidity and, which is most important, the Landau theory of superfluidity [19] made it possible to consider superconductivity as the superfluidity of the electron liquid in metals. However, understanding this fact did not play any particular role at that time because the Landau theory [19] was phenomenological and referred to the Bose–Einstein liquid (it is another thing that at first Landau did not regard the connection with Bose–Einstein statistics to be crucial for superfluidity). And one should have understood superconductivity as a phenomenon in an electron gas (or liquid), i.e. involving particles that obey Fermi–Dirac statistics. No advances were made then in this direction.

Even before that, superconductivity had become the most enigmatic phenomenon in condensed-matter physics and, more concretely, in the physics of metals. Properly speaking, before the creation of quantum theory, the behavior of non-superconducting metals (or, more precisely, metals in the normal, non-superconducting state) had also been absolutely unclear. But the application of quantum mechanics to a degenerate Fermi-gas in works by Pauli, Sommerfeld, Bethe, Bloch, Landau, and many others in the period from 1926 to 1930 changed the situation radically—everything seemed to become, in principle, clear in the theory of metals. Indeed, the advances (see, e. g., [20, 21]) were impressive and, as far as I remember and know, they were practically unconditionally accepted by a wide range of physicists. In actual fact, as Landau used to emphasize, "nobody has abrogated the Coulomb law" and, in this connection, it remained unclear why the electron gas approximation was so successful when applied to metals. Indeed, in normal metals, the kinetic (Fermi) energy E_F is by no means lower than the Coulomb electron–electron interaction energy (e. g., in Ag, the electron concentration is $n = 5.9 \times 10^{22}$ cm^{-3}, the Fermi energy is $E_F = 8.5 \times 10^{-12}$ erg, and the characteristic interaction energy is $e^2 n^{1/3} = 19.3 \times 10^{-12}$ erg). The situation became transparent only when Landau created the Fermi-liquid theory in 1956–58 (see, e. g., [22]). But this is already another epoch.

In the 1930s and, in fact, up to the mid-1950s superconductivity, as has already been said, remained an enigma. So, in 1933, Bethe wrote: "The success in the theory, in the explanation of normal phenomena in conductivity is great, whereas very little has yet been done in solving the problem of superconductivity. Only a number of hypotheses exist which until now have in no way been worked out and whose validity cannot therefore be verified" [20]. These hypotheses are listed in [20, 23] and they all turned out to be erroneous. In the well-known monograph *The Theory of Metals* by A H Wilson [21], published in England in 1936 (and in Russian in 1941), we find the words: 'In spite of all the progress made by the theory of metals over the past years, the phenomenon of superconductivity remains as enigmatic as it was before, and as before it leaves unsuccessful all attempts to explain it.' Interestingly, in the second edition of the book, which appeared in England in 1953, the chapter devoted to superconductivity was dropped completely [21]. There may have been good reason for this: the author had nothing new to say.

Thus, the first period in the study of superconductivity, entitled in the present article "The day before yesterday (1911–41)", ended, in respect of the microtheory of superconductivity, with the understanding of the existence of a real problem and the recognition of obscurity on the way to its solution. Incidentally, this was not for lack of attention or intellectual efforts. Suffice it to say that the attempts of Einstein [24] and Bohr [23] to gain insight into the nature of superconductivity also failed. It was apparent that the interaction between conduction electrons should be taken into account in one or another way. But the key to an effective approach to this issue was only found in 1950, when the previously mentioned isotope effect was found and pointed to the role of the interaction between the conduction electrons and the crystal lattice.

At the same time, great success was achieved in the understanding of the macroscopic behavior of superconductors even at this early stage. After the discovery of the Meissner effect in 1933 [12], it became evident (see [25] and the literature cited there) that the superconducting state is a phase of matter in the thermodynamical sense of this notion. In the depth of a superconductor the magnetic field $H = 0$ (here and in the following we do not distinguish between the field H and the magnetic induction B) and

$$F_{n0} - F_{s0} = \frac{H_{cm}^2(T)}{8\pi} \tag{6.1}$$

where F_{n0} and F_{s0} are the free energies of unit volume of a metal respectively in the normal (n) and superconducting (s) states (phases) and $H_{cm}(T)$ is the critical magnetic field for massive samples. Differentiation of relation (6.1) with respect to T yields a number of thermodynamic relations. It would obviously be out of place to dwell longer on the thermodynamics of superconductors, on the influence of the field, current, etc (see, e.g., books [8, 10, 26–29]).

What is necessary to note when speaking of the history is the previously mentioned paper by Gorter and Casimir [25] and the so-called two-fluid model [30] proposed by them in 1934. According to this model, along with a superconducting current with density j_s, in a superconductor a normal current may flow with density j_n which is due to the flow of 'normal electrons' present in the superconductor at $T > 0$. The total current density is, of course, $j = j_s + j_n$ and it is this density that enters the ordinary electrodynamic equation rot $H = (4\pi/c)j + (1/c)(\partial E/\partial t)$, where E is the electric field strength (the polarization of the medium is neglected).

The normal current in a superconductor does not, in fact, differ from the current in a non-superconducting state and, in the local approximation in the absence of a temperature gradient (and generally in a homogeneous medium), we have

$$j_n = \sigma_n(T)E. \tag{6.2}$$

A significant step forward was the equation for j_s,

$$\text{rot}(\Lambda j_s) = -\frac{1}{c}H \tag{6.3}$$

with Λ as a new constant. This equation was introduced by the Londons [31] in 1935.

The meaning and, so-to-say, the origin of this equation becomes obvious if we consider the hydrodynamic equation for a conducting liquid (gas) consisting of particles with a charge e and a mass m and moving at a velocity $v_s(r, t)$:

$$\begin{aligned}
\frac{\partial v_s}{\partial t} &= -(v_s\nabla)v_s + \frac{e}{m}E + \frac{e}{mc}[v_s H] \\
&\equiv \frac{e}{m}E - \nabla\frac{v_s^2}{2} + \left[v_s\left(\text{rot}\,v_s + \frac{e}{mc}H\right)\right].
\end{aligned} \tag{6.4}$$

Such an equation corresponds to a medium with ideal conductivity [32] and does not prevent a constant magnetic field from being present in this medium. Imposing an additional condition, namely the condition of the absence of vortices extended to the case of a charged liquid

$$\text{rot } \boldsymbol{v}_\text{s} + \frac{e}{mc} \boldsymbol{H} = 0 \tag{6.5}$$

we arrive at the Londons equation (6.3) if we take into account the fact that $\boldsymbol{j}_\text{s} = e n_\text{s} \boldsymbol{v}_\text{s}$ (n_s is the concentration of 'superconducting' charges). Given this, obviously,

$$\Lambda = \frac{m}{e^2 n_\text{s}}. \tag{6.6}$$

Moreover, under condition (6.5), we obtain from (6.4) the second Londons equation

$$\frac{\partial (\Lambda \boldsymbol{j}_\text{s})}{\partial t} = \boldsymbol{E} - \nabla \frac{\Lambda}{2en_\text{s}} j_\text{s}^2. \tag{6.7}$$

The last term on the right-hand side is rather small and is, therefore, typically omitted, although it has quite a real meaning (see, e. g., [33]).

Equation (6.3) together with the field equation rot $\boldsymbol{H} = (4\pi/c)\boldsymbol{j}_\text{s}$ in a homogeneous medium (i. e. for $\Lambda = $ constant) leads to the equations

$$\Delta \boldsymbol{H} - \frac{1}{\delta^2} \boldsymbol{H} = 0 \qquad \Delta \boldsymbol{j}_\text{s} - \frac{1}{\delta^2} \boldsymbol{j}_\text{s} = 0 \qquad \delta^2 = \frac{\Lambda c^2}{4\pi} = \frac{mc^2}{4\pi e^2 n_\text{s}}. \tag{6.8}$$

This implies that the field and the current decrease with the penetration into the superconductor according to a law of the form $H = H_0 \exp(-z/\delta)$, which corresponds to the Meissner effect. At first, however, it was only a qualitative agreement between the theory and experiment, while quantitative measurements [10] remained contradictory for a long time and, in particular, did not confirm the conclusion of the theory concerning the dependence of the critical magnetic field on the thickness of superconducting films (see book [10], paper [34] and the literature cited therein). Furthermore, in the Londons theory, to provide stability of the boundary between the normal and superconducting phases, it was necessary to introduce also the surface energy σ_ns at the interface. But the energy σ_ns is rather high and it remained quite unclear how it could be calculated [34, 35]. Here, however, we pass over to the next period in the study of superconductivity—the period referred to as "Yesterday (1942–86)".

True, the research of thermoelectric phenomena in superconductors which was started in 1927 should also be ascribed to "The day before yesterday (1911–41)". The result of this research was as follows: "Many experiments have shown that all thermoelectric effects disappear in the superconducting state" [10, see p 86]. Indeed, at first glance, this statement seems to reflect the situation correctly. But, in reality, as I pointed out later, in 1944 [36] (see also [33, 37]), thermoelectric effects exist in the superconducting state as well but they are masked to a great extent because of the existence of two currents—j_s and j_n. Strange as it is, thermoelectric effects in the superconducting state have even now been investigated quite insufficiently. This problem, however, stands apart from the magisterial trends in the field of superconductivity and is, therefore, beyond the scope of our further discussion (see paper [37], book [38], and also article 7 of this book).

Finally, it is noteworthy that type II superconductors were, in fact, discovered in the late 1930s, although it took two decades to make things clear. Namely, in 1935–36, L V Shubnikov and his co-authors revealed the behavior of some alloys in a magnetic field, typical of type II superconductors (for the explanation and references see books [8, 10]; previously we have referred, although implicitly, to type I superconductors).[3]

6.3   Yesterday (1942–86)

It is somewhat conditional that the first period in the history of the investigation of superconductivity ("The day before yesterday") is thought of as finished in 1941. One should not, however, forget that this boundary was determined, among other things, by the fact that the Second World War broke out in Western Europe in 1939 and, in the USSR, in 1941. Naturally, the investigation of superconductivity nearly stopped. In the bibliographical index [39], which includes papers devoted to superconductivity over the period 1911–70, only 36 out of the 6579 publications refer to the period 1942–45 (in 1941, according to [39], only nine papers were published, some of them having been written earlier). What a contrast this is with the present-day situation, for which we can point out that, within the period 1989–91, nearly 15 000 papers appeared devoted to high-temperature superconductors, i. e. on average about 15 papers a day were issued.

I was among those few physicists who took an interest in the theory of superconductivity in the war years and this happened under the influence of Landau's paper [19] published not long before. We had been evacuated to the town of Kazan where we were pretty cold and hungry. But people sometimes indulge in research work under any circumstances. All my activities in the field of superconductivity and superfluidity, started in 1943, are described in detail in paper [35] (see article 7 of the present volume). Here I shall dwell on only two directions of my own work which seem of importance in a rather broad context. One is the formulation of the quasi-phenomenological theory of superconductivity, which was then called the Ginzburg–Landau theory (I refer to it as the Ψ-theory of superconductivity). The other direction of this work, which was started in 1964, was the discussion of the possibility of creating high-temperature superconductors.

---

[3] I cannot but mention with bitterness that the outstanding physicist L V Shubnikov fell victim to terror: although completely innocent, he was shot in 1937.

As previously mentioned, the Londons theory correctly reflected the existence of the Meissner effect but was inapplicable to 'strong' magnetic fields comparable with the critical field H_c. In other words, it was only applicable under the condition

$$H \ll H_c. \qquad (6.9)$$

True, condition (6.9) was established with sufficient clarity only later but the impossibility of deriving a correct expression for the critical field H_c on the basis of the Londons theory in the case of thin films became apparent as far back as 1939 (see papers [40, 34]). The question of the surface energy σ_{ns} on the boundary of the superconducting and normal phases also remained absolutely uncertain [34]. All this stimulated the search for a generalization of the Londons' theory: the Ψ-theory published in 1950 [41] may be thought of as such a generalization. In Ψ-theory, the scalar complex function $\Psi(r)$ is introduced as an order parameter to describe superconductivity. This function Ψ is sometimes termed the macroscopic or effective wavefunction and is, in fact, associated with the electron density matrix in a superconductor [41, 35]. The free energy density of the superconductor and the field is written in the form

$$F_{sH} = F_{s0} + \frac{H^2}{8\pi} + \frac{1}{2m} \left| -i\hbar\nabla\Psi - \frac{e^*}{c} A\Psi \right|^2$$

$$F_{s0} = F_{n0} + \alpha|\Psi|^2 + \frac{\beta}{2}|\Psi|^4 \qquad (6.10)$$

where A is the vector potential of the field rot $A = H$ and F_{n0} is the free energy density in the normal state.

Under the condition of thermodynamic equilibrium in the absence of a field, $\partial F_{s0}/\partial|\Psi|^2 = 0$, $\partial^2 F_{s0}/\partial^2|\Psi|^2 > 0$, and $|\Psi|^2 = 0$ at $T > T_c$, while at $T < T_c$ we already have $|\Psi|^2 > 0$. This implies that $\alpha(T_c) \equiv \alpha_c = 0$, $\beta(T_c) \equiv \beta_c > 0$, and $\alpha(T) < 0$ at $T < T_c$. The theory develops in the region near T_c and, within the validity limits of the expansion (6.10), one can put $\alpha(T) = \alpha'_c(T - T_c)$ and $\alpha'_c \equiv (d\alpha/dT)_{T=T_c}$, $\beta(T_c) \equiv \beta_c$. From this, in equilibrium at $T < T_c$ we have

$$|\Psi|^2 = |\Psi_\infty|^2 = \frac{\alpha'_c(T_c - T)}{\beta_c}$$

$$F_{s0} = F_{n0} - \frac{\alpha^2}{2\beta} = F_{n0} - \frac{(\alpha'_c)^2(T_c - T)^2}{2\beta_c} = F_{n0} - \frac{H_{cm}^2}{8\pi}. \qquad (6.11)$$

In the presence of the field, the equation for Ψ is derived upon variation of the free energy $\int F_{sH} dV$ over Ψ^* and has the form

$$\frac{1}{2m} \left(-i\hbar\nabla - \frac{e^*}{c} A \right)^2 \Psi + \alpha\Psi + \beta|\Psi|^2\Psi = 0. \qquad (6.12)$$

Variation of the integral $\int F_{sH}dV$ over $\boldsymbol{A}$ (under the condition div $\boldsymbol{A} = 0$) leads us to the equation (evidently, it is assumed that $\boldsymbol{j}_{\mathrm{n}} = 0$)

$$\Delta \boldsymbol{A} = -\frac{4\pi}{c}\boldsymbol{j}_{\mathrm{s}}$$

$$\boldsymbol{j}_{\mathrm{s}} = -\frac{ie^*\hbar}{2m}(\Psi^*\nabla\Psi - \Psi\nabla\Psi^*) - \frac{(e^*)^2}{mc}|\Psi|^2\boldsymbol{A}. \qquad (6.13)$$

If we put $\Psi = \Psi_\infty = $ constant, then $\boldsymbol{j}_{\mathrm{s}} = -[(e^*)^2/mc]\Psi_\infty^2\,\boldsymbol{A}$, which, as is readily seen, is equivalent to the Londons equation (6.3). In fields comparable with the critical field, the function Ψ is already not constant, which makes the Londons theory inapplicable (and thereby comes the criterion (6.9)). It should be noted that far from T_c, where the Ψ-theory is generally quantitatively inapplicable (at least without some changes), the Londons theory may appear to be inapplicable in the weak field as well. The point is that, for type I superconductors far from T_c, the coupling of the field with the current is non-local. This circumstance was pointed out by Pippard in 1953 [42]. The state of the theory of superconductivity (both macroscopic and microscopic) before the creation of the microtheory of superconductivity by Bardeen, Cooper and Schrieffer (BCS) in 1957 [43] was elucidated in the large review by Bardeen published in 1956 [44]. I refer the reader to this review, in particular, in respect of the allowance for non-locality [42]. The current state of the theory of superconductivity is presented in books [26–29, 38, 128] and elsewhere. But now, in this brief review of the crucial points in the history of the development of this field, we shall only make some more remarks concerning Ψ-theory.

Of interest is the question of the charge e^*, entering the equations of the Ψ-theory [see (6.10), (6.12), and (6.13)]. This charge is involved in the expression for a very important parameter of the theory,

$$\ae = \frac{mc}{e^*\hbar}\sqrt{\frac{\beta_c}{2\pi}} = \frac{\sqrt{2}e^*}{\hbar c}H_{c\,\mathrm{m}}\delta_0^2 \qquad (6.14)$$

where $H_{c\,\mathrm{m}}$ is a thermodynamic critical field [see (6.1) and (6.11)] and δ_0 is the penetration depth of the weak magnetic field, and

$$\delta_0^2 = \frac{mc^2\beta_c}{4\pi(e^*)^2|\alpha|} = \frac{mc^2}{4\pi(e^*)^2\Psi_\infty^2}. \qquad (6.15)$$

The quantities $H_{c\,\mathrm{m}}$ and δ_0 can be measured experimentally and, moreover, a number of measurable quantities (the surface energy σ_{ns} and fields for limiting supercooling and superheating) depend on $\ae$. Thus, using the measurable quantities $H_{c\,\mathrm{m}}, \delta_0$, and $\ae$, one can determine e^* according to (6.14). In this way, one can arrive at the conclusion [45] that $e^* \approx (2\text{–}3)e$, where e is the electron charge. This seemed strange because, as Landau noted, the effective charge e^* must not depend on coordinates (otherwise, the gradient invariance of the theory is violated) and must, therefore, be universal. Only after the creation of BCS theory [43] Gorkov showed [46] that $e^* = 2e$ holds strictly. This result certainly means that we are dealing with Cooper pairs with charge $2e$. Consequently, the charge $e^* = 2e$ is actually universal (in the sense that it does not depend on coordinates) but at the same time it is not equal to e. Interestingly, such a simple idea occurred to no one, in particular neither to me nor to Landau. For Landau, this was not accidental—as mentioned earlier, in his theory of superfluidity [19] Landau did not see any relationship between superfluidity and the Bose–Einstein statistics of ^{4}He atoms. Hence, the idea of electron 'pairing' with so-to-say transformation of fermions into bosons did not suggest itself. As for myself, I cannot find any excuse because I even pointed out that for a charged Bose gas the Meissner effect must take place [47]. Furthermore, I might have known (I do not remember it now) that the idea of electron pairing with subsequent Bose–Einstein condensation and the appearance of superconductivity was suggested by Ogg as far back as 1946 [48], and then by Schafroth [49]. However, Bardeen did not mention Ogg in his extensive 1956 survey [44] and, although he knew the papers by Schafroth, he never even mentioned the possibility of pairing. It was only the paper by Cooper [50] that made the idea of pairing popular and led directly to the creation of the BCS theory [43]. But, and this is interesting, in the BCS paper [43] there is not a single word about Bose–Einstein condensation and, obviously, they failed to recognize the direct relation between this condensation and pairing and its role in the explanation of superconductivity. This can be understood to a certain extent because pairs in the BCS theory have a large size $\xi_0 \sim 10^{-4}$ cm and the condensed (superconducting) state is very far from the condensate of bosons with an atomic size $\xi_0 \sim 10^{-8}$ cm.

However, I have got ahead of my story. The Ψ-theory proved to be very efficient and made it possible to consider a large number of questions and problems (the behavior of films and other superconducting samples in a magnetic field, supercooling and superheating, the calculation of the surface energy, etc). The Ψ-theory turned out to be successful because it lies within the scope of the general phase transition theory and, in this sense, is more general than the BCS theory. At the same time, from the BCS theory near T_c, one certainly obtains the Ψ-theory equations (6.12) and (6.13) with concrete values of the coefficients α and β (see paper [46]). It is, of course, very important that in conventional superconductors the coherence length $\xi = \delta_0/\ae$ (the penetration depth δ_0 is more frequently denoted by the letter λ) is large and, therefore, the fluctuations are small (see [51, 52, 37]). The Ψ-theory is readily extended to the anisotropic case [53] and also holds when more complicated (non-scalar) order parameters are used [54]. In the original work [41], consideration was only given to the case where $\ae < 1/\sqrt{2}$. Then $\sigma_{ns} >$

0 and it was proved that, for æ $= 1/\sqrt{2}$, we already have $\sigma_{\mathrm{ns}} = 0$, and with further growth of æ the energy becomes $\sigma_{\mathrm{ns}} < 0$. In other words, only type I superconductors were dealt with in paper [41], where it was shown that, for æ $> 1/\sqrt{2}$, a certain instability occurs. Only after the work of Abrikosov [55] was it understood that, for æ $> 1/\sqrt{2}$, a vortex lattice is formed and superconductors behave as had been established in the 1930s by Shubnikov and colleagues (this has already been mentioned). In today's terminology, these are type II superconductors which now remain the object of intense investigations and not only in cuprates.

The numerous applications of the Ψ-theory are elucidated in books [22, 27–29] and in many other publications. A huge number of papers are devoted to various applications and generalizations of the Ψ-theory (as an example I may refer the reader to papers [38, 56, 57, 59–61, 141] and book [58]). This is a whole field of research (especially in what concerns vortices and vortex structures) and we cannot dwell on it here.[2*]

The creation of the BCS theory in 1957 [43], i.e. 46 years after the discovery of superconductivity, was a fairly significant event in the history of the study of superconductivity and, properly, for the whole of the physics of condensed media. The BCS work was followed by a series of investigations in which virtually the same results were obtained by other methods, some points were specified, etc (Bogolyubov [62], Valatin [63], Gorkov [46], and others; see the review [64] and the collection of papers [65]).

The most typical result of the BCS theory is the expression for the critical temperature

$$T_{\mathrm{c}} = \theta \exp\left(-\frac{1}{\lambda_{\mathrm{eff}}}\right) \tag{6.16}$$

where $k_{\mathrm{B}}\theta$ is the energy range near the Fermi energy E_{F} where the conduction electrons (more precisely, the corresponding quasi-particles) are attracted, which causes pairing and instability of the normal state; next, in the simplest case $\lambda_{\mathrm{eff}} = \lambda = N(0)V$, where $N(0)$ is the level density near the Fermi surface in the normal state and V is a mean matrix element of the interaction energy corresponding to the attraction.

In the BCS theory, the coupling 'constant' λ is assumed to be small ('weak coupling'), i.e.

$$\lambda \ll 1. \tag{6.17}$$

The theory implies a number of results that can be verified by experiment, for example,

$$\frac{2\Delta(0)}{k_{\mathrm{B}}T_{\mathrm{c}}} = 3.52 \tag{6.18}$$

where $\Delta(0)$ is a superconducting gap (per one quasi-particle) at $T = 0$: for many type I superconductors, the BCS theory proved to be in full agreement with experiment.

Since it is impossible to dwell in more detail here on the development of either experiment or the theory, we shall only mention the Josephson effect [66] and the extension of the BCS theory to the case of strong coupling by Eliashberg [67].

The main landmarks of this stage in the study of superconductivity, which has been described as "Yesterday (1942–86)", are in my opinion the creation of the Ψ-theory (1950), the BCS theory (1957) and, finally, the search for high-temperature superconductors (1964–86). More precisely, we are now mostly speaking of the theory. The experimental research is no less important but it was largely determined by the theory and, in any case, there was no contradiction or contrast between them.

Undoubtedly, the question of why superconductivity is only observed at low temperatures arose long ago. However, before the creation of the BCS theory, no concrete answer to this question could be given, while within the BCS theory the answer is already clear when one applies formula (6.16). The point is that, in the BCS work, the electron–phonon interaction was regarded as the interaction responsible for the attraction between electrons and, thus, for their pairing. This is clear, for the role of this interaction becomes obvious from the isotope effect and, more concretely, from the validity in some cases of the relation

$$T_\mathrm{c} \propto M^{-1/2} \tag{6.19}$$

where M is the ion mass in the lattice (see, for example, [29]).

In the case of electron–phonon interaction, in formula (6.16) for T_c, we have

$$\theta \sim \theta_\mathrm{D} \tag{6.20}$$

where θ_D is the Debye temperature of the corresponding metal. Estimate (6.20) is particularly clear in the language explaining the electron–phonon interaction as a result of the fact that two electrons exchange phonons. But the maximum phonon energy $\hbar\omega_\mathrm{ph}$ is precisely of the order of $k_\mathrm{B}\theta_\mathrm{D}$ and, therefore, $k_\mathrm{B}\theta_\mathrm{D}$ (i.e. the energy range where interelectron attraction occurs) is considered in the BCS theory and, thus, in formula (6.16).

For the majority of metals, $\theta_\mathrm{D} \lesssim 500$ K and $\lambda \lesssim 1/3$. Hence, according to (6.16), $T_\mathrm{c} \lesssim 500\exp(-3) = 25$ K. Such estimates suggest the following value for the phonon mechanism:

$$T_\mathrm{c} \lesssim 30\text{–}40 \text{ K}. \tag{6.21}$$

True, for instance, for lead we have $\lambda \approx 1.5$ but, in this case, $\theta_\mathrm{D} = 96$ K: formula (6.16) will not hold here but analysis based on the expressions for T_c also holding for strong coupling leads to the actual value $T_{\mathrm{c},\mathrm{Pb}} = 7.2$ K (see, e.g., book [68]). For hypothetical metallic hydrogen, where $\theta_\mathrm{D} \sim 3000$ K, one can, however, expect the values $T_\mathrm{c} \sim 200\text{–}300$ K but it was not until 1986 that materials with $T_c > 24$ K were created (for Nb_3Ge synthesized in 1973 the temperature was $T_c = 23.2$–24 K). That is why the opinion that estimate (6.21) was valid became widespread.

Getting somewhat ahead, I should note that, for high-temperature cuprates, the electron–phonon coupling is strong and the Debye temperature is high (e.g. $\lambda \approx 2$, $\theta_D \approx 600$ K; see papers [69–71]). It is generally clear that the electron–phonon mechanism can, in principle, account for high temperature superconductivity (HTSC), too, at least for $T_c \lesssim 200$ K (e.g., for $\lambda \approx 2$ and $\theta_D \approx 1000$ K we already have $T_c \approx 200$ K; see later). However, the temperature T_c is only one of the characteristics of a superconductor and what has been said does not guarantee that the phonon mechanism is the basic one responsible for superconductivity in the known HTSC cuprates. Moreover, in the framework of the phonon mechanism, it is apparently not easy to explain the d-pairing and the high value of $2\Delta(0)/k_BT_c$. Historically, the search for HTSCs by creating materials with simultaneously high θ_D and λ values was at one time unpopular obviously for fear that the lattice would be unstable upon strong coupling.

In any case, the search for HTSCs went at first in a different direction (I shall not mention purely empirical attempts). Namely, as is already apparent from (6.16) and from the aforesaid, to increase T_c one can raise the temperature θ. But this need not necessarily be the electron–phonon interaction and any mechanism that provides electron pairing will do. Not phonons but bound electrons which, of course, also interact with the conduction electrons, may, in principle, be responsible for superconductivity. As far as I know, W Little was the first to pay attention to such a possibility in 1964 [72]. Specifically, he considered a metallic (i.e. conducting) quasi-one-dimensional thread or 'spine' on the side of (or, more precisely, around) which there are 'polarizers', i.e. some molecules interacting with the conduction electrons in the quasi-one-dimensional thread. This interaction (which is obviously a Coulomb interaction by nature) is the one to provide pairing. In the same year, 1964, D A Kirzhnits and I discussed the possibility of the existence of two-dimensional superconductors [73]. That is why it was natural that after the appearance of paper [72] I proposed [74] something analogous but quasi-two-dimensional instead of quasi-one-dimensional, i.e. a metallic film with dielectric layers on both sides (a 'sandwich').

The proposed mechanism of superconductivity, in general, and under the conditions of papers [72, 74], in particular, can be called the exciton mechanism (or, more precisely, the electron–exciton mechanism) meaning that phonons in this case are replaced by electronic excitons, i.e. excitations in a system of bound electrons. Briefly speaking, this is the electronic mechanism of superconductivity. The typical exciton energy $\hbar\omega_{ex} = k_B\theta_{ex}$ of the order of 0.1–1 eV corresponds to the temperature $\theta_{ex} \sim 10^3$–10^4 K. Hence, substituting in the BCS formula $\theta \sim \theta_{ex}$ when $\lambda = 1/3$, we obtain

$T_c \sim$ 50–500 K. They say that 'paper will withstand anything' but how such possibilities could be realized was unclear and remains unclear to the present date. Nevertheless, many points had been clarified and discussed [68, 75] before the discovery of HTSCs in 1986–87 [76, 77] (the history of the early HTSC research is elucidated in papers [35, 78]).

Of what became clear 'yesterday' I shall only linger on two things. First, in quasi-one-dimensional systems fluctuations are particularly large, which causes the lowering of T_c (in a strictly one-dimensional system generally $T_c = 0$). From this point of view, quasi-two-dimensional systems, i.e. 'sandwiches' and layered compounds are much more advantageous. This conclusion [74, 75, 79, 68] has been completely confirmed because all HTSC cuprates are layered compounds. Second, it was established [68] that the doubts as to the possibility of the existence of HTSCs [80], aroused by the lattice instability, were ungrounded. Specifically, a system (a metal) with a negative dielectric permittivity

$$\varepsilon(0, q) < 0 \qquad q \neq 0 \qquad (6.22)$$

can be perfectly stable. Here $\varepsilon(\omega, q)$ is the longitudinal permittivity for the frequency ω and the wavevector q. When $\varepsilon < 0$, the Coulomb interaction $U = e^2/\varepsilon r$ obviously corresponds to attraction. This is precisely what leads to the electron (quasi-particle) pairing. It is interesting that, for large q values, a negative permittivity is realized in many metals [81, 82] owing to the phonon contribution. Negative ε values due to the electron contribution (mechanism) are only attainable in systems with a strong exchange-correlation interaction (allowance for the role of the local field is necessary). But it should be emphasized that no restrictions are generally known for the use of the electron (exciton) mechanism for raising T_c (we do not of course mean temperatures comparable with the degeneration temperature $\theta_F = E_F/k_B$). For more details concerning this point and the derivation of the stability condition (6.22), see papers [71, 81] and the references therein.

I have permitted myself to discuss here in some more detail the HTSC research carried out by myself and my colleagues working at the P N Lebedev Physical Institute in Moscow [68, 71, 73, 74, 78, 81, 82] not of course to claim priority. We have neither obtained HTSC materials nor given exact prescriptions for the synthesis of such materials which were obtained in 1986–87 [76, 77].[4] However, I do not think that a total disregard (see, in particular, [83]) of everything done earlier in the HTSC research by Little [72], by us in Moscow [35, 68], and by a number of other authors (see, e.g., paper [79] and the references in [78]) is justified. Another standpoint is also possible if cuprates and their superconductivity are assumed

---

[4] I would like to note here that the importance of studying oxides, carbides, and nitrides was pointed out in book [68] and paper [82].

to be something quite special, not related to low-temperature physics. I do not think such an opinion is grounded, although the distinguished position of cuprates has now become particularly obvious (this fact has even been reflected in the title of the book [29]: *Superconductivity of Metals and Cuprates*).

6.4   Today (1987–?)

The study of conventional superconductors has continued on a large scale. Various non-stationary processes, including thermoelectric effects [38, 37, 28, 29], the study of vortices and different vortex structures [28, 58–61, 84, 85, 127], and the co-existence of superconducting and magnetic ordering [86, 142][5] can be mentioned as especially topical problems [109]. However, many other interesting questions might also be mentioned. But in the broad context, the most important current problem is everything related to HTSCs. Their discovery in 1986–87, as is well known, gave rise to a real 'boom', they attracted unprecedented attention, and a huge number of papers began to appear (see, e. g., [83]). But I think nobody could then believe that synthesized oxides—cuprates—would turn out to be so radically different from conventional superconductors. Now it has become clear that HTSC cuprates, although I do not think that they should be separated from other superconducting metals by a Chinese wall, represent a manifestly distinguished class of superconductors. Their properties are elucidated in monographs [28, 29] devoted to superconductivity, in special collections of papers [87, 88] and in numerous reviews (I shall only mention a few of them [89–93]; see also a number of papers cited later and in the subsequent article 7).

In spite of the fact that HTSC materials have been investigated for 13 years and much effort has been made (tens of thousands of publications have appeared), the picture in the early 2000 remains on the whole fairly cloudy[6]. This is largely due to the complexity of cuprate structure and, mainly, to the difficulties in obtaining perfect single crystals and controlling the degree of doping, homogeneity of samples, etc. That is why a number of experimental results turned out to be unreliable or insufficiently clear. I do not even want to try to outline the present situation. But, at the same time, I cannot but mention the perfection of some of the applied experimental methods, for instance observations, using a scanning tunnel microscope, of the electron density distribution near individual impurity atoms in HTSC cuprates [94]. See also papers [153, 154]. In respect of the theory, it suffices to say that there exists no generally accepted view of the mechanism of superconductivity lead-

---

[5] See also several papers in 2000 *J. Superconductivity* **13** nos 5, 6.
[6] The situation had not changed by May 2004.

ing to high T_c values in cuprates. Here, I shall restrict myself to a few remarks.

The great success of the BCS theory has led to the enduring dominance of the ideology associated with this theory. Thus, at the early stages, the HTSC problem was discussed in the framework or, better to say, on the basis of the conceptions of the BCS theory and its extension to the case of strong coupling. Indeed, there were attempts to raise T_c by increasing the temperature θ in the BCS formula (6.16) through a replacement of $\theta \sim \theta_D$, which holds within the phonon mechanism, by $\theta \sim \theta_{ex}$ for the exciton mechanism of superconductivity. Another way, which is also apparent from (6.16), is an increase of the coupling constant λ_{eff} in passing over to strong coupling, when in the simplest version we have

$$T_c = \theta \exp\left\{ -\frac{1+\lambda}{\lambda - \mu^*} \right\} \qquad (6.23)$$

where $\lambda - \mu^* = \lambda_{eff}$ is the coupling constant involved in the BCS formula (6.16), λ is the strength of the coupling due to the phonon interaction (the phonon mechanism) or excitons (the exciton mechanism), and $\mu^* = \mu[1 + \mu \ln(\theta_F/\theta)]^{-1}$ reflects the role of Coulomb repulsion (for more details, see, e.g., books [29, 68]). If, in (6.23), we put $\mu^* = 0.1$, then for $\lambda = 3$ we already have $T_c = 0.25\theta$ and, for the phonon mechanism with $\theta = \theta_D = 400$ K, we obtain $T_c = 100$ K. This example only demonstrates that for cuprates with $\theta_D \sim 600$ K even for $\lambda = 2$ the temperature is already $T_c \approx 130$ K. Thus, to obtain values $T_c \sim 100$ K in cuprates is no problem in itself. But this does not prove that in cuprates we are dealing with the phonon mechanism of superconductivity because T_c is only one of the characteristic quantities. The behavior of cuprates in the superconducting state, for example the observed ratio $2\Delta(0)/k_B T_c$, does not allow us to identify the mechanism of superconductivity in cuprates as purely phonon. At the same time, a whole number of typical features in the behavior of cuprates in the normal state, which are customarily assumed to be specific, can, in fact, be well explained within the phonon mechanism [95, 115]. I refuse to understand how the role of the phonon mechanism in cuprates can be ignored,[7] although this mechanism is not, in fact, the only one to determine all the properties of these substances.

Undoubtedly, the study of HTSC in cuprates made us realize or rather recall that the phonon mechanism and the BCS approach itself are not the only possible ones for the understanding of superconductivity. Indeed, superfluidity in ^{3}He and in neutron stars definitely has nothing in common with the phonon mechanism. Bose–Einstein condensation (BEC) leading

---

[7] A confirmation of the important role of phonons in cuprates is paper [112] reporting a strong isotope effect in one of the HTSC cuprates (not T_c but the temperature T^* at which a so-called pseudo-gap appears in a normal-state underdoped crystal). See, however, [113] and comment 4 at the end of this article.

generally to superfluidity in the case of neutral particles and to superconductivity for charged particles does not depend on the boson formation
mechanism (we mean the formation of bosons as a result of the pairing of
two electrons).[8] In my opinion, it would be correct to call such a mechanism the Schafroth mechanism [49, 96], although Schafroth had a predecessor [48].

For an ideal Bose gas of zero-spin particles of mass m^*, the temperature
of the beginning of BEC is

$$T_c = \frac{3.31\hbar^2 n^{2/3}}{m^* k_B} = 2.9 \times 10^{-11} \left(\frac{m}{m^*}\right) n^{2/3} \text{ (K)} \qquad (6.24)$$

where $m = 9.1 \times 10^{-28}$ g is the free electron mass and n is the concentration
of bosons (in cm^{-3}).

It is interesting that when applied to liquid ^4He, formula (6.24) leads
to the value $T_c = 3.1$ K for the temperature of the λ-point, whereas actually $T_\lambda = 2.17$ K. In view of the fact that liquid helium is rather far
from an ideal gas, this closeness of T_c and T_λ is evidence of the determining role of the Bose statistics of particles for the superfluidity of ^4He.
According to (6.24), even for $n \sim 10^{21}$ cm^{-3} and $m^* \sim m$, we obtain
$T_c \sim 3 \times 10^3$ K, and so, from this point of view, obtaining the values
$T_c \sim 100$ K is no problem. In HTSC cuprates, the coherence lengths are
small and, thus, the pairs are obviously much smaller than the typical sizes
of Cooper pairs in conventional superconductors. Hence, the Schafroth
mechanism or BEC of local pairs in application to HTSC was repeatedly
mentioned from the very beginning (see, e. g., paper [78]) and was then
developed in detail [98]. But such an approach to HTSC cuprates meets
with serious objections, as all the other theories of superconductivity in
cuprates known from the literature. Among them, the spin mechanism is
worthy of note, in which pairing is due to spin interaction (I shall restrict
myself to mentioning the first paper [99] and review [100] in this field). The
exciton mechanism has already been pointed out [68, 72, 74, 75, 78, 79];
the basis of it is the BCS theory. Let us also mention the electron mechanisms [110, 111].

In a real substance, there certainly simultaneously exist electron–
phonon, spin, and electron–electron (electron–exciton) interactions. So,
strictly speaking, consideration may be restricted to one of these interactions only in some limiting cases. For example, in conventional superconductors, the electron–phonon interaction prevails. But, in cuprates,
both electron–phonon and electron–electron (and maybe also spin) interactions are probably significant (in this connection see, in particular, paper [114]).

---

[8] The existence of BEC was pointed out by Einstein as far back as 1925 [97] (see also
[15, 52, 155]).

The possibility of applying the notions of a Fermi-liquid and a BCS-type theory to HTSC cuprates is called into question [101]. The hypothesis of an electron liquid other than a Fermi-liquid is beyond doubt deep and important [102]. It is, in principle, possible that non-Fermi-liquid effects are significant and even decisive in HTSC cuprates. The future will show if this is so. Of particular interest in this respect are the experiments [139] but they need confirmation and verification (in the experiments in [139], electrical and thermal conductivities in cuprates are measured at a low temperature and in a sufficiently strong magnetic field that destroys the superconductivity; however, it remains unclear to me how the effect of this field on the transfer processes is involved). It is of interest to analyze what the thermoelectric effects would be in the case of a non-Fermi-liquid model (see article 7, section 7.5). But I am not afraid to express my intuitive opinion that the resources of the BCS type theory (including its extension to the case of strong coupling) are far from being exhausted. It is, therefore, quite possible that cuprates, too, (to say nothing of fullerenes and perovskite type substances like $Ba_{1-x}K_xBiO_3$) are mostly described using the concepts of a Fermi-liquid, the formation of pairs with charge $2e$ and their collectivization.

Here, however, I proceed to the section 'Tomorrow'. Today, the situation in the field of superconductivity is primarily characterized by obscurity of the picture of HTSC cuprates [115]. This is the main thing now (however, some new interesting issues have appeared, which are mentioned in section 6.6).

6.5   Tomorrow

In the title of the previous section 'Today (1987–?)' there stands a question mark, since as in the sections 'The day before yesterday (1911–41)' and 'Yesterday (1942–86)' some landmarks are understood, for example the discovery of superconductivity in 1911 and the discovery of HTSC materials in 1986–87. For this reason, the landmark of "Today" should not be March of 2000 when this paper is being written, and nor in mid-2004 when its translation is being prepared but it should be some event. What event will it be? It is desirable that this landmark be an insight into the mechanism of superconductivity in HTSC cuprates. So many experimental results on cuprates have been obtained in the past 15 years and the experimental methods employed have so much advanced (I judge, say, by papers [92, 94, 103, 104, 112, 153, 154, 156]), that a certain clarity may be expected in the experiment in the near future. If this happens, the theoretical comprehension will hardly keep us waiting much longer. Now I can only make some remarks.

If the phonon mechanism with strong coupling, though not the only one, is still determining, the value of the critical temperature T_c is unlikely to exceed approximately 200 K. [The maximum temperature attained to date (this happened in 1994) was $T_c \approx 164$ K for the cuprate $HgBa_2Ca_2Cu_3O_{8+x}$ under high pressure; at atmospheric pressure for this material $T_c = 135$ K.] This is clear from expressions of the type (6.23) for T_c and from the fact that $\theta \sim \theta_D \lesssim 10^3$ K. For spin mechanisms, the role of θ is played by Curie or Néel temperatures and they are no higher than 10^3 K either. For the exciton and some related electron mechanisms (the plasmon mechanism is sometimes discussed), $\theta \sim \theta_{ex}$. The natural upper boundary for θ_{ex} is the Fermi temperature $\theta_F = E_F/k_B \lesssim 10^5$ K. The electron mechanism is not known to meet with objections of principal character (see [68, 71] and article 7 of this book), and from this point of view it may 'do its best' under some conditions. And in this case the dream, i.e. room-temperature superconductivity (RTSC) with $T_c \sim 300$–400 K, would come true.

Undoubtedly, many laboratories are engaged in the search for materials with increasingly high T_c values. The lack of advances in this direction over a number of years (since 1994) testifies to the fact that the T_c of cuprates and many other tested compounds can hardly be raised. But the number of possible compounds is huge, which naturally gives hope that a substance with higher T_c values will be found. I believe, as before, that quasi-two-dimensional (layered) structures are promising; for some ideas concerning this issue, see [105]. Wide possibilities for experiments in this field are known [106, 118, 154].

It should be mentioned here that the current situation in solid state theory cannot yet be thought of as satisfactory. The progress made over the past century is great if we look at the distance passed from the idea of electron motion in conductors, suggested by Drude in 1900, up to today's condition of the physics of metals. But, in contrast, the properties of even what seems to be the simplest system, namely metallic hydrogen, still cannot be predicted from 'first principles' [107]. The judgement, sometimes encountered, that almost all the principal things in physics have already been done are simply absurd (see, e.g., [108]). There is no doubt that the theory of many-particle systems faces unsolved problems of great difficulty. Not until it becomes possible to calculate the parameters and characteristics of compounds of any prescribed composition and structure will one be able to think of condensed-matter theory as practically accomplished. Certainly, this also refers to superconductors (true, it should be pointed out that for simple metals like Al and Pb, advances have already been made [95]). It is difficult to say how many decades we shall wait for the achievement of this goal. We now have only one natural landmark in view—the year 2011, the centenary of the discovery of superconductivity. Unfortunately, we are unable to make a definite prognosis even for the decade left before this centenary. But I would not be very much surprised if room-temperature superconductors were created by 2011. This is, however, no more than a dream. But high-temperature superconductivity was also only a dream before 1986.

Supplement[9]

The Conference MTSC 2000 (Major Trends in Superconductivity in the New Millennium) and the Symposium "Itinerant and Localized States in HTSC" held on 1–9 April, 2000, i. e. immediately after the conference, were rather representative assemblies (nearly 130 scientists were present at the conference). Many experimental data were reported and various theoretical issues associated with superconductivity in cuprates and in some other substances were discussed. However, no essentially new insight into the HTSC in cuprates was proposed. It is surprising that the long discussion of the problem did not contribute to the clarification of the theory of cuprate superconductivity. There exist different points of view but I hope a consensus will be reached in the near future. The situation in mid-2000 will largely be elucidated in the Proceedings of the MTSC 2000 Conference (published in 2000 *Journal of Superconductivity* nos 5 and 6) and also in the paper by E G Maksimov [115].

Here I only want to make a few remarks not connected with cuprates. The data [116] reported at the conference testify to the probable existence of superconductivity with $T_c = 91$ K on the surface of the compound WO_3 doped with sodium (Na). We are speaking of strongly diamagnetic (at $T < T_c = 91$ K) small regions localized on the surface. The most significant point here is the high T_c value in the absence of copper. At the same time, it is natural to recall here two-dimensional surface superconductivity which has long been discussed [73, 117]. A two-dimensional conductor may pass over to a superconducting state and there are different possibilities for this. One of them occurs if a substance (a metal, a semiconductor, or a dielectric in the case of volume effects) possesses surface levels. In the case of an appropriate position and occupation of these levels (e. g. Tamm levels [117]), the surface may appear to be metallic and then also superconducting. Another possibility is the coating of a non-superconductive material with a monolayer (of, e. g., CuO_2) which may turn out to be superconductive (the technological process is well known, see [106, 118]).

The report of a possible high-temperature superconductivity in the $WO_3 + Na$ system makes us emphasize once again that there is no reason to assume high-temperature superconductivity (we shall define it as superconductivity with $T_c > T_{b, N_2} = 77.4$ K) to exist in cuprates only. The search for HTSC in various substances has been and is now being con-

---

[9] This supplement was written after the MTSC 2000 Conference, and all this article, 6, presents the discussions given at that Conference.

ducted but the earlier positive results not related to cuprates have not been reproduced and thus confirmed. I am afraid that the fate of the results of experiments with WO_3+Na has been the same, for I had not heard of their confirmation by May, 2004. Nevertheless, I believe that there are insufficient grounds to think of all the reported past observations of HTSC as erroneous. This particularly concerns CuCl (see, for example, [119, 120]; for some other references see also [78]). Since then, the crucial significance in some cases of even a small doping, i. e. the presence of impurities, especially oxygen, has become clear. Moreover, it has become possible to state the appearance of small superconducting regions not only from measurements of diamagnetic susceptibility (see, e. g., [116]). That is why some earlier observations, especially with CuCl, should be repeated. At the same time, the widescale search for HTSC in various substances is particularly promising when a material is created using layer-by-layer sputtering (see [106, 118]). The study of non-damping currents in carbon nanotubes [121] also deserves attention.[2*] (Some new results are also mentioned in comments 3 and 4.)

Comments

1*. This paper was published in 2000 *Usp. Fiz. Nauk* **170** 619 [2000 *Phys. Usp.* **43** 573]. It was the talk delivered on 29 March, 2000 at the Scientific Session of the Division of General Physics and Astronomy of the Russian Academy of Sciences and on 1 April at the MTSC 2000 Conference (Major Trends in Superconductivity in the New Millennium) held in Klosters, Switzerland. The proceedings of this conference, including the present paper, were published in the *Journal of Superconductivity* (2000).

2*. A whole number of interesting results has recently been obtained by G F Zharkov [127] (see also [166, 167]). I also take the opportunity to mention the second edition of the book by V V Schmidt [128] which is a good introduction to the physics of superconductors.

3*. The resumption of interest in conductivity and even a possible superconductivity in sodium–ammonium solutions [122] seems to be well justified. Recall that such studies were begun by Ogg as far back as 1946 [48]. A possible superconductivity in nanotubes is discussed, along with [121], also in paper [123]. I would like to mention the unexpected discovery of superconductivity in MgB_2 with $T_c \approx 40$ K [124–126]. Although this substance does not belong, according to the terminology adopted here, to HTSC materials, this discovery is significant because the long known metal proved to be superconducting and, in addition, to have a rather high critical temperature. Incidentally, the T_c of MgB_2 is even slightly higher than that of the first superconductor $La_{2-x}Ba_xCuO_4$ [76] which is usually thought of as an HTSC (what is more important, of course, is that MgB_2 is not a cuprate). In MgB_2, the isotope effect is clearly pronounced (I mean in comparison with $Mg^{10}B_2$ and $Mg^{11}B_2$ [125]) and it is obviously a "conventional" superconductor (although a layered one) with a phonon mechanism of superconductivity. Investigations of MgB_2 have attracted much attention [124–126, 132, 152, 157–161, 163–165].

4*. A vivid confirmation of the role of the electron–phonon interaction in HTSC seemed to have been obtained in [129] (see, however, [145]). In 2001, a number of papers appeared devoted to the coexistence of superconductivity and ferromagnetism (see, in particular, [130, 131, 142, 146, 147, 162]) and to many other problems, for instance, superconductivity in organic metals [133, 144] and to the study of the "pseudogap" [134–136, 148, 156]. The origin of the "pseudogap" (it emerges in underdoped cuprates, i.e. when cuprate doping is less than needed to provide the maximum T_c value) remains unclear. The issue of the non-uniformity of the metallic state, the so-called stripes which are observed in a number of cuprates (especially also in regions of underdoping) [137, 138, 136] is also unclear. On the whole, some lull was observed in the study of superconductivity in 2001 and 2002 (see, however, papers [124–126, 168]). Nevertheless, there is an impression that it is most likely to be the calm before the storm. The impression is grounded because an intensive study of superconductors is under way. This, in particular, was demonstrated at the Scientific Session of the Division of General Physics and Astronomy of the Russian Academy of Sciences held on 19 December, 2001 [140] and in papers [141–152].

The subsection 'Tomorrow' in the present chapter undoubtedly arouses dissatisfaction, although I hope I have described the current situation in mid-2003 or even mid-2004 adequately. It would be pertinent to stress that future advances in the study of superconductivity cannot be associated with the creation of some giant installations such as the accelerator LHC or the thermonuclear reactor ITER. Hence, new significant results may be expected any moment.

References[10]

[1] Kamerlingh Onnes 1911 *Commun. Phys. Lab. Univ. Leiden* **119b** (cited in the following as *Commun. Leiden*)
[2] de Nobel J 1996 *Phys. Today* **49**(9) 40
[3] Kamerlingh Onnes H 1911 *Commun. Leiden* **120b; 122b**
[4] Kamerlingh Onnes H 1911 *Commun. Leiden* **124c** (This paper is placed as an appendix in paper [6])
[5] Kamerlingh Onnes H 1913 *Commun. Leiden* **133a, b, c, d**

---

[10] This list of references does not at all claim completeness even in respect of the problems discussed here. Moreover, in some cases, the papers cited accidentally appeared in the field of vision of the author but they contain a large number of other references. In particular, the citation of many papers by the present author serve only as information and are not intended to show some priority claims.

[6] Ginzburg V L 1992 *Sverkhprovodimost'*: *Fiz. Khim. Tekhn.* **5** 1 [1992 *Superconductivity: Phys. Chem. Technol.* **5** 1]

[7] Kamerlingh Onnes H 1913 *Commun. Leiden. Suppl.* **34b** (This is the Nobel lecture. It is undoubtedly published elsewhere)

[8] Ginzburg V L 1946 *Sverkhprovodimost'* (*Superconductivity*) (Moscow: Izd. AN SSSR)

[9] Kamerlingh Onnes H and Tuyn W 1922 *Commun. Leiden* **160a**

[10] Shoenberg D 1965 *Superconductivity* (Cambridge: Cambridge University Press) (The first edition of this book was published in 1938 (in the series Cambridge Physical Tracts) and the second edition in 1952 (in the series Cambridge Monographs on Physics) [Translated into Russian 1955 *Sverkhprovodimóst'* (Moscow: IL)]

[11] Kamerlingh Onnes H 1924 *Commun. Leiden Suppl.* **50a**

[12] Meissner W and Ochsenfeld R 1933 *Naturwissensch.* **21** 787

[13] Kamerlingh Onnes H 1911 *Commun. Leiden* **119a**; *Proc. R. Acad. Amsterdam* **13** 1903

[14] Keesom W H 1942 *Helium* (Amsterdam: Elsevier) [Translated into Russian 1949 (Moscow: IL)]

[15] London F 1954 *Superfluids* vol 2 *Macroscopic Theory of Superfluid Helium* (New York: Wiley)

[16] Kapitza P L 1938 *Nature* **141** 74

[17] Allen F and Missner A D 1938 *Nature* **141** 75; 1939 *Proc. R. Soc. London* **172A** 467

[18] Kapitza P L 1941 *Zh. Eksp. Teor. Fiz.* **11** 581; *J. Phys. USSR* **4** 181; **5** 59

[19] Landau L D 1941 *Zh. Eksp. Teor. Fiz.* **11** 592; *J. Phys. USSR* **5** 71

[20] Sommerfeld A and Bethe H 1933 in *Handbuch der Physik* vol 24, pt 2 (Berlin: Springer) p 333 [Translated into Russian 1938 *Elektronnaya Teoriya Metallov* (Leningrad: ONTI)]

[21] Wilson A H 1936 *The Theory of Metals* (Cambridge: Cambridge University Press); 1953 2nd edn (Cambridge: Cambridge University Press) [Translated into Russian 1941 *Kvantovaya Teoriya Metallov* (*Quantum Theory of Metals*) (Moscow, Leningrad: OGIZ)]

[22] Lifshitz E M and Pitaevskii L P 1978 *Statisticheskaya Fizika* (*Statistical Physics*) Part II *Teoriya Kondensirovannogo Sostoyaniya* (*The Theory of the Condensed State*) (Moscow: Nauka) [Translated into English 1980 (New York: Pergamon Press)]

[23] Hoddeson L, Baym G, and Eckert M 1987 *Rev. Mod. Phys.* **59** 287

[24] Einstein A 1922 *Gedankbook Kamerlingh Onnes* ed C A Crommelin (Leiden) p 429 [Translated into Russian Einstein A 1966 *Sobranie Nauchnykh Trudov* (*Collection of Scientific Papers*) vol 3 (Moscow: Nauka) p 432]

[25] Gorter C J and Casimir H 1934 *Physica* **1** 306

[26] Landau L D and Lifshitz E M 1992 *Elektrodinamika Sploshnykh Sred* (*Electrodynamics of Continuous Media*) (Moscow: Nauka) See ch XI [Translated into English 1984 (New York: Pergamon Press)]

[27] Tilley D R and Tilley R 1986 *Superfluidity and Superconductivity* 2nd edn (Bristol: A Hilger)

[28] Tinkham M 1996 *Introduction to Superconductivity* 2nd edn (New York: McGraw-Hill)

[29] Waldram J R 1996 *Superconductivity of Metals and Cuprates* (Bristol: Institute of Physics)

[30] Gorter C J and Casimir H 1934 *Phys. Z.* **35** 963

[31] London F and London H 1935 *Proc. R. Soc. London* **149A** 71; *Physica* **2** 341

[32] Becker R, Heller G and Sauter F 1933 *Z. Phys.* **85** 772

[33] Ginzburg V L and Zharkov G F 1978 *Usp. Fiz. Nauk* **125** 19 [1978 *Sov. Phys. Usp.* **21** 381]

[34] Ginzburg V L 1946 *Zh. Eksp. Teor. Fiz.* **16** 87; 1945 *J. Phys. USSR* **9** 305

[35] Ginzburg V L 1997 *Usp. Fiz. Nauk* **167** 429 [1997 *Phys. Usp.* **40** 407]; see also article 7 in this book

[36] Ginzburg V L 1944 *Zh. Eksp. Teor. Fiz.* **14** 177; *J. Phys. USSR* **8** 148

[37] Ginzburg V L 1998 *Usp. Fiz. Nauk* **168** 363 [1998 *Phys. Usp.* **41** 307]

[38] Gulian A M and Zharkov G F 1999 *Nonequilibrium Electrons and Phonons in Superconductors* (New York: Kluwer Academic, Plenum)

[39] Karasik V R (ed) 1975 *Sverkhprovodimost'. Bibliograficheskii Ukazatel' 1911–1970* [*Superconductivity. Bibliographical Index* 1911–1970] (Moscow: Nauka)

[40] Appleyard E *et al* 1939 *Proc. R. Soc. London* **172A** 540

[41] Ginzburg V L and Landau L D 1950 *Zh. Eksp. Teor. Fiz.* **20** 1064 [Soviet journals were not translated into English in that period and *Journal of Physics USSR* was closed in 1947. The English translation of the article appeared in 1965 *Collected Papers of L D Landau* (Oxford: Pergamon Press) p 546]

[42] Pippard A B 1953 *Proc. R. Soc. London* **216A** 547

[43] Bardeen J, Cooper L N and Schrieffer J R 1957 *Phys. Rev.* **108** 1175

[44] Bardeen J 1956 in *Kältephysik* (*Handbuch der Physik* **15**) ed S von Flügge (Berlin: Springer) p 274

[45] Ginzburg V L 1955 *Zh. Eksp. Teor. Fiz.* **29** 748 [1956 *Sov. Phys. JETP* **2** 589]

[46] Gorkov L P 1958 *Zh. Eksp. Teor. Fiz.* **36**; 1959 **37** 1407 [1959 *Sov. Phys. JETP* **9** 1364; 1960 **10** 998]

[47] Ginzburg V L 1952 *Usp. Fiz. Nauk* **48** 26; 1953 *Fortschr. Phys.* **1** 101

[48] Ogg R A Jr 1946 *Phys. Rev.* **69** 243; **70** 93

[49] Schafroth M R 1954 *Phys. Rev.* **96** 1149, 1442; 1955 **100** 463

[50] Cooper L N 1956 *Phys. Rev.* **104** 1189

[51] Ginzburg V L 1960 *Fiz. Tverd. Tela* **2** 2031 [1961 *Sov. Phys. Solid State* **2** 1824]

[52] Landau L D and Lifshitz E M 1955 *Statisticheskaya Fizika* (*Statistical Physics*) pt I (Moscow: Fizmatlit) ch 14 [Translated into English 1980 (Oxford: Pergamon Press)]

[53] Ginzburg V L 1952 *Zh. Eksp. Teor. Fiz.* **23** 236

[54] Sigrist M and Ueda K 1991 *Rev. Mod. Phys.* **63** 239

[55] Abrikosov A A 1957 *Zh. Eksp. Teor. Fiz.* **32** 1442 [1957 *Sov. Phys. JETP* **5** 1174]

[56] Weinan E 1994 *Phys. Rev.* **50B** 1126; *Physica* D **77** 383

[57] Sakaguchi H 1995 *Prog. Theor. Phys.* **93** 491

[58] Bethuel F, Brezis H and Helein F 1994 *Ginzburg–Landau Vortices* (Boston:

Birkhauser)

[59] Ovchinnikov Yu N 1999 *Zh. Eksp. Teor. Fiz.* **115** 726 [1999 *JETP* **88** 398]

[60] Aranson I S and Pismen L M 2000 *Phys. Rev. Lett.* **84** 634

[61] Zharkov G F, Zharkov V G and Zvetkov A Yu 2000 *Phys. Rev. B* **61** 12293

[62] Bogolyubov N N 1958 *Zh. Eksp. Teor. Fiz.* **34** 58 [1958 *Sov. Phys. JETP* **7** 41]

[63] Valatin J 1958 *Nuovo Cimento* **7** 843

[64] Kuper C 1959 *Adv. Phys.* **8**(29) 1

[65] Bogolyubov N N (ed) 1960 *Teoriya Sverkhprovodimosti* [*Theory of Superconductivity. Collected papers*] (Moscow: IL)

[66] Josephson B D 1962 *Phys. Lett.* **1** 251

[67] Eliashberg G M 1960 *Zh. Eksp. Teor. Fiz.* **38** 966; **39** 1437 [*Sov. Phys. JETP* 1960 **11** 696; 1961 **12** 1000]

[68] Ginzburg V L and Kirzhnits D A (eds) 1982 *High-Temperature Superconductivity* (New York: Consultants Bureau)

[69] Ginzburg V L and Maksimov E G 1994 *Physica C* **235–240** 193
    Ginzburg V L 1993 *Physica C* **209** 1

[70] Maksimov E G 1995 *J. Supercond.* **8** 433

[71] Ginzburg V L and Maksimov E G 1992 *Sverkhprovodimost': Fiz., Khim., Tekhn.* **5** 1543 [1992 *Superconductivity: Phys., Chem., Technol.* **5** 1505]

[72] Little W A 1964 *Phys. Rev.* **134** A1416

[73] Ginzburg V L and Kirzhnits D A 1964 *Zh. Eksp. Teor. Fiz.* **46** 397 [1964 *Sov. Phys. JETP* **19** 269]

[74] Ginzburg V L 1964 *Zh. Eksp. Teor. Fiz.* **47** 2318 [1964 *Sov. Phys. JETP* **20** 1549]; 1964 *Phys. Lett.* **13** 101

[75] Little W (ed) 1970 *Proc. Int. Conf. on Organic Superconductors* (New York: Wiley)

[76] Bednorz J G and Müller K A 1986 *Z. Phys.* **64B** 189

[77] Wu M K *et al* 1987 *Phys. Rev. Lett.* **58** 908

[78] Ginzburg V L 1989 *Prog. Low Temp. Phys.* **12** 1

[79] Allender D and Bardeen J 1973 *Phys. Rev.* **7B** 1020
    Allender D, Bray J, and Bardeen J 1973 *Phys. Rev.* **8B** 4443

[80] Cohen M L and Anderson P W 1972 *Superconductivity in d- and f- Band Metals* (*American Institute of Physics Conf. Ser.* 4) ed D H Douglass (New York: American Institute of Physics) p 17

[81] Dolgov O V, Kirzhnits D A and Maksimov E G 1981 *Rev. Mod. Phys.* **53** 81

[82] Dolgov O V and Maksimov E G 1982 *Usp. Fiz. Nauk* **138** 95 [1982 *Sov. Phys. Usp.* **25** 688]

[83] Nagaoka Y (ed) 1987 *Proc. 18th Int. Conf. on Low Temperature Physics, Kyoto, Japan, 1987* (*Japan J. Appl. Phys.* **26**, Suppl. 26-3) pt 3

[84] Crabtree G W and Nelson D R 1997 *Phys. Today* **50**(4) 38

[85] Palacios J J 2000 *Phys. Rev. Lett.* **84** 1796

[86] Amici A, Thalmeier P, and Fulde P 2000 *Phys. Rev. Lett.* **84** 1800

[87] Lynn J W *et al* (eds) 1990 *High Temperature Superconductivity* (New York: Springer)

[88] Ginsberg D M (ed) *Physical Properties of High Temperature Superconductors* (Singapore: World Scientific) [Several volumes, the first of which

was published in 1989]

[89] Ruvalds J 1996 *Supercond. Sci. Technol.* **9** 905

[90] Ford P J and Saunders G A 1997 *Contemp. Phys.* **38** 1

[91] Goldman A M and Markovic N 1998 *Phys. Today* **51**(11) 39

[92] Timusk T and Statt B 1999 *Rep. Prog. Phys.* **62** 61

[93] Batlogg B and Varma C 2000 *Phys. World* **13**(2) 33

[94] Balatsky A V 2000 *Nature* **403** 717
       Pan S H *et al* 2000 *Nature* **403** 746

[95] Maksimov E G, Savrasov D Yu and Savrasov S Yu 1997 *Usp. Fiz. Nauk*
       167 353 [1997 *Phys. Usp.* **40** 337]

[96] Schafroth M R, Burler S T and Blatt J M 1957 *Helv. Phys. Acta* **30** 93

[97] Einstein A 1925 *Sitzungsber. Preuss. Akad. Wiss., Phys.-Math. Kl.*, 3
       [Translated into Russian Einstein A 1965 *Sobranie Nauchnykh Trudov
       (Collection of Scientific Papers)* vol 3 (Moscow: Nauka) p 489]

[98] Alexandrov A S and Mott N F 1994 *High Temperature Superconductors
       and Other Superfluids* (London: Taylor & Francis)

[99] Akhiezer A I and Pomeranchuk I Ya 1959 *Zh. Eksp. Teor. Fiz.* **36** 859 [1959
       Sov. Phys. JETP **9** 605]

[100] Izyumov Yu A 1999 *Usp. Fiz. Nauk* **169** 225 [1999 *Phys. Usp.* **42** 215]

[101] Anderson P W 1997 *The Theory of Superconductivity in the High-T*$_c$
       Cuprates (Princeton, NJ: Princeton University Press)

[102] Schofield A J 1999 *Contemp. Phys.* **40** 95

[103] Grüninger M *et al* 2000 *Phys. Rev. Lett.* **84** 1575

[104] Lake B *et al* 1999 *Nature* **400** 43

[105] Geballe T H and Moyzhes B Y *Proc. Int. Conf. on Materials and Mecha-
       nisms of Superconductivity High Temperature Superconductors VI*; 2000
       Physica C **341–348** 1821

[106] Bozovic I and Eckstein J N 1996 See [88] vol 5

[107] Maksimov E G and Shilov Yu I 1999 *Usp. Fiz. Nauk* **169** 1223 [1999 *Phys.
       Usp.* **42** 1121]

[108] Ginzburg V L 1999 *Usp. Fiz. Nauk* **169** 419 [1999 *Phys. Usp.* **42** 353]; see
       also article 1 in this book

[109] Tallon J 2000 *Phys. World* **13**(3) 27

[110] Kocharovsky V V and Kocharovsky Vl V 1992 *Physica C* **200** 385

[111] Belyavsky V I, Kapaev V V and Kopaev Yu V *Proc. Int. Conf. on Materials
       and Mechanisms of Superconductivity High Temperature Superconductors
       VI*; 2000 *Physica C* **341–348** 185
       Belyavskii V I, Kapaev V V and Kopaev Yu V 2000 *Zh. Eksp. Teor. Fiz.*
       118 941 [2000 *JETP* **91** 817]
       Belyavskii V I and Kopaev Yu V 2000 *Pis'ma v JETP* **72** 734 [2000 *JETP
       Lett.* **72** 511]; 2001 **73** 87 [2000 *JETP Lett.* **73** 82]

[112] Tamprano D R *et al* 2000 *Phys. Rev. Lett.* **84** 1990

[113] Ohno T and Asayama K 1999 *Phys. Lett.* **258A** 367

[114] Bill A, Morawitz H and Kresin V Z 2000 *J. Supercond.* **13** 907

[115] Maksimov E G 2000 *Usp. Fiz. Nauk* **170** 1033 [2000 *Phys. Usp.* **43** 965]

[116] Reich S and Tsabba Y 1999 *Eur. Phys. J.* **B9** 1
       Reich S *et al* 2000 *J. Supercond.* **13** 855

[117] Ginzburg V L 1989 *Phys. Scripta* **T27** 76

178   References

[118] Bozovic I 2001 *IEEE Trans. Appl. Supercond.* **11** 2686
[119] Chu C W *et al* 1978 *Phys. Rev.* **18B** 2116
[120] Brandt N B *et al* 1978 *Pis'ma Zh. Eksp. Teor. Fiz.* **27** 37 [1978 *JETP Lett.* **27** 33]
[121] Tsebro V I, Omel'yanovskii O E, and Moravskii A P 1999 *Pis'ma Zh. Eksp. Teor. Fiz.* **70** 457 [1999 *JETP Lett.* **70** 462]; see also Tsebro V I and Omel'yanovskii O E 2000 *Usp. Fiz. Nauk* **170** 906 [2000 *Phys. Usp.* **43** 847]
[122] Edwards P P 2000 *J. Supercond.* **13** 933
[123] Pokropivny V V 2000 *J. Supercond.* **13** 607
[124] Nagamatsu J *et al* 2001 *Nature* **410** 63
[125] Bud'ko S L *et al* 2001 *Phys. Rev. Lett.* **86** 1877
[126] Uchiyama H *et al* 2002 *Phys. Rev. Lett.* **88** 157002
[127] Zharkov G F 2002 *Zh. Eksp. Teor. Fiz.* **122** 600 [2002 *JETP* **95** 517]; 2002 *J. Low Temp. Phys.* **128** 87; 2003 **130** 45
[128] Schmidt V V 2000 *Vvedenie v Fiziku Sverkhprovodnikov* [*Introduction to the Physics of Superconductors*] (Moscow: Izd. MTsNMO) [Translated into English 1997 *The Physics of Superconductors: Introduction to Fundamental and Applications* (Berlin: Springer)]
[129] Lanzara A *et al* 2001 *Nature* **412** 510
     Allen P B 2001 *Nature* **412** 494
[130] Pfleiderer C *et al* 2001 *Nature* **412** 58
[131] Aoki D *et al* 2001 *Nature* **413** 613
[132] Bohnen K-P, Heid R and Renker B 2001 *Phys. Rev. Lett.* **86** 5771
[133] Singleton J 2001 *Phys. World* **14**(7) 23
[134] Chaban I A 2001 *J. Supercond.* **14** 481
[135] Vyas A and Lam C C 2001 *J. Supercond.* **14** 487
[136] Sadovskii M V 2001 *Usp. Fiz. Nauk* **171** 539 [2001 *Phys. Usp.* **44** 515]
     Loktev V M, Quick R M, and Sharapov S G 2001 *Phys. Rep.* **349** 1
[137] Nagaev E L 1995 *Usp. Fiz. Nauk* **165** 529 [1995 *Phys. Usp.* **38** 497
[138] Zaanen J 1999 *Physica C* **317–318** 217; *Science* **286** 251
[139] Hill R W *et al.* 2001 *Nature* **414** 711; see also 2001 *Nature* **414** 697
[140] 2002 *Usp. Fiz. Nauk* **172** 701 [2002 *Phys. Usp.* **45** 645]
[141] Babaev E 2002 *Phys. Rev. Lett.* **89** 067001
[142] Flouqnet J and Buzdin A 2002 *Phys. World* **15**(1) 41
[143] Canfield P C and Budko S L 2002 *Phys. World* **15**(1) 29
[144] Singlton J and Mielke C 2002 *Phys. World* **15** 35
[145] Wang N L *et al* 2002 *Phys. Rev. Lett.* **89** 087003
[146] Walker M B and Samokhin K V 2002 *Phys. Rev. Lett.* **88** 207001
[147] Singh D J and Mazin I I 2002 *Phys. Rev. Lett.* **88** 187004
[148] Kaminski A *et al* 2002 *Nature* **416** 610
[149] Zhu Y *et al* 2002 *Phys. Rev. Lett.* **88** 247002
[150] Blanchard S *et al* 2002 *Phys. Rev. Lett.* **88** 177201
[151] Arumugam S *et al* 2002 *Phys. Rev. Lett.* **88** 247001
[152] Choi H J *et al* 2002 *Nature* **418** 758
     Pickett W 2002 *Nature* **418** 733
[153] McElroy K *et al* 2003 *Nature* **422** 592
[154] Bozovic I *et al* 2003 *Nature* **422** 873

[155] Pitaevskii L and Stringari S 2003 *Bose–Einstein Condensation* (Oxford: Clarendon Press)

[156] Alff L *et al* 2003 *Nature* **422** 698

[157] Seneor P *et al* 2002 *Phys. Rev.* **65B** 012505

[158] Gonnelli R S *et al* 2002 *Phys. Rev. Lett.* **89** 247004

[159] Iavarone M *et al* 2002 *Phys. Rev. Lett.* **89** 187002

[160] Eskildsen M R *et al* 2002 *Phys. Rev. Lett.* **89** 187003

[161] Shukla A *et al* 2003 *Phys. Rev. Lett.* **90** 095506

[162] Ovchinnikov S G 2003 *Usp. Fiz. Nauk* **173** 27 [2003 *Phys. Usp.* **46** 21]

[163] Kim M-S *et al* 2002 *Phys. Rev.* **66B** 064511

[164] Kotegawa H *et al* 2002 *Phys. Rev.* **66B** 064516

[165] Souma S *et al* 2003 *Nature* **423** 65

[166] Hove J, Mo S, and Sudbø A 2002 *Phys. Rev.* **66B** 064524

[167] Li D and Rosenstein B 2003 *Phys. Rev. Lett.* **90** 167004

[168] Zhou X J *et al* 2003 *Nature* **423** 398

Article 7

Superconductivity and superfluidity (what was done and what was not)[1*]

7.1 Introduction: early works

I, the author of the present paper, am 80 years old (the paper was written in 1997) and I cannot hope to obtain new important scientific results. At the same time, I feel I need to summarize my work of over 50 years. I do not mean now my work in general (I have been engaged in solving quite a variety of physical and astrophysical problems, see [1, p 312]) but my activity in the field of superconductivity and superfluidity. In general, it is not traditional to write such papers. In my opinion, however, this comes from a certain prejudice. In any case, I decided to try and write such a paper, something like a scientific autobiography, but devoted only to two related problems—superconductivity and superfluidity. I may say that it is not associated with some priority or any other claims: it is only a desire to continue my work, though in an unusual form. I leave it to the reader to judge whether this attempt has been pertinent and successful.

I began working, i.e. obtaining some results in physics in 1938–39, when I graduated from the Physics Faculty of Moscow State University. Before the War, i.e. up to the mid-1941, I was engaged in classical and quantum electrodynamics, as well as the theory of higher-spin particles. We somehow felt that war would break out and were scared of it but were unprepared and lived with the hope that the danger would pass. I am not going to generalize but this atmosphere reigned in the Department of Theoretical Physics of FIAN (P N Lebedev Physical Institute of the USSR Academy of Sciences). When it did not pass by, we began looking, while waiting for the call-up or some other changes in our lives, for an application of our abilities which might be of use for defense. I, for one, was engaged in problems of radio-wave propagation in the ionosphere (see [1, 2]). But these and similar subjects remained, at least in my case, far from finding an application in defense. Therefore, I went on working in various fields under these or other influences.

The most important such influence, not to mention the continuation of research in the field of relativistic theory of spin particles, was exerted by L D Landau. In 1939, after a year's confinement in prison, Landau started working on the theory of superfluidity of helium II.[1] I was present, if I am not mistaken, in 1940, at Landau's talk devoted to this theory (the corresponding paper [4] was submitted for publication in 15 May, 1941). At the end of the paper [4], he also considered superconductivity interpreted as the superfluidity of an electron liquid in a metal. I do not know whether an assertion of the kind had ever been expressed before but it is hardly probable. (Some hint in this respect was made in [5].) The point is that superfluidity in the proper sense of the word was discovered only in 1938 independently and simultaneously by Kapitza [5] and Allen and Misener [6].

We mean here a frictionless flow through capillaries and gaps. As to the anomalous behavior of liquid helium (^{4}He) below the λ-point, i. e. at a temperature $T < T_\lambda = 2.17$ K, the study of this issue began, in effect, in 1911. Precisely in the year when superconductivity was discovered [7] (for more details, see [8, 9]; paper [7] is also included in [9] as an appendix), Kamerlingh Onnes reported a helium density maximum at T_λ [10, 11]. It was only in 1928 that the existence of two phases—helium I and helium II—became obvious and, in 1932, a clear λ-shaped curve for the temperature dependence of the specific heat near the λ-point was obtained. The superhigh thermal conductivity of helium II was discovered by W Keezom and A Keezom (see the references in [11, 12]) in 1936 and, finally, superfluidity was revealed [5, 6] in 1938. One can thus say that it took 27 years (from 1911 to 1938) to discover superfluidity [127]. Such a long process is in obvious contrast with the discovery of superconductivity which was practically a one-stroke occurrence [7] (for details, see [8, 9] and article 6 in this volume). One can hardly doubt that the reason lies in the different methods. Superconductivity was discovered when the electrical resistance of a wire (or, more precisely, a capillary filled with mercury) was being measured. It is a much more difficult task to investigate the character of liquid flow (concretely, helium II) through gaps and capillaries and, besides, one must hit upon the idea of carrying out such experiments.

At the same time, the origin of superfluidity remained obscure. Landau believed [4] that the responsibility rested with the spectrum of 'elementary excitations' in a liquid while Bose statistics and Bose–Einstein condensation had nothing to do with it. F London and L Tisza [12] on the contrary, associated superfluidity with Bose–Einstein condensation. The validity of the latter opinion became obvious in 1949 after liquid ^{3}He with atoms obey-

---

[1] As is well known, P L Kapitza's plea for Landau's discharge from prison was motivated by the very wish to have his assistance in the field of superfluidity theory (see [3, p 345]).

ing Fermi statistics, the properties differing radically from those of liquid ^{4}He, had been obtained. Theoretically, the same conclusion was drawn by Feynman (see [13]). But nothing could be derived from it in respect of superconductivity because electrons obey Fermi statistics. As we know today, the solution of the problem (or rather the puzzle) lies in the fact that electrons in a superconductor form 'pairs' with zero spin. Such pairs can undergo Bose–Einstein condensation with which the transition to a superconducting state is associated. My fairly modest contribution to this subject consists in pointing out that, in a Bose gas of charged particles, the Meissner effect must be observed [14]. The idea of 'pairing' itself did not occur to me. To the best of my knowledge, Ogg was the first to suggest it in 1946 [15]. This viewpoint was supported and further developed by Schafroth [16]. However, the cause and mechanism of pairing remained absolutely vague; and it was only in 1956 that Cooper [17] pointed out a concrete mechanism of pairing in a Fermi gas with attracting particles. This was the basis on which Bardeen, Cooper, and Schrieffer (BCS) finally formulated the first consistent, though model-type microtheory of superconductivity [18] in 1957. It is curious that paper [18] contains no indications of Bose–Einstein condensation, while it is, in fact, the crucial point.

However, I am running many years ahead as far as my own work is concerned. Concretely, in 1943, I tried [19], on the basis of the Landau theory [4] of superfluidity, to construct a quasi-microscopic theory of superconductivity. The paper postulated a spectrum of electrons (charged 'excitations') in a metal with a gap Δ. For such a spectrum, superconductivity (superfluidity of a charged liquid) must be observed. The introduction of a gap provided the critical field with a dependence on temperature and penetration depth into a superconductor which approximately corresponded to the actual one. Comparison between theory and experiment gave the value $\Delta/k_B T_c = 3.1$. As is well known, in BCS theory $2\Delta_0/k_B T_c = 3.52$ but the most important point is that $\Delta_0 \equiv \Delta(0)$ is the value of the gap at $T = 0$ and, with increasing temperature, the gap decreases to yield $\Delta(T_c) = 0$. In my paper, the gap Δ was assumed to be constant and a satisfactory agreement with experiment is possibly explained by the inaccuracy of the experimental data employed. I do not think that a more detailed analysis of this question is pertinent because model [19] is of no more than historical value now. Nonetheless, article [19] did have some ideas that could have been of interest, for example the occurrence of resonance phenomena for incident radiation at a frequency $\nu = \Delta/h$ was mentioned. In any case, the fact is that in his well-known review [20], published in 1956, Bardeen covered the results of paper [19] rather extensively. Notice that paper [19] also presented a survey of the macrotheory of superconductivity. It was followed by note [21] considering gyromagnetic and electron inertia experiments with superconductors. Finally, in the same year, 1944, paper [22], devoted to thermoelectric phenomena in superconductors, was published.[2] This latter paper remains topical even now and we shall return to it in section 7.5. The previously mentioned papers [19, 21, 22] were included in the monograph *Superconductivity* [24] written in 1944. Before taking up superconductivity, I analyzed [23], on the basis of the Landau theory, the problem of light scattering in helium II. In what follows, I shall consider this and some other papers devoted to superfluidity (see section 7.6).

7.2 Ψ-theory of superconductivity (Ginzburg–Landau theory)

Within the first two decades after the discovery of superconductivity, its study went rather slowly compared to today's standards. This does not seem strange if we remember that liquid helium, which was first obtained in Leiden in 1908, became available elsewhere only after 15 years, i.e. in 1923. Without plunging into the history (see [8, 9, 11]; see also article 6 in this volume), I shall restrict myself to the remark that the Meissner effect was only discovered [25] in 1933, i.e. 22 years after the discovery of superconductivity. Only after that did it become clear that a metal in normal and superconducting states can be treated as two phases of a substance in the thermodynamic sense of this notion. As a result, in 1934, there appeared [26, 20] the so-called two-fluid approach to superconductors and also the relation

$$F_{n\,0}(T) - F_{s\,0}(T) = \frac{H_{c\,m}^2(T)}{8\pi} \tag{7.1}$$

where $F_{n\,0}$ and $F_{s\,0}$ are free-energy densities (in the absence of a field) in the normal and superconducting phase, respectively, and $H_{c\,m}$ is the critical magnetic field destroying superconductivity. Differentiation of expression (7.1) with respect to T leads to expressions for the differences of entropy and specific heat.

According to the two-fluid picture, the total electric current density in a superconductor is

$$\boldsymbol{j} = \boldsymbol{j}_s + \boldsymbol{j}_n \tag{7.2}$$

where $\boldsymbol{j}_s$ and $\boldsymbol{j}_n$ are the densities of the superconducting and normal current.

The normal current in a superconductor does not, in fact, differ from the current in a normal metal and, in the local approximation, we have

$$\boldsymbol{j}_n = \sigma_n(T)\,\boldsymbol{E} \tag{7.3}$$

---

[2] It should be noted that all the three papers [19, 21, 22] were submitted for publication on the same date (23 November, 1943). I do not remember why this happened. Most probably it was connected with some special conditions pertaining during the war.

where E is the electric field strength and σ_n is conductivity of the 'normal part' of the electron liquid; for simplicity, we henceforth take $j_n = 0$, unless otherwise specified.

In 1935, F London and H London proposed [27] for j_s the equations (now referred to as Londons' equations)

$$\mathrm{rot}\,(\Lambda j_s) = -\frac{1}{c}\,H \tag{7.4}$$

$$\frac{\partial(\Lambda j_s)}{\partial t} = E \tag{7.5}$$

where Λ is a constant and the magnetic field strength H here and later does not differ from the magnetic induction B.

We arrive at such equations, for example, proceeding from the hydro-dynamic equations for a conducting 'liquid' which consists of particles with charge e, mass m, and velocity $v_s(r, t)$:

$$\begin{aligned}\frac{\partial v_s}{\partial t} &= -(v_s\nabla)v_s + \frac{e}{m}\,E + \frac{e}{mc}\,[v_s H] \\ &\equiv \frac{e}{m}\,E - \nabla\frac{v_s^2}{2} + \left[v_s\left(\mathrm{rot}\,v_s + \frac{e}{mc}\,H\right)\right].\end{aligned} \tag{7.6}$$

Such an equation corresponds to infinite (ideal) conductivity [28] and is not an obstruction to the presence of a constant magnetic field in a supercon-ductor, which contradicts the existence of the Meissner effect. Therefore, the Londons imposed, so to say, an additional condition $\mathrm{rot}\,v_s + eH/mc = 0$ interpreted as the condition of a vortex-free motion for a charged liquid. If j_s is written in the form $j_s = en_s v_s$, where n_s is the charge concentration, the additional condition for $n_s = $ constant assumes precisely form (7.4) and

$$\Lambda = \frac{m}{e^2 n_s}. \tag{7.7}$$

Equation (7.6) transforms to (7.5) up to a small term proportional to ∇v_s^2 (see section 7.5). Within such an approach, the principal Londons' equa-tion (7.4) is, of course, merely postulated. This condition is an effect of quantum nature and follows from the Ψ-theory of superconductivity [29] considered later and from the microtheory of superconductivity [18, 30] which, in turn, transforms near T_c to Ψ-theory [31]).

Londons' equation (7.4), along with the Maxwell equation

$$\mathrm{rot}\,H = \frac{4\pi}{c}\,j_s \tag{7.8}$$

at $\Lambda = $ constant (we are obviously dealing with the quasi-stationary case), leads to the equations

$$\Delta H - \frac{1}{\delta^2}\,H = 0 \qquad \Delta j_s - \frac{1}{\delta^2}\,j_s = 0 \tag{7.9}$$

$$\delta^2 = \frac{\Lambda c^2}{4\pi} = \frac{mc^2}{4\pi e^2 n_{\mathrm{s}}}. \tag{7.10}$$

Equation (7.9) implies that the magnetic field H and the current density j_{s} exponentially decay through the superconductor depth (for example, in the field parallel to and near a flat boundary, we have $H = H_0 \exp(-z/\delta)$, where z is the distance from the boundary), i.e. the Meissner effect arises. The Londons' equations still hold true but only in the case of a weak field

$$H \ll H_{\mathrm{c}} \tag{7.11}$$

where H_{c} is the critical magnetic field destroying superconductivity (in the case of non-local coupling between the current and the field, Londons' equations do not hold either [20, 30] but we do not consider such cases here). We mean here type I superconductors. For type II superconductors, the Londons' theory has a wider limit of applicability, including the vortex phase for $H \ll H_{\mathrm{c}2}$ at any temperature. But if the field is strong, i.e. comparable with H_{c}, Londons' theory may be invalid or otherwise insufficient. So, from Londons' theory, it follows that the critical magnetic field H_{c}, in which the superconductivity of a flat film of thickness $2d$ is destroyed (in the field parallel to it), is

$$H_{\mathrm{c}} = \left(1 - \frac{\delta}{d}\,\mathrm{th}\frac{d}{\delta}\right)^{-1/2} H_{\mathrm{c\,m}}$$

where $H_{\mathrm{c\,m}}$ is the critical field for a massive specimen (see [32, 33, 24] and references therein). This expression for H_{c}, however, contradicts experimental data. The situation can be improved by introducing different surface tensions σ_{n} and σ_{s} on the boundaries of the normal and the superconducting phases with a vacuum [32]. It turns out, however, that

$$\frac{\sigma_{\mathrm{n}} - \sigma_{\mathrm{s}}}{H_{\mathrm{c\,m}}^2/8\pi} \sim \delta \sim 10^{-5} \ (\mathrm{cm}).$$

At the same time, it might be expected that $(\sigma_{\mathrm{n}} - \sigma_{\mathrm{s}}) \sim (10^{-7} - 10^{-8})\,H_{\mathrm{c\,m}}^2/8\pi$, i.e. is of the order of the volume energy $H_{\mathrm{c\,m}}^2/8\pi$ multiplied by an atomic scale length. Moreover, in the theory based on Londons' equations, on the boundary between the normal and superconducting phases, the surface tension (surface energy) connected with the field and the current is $\sigma_{\mathrm{ns}}^{(0)} = -\delta\,H_{\mathrm{c\,m}}^2/8\pi$. Consequently, to obtain a positive surface tension $\sigma_{\mathrm{ns}} = \sigma_{\mathrm{ns}}^{(0)} + \sigma_{\mathrm{ns}}^{(\prime)}$ observed for a stable boundary, it is necessary to introduce a certain surface energy $\sigma_{\mathrm{ns}}^{(\prime)} > \delta\,H_{\mathrm{c\,m}}^2/8\pi$ of non-magnetic origin. The introduction of such a comparatively high energy is totally ungrounded. On the contrary, one can think that a rational theory of superconductivity must automatically lead to the possibility of expressing the energy σ_{ns} in terms of parameters characterizing the superconductor.

Such a theory that generalized the Londons' theory, eliminated the indicated difficulties, and suggested some new conclusions was the Ψ-theory [29] formulated in 1950.[3] In the same year, I wrote a review [33] devoted to the macro-theory of superconductivity including the Ψ-theory.

In the absence of a magnetic field, the superconducting transition is a second-order transition. The general theory of such transitions always includes [34] a certain parameter (the order parameter) η which, when in equilibrium, differs from zero in the ordered phase and equals zero in the disordered phase. For example, in the case of ferroelectrics, the role of η is played by the spontaneous electric polarization P_s and, in the case of magnetics, by the spontaneous magnetization M_s (not long before the appearance of our paper [29], both these cases were discussed in the review [35]). In superconductors, where the ordered phase is superconducting, for the order parameter we chose a complex function Ψ which plays the role of an 'effective wavefunction of superconducting electrons'. This function can be so normalized that $|\Psi|^2$ is the concentration n_s of 'superconducting electrons'.

The free energy density of a superconductor and the field was written in the form

$$F_{sH} = F_{s0} + \frac{H^2}{8\pi} + \frac{1}{2m}\left| -i\hbar\nabla\Psi - \frac{e}{c}A\Psi \right|^2$$

$$F_{s0} = F_{n0} + \alpha|\Psi|^2 + \frac{\beta}{2}|\Psi|^4 \qquad (7.12)$$

where A is vector potential of the field $H = \operatorname{rot} A$. Without the field, in the state of thermodynamic equilibrium $\partial F_{s0}/\partial|\Psi|^2 = 0$, $\partial^2 F_{s0}/\partial^2|\Psi|^2 > 0$ and we must have $|\Psi|^2 = 0$ for $T > T_c$ and $|\Psi|^2 > 0$ for $T < T_c$. This implies that $\alpha_c \equiv \alpha(T_c) = 0$ and $\beta_c \equiv \beta(T_c) > 0$, and $\alpha < 0$ for $T < T_c$. Within the validity limits of expansion (7.12) in $|\Psi|^2$, one can put $\alpha = \alpha'_c(T - T_c)$ and $\beta(T) = \beta_{T_c} \equiv \beta_c$. From this, at $T < T_c$ (see also equation (7.1)), we have

$$|\Psi|^2 \equiv |\Psi_\infty|^2 = -\frac{\alpha}{\beta} = \frac{\alpha'_c(T_c - T)}{\beta_c}$$

$$F_{s0} = F_{n0} - \frac{\alpha^2}{2\beta} = F_{n0} - \frac{(\alpha'_c)^2(T_c - T)^2}{2\beta_c} = F_{n0} - \frac{H^2_{cm}}{8\pi}. \qquad (7.13)$$

In the presence of the field, the equation for Ψ is derived upon varying the

---

[3] This theory is usually called the Ginzburg–Landau theory. I try, however, to avoid this term, not out of false modesty but rather because in such cases the use of one's own name is not conventional in Russian. Furthermore, in its application to superfluidity (not superconductivity) the Ψ-theory was developed not with L D Landau but with L P Pitaevskii and A A Sobyanin (see section 7.4).

free energy $\int F_{s\,H} dV$ with respect to Ψ^* and, obviously, has the form

$$\frac{1}{2m}\left(-i\hbar\nabla - \frac{e}{c} \boldsymbol{A}\right)^2 \Psi + \alpha\Psi + \beta|\Psi|^2\Psi = 0. \qquad (7.14)$$

If, on the superconductor boundary, the variation $\delta\Psi^*$ is arbitrary, i. e. no additional condition is imposed on Ψ and no additional term corresponding to the surface energy is introduced in equation (7.12), then the condition of minimal free energy is the so-called natural boundary condition on the superconductor boundary,

$$\boldsymbol{n}\left(-i\hbar\nabla\Psi - \frac{e}{c} \boldsymbol{A}\Psi\right) = 0 \qquad (7.15)$$

where $\boldsymbol{n}$ is the normal to the boundary (for more details, see [29] and section 7.3). Condition (7.15) refers to the case of a boundary between a superconductor and a vacuum or a dielectric. As regards the equation for $\boldsymbol{A}$, under the condition $\mathrm{div}\,\boldsymbol{A} = 0$ and after variation of the integral $\int F_{s\,H} dV$ over $\boldsymbol{A}$, it becomes

$$\Delta\boldsymbol{A} = -\frac{4\pi}{c}\boldsymbol{j}_s \qquad \boldsymbol{j}_s = -\frac{ie\hbar}{2m}(\Psi^*\nabla\Psi - \Psi\nabla\Psi^*) - \frac{e^2}{mc}|\Psi|^2\boldsymbol{A}. \qquad (7.16)$$

Here, of course, we assume that $\boldsymbol{j}_n = 0$, i. e. the total current is superconducting. An expression similar to (7.14) is, of course, also obtained for Ψ^* and, as expected, we have $\boldsymbol{j}_s\boldsymbol{n} = 0$ on the boundary (see equation (7.15)). The solution of the problem of the distribution of the field, current, and function Ψ in a superconductor is reduced to the integration of the system of equations (7.14) and (7.16). An expression similar to (7.14) is, of course, also obtained for Ψ^* and, as expected, we have $\boldsymbol{j}_s\boldsymbol{n} = 0$ on the boundary (see equation (7.15)). The solution of the problem of the distribution of the field, current, and function Ψ in a superconductor is reduced to the integration of the system of equations (7.14) and (7.16). Assuming $\Psi = \Psi_\infty = $ constant, the superconducting current density is $\boldsymbol{j}_s = -e^2|\Psi_\infty|^2\boldsymbol{A}/mc = -e^2 n_s \boldsymbol{A}/mc$ (with normalization $|\Psi_\infty|^2 = n_s$). Applying the operation rot to this expression, we obtain Londons' equation (7.4) (see also equation (7.7)). Thus, the Ψ-theory generalizes the Londons theory and passes over into it in the limiting case $\Psi = \Psi_\infty = $ constant.

Paper [29] is rather long (19 pages) and solves several problems to which we shall return in what follows. After that, I myself, sometimes with co-authors, devoted a number of papers to the development of the Ψ-theory of superconductivity. These papers are mentioned later. Moreover, this theory was further promoted and accounted for in a lot of papers and books (see, for example, [20, 30, 33, 36–41, 229, 236–239]). I do not follow the corresponding literature now, the more so as equation (7.14) and its extensions are widely used outside superconductivity or only in application to superconductors (see, for example, [42–44]). This equation is also being investigated by mathematicians whose works are incomprehensible to me (see, for example, [45]). The relativistic generalization of the equations of the Ψ-theory and some of the concepts associated with this theory also enjoy wide application in quantum field theory (spontaneous symmetry breaking, etc; see, for example, [46]).[4] In such a situation, it seems absolutely impossible to elucidate here the present-day state of the Ψ-theory or even focus in detail on the original paper [29] and my subsequent papers.

But what I think necessary is to tell the story of the appearance of paper [29] and to speak about the role of Landau and myself. Nobody else can do this because regretfully Lev Davidovich Landau passed away long ago (he stopped working in 1962 and died in 1968). At the same time, this is, of course, a very delicate question. That is why, when 20–25 years ago I was approached by the bibliographical magazine *Current Contents* with a request to elucidate the history of the appearance of paper [29], I refused. My refusal was motivated by the fact that my story might be interpreted as an attempt to exaggerate my role. And, in general, I had no desire to prove that I was indeed a full co-author and not a student or a postgraduate to whom Landau 'had set a task', whilst actually doing everything himself. Without such a premise it is difficult to explain why our paper has been frequently cited as Landau and Ginzburg, although it is known to have Ginzburg and Landau in the title. Of course, I have never made protestations concerning this point and, in general, consider it to be a trifle but still I believe that such a citation with a wrong order of authors is incorrect. It would certainly still be incorrect even if my role had indeed been a secondary one. But I do not think so, neither did Landau and this fact was well known to his circle and generally in the USSR. As to foreigners, they really did not know much of scientific research in the USSR at that time, for in 1950 the Cold War was at its height. As far back as 1947, the *USSR Journal of Physics*, which was a good journal, stopped being published and paper [29] appeared only in Russian. We could not go abroad at that time. Perhaps we sent a reprint to D Shoenberg or he himself came across this article in *Zh. Eksp. Teor. Fiz.* In any event, Shoenberg translated the paper into English on his own initiative, distributed it among some people and it became available at least to some colleagues. Landau's name played, of course, a positive role and stimulated a lively interest in the paper.

One way or another, I decided to dwell on the history of the appearance

[4] To confirm this, I would cite paper [46] (see p 184; p 480 in the English translation): "It is easy to see that the Higgs model is fully analogical to the Ginzburg–Landau theory and is its relativistic generalization. It turned out that this conclusion bears an important heuristic value by allowing to establish direct analogs between superconductivity theory and theories of elementary particles, including the Higgs model" (see also [265]).

of the work [29] because the present paper would be incomplete if I did not.

I regard the already mentioned paper [32], being accomplished as far back as 1944 (it was submitted for publication on 21 December, 1944), as initial. From [32], it is quite clear that the Londons theory is invalid for the description of the behavior of superconductors in strong enough fields and, in particular, for the calculation of the critical field in the case of films. The introduction of the surface energies σ_n and σ_s was an artificial technique and these quantities were absurdly large new constants whose values were not predicted by the theory. The same applies to the surface energy σ_{ns} on the boundary between the normal and superconducting phases. It was also absolutely unclear how the critical current should be calculated in the case of small-sized superconductors. Therefore, it was necessary somehow to generalize the Londons' theory to overcome its limits. Unfortunately, advancement in this direction was slow. One of the possible explanations is that like many theoretical physicists of my generation and the previous, I was simultaneously engaged in the solution of various problems and did not concentrate on anything definite (it can be seen, for instance, from the bibliographical index [47]). But there was gradual progress. So, on the basis of the conception of the Landau theory [4], I came to the conclusion [48] that electromagnetic processes in superconductors must be nonlinear and, incidentally, suggested a possible experiment for revealing such nonlinearity. The main point is that, in note [48,] I made the following remark: 'The indication of a possible inadequacy of the classical description of superconducting currents consists in the fact that the zero energy of excitation in a superconductor is equal in order of magnitude to $\hbar^2 n/m\delta \sim 1$ erg cm^{-2} (for $\delta \sim 10^{-5}$ cm and $n \sim 10^{22}$ cm^{-3}) and is thus higher than the magnetic energy $\delta H^2/8\pi \sim 0.1$ erg cm^{-2} (for $H \sim 500$ Oe)'. The feeling that the theory of superconductivity should take into account quantum effects was also reflected in note [49] devoted for the most part to critical velocity in helium II. At the same time, in that paper I also tried to apply the theory of second-order phase transitions to the λ-transition in liquid helium.

It seems surprising and, unfortunately, it did not occur to me at that time to ask why Landau, the author of the theory of phase transitions [34] and the theory of superfluidity [4], had never posed the question of the order parameter η for liquid helium. In [49], I chose as such a parameter ρ_s, i.e. the density of the superfluid phase of helium II. However, this choice raises doubts because the expansion of the free energy (thermodynamic potential) begins with the term $\alpha\rho_s$, whereas, in the general theory, the first term of the expansion has the form $\alpha\eta^2$. Hence, $\sqrt{\rho_s}$ is a more pertinent choice as the order parameter. But $\sqrt{\rho_s}$ is proportional to a certain wavefunction Ψ so far as it is precisely the quantity $|\Psi|^2$ that is proportional to the particle concentration. Unfortunately, I do not remember exactly whether it was these arguments alone that prompted me to introduce the order parameter $\eta = \Psi$ and nothing is said about it in [49]. More important for me was the desire to explain the surface tension σ_{ns} by the gradient term $|\nabla\Psi|^2$. In quantum mechanics, this term has the form of kinetic energy $\hbar^2|\nabla\Psi|^2/2m$. It was precisely this idea that I suggested to Landau, probably in late 1949 (paper [29] was submitted on 20 April, 1950 but it had taken much time to prepare it). I was on good terms with Landau, I attended his seminars and often asked his advice on various problems. Landau supported my idea of introducing the 'effective wavefunction Ψ of superconducting electrons' as the order parameter, and so we were immediately led to the free energy (7.12). The thing I do not remember exactly (and certainly do not want to contrive) is whether I came to him with the ready expression

$$\frac{1}{2m}\left| -i\hbar\nabla\Psi - \frac{e}{c}\boldsymbol{A}\Psi \right|^2$$

or with an expression without the vector potential. The introduction of the latter is obvious by analogy with quantum mechanics but perhaps this was made only during a conversation with Landau. I feel I should present my apologies to the reader for such reservations and uncertainty but since that time nearly 50 years have passed (!), no notes have remained, and I never thought that I would have to recall those remote days.

After the basic equations (7.12), (7.14), and (7.16) of the Ψ-theory were derived, one had to solve various problems on their basis and compare the theory with experiment. Naturally, it was I who was mostly concerned with this but I regularly met with Landau to discuss the results. What has been said may produce the impression that my role in the creation of the Ψ-theory was even greater than that of Landau. But this is not so. One should not forget that the fundamental basis was the theory [50, 34] of second-order phase transitions developed by Landau in 1937 which I had employed in a number of cases [35, 49] and applied to the theory of superconductivity in paper [29]. Moreover, I find it necessary to note that the important remark made in [29] concerning the meaning of the Ψ-function used as an order parameter was due to Landau himself. I shall cite the relevant passage from [29]:

> Our function $\Psi(\boldsymbol{r})$ may be thought of as immediately related to the density matrix $\rho(\boldsymbol{r}, \boldsymbol{r}') = \int \Psi^*(\boldsymbol{r}, \boldsymbol{r}_i')\Psi(\boldsymbol{r}', \boldsymbol{r}_i')d\boldsymbol{r}_i'$, where $\Psi(\boldsymbol{r}, \boldsymbol{r}_i')$ is the true Ψ-function of electrons in a metal which depends on the coordinates of all the electrons $\boldsymbol{r}_i$ ($i = 1, 2, \ldots, N$) and $\boldsymbol{r}_i'$ are the coordinates of all the electrons except a distinguished one (its coordinates are $\boldsymbol{r}$ and at another point $\boldsymbol{r}'$). One may think that for a non-superconducting body, where the long range order is absent, as $|\boldsymbol{r} - \boldsymbol{r}'| \to \infty$ we have $\rho \to 0$, while in the superconducting state $\rho(|\boldsymbol{r} - \boldsymbol{r}'| \to \infty) \to \rho_0 \neq 0$. In this case it is natural to assume the density matrix to be related to the introduced Ψ-function as $\rho(\boldsymbol{r}, \boldsymbol{r}') = \Psi^*(\boldsymbol{r})\Psi(\boldsymbol{r}')$.

Accordingly, the superconducting (or superfluid) phase is characterized by a certain long-range order which is absent in ordinary liquids (see also [30, section 26]; [51, 52], [53, section 9.7]). This result is usually ascribed to Yang [51] and is referred to as ODLRO (off-diagonal long-range order) [53]. However, as we can see, Landau realized the possibility of the existence of this long-range order 12 years before Yang. I mentioned this fact in [54].

In expression (7.12) and subsequent ones, the coefficients e and m appear. These designations were of course chosen by analogy with the quantum-mechanical expression for the Hamiltonian of a particle with charge e and mass m. Our Ψ-function is, however, not the wavefunction of electrons. The coefficient m can be taken arbitrarily [29] because the Ψ-function is not an observed quantity: an observed quantity is the penetration depth δ_0 of a weak magnetic field (see equations (7.12), (7.13), and (7.16)):

$$\delta_0^2 = \frac{mc^2 \beta_c}{4\pi e^2 |\alpha|} = \frac{mc^2}{4\pi e^2 |\Psi_\infty|^2}. \tag{7.17}$$

Since the Ψ-theory in a weak field (7.11) transforms to the Londons theory (though, a number of problems cannot be stated in Londons theory even in this case), the penetration depth δ_0 is frequently called the London penetration depth and is denoted by δ_L or λ_L. If we assume [29] e and m to correspond to a free electron ($e_0 = 4.8 \times 10^{-10}$ CGS, $m_0 = 9.1 \times 10^{-28}$ g), then $|\Psi_\infty|^2 = n_s$, where n_s is the 'superconducting electron' concentration thus defined. In fact, one can choose any arbitrary value of m [29, 37] which will only affect the normalization of the observed quantity $|\Psi_\infty|^2$. In the literature, $m = 2m_0$ occasionally occurs, which corresponds to the mass of a 'pair' of two electrons. As to the charge e in equation (7.12) and subsequent expressions, it is an observed quantity (see later). It seemed to me from the very beginning that one should regard the charge e in equation (7.12) as a certain 'effective charge' e_{eff} and take it as a free parameter. But Landau objected and, in paper [29], it is stated as a compromise that "there is no reason to assume the charge e to be other than the electron charge". Running ahead I shall note that I still went on thinking of the question of the role of the charge $e \equiv e_{\text{eff}}$ as open and pointed out the possibility of clarifying the situation by comparing the theory with experiment (see [14, p 107]). The point is that the essential parameter involved in the Ψ-theory is the quantity

$$æ = \frac{mc}{e\hbar} \sqrt{\frac{\beta_c}{2\pi}} = \frac{\sqrt{2}e}{\hbar c} H_{c\,m} \delta_0^2. \tag{7.18}$$

In paper [29], we set $e = e_0$ and could, therefore, determine $æ$ from experimental data on $H_{c\,m}$ and δ_0. At the same time, the parameter $æ$ enters the expressions for the surface energy σ_{ns}, for the penetration depth in a strong field ($H \gtrsim H_{cm}$) and the expressions for superheating and supercooling limits. Using the approximate data of measurements available at the time, I came to the conclusion [55] (this paper was submitted for publication on 12 August, 1954) that the charge $e \equiv e_{\text{eff}}$ in equation (7.18) is two to three times greater than e_0. When I discussed this result with Landau, he put forward a serious objection to the possibility of introducing an effective charge (he had apparently had this argument in mind before, when we discussed paper [29] but did not then advance it). Specifically, the effective charge might depend on the composition of a substance, its temperature and pressure and, therefore, might appear to be a function of coordinates. But, in that case, the gradient invariance of the theory would be broken, which is inadmissible. I could not find arguments against this remark and, with the consent of Landau, I included it in paper [55]. The explanation seems now to be quite simple. No, an effective charge e_{eff}, which might appear to be coordinate-dependent, should not have been introduced. But it might well be supposed that, say, $e_{\text{eff}} = 2e_0$. And this was exactly the case but it became obvious only after the creation of BCS theory [18] in 1957 and the appearance of the paper by Gorkov [31] who showed that the Ψ-theory near T_c follows from the BCS theory. More precisely, the Ψ-theory near T_c is certainly wider than the BCS theory in the sense that it is independent of some particular assumptions used in the BCS theory. But this is a different subject. The formation of pairs with charge $2e_0$ is a very general phenomenon, too. I have already emphasized that the idea of pairing and, what is important, the realistic character of such pairing was far from trivial.

So, in the Ψ-theory, we have $e = 2e_0$ and, consequently (see equation (7.18)),

$$\ae = \frac{2\sqrt{2}e_0}{\hbar c} H_{\text{c m}} \delta_0^2. \tag{7.19}$$

As can be seen from the calculations, the surface tension $\sigma_{\text{n s}}$ is positive only for $\ae < 1/\sqrt{2}$. An analytical calculation of $\sigma_{\text{n s}}$ encounters difficulties. In paper [29], this was only done for sufficiently small $\ae$:

$$\sigma_{\text{n s}} = \frac{\delta_0 H_{\text{c m}}^2}{\sqrt{2} \cdot 3\pi\ae} \qquad \Delta = \frac{\sigma_{\text{n s}}}{H_{\text{c m}}^2/8\pi} = \frac{1.89\delta_0}{\ae} \qquad \sqrt{\ae} \ll 1. \tag{7.20}$$

From this, it is already seen that the Ψ-theory leads to $\sigma_{\text{n s}}$ values of the required order of magnitude. It is only in paper [56] that the energy $\sigma_{\text{n s}}$ is calculated analytically up to terms of the order of $\ae\sqrt{\ae}$. The result is as follows [the value $\Gamma = 2\sqrt{2}/3$ corresponds to expression (7.20)]:

$$\sigma_{\text{n s}} = \frac{\delta_0 H_{\text{c m}}^2}{4\pi\ae} \Gamma \qquad \Gamma = \frac{2\sqrt{2}}{3} - 1.02817\sqrt{\ae} - 0.13307\ae\sqrt{\ae} + \cdots . \tag{7.21}$$

As $\ae$ increases, the energy $\sigma_{\text{n s}}$ decreases and, in [29], it was pointed out that, according to numerical integration,

$$\sigma_{ns} = 0 \qquad æ = \frac{1}{\sqrt{2}}. \tag{7.22}$$

But it was also shown that for $æ > 1/\sqrt{2}$, there occurs some specific instability of the normal phase, namely nuclei of the superconducting phase are formed in it. Concretely, this instability arises in the field

$$H_{c2} = \sqrt{2}æH_{cm}. \tag{7.23}$$

(It should be noted that formula (7.23) is present in [29] in an implicit form; and it was written explicitly in [57].) In the case $æ < 1/\sqrt{2}$, the field H_{c2} corresponds to the limit of a possible supercooling of the normal phase (for $H < H_{c2}$, this phase becomes metastable; see also [57], where, as in some of my other papers, the field H_{c2} is denoted by H_{k1}). When $æ > 1/\sqrt{2}$, it is clear from equation (7.23) that superconductivity is preserved in some form in the field $H > H_{cm}$ too and vanishes only in the field H_{c2}. Generally, it is just for $æ = 1/\sqrt{2}$ that the change in the behavior of a superconductor becomes pronounced. Hence, there were no doubts in the validity of the result (7.22). Analytically this is proved, for example, in [30, 37, 38]. It turns out that for pure superconducting metals we typically have $æ < 1/\sqrt{2}$ or even $æ \ll 1/\sqrt{2}$ (for instance, according to [30], $æ$ is equal to 0.01 for Al, 0.13 for Sn, 0.16 for Hg, and 0.23 for Pb). Such superconductors are called type I superconductors. If $æ > 1/\sqrt{2}$, the surface tension σ_{ns} is negative and we then deal with type II superconductors (for the most part alloys) whose behavior was first investigated thoroughly in experimental studies by L V Shubnikov[5] and co-authors as far back as 1935–36 (for references and explanations see [58, 24]). In [29], we considered only type I superconductors and we read such a phrase there: "For sufficiently large $æ$, on the contrary, $\sigma_{ns} < 0$, which is indicative of the fact that such large $æ$ do not correspond to the typically observed picture". So we, in fact, overlooked the possibility of the existence of type II superconductors. Neither was I engaged in the study of type II superconductors later on. In this respect, I only made a remark in [57]. The theory of the behavior of type II superconductors based on the Ψ-theory was constructed in 1957 by Abrikosov [59] (see also [30, 41]). As indicated in [59] and [30, p 191], Landau was the first to suggest that in alloys $æ > 1/\sqrt{2}$.

Allowing for equations (7.13) and (7.17), one can write

$$H_{cm} = \left(\frac{4\pi(\alpha_c')^2}{\beta_c}\right)^{1/2}(T_c - T) \qquad \delta_0 = \left(\frac{m_0 c^2 \beta_c}{16\pi e_0^2 \alpha_c'}\right)^{1/2}(T_c - T)^{-1/2}. \tag{7.24}$$

---

[5] L V Shubnikov, a prominent experimental physicist, was guiltlessly executed in 1937.

These expressions, the same as the whole Ψ-theory, are strictly speaking valid only in the vicinity of T_c, i. e. the condition $(T_c - T) \ll T_c$ is needed. However, the condition of applicability of the theory for small æ is, in fact, more rigorous because to satisfy the local approximation, the penetration depth δ_0 must significantly exceed the size ξ_0 of the Cooper pair (the corresponding condition written in [30, section 45] has the form $(T_c - T) \ll æ^2 T_c$ but in [29] this, of course, could not yet be discussed). Along with the penetration depth δ_0, the Ψ-theory involves one more parameter which has the dimension of length—the so-called coherence length or the correlation radius (length)

$$\xi = \frac{\hbar}{\sqrt{2m_0|\alpha|}} = \frac{\hbar}{\sqrt{2m_0\alpha'_c(T_c - T)}} = \frac{\hbar \tau^{-1/2}}{\sqrt{2m_0\alpha'_c T_c}} = \xi(0)\tau^{-1/2} \quad (7.25)$$

where $\tau = (T_c - T)/T_c$ and $\xi(0) = \hbar/\sqrt{2m_0\alpha'_c T_c}$ is a conditional correlation radius for $T = 0$ (we call it conditional because the Ψ-theory is strictly speaking applicable only in the vicinity of T_c). To compare the formulae written here with those of [30], one should bear in mind that in our expression (7.12) in [30] $m = 2m_0$ and, of course, $e = 2e_0$.

As is readily seen (see equations (7.18), (7.19), (7.24)),

$$æ = \frac{m_0 c}{2e_0 \hbar}\sqrt{\frac{\beta_c}{2\pi}} = \frac{\delta_0(T)}{\xi(T)}. \quad (7.26)$$

In addition to these mentioned problems, some more points were considered in [29], namely the field in a superconducting half-space and critical fields for plates (films) in the case where superconductivity is destroyed by the field and current. The penetration depth of the field in a superconducting half-space adjoining a vacuum has the form

$$\delta = \delta_0\left[1 + f(æ)\left(\frac{H_0}{H_{cm}}\right)^2\right] \qquad f(æ) = \frac{æ(æ + 2\sqrt{2})}{8(æ + \sqrt{2})^2} \quad (7.27)$$

where H_0 is the external field (the field for $z = 0$) and, by definition, $\delta = \int_0^\infty H(z)\mathrm{d}z/H_0$. The nonlinearity of the electrodynamics of superconductors, which was assumed already in [48] and is reflected in the dependence of δ on H_0, is fairly small. So, even for $æ = 1/\sqrt{2}$ and $H_0 = H_{cm}$, the depth is $\delta = 1.07\delta_0$. In 1950, there were no accurate enough experimental measurements of $\delta(H)$. I am not sure that they have yet been carried out, though it is probable.

Now I should make or, rather, repeat one general remark. I was never long engaged in studying only superconductivity but researched various fields (see [1, p 309] and [47]). As to the macroscopic theory of superconductivity (the Ψ-theory and its development), it was generally beyond the scope of my interest from a certain time (see section 7.3). As a result, I am ignorant of the current state of the problem as a whole. Unfortunately, neither am I aware of the existence of a monograph compiling all the material (I am afraid there is no such book). Moreover, I forgot much of what I had done myself and now recollect the old facts, sometimes with surprise, when reading my own papers. That is why I cannot be convinced that my old calculations were unerring, I do not know the subsequent calculations and the results of their comparison with experiment. But the present paper does not even claim to make a current review, it is only an attempt to elucidate some problems of the history of studies of superconductivity and superfluidity in an autobiographical context. Those uninterested will just not read it and, in this, I find some consolation.

The concluding part of paper [29] is devoted to a consideration of superconducting plates (films) of thickness $2d$ in an external magnetic field H_0 parallel to the film and also in the presence of a current $J = \int_{-d}^{+d} j(z)dz$ (where $j(z)$ is the current density) flowing through the film. Instead of J, it is convenient to work in terms of the field $H_J = 2\pi J/c$ created by the current outside the film.

In the absence of current, the critical field H_c destroying superconductivity for thick films with $d \gg \delta_0$ is (see equation (7.27))

$$\frac{H_c}{H_{cm}} = 1 + \frac{\delta_0}{2d}\left(1 + \frac{f(æ)}{2}\right) \qquad d \gg \delta_0. \tag{7.28}$$

For sufficiently thin films a transition to the normal state is a second-order transition (i.e. for $H_0 = H_c$, the function Ψ is equal to zero) and, for small $æ$, we have

$$\left(\frac{H_c}{H_{cm}}\right)^2 = 6\left(\frac{\delta_0}{d}\right)^2 - \frac{7}{10}æ^2 + \frac{11}{1400}æ^4\left(\frac{d}{\delta_0}\right)^2 + \cdots \qquad d \ll \delta_0. \tag{7.29}$$

For films with half-thickness $d > d_c$, where

$$d_c^2 = \frac{5}{4}\left(1 - \frac{7}{24}æ^2 + \cdots\right)\delta_0^2 \tag{7.30}$$

we are already dealing with first-order transitions with a release of latent transition heat (in other words, d_c is a tricritical point or, as it was termed before, a critical Curie point).

In the presence of a current and field (for $æ = 0$),

$$\frac{H_{J_c}}{H_{cm}} = \frac{2\sqrt{2}}{3\sqrt{3}}\frac{d}{\delta_0}\left[1 - \left(\frac{H_0}{H_c}\right)^2\right]^{3/2} \qquad d \ll \delta_0 \tag{7.31}$$

where H_c is the critical field for a given film in the absence of a current (see equation (7.29)), H_0 is the external field, and J_c is the critical current destroying superconductivity ($H_{J_c} = 2\pi J_c/c$).

The field H_c for such films is much larger than the critical field H_{cm} for bulk samples and $H_{J_c} \ll H_{cm}$. It is interesting, however, that according to equations (7.29) and (7.31) (for $æ = 0$ and $H_0 = 0$), we are led to

$$H_c H_{J_c} = \frac{4}{3} H_{cm}^2. \tag{7.32}$$

In [29] we certainly tried to compare the theory with the then available experimental data. But the latter were not numerous and, particularly importantly, their accuracy was low. To the best of my knowledge, all the results of the theory were later confirmed by experiment.

7.3   The development of the Ψ-theory of superconductivity

In paper [29], we did not solve all the problems nor even those which were easy to formulate. Therefore, I naturally continued, although with some intervals, to develop the Ψ-theory for several years. For example, in paper [60] (see also [14]), I considered in more detail than in [29] the destruction of superconductivity of thin films having half-thickness $d > d_c$ (see equation (7.30): the condition $(æd/\delta_0)^2 \ll 1$ was used). Critical fields were found for supercooling and superheating. I note that not for films but for cylinders and balls critical fields were calculated (on the basis of the Ψ-theory) by Silin in [61] and myself in [62] (see also [229]). The critical current for superconducting films deposited onto a cylindrical surface was found in paper [63]. The question of normal phase supercooling (see equation (7.23)) was discussed in paper [57], which has already been mentioned, and the critical field for superheating of the superconducting phase in bulk superconductors was calculated in paper [62]. So, for small $æ$, the critical field for superheating (denoted as the field H_{k2} in [62]) is

$$\frac{H_{c1}}{H_{cm}} = \frac{0.89}{\sqrt{æ}} \qquad \sqrt{æ} \ll 1 \tag{7.33}$$

where the coefficient is obtained from numerical integration.[2*]

In several papers (see [14, 32, 55, 64]), I discussed, in particular, the behavior of superconductors in a high-frequency field but later on showed no interest in this issue and am now unaware whether these papers were of interest and importance for experiments (in respect of the behavior in a high-frequency field).

As I have already emphasized, the Ψ-theory can be immediately applied only in the vicinity of T_c. Naturally, I wished to extend the theory to the case of any temperature. In the framework of the phenomenological approach this goal can be achieved in different ways. So, Bardeen [65] suggested replacing the expression for the free energy F_{s0} from equation (7.12) with another expression involving a more complicated dependence of $F_{s0}(|\Psi|^2)$ on $|\Psi|^2$. The same object can, however, be attained [66] without changing expression (7.12) but by assuming a certain dependence of the coefficients α and β on temperature or, more precisely, on the ratio T/T_c. A somewhat different approach to the problem consists [67] not in assuming the dependence $F_{s0}(|\Psi|^2)$ in advance but rather in finding it from comparison with experiment.

After the creation of the BCS theory in 1957 and the papers [31] by Gorkov, I almost lost interest in the theory of superconductivity. Superconductivity was no longer an enigma (it had been an enigma for a long 46 years after its discovery in 1911). Quite a lot of other attractive problems existed and I thought that I would drop superconductivity for ever. It was merely by inertia that, in 1959, when it became finally clear that the effective charge in the Ψ-theory was $e_{\text{eff}} \equiv e = 2e_0$, I compared [68] the Ψ-theory with the available experimental data and made sure that everything was all right. I will also mention the note [69] devoted to the allowance for pressure in the theory of second-order phase transitions as applied to a superconducting transition.

It was F London [70] who had already pointed out that a magnetic flux through a hollow massive superconducting cylinder or a ring must be quantized and that the flux quantum must be $\Phi_0 = hc/e$ and the flux $\Phi = k\Phi_0$, where k is an integer and e is the charge of the particles carrying the current. Naturally, London assumed $e = e_0$ to be a free electron charge. It was only in 1961 that the corresponding experiments were carried out (for references and a description of the experiments see, for example, [71]) demonstrating that, in fact, $e = 2e_0$. The latter is quite clear from the point of view of the BCS theory according to which it is pairs of electrons that are carried over. Thus,

$$\Phi = \frac{hck}{2e_0} = \frac{\pi \hbar ck}{e_0} = \Phi_0 k \qquad \Phi_0 = 2 \times 10^{-7} \, \text{G cm}^2 \, (k = 0, 1, 2, \ldots). \quad (7.34)$$

This result (7.34) refers, however, only to the case of doubly connected bulk samples, for instance hollow cylinders with wall thickness substantially exceeding the magnetic field penetration depth δ in a superconductor. And yet, samples of any size, as well as those located in an external magnetic field, etc, are also of interest. Within the framework of the Ψ-theory, I solved this problem in paper [72]. A similar but less thorough and comprehensive analysis appeared nearly simultaneously in papers [73, 74] (all the papers [72–74] were submitted for publication in mid-1961).

I have not yet mentioned my papers [75] and [76] which were written before the creation of the BCS theory which, however, fell out of the scope of direct application of the Ψ-theory [29]. So, in [75], the Ψ-theory was extended to the case of anisotropic superconductors. In the 'low-temperature' (conventional) superconductors known at that time, anisotropy is either absent altogether (isotropic and cubic materials) or is fairly small. It was apparently for this reason that in [29] we assumed, even without reservations, that metals are isotropic. But as early as in paper [22], when I considered thermoelectric phenomena, I had to examine an anisotropic (i. e. non-cubic) crystal and, in view of this, I generalized the Londons' theory (7.4), (7.5) by introducing a symmetric tensor of rank two, Λ_{ik}, instead of the scalar Λ so that rot $\Lambda(\boldsymbol{j}) = -\boldsymbol{H}/c$, $\Lambda_i(\boldsymbol{j}) = \Lambda_{ik}\,j_k$, (here $\boldsymbol{j} = \boldsymbol{j}_s$ is the superconducting current density). Such a generalization is, of course, obvious enough but I mention it here because in the extensive review [20] Bardeen refers in this connection only to papers [78, 79] by Laue which appeared later.

In [75], the complex scalar function $\Psi(\boldsymbol{r})$ for anisotropic material is introduced as before but the free energy is written not in the form (7.12) but as

$$F_{\mathrm{s}H} = F_{\mathrm{s}0} + \frac{H^2}{8\pi} + \frac{1}{2m_k}\left| -\mathrm{i}\hbar\frac{\partial\Psi}{\partial x_k} - \frac{2e_0}{c}A_k\Psi \right|^2 \qquad (7.35)$$

where doubly occurring indices are summed and, in [75], the charge e is taken instead of $2e_0$; and for an isotropic or cubic material, $m_1 = m_2 = m_3 = m$, and we obtain equation (7.12).

As mentioned previously, the anisotropy in 'conventional' superconductors is not large, i. e. the 'effective masses' m_k differ little from one another. But in the majority of high-temperature superconductors, in contrast, the anisotropy is very large and it is expression (7.35) and the corollaries to it, partially mentioned already in [75], that are widely used. An interesting effect related to anisotropy of a superconductor is noted in [238].

Among the superconductors known in the 1950s, there was not a single ferromagnetic. This is, of course, not accidental. The point is that even digressing from microscopic reasons, the presence of ferromagnetism hampers the occurrence of superconductivity [76]. Indeed, one can see that in the depth of a ferromagnetic superconductor the magnetic induction $\boldsymbol{B}$ must also be zero. But spontaneous magnetization $\boldsymbol{M}_{\mathrm{s}}$ causes induction $\boldsymbol{B} = 4\pi\boldsymbol{M}_{\mathrm{s}}$. Consequently, in a ferromagnetic superconductor, even in the absence of an external magnetic field, there must flow a surface superconducting current compensating for the 'molecular' current responsible for magnetization. From this, it follows that a thermodynamic critical magnetic field for a ferromagnetic superconductor is

$$H_{\mathrm{c\,m}}(T) = \frac{H_{\mathrm{c\,m}}^{(0)}(T)}{\sqrt{\mu}} - \frac{4\pi M_{\mathrm{s}}}{\mu}$$

$$H_{\mathrm{c\,m}}^{(0)} = \sqrt{8\pi(F_{\mathrm{n}0} - F_{\mathrm{s}0})} \qquad (7.36)$$

where the ferromagnetic is assumed to be 'ideal', i. e. for it $\boldsymbol{B} = \boldsymbol{H} + 4\pi\boldsymbol{M} = \mu\boldsymbol{H} + 4\pi\boldsymbol{M}_{\mathrm{s}}$ (μ is magnetic permittivity) and $F_{\mathrm{n}0}$ and $F_{\mathrm{s}0}$ are free energies for the normal and superconducting phases of a given metal in the absence of magnetization and a magnetic field. Obviously, superconductivity is only possible under the condition $H_{cm}^{(0)}(0) > 4\pi M_s/\sqrt{\mu}$ which can hold, in fact, only for ferromagnetics with a not very large spontaneous magnetization M_s. With the appearance of the BCS theory, it became clear that superconductivity and ferromagnetism obstruct each other even irrespective of the previously mentioned so-called electromagnetic factor. Indeed, conventional superconductivity is associated with the pairing of electrons with oppositely directed spins, while ferromagnetism corresponds to parallel spin orientation. Thus, the exchange forces that lead to ferromagnetism obstruct the appearance of superconductivity. Nevertheless, ferromagnetic superconductors were discovered but naturally with fairly low values of T_c and the Curie temperature T_M (see [77, 217, 266] and also article 6 in this volume).

Unfortunately, I am not aware of corresponding experiments and wish to emphasize here that the 'electromagnetic factor' was allowed for in only the simplest, trivial case of an equilibrium uniform magnetization of bulk metal. However, there exist alternative possibilities [76].[8*]

For example, let us assume that a ferromagnetic metal possesses a large coercive force and that in the external field $H_c < H_{coer}$ magnetization can remain directed opposite to the field (for simplicity we consider cylindrical samples in a parallel field). Then, for $M_s < 0$ (the magnetization is directed oppositely to the field), superconductivity may exist under the condition $H_{cm}^{(0)}(0) > 4\pi |M_s|/\sqrt{\mu} - \sqrt{\mu} H_{coer}$, i.e. in principle, the 'electromagnetic factor' may be absolutely insignificant. Of even greater interest are possibilities arising in the case of thin films and generally small-size samples. For them, the critical field $H_c^{(0)}$, as is well known and has already been mentioned, may substantially exceed the field $H_{cm}^{(0)}$ for bulk metal. At the same time, a critical field for a ferromagnetic superconducting film, even for $M_s > 0$ (when the magnetization is directed along the field), has, as before, the form (7.36) but with $H_{cm}^{(0)}$ replaced by $H_c^{(0)}$. Now, the presence of magnetization M_s may already be of no importance. Thus, additional possibilities open up for investigating ferromagnetic superconductors. I do not know if these possibilities have ever been considered.[6]

We have up to now discussed only equilibrium or metastable (superheated or supercooled) states of superconductors, fluctuations being totally ignored. Meanwhile, fluctuations near phase transition points, especially for second-order transitions, generally speaking play an important role (see, for example, [34, section 146]). In the case of superconductors, one should expect fluctuations of the order parameter Ψ both below and above T_c. I can tell the reader about my activity in this field. In 1952, at the end of

---

[6] In recent years, the superconductivity in ferromagnetics has attracted much attention [253–255, 258, 259].

paper [80], it was noted that fluctuations of the 'concentration of superconducting electrons' n_s must also be present above T_c and this must affect first
of all the complex dielectric constant of a metal. At the end of review [14],
this remark was made again with emphasis on the fact that as $T \to T_c$
the fluctuations must be large. However, I never elaborated upon this observation later. Fourteen years had passed before V V Schmidt [81] (who
untimely demised in 1985) went farther and considered (with a reference to
the paper [80]) the question of the fluctuational specific heat of small balls
above T_c and also mentioned the possibility of observing the fluctuational
diamagnetic moment of such balls. It is curious that another two physicists with this name investigated [82, 83] the same issue and, moreover,
considered fluctuational conductivity above T_c (for the fluctuation effects
see also [30, 84, 85]).

Let us now turn to a very important question of the applicability limits
of Landau's phase transition theory both in the general context and in its
application to superconductors [86].

Landau's phase transition theory [50, 34] is well known to be the mean
field theory (or, as it is sometimes referred to, molecular or self-consistent
field theory). This means that the free energy (or a corresponding thermodynamic potential) of the type

$$F = F_0 + \alpha\eta^2 + \frac{\beta}{2}\eta^4 + \frac{\gamma}{6}\eta^6 + g(\nabla\eta)^2 \qquad (7.37)$$

does not allow for the contribution from the fluctuations of η.

As we have seen in the example of a superconductor, when $\eta = \Psi$ (see
equations (7.12), (7.13)), below the second-order transition point (we set
$\gamma = 0$), the equilibrium value is

$$\eta_0^2 = -\frac{\alpha}{\beta} = \frac{\alpha'_c(T_c - T)}{\beta_c}. \qquad (7.38)$$

Taking the Landau theory as the first approximation and using it as a basis,
one can find the fluctuations of various quantities, in particular, the parameter η itself. Naturally, the Landau theory holds true and the fluctuations
calculated on its basis hold true only as long as they are small compared to
the mean values obtained within the Landau theory. In application to η,
this means that the condition

$$\overline{(\Delta\eta)^2} \ll \eta_0^2 \qquad (7.39)$$

must hold, where obviously $\overline{(\Delta\eta)^2}$ is the statistical mean of the fluctuation
of the quantity η (the fluctuation $\overline{(\Delta\eta)}$ is zero because we calculate the
deviations from the value η_0 corresponding to the minimum free energy).

The use of criterion (7.39) leads to the following condition of applicability of the Landau theory (see equations (7.37), (7.38))

$$\tau \equiv \frac{T_c - T}{T_c} \gg \frac{k_B^2 T_c \beta_c^2}{32\pi^2 \alpha_c' g^3} \tag{7.40}$$

where k_B is the Boltzmann constant. This means that the Landau theory can be exploited within the temperature range in the vicinity of the transition point T_c satisfying inequality (7.40). A condition of type (7.40) or similar was derived in different but close ways in [86–88, 34]. For example, in [88] the condition of applicability of the Landau theory is written in the form (in our notation; moreover, in [88, 34] k_B was set unity)

$$\text{Gi} = \frac{T_c \beta_c^2}{\alpha_c' g^3} \ll \tau \ll 1 \qquad \tau = \frac{T_c - T}{T_c}. \tag{7.41}$$

The number Gi in [88] was called the Ginzburg number but I never employ this terminology for the reason mentioned earlier in respect of the Ψ-theory. In my opinion it is more appropriate to employ a criterion of the form (7.40) because the coefficient $1/32\pi^2$ is fairly small and this extends, in fact, the limits of applicability of the Landau theory (note that in [86] the coefficient $1/32\pi^2$ in the final expression (5b) is omitted but it is clear from formula (7.4) for $\overline{(\Delta\eta)^2}$).

Obviously, the smaller the number Gi is, the closer to the transition point the Landau theory can be used, in which, in particular, the specific heat simply undergoes a jump (without λ-singularity) and $\eta_0^2 \sim (T_c - T)$. This immediately implies, for example, that in liquid helium (^{4}He) the parameter Gi is large and this results in the existence of the λ-singularity. In [86], various transitions are discussed, the most detailed consideration being given to ferroelectrics to which the Landau theory is generally well applicable, as to other structure phase transitions. This subject was discussed many years later in paper [89] but we shall not touch upon it here. In the present paper, we are concerned with superconducting transitions and the λ-transition in liquid helium. The latter is dealt with in section 7.4. As far as superconductors are concerned, from comparison of the expressions (7.12) with $e = 2e_0$ and $m = m_0$, (7.25), (7.26), (7.37), and (7.40), it follows that condition (7.40) takes on the form

$$\tau \equiv \frac{T_c - T}{T_c} \gg \tau_G \equiv \frac{(k_B \beta_c)^2}{32\pi^2 (\alpha_c')^4 T_c^2 [\xi(0)]^6}. \tag{7.42}$$

This expression, however, bears no specific features for superconductors and refers to any second-order transition described by the Landau theory. In the framework of this theory, as is clear from [34] and, for example, from equations (7.13) or (7.37), the jump ΔC of specific heat $C = TdS/dT$, where $S = -\partial F/\partial T$ is entropy, at transition is

$$\Delta C = \frac{(\alpha_c')^2 T_c}{\beta_c}. \tag{7.43}$$

From equation (7.43), it is clear that condition (7.42) involves, in particular, the directly measurable quantity ΔC. Next, for superconductors (see equations (7.13), (7.23), (7.25), (7.26), and (7.34)),

$$H_{cm}^2 = \frac{4\pi(\alpha_c')^2}{\beta_c}(T_c - T)^2 = \frac{4\pi(\alpha_c')^2 T_c^2}{\beta_c}\tau^2 \equiv H_{cm}^2(0)\tau^2$$

$$H_{c2}^2 = 2\ae^2 H_{cm}^2 \qquad \xi^2 = \frac{\hbar^2}{2m_0\alpha_c' T_c}\tau^{-1} \equiv \xi^2(0)\tau^{-1} \tag{7.44}$$

$$\ae^2 = \frac{m_0^2 c^2 \beta_c}{8\pi e_0^2 \hbar^2} \qquad \xi^{-2}(0) = \frac{2e_0}{\hbar c}H_{c2}(0) = \frac{2\pi H_{c2}(0)}{\Phi_0}$$

$$H_{c2}^2(0) = 2\ae^2 H_{cm}^2(0).$$

To avoid misunderstanding, we shall stress that all our consideration, as well as the Ψ-theory itself, refers directly only to the region near T_c. Consequently, the quantities $H_{cm}(0)$ and $H_{c2}(0)$ are somewhat formal and are not at all the true values of the fields $H_{cm}(T)$ and $H_{c2}(T)$ at $T = 0$. In view of this, it would be more correct to employ the derivatives $(dH_{cm}/dT)_{T=T_c} = -H_{cm}(0)/T_c$ and $(dH_{c2}/dT)_{T=T_c} = -H_{c2}(0)/T_c$ which can be measured in experiment.

Allowing for equations (7.43) and (7.44), one can rewrite condition (7.42) in the form

$$\tau \gg \tau_G = \left(\frac{2\pi}{\Phi_0}\right)^3 \frac{H_{c2}^3(0)}{32\pi^2(\Delta C)^2} \qquad \Phi_0 = \frac{\pi\hbar c}{e_0}. \tag{7.45}$$

For type I superconductors, the substitution in equations (7.42) and (7.45) of the values of $\xi(0)$ (or $H_{c2}(0)$) and ΔC known from experiment, even without account of the factor $1/32\pi^2 \sim 3 \times 10^{-3}$, yields the estimate $\tau_G \sim 10^{-15}$ (see [86] for $T_c \sim 1$ K) or, on the basis of the BCS model, the estimate $\tau_G \sim (k_B T_c/E_F)^4 \sim 10^{-12}$–$10^{-16}$ (here E_F is the Fermi energy; see [88, 30, section 45]). Physically it is obvious that the smallness of the value τ_G for superconductors is due to the high value of the correlation radius $\xi(0)$ in type I superconductors. In this case, the characteristic value $\xi(0) \sim \xi_0 \sim 10^{-4}$–$10^{-5}$ cm is of the order of the size of a Cooper pair. For structure phase transitions, $\xi(0) \sim d \sim 3 \times 10^{-8}$ cm and is of the order of interatomic length and the fluctuation region must be seemingly large. But, in this case (in particular, in ferroelectrics), the relative smallness of τ_G is caused by other factors (see [86, 89]).

Thus, the Ψ-theory is generally speaking well applied to superconductors. The words 'generally speaking' refer to several circumstances.

Firstly, we have considered here the three-dimensional case. For quasi-two-dimensional (thin films), quasi-one-dimensional (thin wires, etc), and quasi-zero-dimensional (small seeds, say, balls) superconductors, the conditions of applicability of the theory are different: the fluctuation region is wider than for a three-dimensional system. Unfortunately, I do not know all aspects of the problem (see, however, [90]). Secondly, as has already been emphasized, good applicability of the mean field approximation (the Landau theory and, in particular, the Ψ-theory) is in no way an obstruction to the calculation of various fluctuation effects as long as they are sufficiently small (see, for example, [81–85, 90, 91]). It is of importance, especially in application to high-temperature superconductors (HTSCs), that paper [90] analyses, on the basis of the expression (7.35), the anisotropic case. Third, in a number of superconductors (dirty alloys, HTSCs), the parameter æ is large or even very large (reaching hundreds) while the correlation length is small. Then the fluctuation region, i. e. the temperature range in which inequality (7.42), (7.45) is violated, is not so small. So, in [90], we present the values $\tau_G = (0.2-2) \times 10^{-4}$ for HTSCs. Somewhat lower values are reported in [92]. For $\tau_G \sim 10^{-4}$ and $T_c \sim 100$ K, the width of the fluctuation region is $\Delta T \sim 10^{-2}$ K (in this region the fluctuations are already high and are, therefore, not a small correction). This region does not seem to be so very large but in experiments the variation of the specific heat of some HTSCs near T_c has a clearly pronounced λ-shaped form similar to the one we observe in helium II (see [93, p 2; 132]), where the original literature is cited).

In view of the latter circumstance, it seems interesting to extend the Ψ-theory to the fluctuation region. We shall touch upon this issue in section 7.4 because this extension was proposed in application to liquid helium. But after the discovery of HTSCs in 1986–87, such a 'generalized Ψ-theory' was suggested in application to superconductors as well [94, 90, 54].

Underlying the 'generalized' Ψ-theory of superconductivity is the following expression:

$$\widetilde{F} = \widetilde{F}_{n0} + \frac{C_0 T_c}{2} \tau^2 \ln \tau + \int \left[-a_0 \tau^{4/3} |\Psi|^2 + \frac{b_0}{2} \tau^{2/3} |\Psi|^4 \right.$$
$$\left. + \frac{g_0}{3} |\Psi|^6 + \frac{\hbar^2}{4m_k} \left| \left(\nabla_k - i \frac{2e_0}{\hbar c} A_k \right) \Psi \right|^2 \right] dV \qquad (7.46)$$

for the free energy which leads to the following equation for Ψ:

$$-\frac{\hbar^2}{4m_k} \left(\nabla_k - i \frac{2e_0}{\hbar c} A_k \right)^2 \Psi + \left(-a_0 \tau^{4/3} + b_0 \tau^{2/3} |\Psi|^2 + g_0 |\Psi|^4 \right) \Psi = 0. \quad (7.47)$$

If one neglects anisotropy and sets $m_k = m_0/2$, then equation (7.47) will differ from (7.14) by a transformed temperature dependence of the coefficients and by the presence of the term proportional to $|\Psi|^4 \Psi$. Taking the example of helium II, we shall see in section 7.4 that the 'generalized' Ψ-theory entails a number of consequences near T_c which correspond in reality in the case of liquid helium. One might think that this could also be extended to superconductors with a very small correlation length. Such a case corresponds in a certain measure to the Schafroth model [16] which involves small-sized pairs. One of the directions of HTSC theory is based precisely on this model [93].

An important point in the 'generalized' Ψ-theory is the problem of boundary conditions. Condition (7.15) is, generally speaking, already insufficient here and should be replaced [37, 90, 95] by a more general condition

$$n_k \Lambda_k \left[\frac{\partial \Psi}{\partial x_k} - \mathrm{i} \frac{2e_0}{\hbar c} A_k \Psi \right] = -\Psi \qquad (7.48)$$

on the boundary with a vacuum or a dielectric, where all the quantities are, of course, taken on the boundary, n_k are the components of the unit vector n perpendicular to the boundary, and Λ_k are some coefficients having dimensions of length, sometimes referred to as extrapolation lengths. For the isotropic case, when $\Lambda_k = \Lambda$, equation (7.48) takes on the form

$$n \left(\nabla \Psi - \mathrm{i} \frac{2e_0}{\hbar c} A \Psi \right) = -\frac{1}{\Lambda} \Psi \qquad (7.49)$$

(this Λ should not be confused with the coefficient (7.17) involved in the Londons theory (7.4), (7.5)).

For $\Lambda_k \gg \xi_k(T)$, condition (7.49) becomes condition (7.15) because, generally speaking, $\partial \Psi / \partial x_k \sim \Psi / \xi_k$. In the case $\Lambda_k \ll \xi_k(T)$, however, we arrive at the boundary condition

$$\Psi = 0. \qquad (7.50)$$

This condition on a rigid wall was chosen in the initial Ψ-theory of superfluidity [96]. As far as I know, the 'generalized' Ψ-theory of superconductivity was never used after paper [90]. Two reasons for this are possible. On the one hand, the 'generalized' Ψ-theory has no reliable microscopic grounds (as distinct from the conventional Ψ-theory of superconductivity considered earlier). On the other hand, the investigations of HTSC are obviously at such a stage now that it has probably not yet become necessary to solve problems requiring the application of the 'generalized' Ψ-theory. As far as the conventional Ψ-theory is concerned, its application to HTSC is also now only rather small scale.

I have dwelt on the development of the initial Ψ-theory [29] in three directions: allowing for anisotropy [75], for ferromagnetic superconductors [76], and in a fluctuation region [90]. Also of importance are extensions in another two directions, namely to the non-stationary case, when the function Ψ is time dependent, and to superconductors in which the order parameter is not reduced to the scalar complex function $\Psi(r)$. I obtained no results in either of these two directions. True, in what concerns the non-stationary generalization of the Ψ-theory, I already understood [64] in 1950 that this task did exist but restricted myself to the remark that equation (7.14) might be supplemented with the term $i\hbar\partial\Psi/\partial t$. Meanwhile, an allowance for relaxation is more significant. The corresponding equations for $\Psi(r,t)$ are discussed in reviews [85, 97, 237]. As to the so-called unconventional superconductors in which Cooper (or analogous) pairs are not in the s-state, I not only failed to contribute to this field, I also have a poor knowledge of it. By the way, the possibility of 'unconventional' pairing was first pointed out [98] for superfluid ^{3}He and this fact was later confirmed. In the case of superconductivity, the 'unconventional' pairing takes place for at least several superconductors with heavy fermions (UB_{13}, $CeCu_2Si_2$, UPt_3) and, apparently, several HTSCs—the cuprates. I shall restrict myself only to pointing to one of the pioneering papers in this field [99] and reviews [100–103]. It is a pleasure to me to note also that 'unconventional' superconductors are now the subject of successful research by Yu S Barash [104], my immediate colleague (our joint research was, however, conducted in quite a different field—the theory of Van der Waals forces [105]). It is noteworthy that an appropriately extended Ψ-theory is extensively used for 'unconventional' superconductors as well [99–102].

7.4 Ψ-theory of superfluidity

As I have already mentioned, the behavior of liquid helium near the λ-point was beyond the scope of Landau's interests. He also remained indifferent to the behavior of superfluid helium near a rigid wall. As for me, I was for some reason interested in both these questions from the very beginning of my work in the field of superfluidity, i.e. from 1943 [19]. I have already mentioned the attempt [49] to introduce the order parameter ρ_s near the λ-point. As regards the behavior of helium near the wall, it looks like this. Helium atoms stick to the wall (they wet it, so to say). How can it be combined with a flow along the wall of the superfluid part of the liquid with density ρ_s and a velocity v_s? We know that in the Landau theory of superfluidity [4] the velocity v_s along the wall (as distinct from the velocity v_n of a normal liquid) does not become zero on the wall. This means that, on the wall, the velocity v_s must become discontinuous (the velocity v_s cannot tend gradually to zero because of the condition rot $v_s = 0$). This velocity discontinuity must be associated with a certain surface energy σ_s [106]. Estimates show that this energy σ_s is rather high ($\sigma_s \sim 3 \times 10^{-2}$ erg cm^{-2}) and its existence must have led to a pronounced effect. Specifically, something like dry friction must have been observed—to move a rigid body placed in helium II, the energy $\sigma_s S$ must have been expended, where S is the body (say, plate) surface area. However, specially conducted experiments showed [107] that no energy $\sigma_s S$ is actually needed and a possible value of σ_s is at least by many orders of magnitude smaller than the previously mentioned estimates [106]. How can this contradiction be eliminated? The solution of the problem I saw in the assumption that the density ρ_s decreases on approaching the wall and, on the wall itself, $\rho_s(0) = 0$. Thus, the discontinuity of the velocity v_s on the wall is of no importance because the flow $j_s = \rho_s v_s$ tends gradually to zero on the wall itself even without a change of velocity v_s. By that time (1957), the Ψ-theory of superconductivity [29] had long since been constructed and there was no problem in extending it to the case of superfluidity and with the boundary condition $\Psi(0) = 0$ on the wall (see equation (7.50)), which provided the condition $\rho_s(0) = 0$, as well.

Unfortunately, I do not at all remember how far I had advanced in constructing the Ψ-theory of superfluidity before I learnt that L P Pitaevskii was engaged in solving the same problem. We, naturally, joined our efforts and the outcome was our paper [96] which we submitted for publication on 10 December, 1957.

The Ψ-theory of superfluidity constructed in [96] will, henceforth, be referred to as the initial Ψ-theory of superfluidity. The point is that this theory was later found to be inapplicable to helium II in the quantitative respect and we had to generalize it. Such a generalized Ψ-theory of superfluidity, developed by A A Sobyanin and myself [108–112], is far from being so well grounded as the Ψ-theory of superconductivity. In this connection and, I think, in view of an insufficient awareness of the distinction between the generalized theory and the initial one [96], the Ψ-theory of superfluidity has not attracted attention and, at the present time, remains undeveloped[7] and not systematically verified. Meanwhile, the microtheory of superfluidity is not nearly so well developed as the microtheory of superconductivity and the role of the macrotheory of superfluidity is particularly high. This has led Sobyanin and myself to the conviction that the development of the Ψ-theory of superfluidity and its comparison with experiment would be highly appropriate.

The most comprehensive of the cited reviews devoted to the generalized Ψ-theory of superfluidity [110] amounts to 78 pages. This alone makes it clear that, in this article, I have no way of giving an in-depth consideration to the Ψ-theory of superfluidity. Here we shall restrict ourselves to brief remarks.

We shall begin with the initial theory [96]. It is constructed in much

---

[7] One of the reasons, and perhaps even the main one, is the fact that A A Sobyanin has become a politician and for several years now has not been working practically as a physicist.[3*]

the same manner as the Ψ-theory of superconductivity [29]. As the order parameter, we took the function $\Psi = |\Psi| \exp i\varphi$ acting as an 'effective wavefunction of the superfluid part of a liquid' and so the density ρ_s and the velocity v_s are expressed as

$$\rho_s = m|\Psi|^2 \qquad v_s = \frac{\hbar}{m}\nabla\varphi$$

$$j_s = \rho_s v_s = -\frac{i\hbar}{2}(\Psi^*\nabla\Psi - \Psi\nabla\Psi^*) = \hbar|\Psi|^2\nabla\varphi \tag{7.51}$$

where $m = m_{\mathrm{He}}$ is the mass of a helium atom and a convenient normalization of Ψ is chosen; in [96] it is shown (see also later) that in the expression for v_s we have $m = m_{\mathrm{He}}$ irrespective of the manner in which Ψ is normalized. Then there come the expressions

$$F = F_0 + \frac{\hbar^2}{2m}|\nabla\Psi|^2$$

$$F_0 = F_{\mathrm{I}} + \alpha|\Psi|^2 + \frac{\beta}{2}|\Psi|^4 \qquad \alpha = \alpha'_\lambda(T - T_\lambda) \qquad \beta = \beta_\lambda \tag{7.52}$$

which are usual for the mean field theory (the Landau phase transitions theory), where $F_{\mathrm{I}}(\rho, T)$ is the free energy of helium I and T_λ is the temperature of the λ-point. In equilibrium, homogeneous helium II

$$|\Psi_0|^2 = \frac{\rho_s}{m} = \frac{|\alpha|}{\beta_\lambda} = \frac{\alpha'_\lambda(T_\lambda - T)}{\beta_\lambda} \qquad \Delta C_p = C_{p,\mathrm{II}} - C_{p,\mathrm{I}} = T_\lambda \frac{(\alpha'_\lambda)^2}{\beta_\lambda}. \tag{7.53}$$

In inhomogeneous helium II, the function Ψ obeys the equation

$$-\frac{\hbar^2}{2m}\nabla\Psi + \alpha\Psi + \beta_\lambda|\Psi|^2\Psi = 0 \tag{7.54}$$

which should be solved with the boundary condition (7.50) on a rigid wall.

As in equation (7.25), we introduce the correlation length (in [96] it is denoted by l)

$$\xi(T) = \frac{\hbar}{\sqrt{2m|\alpha|}} = \frac{\hbar\tau^{-1/2}}{\sqrt{2m\alpha'_\lambda T_\lambda}} = \xi(0)\tau^{-1/2} \qquad \tau = \frac{T_\lambda - T}{T_\lambda} = \frac{t}{T_\lambda}. \tag{7.55}$$

The estimate presented in [96] and based on the data of ΔC_p and ρ_s measurements (see equation (7.54)) gives approximately $\xi(0) \sim 3 \times 10^{-8}$ cm. At the same time, the Ψ-theory is applicable only provided the macroscopic Ψ-function changes little on atomic scales. This implies the condition $\xi(T) \gg a \sim 3 \times 10^{-8}$ cm (here a is the mean interatomic distance in liquid helium). Consequently, the Ψ-theory can only hold near the λ-point for $\tau \ll 1$, say, for $(T_\lambda - T) < (0.1\text{--}0.2)$ K. Of course, proximity to T_λ is also the condition of applicability of expansion (7.52) in $|\Psi|^2$. The small magnitude of the length $\xi(0)$ in helium leads at the same time to considerable dimensions of the fluctuation region [86]. Indeed, applying criterion (7.42), we arrive at the value $\tau_G \sim 10^{-3}$ for helium (see [108, formula (2.46)]). Thus, it turns out that the initial Ψ-theory of superfluidity can only hold under the condition 10^{-3} K $\ll (T_\lambda - T) \lesssim 0.1$ K, i.e. it is practically inapplicable because in the studies of liquid helium, of particular interest is exactly the range of values $(T_\lambda - T) \ll 10^{-3}$ K. The fact that the mean field theory leading to the jump in specific heat (7.53) does not hold for liquid helium (we certainly mean ^{4}He) is attested by the existence of a λ-singularity in the specific heat as well as the circumstance that the density ρ_s near T_λ does not behave at all like $(T_\lambda - T)$ according to (7.53) but rather changes by the law

$$\rho_s(\tau) = \rho_{s0}\tau^\zeta \qquad \zeta = 0.6705 \pm 0.0006 \qquad (7.56)$$

where the value of ζ is borrowed from the most recently reported data [113]. Note that, in [108], we gave the value $\zeta = 0.67 \pm 0.01$ and, in [110], the values $\zeta = 0.672 \pm 0.001$ and $\rho_s = 0.35\tau^\zeta$ g cm^{-3}. Hence, to a high accuracy, we have

$$\zeta = \frac{2}{3}. \qquad (7.57)$$

I cannot judge whether ζ actually differs from 2/3 but if it does, the difference does not exceed 1%. It is noteworthy that, in 1957, when paper [96] was accomplished, the variation of ρ_s by the law (7.56) was not yet known. We, therefore, did not raise an alarm immediately (the λ-type behavior of specific heat is less crucial in this respect because it may not be associated with variations of Ψ, whereas the density ρ_s is proportional to $|\Psi|^2$).

Thus, the initial Ψ-theory of superfluidity [96] is inapplicable to liquid helium (^{4}He). However, owing to its simplicity, it has a qualitative and occasionally even quantitative significance for ^{4}He as well. The main thing is that liquid ^{4}He is not the only existing superfluid liquid, suffice it to mention liquid ^{3}He at very low temperatures, ^{3}He–^{4}He solutions, non-dense ^{4}He films and neutron liquid in neutron stars, as well as possible superfluidity in an exciton liquid in crystals, in supercooled liquid hydrogen [114], and in the Bose–Einstein condensate of the gas of various atoms (it is this very question that is presently commanding the attention of physicists; see, for example, [115] and references therein).[4*] In some of these cases, the fluctuation region may appear to be small enough, so that the initial Ψ-theory of superfluidity may prove sufficient. This is apparently the situation in the particularly important case of superfluidity in ^{3}He (see comment 9). We shall, therefore, dwell briefly on the results obtained in paper [96].

We found the distribution $\rho_s(z)$ near a rigid wall and in a liquid helium film of thickness d. The function $\Psi(z)$ and, of course, $\rho_s = m|\Psi|^2$, where z is the coordinate perpendicular to the film, has a dome-like shape because on the boundaries of the film we have $\Psi(0) = \Psi(d) = 0$ (see equation (7.50)). Naturally, for a sufficiently small thickness d, the equilibrium value is $\Psi = 0$, i.e. the superfluidity vanishes. The corresponding critical value d_c (for $d < d_c$ a film is not superfluid) is equal to

$$d_c = \pi\xi(T) = \frac{\pi\hbar\tau^{-1/2}}{\sqrt{2m\alpha'_\lambda T_\lambda}} \qquad \tau = \frac{T_\lambda - T}{T_\lambda}. \tag{7.58}$$

This result implies that, for a film, the λ-transition temperature is lower than that for 'bulk' helium. Concretely, from equation (7.58), it follows that, for a film, the λ-transition takes place at a temperature ($T_\lambda \equiv T_\lambda(\infty)$)

$$T_\lambda(d) = T_\lambda - \frac{\pi^2\hbar^2}{2m\alpha'_\lambda d^2} = T_\lambda - \frac{\pi^2 T_\lambda \xi^2(0)}{d^2}. \tag{7.59}$$

The specific heat of the film changes with varying d, too. Such effects in small samples are observed experimentally. In [96] we also solved the problem of the vortex line, the value of Ψ on its axis being equal to zero and the velocity circulation around the line being

$$\oint v_s ds = \frac{2\pi\hbar k}{m} \qquad k = 0, 1, 2, \dots . \tag{7.60}$$

In this formula, the ^{4}He atom mass $m = m_{He}$ should be used considering that the circulation cannot change with temperature and, as was shown by Feynman [116], at $T = 0$ it is the mass m_{He} that enters into equation (7.60). Finally, in [96], we found the surface energy on the boundary between helium II and a rigid body and the vortex line energy.

The fact that for liquid helium and a number of other transitions the mean field (Landau theory) does not hold led to the appearance of the generalized theory in which the free energy is written in the form (7.37) but with a different temperature dependence of the coefficients. Specifically, for the order parameter Ψ, we write

$$F_{II} = F_I - a_0\tau|\tau|^{1/3}|\Psi|^2 + \frac{b_0}{2}|\tau|^{2/3}|\Psi|^4 + \frac{g_0}{3}|\Psi|^6. \tag{7.61}$$

Since for small $|\Psi|^2$ in equilibrium (see equation (7.53)) $|\Psi_0|^2 = |\alpha|/\beta = a_0\tau^{2/3}/b_0$, this result is in agreement with (7.56), (7.57). Expression (7.61) is naturally so chosen as to correspond to experiment. Parenthetically, the same method in application to the Ψ-theory of superconductivity was employed in paper [67], only not near but far from T_c. As far as I know, expression (7.61) was first applied by Yu G Mamaladze [117]. Some other authors also discussed a generalization of the phase transition theory in the spirit of involving an equation of the type (7.61) (see references in [108]).

Sobyanin and I developed the generalized Ψ-theory of superfluidity [108–112] on the basis of expression (7.61) which in turn underlay the 'generalized' Ψ-theory of superconductivity (see [90] and section 7.3). But while the latter is of limited significance, the generalized Ψ-theory of superfluidity is a unique scheme capable of describing the behavior of liquid helium near the λ-point, not counting the incomparably more sophisticated approach based on the renormalization group theory (see [118] and references therein). In addition, this approach [118] is either of no or limited validity for the inhomogeneous and non-stationary cases.

Without going into details, we shall immediately present the expression for the involved free-energy density in some reduced units (instead of free energy, other thermodynamic potentials were used in [108–112] but this is of no importance):

$$F_{\text{II}} = F_{\text{I}} + \frac{3\Delta C_p}{(3+M)T_\lambda}\left[-t|t|^{1/3}|\Psi|^2 + \frac{(1-M)|t|^{2/3}}{2}|\Psi|^4 \right.$$
$$\left. + \frac{M}{3}|\Psi|^6 + \frac{\hbar^2}{2m}|\nabla\Psi|^2\right]. \tag{7.62}$$

Here $t = T_\lambda - T$, ΔC_p is the jump of specific heat determined by expression (7.53), M is the constant introduced in the theory, $\Psi = \Psi/\Psi_{00}$, $\Psi_{00} = \sqrt{1.43\rho_\lambda/m}$, $\rho_s = 1.43\rho_\lambda \times (T_\lambda - T)^{2/3}$. In the simplest version of the theory, we have $M = 0$ and, irrespective of this fact, the reduced order parameter Ψ is sometimes (for instance, in the vicinity of the axis of a vortex line) rather small and the term $|\Psi|^6$ in equation (7.62) can be ignored. Comparison with experiment for helium II leads to the estimate $M = 0.5 \pm 0.3$ (see [112]). The transition is second order for $M < 1$ and first order for $M > 1$.

For a shift of the λ-transition temperature in a film (for $M < 1$), we have

$$\Delta T_\lambda = T_\lambda - T_\lambda(d) = 2.53 \times 10^{-11}\left(\frac{3+M}{3}\right)d^{-3/2}\,(\text{K}) \tag{7.63}$$

which generalizes expression (7.59) and corresponds to experimental data; and for a capillary with diameter d, the coefficient 2.53 in equation (7.63) is replaced by 4.76. Expressions for a number of other quantities (density, specific heat, etc) are obtained and the effect of the external (gravitational, electric) fields as well as van der Waals forces are taken into account. The behavior of ions in helium II, the dependence of the density ρ_s on velocity v_s, and the vortex line structure are considered [119]. Furthermore, the theory is extended to the case of the presence of a flow of the normal part of a liquid (density ρ_n, velocity v_n) and the presence of dissipation and relaxation (for a non-stationary flow; for the initial Ψ-theory, this was done partially in paper [120]). The problem of vortex creation in a superfluid liquid (see [110] where the corresponding literature is cited) is very interesting. We note that, somewhat unexpectedly, this question proved to be of interest for simulating the process of creation of so-called topological defects in cosmology [121]. I believe that in an analysis of corresponding experiments the Ψ-theory of superfluidity may turn out to provide quite suitable methods.

The generalized Ψ-theory of superfluidity was not developed 'from first principles' or on the basis of a certain reliable microtheory (as in the situation with the Ψ-theory of superconductivity). This is a phenomenological theory that rests on the general theory of second-order phase transitions (Landau theory and scaling theory) and on experimental data [110, 111]. Such data are unfortunately quite insufficient for drawing a vivid conclusion concerning the region of applicability of the Ψ-theory. In the papers [122, 123], we find rather pessimistic judgements in this respect but Sobyanin was of the opinion that such a criticism is groundless. I do not hold any particular viewpoint here but my intuition suggests a very positive role of both the initial [96] and the generalized [108–112] Ψ-theories of superfluidity. In any case, clarification of the precision and the role of the Ψ-theory is currently pressing because experimental studies of superfluidity in helium II are continued (see, for example, [124, 125]; see also comments 4, 10, and 13).

7.5   Thermoelectric phenomena in superconductors

Different papers have their own fate. My first paper [19] on superconductivity now seems dull to me and this is all bygone times. And what concerns the second paper [22] accomplished in the same year, 1943, remains topical up to the present date. It was devoted to thermoelectric phenomena in superconductors. Before that, thermoelectric effects had been considered (see, for example, [58, 126]) to disappear completely in the superconducting state. Specifically, when a superconducting current passes through a seal of two superconductors, the Peltier effect is absent, the same as a noticeable thermoelectric current is absent upon heating one of the seals of a circuit consisting of two superconductors. But as a matter of fact, thermoelectric phenomena in superconductors do not vanish completely, although they can manifest themselves only under special conditions [22, 24]. The point is that in a superconductor one should take into account the possibility of the appearance of two currents—superconducting (the density j_s) and normal (the density j_n). In a non-superconducting (normal) state in a metal there may flow only one current j, Ohm's law $j = \sigma E$ holding in the simplest case. If there exists a gradient of chemical potential μ of electrons in a metal and a temperature gradient, then

$$j = \sigma \left(E - \frac{\nabla \mu}{e_0} \right) + b \nabla T. \qquad (7.64)$$

In the superconducting state, as can readily be seen (see, for example, [128]), for the normal current, we have

$$j_{\mathrm{n}} = \sigma_{\mathrm{n}}\left(E - \frac{\nabla\mu}{e_0}\right) + b_{\mathrm{n}}\nabla T \tag{7.65}$$

instead of equation (7.3), and in the Londons' approximation equation (7.4) is preserved; instead of equation (7.5), we obtain

$$\frac{\partial(\Lambda j_{\mathrm{s}})}{\partial t} = E - \frac{\nabla\mu}{e_0} + \nabla\frac{\Lambda j_{\mathrm{s}}^2}{2\rho_e} \tag{7.66}$$

where μ is the chemical potential of electrons and $\rho_e = e_0 n_{\mathrm{s}}$, n_{s} is the concentration of 'superconducting electrons' ($j_{\mathrm{s}} = e_0 n_{\mathrm{s}} v_{\mathrm{s}}$). Here we omit the detail connected with the necessity of introducing different chemical potentials μ_{n} and μ_{s} in non-equilibrium conditions for a normal and superconducting electron subsystems (see [128]). Note that the last term on the right-hand side of equation (7.66) is of hydrodynamic character (see equation (7.6)) and in equation (7.5), it was omitted because of its small magnitude. However, the contribution of this term can be observed experimentally (see [128, 243] and references therein). Forgetting again about the last term in equation (7.66) in the stationary case for a superconductor, we have

$$E - \frac{\nabla\mu}{e_0} = 0 \tag{7.67}$$

from which it follows that (see equation (7.65))

$$j_{\mathrm{n}} = b_{\mathrm{n}}(T)\nabla T. \tag{7.68}$$

Thus, in a superconductor, the thermoelectric current j_{n} does not vanish completely. Why then is it not observed? As has already been mentioned, under particularly simple conditions, a normal current is totally compensated for by a superconducting current, i.e.

$$j = j_{\mathrm{s}} + j_{\mathrm{n}} = 0 \qquad j_{\mathrm{s}} = -j_{\mathrm{n}}. \tag{7.69}$$

By 'particularly simple conditions', we understand a homogeneous and isotropic superconductor, say, a non-closed small cylinder (a wire) on one end of which the temperature is T_1 and on the other end T_2 (we assume that $T_{1,2}$ is less than T_{c}).[8] In such a specimen, in the normal state (for $T_{1,2} > T_{\mathrm{c}}$), we certainly have $j = 0$ and $E = \nabla\mu/e_0 - b\nabla T/\sigma$ (see equation (7.64));

---

[8] I did not want to place figures in this paper, although perhaps they would not be out of place here. But all the necessary illustrations concerning thermoeffects can be found in the readily available papers [128, 129, 213] and also in my Nobel lecture [264].

in the superconducting state, we of course also have $j = 0$ but (see equations (7.68) and (7.69))

$$j_s = -j_n = -b_n \nabla T \qquad E - \frac{\nabla \mu}{e_0} = 0. \qquad (7.70)$$

If a superconductor is inhomogeneous and (or) anisotropic, then, generally speaking, the total compensation (7.69) does not occur and a certain, although weak, thermoelectric current must be [22] and is, in fact, observed [128, 129, 213]. But one should not think that in the simple case considered earlier, when $j = 0$, all thermoelectric effects disappear. Indeed, the thermoelectric current j_n must be associated with some heat transfer, i.e. in superconductors, there must occur an additional (say, circulational or convective) heat transfer mechanism similar to the one that exists in a superfluid liquid.[9] This analogy was, properly speaking, the starting point for me in paper [22]. However, in [22], I made no estimate of the additional (circulational) thermal conductivity. Later I decomposed [64] the total heat conductivity κ involved into the heat transfer equation $q = -\kappa \nabla T$ into three parts: $\kappa = \kappa_{ph} + \kappa_e + \kappa_c$. Here κ_{ph} stands for the contribution due to phonons (the lattice), κ_e is due to electron motion such that there is no circulation (i.e. under the condition $j_n = 0$), and κ_c is due to circulation (convection). The estimates done in [64] indicated that κ_c must be negligibly small compared to κ_e but now, unfortunately, I do not understand these estimates.

After the BCS theory was created, it became possible to carry out a microscopic evaluation of κ_e and κ_c. According to [130], at $T \sim T_c$

$$\frac{\kappa_c}{\kappa_e} \sim \frac{k_B T_c}{E_F} \qquad (7.71)$$

where E_F is the Fermi energy of electrons in a given metal.

This estimation was obtained earlier [222] on the basis of the two-fluid model and some assumptions. Finally, I also came to result (7.71) by estimating the thermal flux (heat transfer) due to creation of Cooper pairs in the colder end of a sample and their decay in the hot end [129, 132].[10] I had some doubts of whether the heat flux calculated on the basis of the kinetic equation [130] and an allowance of the effect on the boundaries [129, 132] should be summed up. Such an assumption is, however, erroneous: whenever the kinetic equation holds (i.e. the free path of 'normal electrons' is small compared to the sample length), the kinetic calculation, and the allowance for pair creation and breakdown on the boundaries are equivalent.

---

[9] Such heat transfer is also possible in semiconductors that possess the corresponding electron and hole conductivities simultaneously (see [131]).

[10] This result was also presented in section 5 of the paper 1997 *Phys.–Usp.* **40** 407 before its modification.

However, the doubts in the validity of estimate (7.71) appeared to be useful since a more consistent estimation gave another result [213]:

$$\frac{\kappa_c}{\kappa_e} \sim \left(\frac{k_B T_c}{F_F}\right)^2. \tag{7.72}$$

Apparently, kinetic calculations in [130] contained an error. The previously mentioned referred to isotropic superconductors but, in this case, the Seebeck coefficient $S = b/\sigma$ is known to be underestimated for the well-known reason by a quantity of the order of $E_F/k_B T$ (see [133, 134, 223]). Hence, for anisotropic and unconventional superconductors, estimate (7.71) is likely to be reasonable. For conventional isotropic superconductors at $T_c \sim (1\text{--}10)$ K and $E_F \sim (3\text{--}10)$ eV, the convective thermal conductivity is quite negligible according to (7.72) because $\kappa_c/\kappa_e \lesssim 10^{-7}$. But for high-temperature superconductors at $T_c \sim 100$ K and $E_F \sim 0.1$ eV, we already have $\kappa_c/\kappa_e \sim 0.1$ according to (7.71). The roughness of the estimate allows the suggestion that, in some cases, convective thermal conductivity may be appreciable. Therefore, I tried to explain [132] in this way the observed peak of thermal conductivity coefficient in HTSCs at $T \sim T_c/2$ (see [135--138]). However, this effect can also be explained by the corresponding temperature dependence of the coefficients κ_{ph} and κ_c. This issue was discussed in the literature. Observation of the Righi--Leduc effect, also referred to as the thermal Hall effect [224], led to the conclusion [225] that the phonon part of thermal conductivity makes no case here (i. e. the contribution of the coefficient κ_{ph} is insignificant; see [224, 138]). At the same time, it is impossible to separate directly the contributions from κ_e and κ_c and I am not aware whether it can generally be done (see, however, later; an analysis is needed that would involve the role of anisotropy and the external magnetic field; see [213]).[11]

I have dwelt on the convective thermal conductivity (heat transfer) in superconductors at such length because I feel somewhat particularly unsatisfied in this respect. I have never properly investigated the microtheory or, as it is more often called, the electron theory of metals. That is why I was unable, and never even tried, to construct a consistent microtheory of convective heat transfer. Now it is certainly late for me. But I hope that someone will investigate this problem sooner or later.

If a superconductor is not homogeneous and isotropic, as has already been mentioned, no complete compensation of currents j_n and j_s occurs and some thermal currents must generally flow. The simplest cases are as follows: an isotropic but inhomogeneous superconductor and a homogeneous but anisotropic superconductor (monocrystal). More than 50

---

[11] See also recent papers [244, 245] devoted to heat transfer and related problems.

years ago (!), when paper [22] was written, alloys and generally inhomogeneous superconductors were thought of as something 'polluted', and it was not even clear whether the Londons' equations can be used in these conditions. For this reason, the case of inhomogeneous superconductors was only slightly touched upon in [22]. Concretely, it was pointed out that if in a bimetallic plate (say, different superconductors sealed or welded to each other), there is a temperature gradient perpendicular to the seal plane, an uncompensated current j is excited along the seal line, which runs around the seal: this generates a magnetic field perpendicular to both the plate and the seal line (see figure 3(a) in [128] and figure 3 in [129]). As I have said, such a version does not seem interesting. Attention was, therefore, given to a monocrystal with non-cubic symmetry when the tensor Λ_{ik} does not degenerate into a scalar (for cubic and isotropic superconductors $\Lambda_{ik} = \Lambda\delta_{ik}$). If in such a plate-like crystal, the temperature gradient ∇T is not directed along the symmetry axis, a current j flowing round the plate is excited and a magnetic field H_T, proportional to $|\nabla T|^2$, is generated perpendicular to the plate. This field can easily be measured using modern methods. For details, see [22, 128, 129, 140]. Unfortunately, attempts to observe the thermoelectric effect in question were made only in paper [141], the results of which remain ambiguous [128, 140].

As it turned out, the thermoeffect for inhomogeneous isotropic superconductors is easier to analyze and easier to observe. For this purpose, it is most convenient to consider not a bimetallic plate but rather a superconducting ring (a circuit) consisting of two superconductors (with one seal at a temperature T_2 and the other at a temperature $T_1 < T_2$; see figure 3(b) in [128], figure 7 in [129], or figure 3 in [213]). The pertinence of the choice of this particular version was indicated in papers [142, 143]. Paper [142] argued that this effect was quite different from that considered in [22] but this was a misunderstanding [128, 144]. Indeed, a bimetallic plate and a circuit of two superconductors differ topologically because of the presence of a hole in the latter case, which leads to the possibility of the appearance of a quantized magnetic-field flux through the hole (see figure 3 in [128]). A simple calculation (see [128, 129, 142–145]) shows that the flux through the indicated hole is equal to

$$\Phi = k\Phi_0 + \Phi_T \qquad \Phi_T = \frac{4\pi}{c}\int_{T_1}^{T_2}(b_{\text{n, II}}\,\delta_{\text{II}}^2 - b_{\text{n, I}}\,\delta_{\text{I}}^2)\mathrm{d}T$$

$$\Phi_0 = \frac{hc}{2e_0} = 2\times10^{-7}\ \text{G cm}^2 \qquad k = 0, 1, 2, \ldots \tag{7.73}$$

where the indices I and II refer to metals I and II forming the superconducting circuit, respectively, $\delta \equiv \delta_0$ is the penetration depth: for $k = 0$, we obtain the result for a bimetallic plate. If we assume for simplicity that $(b_n\delta^2)_{II} \gg (b_n\delta^2)_I$ and $\delta_{II}^2 = \delta_{II}^2(0)(1 - T/T_{c,II})^{-1}$, then from equation (7.73), we obtain $(T_c = T_{c,II})$

$$\Phi_T = \frac{4\pi}{c} b_{n,II}\delta_{II}^2(0) T_c \ln\left(\frac{T_c - T_1}{T_c - T_2}\right). \tag{7.74}$$

Estimates for tin $(b_n(T_c) \sim 10^{11}\text{--}10^{12}$ CGSE, $\delta(0) \approx 2.5 \times 10^{-6}$ cm) when $(T_c - T_2) \sim 10^{-2}$ K, $(T_c - T_1) \sim 0.1$ K, and generally $\ln[(T_c - T_1)/(T_c - T_2)] \sim 1$ lead to the value $\Phi_T \sim 10^{-2}\Phi_0$. Such a flux can readily be measured and this was done in a number of papers as far back as 20 years ago (for the references, see [128, 145]). Here I will only refer explicitly to paper [146], which also confirmed the result (7.74).

As far as the thermoelectric current in a superconducting circuit is concerned, everything seems to be clear in principle but this is not so. The point is that for a sufficiently massive and closed toroidal-type circuit (a hollow cylinder made of two superconductors), the measured flux $\Phi(T)$ appeared [145] to be several orders of magnitude higher than the flux (7.74) and, moreover, to possess a different temperature dependence. The origin of such an 'enormous' thermoeffect in superconductors has not yet been clarified. A probable explanation was suggested by R M Arutyunyan and G F Zharkov [147], although it has not yet been confirmed by experiment. In this case, the measured flux through the hole is equal to $\Phi_T + k\Phi_0$ rather than Φ_T. As the critical temperature of the hottest seal T_2 approaches the temperature T_c of one of the superconductors, the resulting increase in thermoelectric current increases the entrapped flux $k\Phi_0$, i.e. a growth of k, energetically advantageous. This question was discussed in a number of papers [147–150] but the mechanism responsible for the increase in the flux $\Phi(T)$ still remained unclear and no new experiments have been carried out. The mechanism of vortex formation in the walls of a superconducting cylinder that leads to an increase of an entrapped flux with increasing thermoelectric current has been proposed only recently in [151]. There are other explanations [246] of the results obtained in [145].

It should be noted that expression (7.73), which also implies formula (7.74), is derived on the assumption that the total current density $\boldsymbol{j} = \boldsymbol{j}_s + \boldsymbol{j}_n$ is zero throughout the entire circuit thickness. Meanwhile, near T_c, when the field penetration depth δ increases (more than this, $\delta \to \infty$ as $T \to T_c$), the current density j tends to the value corresponding to the current in the normal state (i.e. at $T > T_c$). Clearly, the flux Φ must then increase. Under such conditions, allowance should be made for the appearance in a superconductor of some charges (the so-called charge imbalance effect; see [145, 226] and other literature cited in [145, 213, 226]). It is only allowance for the role of these charges that provides continuity for the transition from the normal to the superconducting state. By the way, near T_c, particularly when the coherent length ξ is small, the fluctuation effects also deserve attention. The influence of the charge imbalance effect upon the temperature dependence of the flux Φ in a superconducting ring was discussed in [227], where the effect was found to be small but the physical meaning of this result is not clear to me. I believe, in particular, that the allowance for flux entrapment (i.e. an increase in the number of trapped quanta of the flux with temperature) should be analyzed simultaneously with the allowance for the charge imbalance effect. The latter would also provide clear insight (which in my opinion has not yet been attained) into the character and the results of measurements of thermal e.m.f. in the circuit upon a superconducting transition of one and then both of its units.

I turned [129, 132, 139] to the convective mechanism of thermal conductivity in superconductors many times and could not then understand why this issue was being ignored. Now (after paper [213]), the most probable explanation seems to be the fact that within a correct kinetic calculation the convective mechanism is involved automatically. Therefore, the contributions of κ_e and κ_c need not be separated from the observed coefficient of the electron component of thermal conductivity $\kappa_e^{tot} = \kappa_e + \kappa_c$. But is it always (when anisotropy and the action of external forces are involved) impossible to separate κ_e and κ_c? This remains unclear to me. Furthermore, the coefficient κ_e can perhaps be determined by measuring the conductivity σ_n according to the Wiedemann–Franz law. Then the coefficient κ_c will be determined as the difference $\kappa_e^{tot} - \kappa_e$.

I hope, although not very much, that thermoeffects in superconductors (in a superconducting state) will no longer be ignored and there will finally appear corresponding experiments involving, in particular, HTSCs. In my opinion, it is nevertheless conceivable that the convective heat conduction mechanism plays a part in some cases.[11*]

Concluding this section, I would like to emphasize that, in accordance with the general context of this paper, I have only concentrated on those thermoelectric phenomena in superconductors which I investigated myself. Nevertheless, there exist some other related aspects of the problem. In this respect, I shall restrict myself by referring the reader to reviews [128, 129, 145] and the references therein, as well as the books [40, 228] and the papers [152–154, 225].

7.6 Miscellanea (superfluidity, astrophysics, and other things)

As mentioned in section 7.1, my first work [23] in the field of low-temperature physics, which was accomplished at the beginning of 1943, was devoted to light scattering in helium II. This question was rather topical at that time because, when comparing the transition in helium and the Bose–Einstein gas condensation, one might expect very strong scattering near the λ-point. At the same time, the Landau theory [4] suggested no anomaly. But this was, so to say, a trivial result. The most interesting thing is that the scattering spectrum must consist not of the central line and a Mandelstam–Brillouin doublet as in usual liquids but of two doublets. Indeed, the Mandelstam–Brillouin doublet is associated with scattering on sound (or, more precisely, hypersonic) waves, while the central line is associated with scattering on entropy (isobaric) fluctuations. In the case of helium II and generally superfluid liquids, entropy fluctuations propagate (or, more precisely, dissipate) in the form of a second sound. This is the reason why instead of a central peak a doublet must be observed that corresponds to scattering on second sound waves. In paper [23], I noted, however, that 'the inner anomalous doublet cannot be actually observed because on the one hand the corresponding splitting is too small ($\Delta\omega_2/\omega_2 \sim u_2/c \lesssim 10^{-7}$) and on the other hand, and this is particularly important, the intensity of this doublet is quite moderate'. Indeed, the inner-to-outer doublet intensity ratio is $I_2/I_1 \approx C_p/C_V - 1$ ($C_{p,V}$ is the specific heat at a constant pressure or for a constant volume). Even near the λ point, at low pressure in helium II, we have $C_p/C_V = 1.008$. However, as in many other cases in physics, the pessimistic prediction did not prove to be correct. Firstly, the intensity of the inner doublet increases greatly with pressure and, secondly, and this is especially significant, the use of lasers promoted great progress in the study of light scattering. As a result, the inner doublet could be observed and investigated (see [155] and review [156, p 830]).

I have already mentioned papers [49, 106] devoted to superfluidity, to say nothing of papers [96, 108–112, 119] on the Ψ-theory of superfluidity. I would like also to mention the notes [157, 158] whose titles cast light on their contents. Finally, I shall dwell on the thermomechanical circulation effect in a superfluid liquid [144, 159]. In a ring-shaped vessel filled with a superfluid liquid (concretely, helium II was discussed) and having two 'weak links' (for example, narrow capillaries), in the presence of a temperature gradient there must occur a superfluid flow spreading to the entire vessel. Curiously, the conclusion concerning the existence of such an effect was suggested [144] by analogy with the thermoelectric effect in a superconducting circuit. At the same time, the conclusion was drawn concerning the existence of thermoelectric effects in superconductors [22], in turn, by analogy with the 'inner convection' occurring in helium II in the presence of temperature gradient.

The effect under discussion was observed [160] but the accuracy of measurements of the velocity v_s was not enough to fix the jumps of circulation in superfluid helium (the circulation quantum is $2\pi\hbar/m_{\text{He}} \approx 10^{-3}$ cm^2 s^{-1}) which had been predicted by the theory [159]. Meanwhile, there exist interesting possibilities of observing not only jumps of circulation of a superfluid flow but also peculiar quantum interference phenomena (to this end, 'Josephson contacts' must be present in the 'circuit', for example, narrow-slit diaphragms). In my opinion, the circulation effect in a non-uniformly heated ring-shaped vessel is fairly interesting and not only for ^{4}He or solutions of ^{4}He with ^{3}He but perhaps also in the case of superfluidity of pure ^{3}He. Considering an extensive front of research in the field of superfluidity all over the world, I cannot understand why this effect is totally neglected. I do not know whether this is a matter of fashion, a lack of information, or something else.[12]

To save space in the other sections of the present paper, I shall mention here the works [114, 161–163]. The first of them [114] stresses the fairly obvious fact that molecular hydrogen H_2 does not become superfluid only for the reason that, at a temperature T_m exceeding the λ-transition temperature T_λ, it solidifies. As is well known, for H_2 the temperature T_m is 14 K, whereas by estimation T_λ should be nearly 6 K. Perhaps liquid hydrogen may be supercooled, for example, by way of expansion (a negative pressure), application of some fields and the use of films on different substrates as well as in the dynamical regime.

The possibility of observing the secondary sound and convective heat transfer in superconductors, in the first place accounting for exciton-type excitations (we mean bosons) was considered in paper [161]. I should say that paper [161] was written in 1961 and I am unaware of the present state of the questions discussed in it.

In 1978, there appeared reports on the observation of a very strong diamagnetism (superdiamagnetism) in CuCl, when the magnetic susceptibility χ is negative, and $|\chi| \sim 1/4\pi$ (of course, $|\chi| < 1/4\pi$ because $\chi = -1/4\pi$ corresponds to an ideal diamagnetism). After that (in 1980), there appeared indications of the existence of superdiamagnetism in CdS too. What it was that was actually observed in the corresponding experiments (for references see [162]) remains still unclear and this question was somehow 'drawn in the sand'.[5*] Many physicists believe that the measurements were merely erroneous. In any case, attempts were made to associate the observations with the possibility of the existence of superdiamagnetics other than superconductors.[13]

The last study in this direction in which I took part was reported in paper [162]. Further on, the question of superdiamagnetism somehow ·"faded away' (see, however, [164]) and I am unacquainted with the progress

---

[12] A A Sobyanin has recently pointed out the interesting possibility of 'spinning-up' the normal component of helium II inside a vessel by means of electric and magnetic fields acting on the helium ions.[3*]

[13] In these experiments, a very strong diamagnetism was observed but the conductivity of the samples was not at all anomalously large. Such a situation is also possible for superconductors in the case of superconducting seeds (granules) are separated by non-superconducting layers. The question, however, arose as to whether or not superdiamagnetism can be observed in dielectrics and non-superconductor, in general.

in this field. When seeking ways of explaining superdiamagnetism, I made an attempt to generalize the Ψ-theory of superconductivity [163]. It is unknown to me whether this paper is of any value now.

Concluding this section, I shall dwell on an astrophysical problem, namely the possibility of the existence of superconductivity and superfluidity in space. It seems to me that a small digression will not lead us beyond the scope of the general context of the paper. When I was young and then middle-aged, I used to entertain myself or, maybe, to do an exercise which I called then a brainstorming (I wrote about it in my book [1, p 305]).[6*] The procedure of the 'attack' was as follows: looking at my watch, I set myself a task to think up some effect within a certain time interval, say, within 15–20 min. Here is a concrete example. If I am not mistaken, it was 1962, I was travelling by train from Kislovodsk to Moscow. I was alone in the compartment with no book to read and so decided to conceive of something. I had been engaged in low-temperature physics and astrophysics for a number of years and, therefore, a natural question for me was where and under what conditions superfluidity and superconductivity could be observed in space. To formulate a question is frequently equivalent to doing half the work. It actually took me no more than the prescribed time to think that the existence of superfluidity is possible in neutron stars and superconductivity in the atmosphere of white dwarfs and that there may exist superfluidity of the neutrino 'sea'. On returning to Moscow, I took up all three problems—the first two together with D A Kirzhnits [165, 166] and the third in collaboration with G F Zharkov [167].

The interaction between neutrons with antiparallel spins in the s-state corresponds to attraction and, therefore, in a degenerate neutron gas there will appear pairing in the spirit of BCS theory. For the gap width $\Delta(0) \sim k_B T_c$, we obtained the estimate $\Delta(0) \sim (1-20)$ MeV, i.e. in the center of a neutron star (for a density $\rho \sim 10^{14}-10^{15}$ g cm^{-3}) we obtained $T_c \sim 10^{10}-10^{11}$ K, while, on the neutron phase boundary (for $\rho \sim 10^{11}$ g cm^{-3}), we had $T_c \sim 10^7$ K. It was also indicated that the rotation of a neutron star results in the formation of vortex lines. The fact that, in nuclear matter, superfluidity may occur had actually been known before but applied to neutron stars (at that time, in 1964, they had not yet been discovered), as far as I know, our paper was pioneering. Incidentally, in paper [168], in which I summarized my activity in the field of superfluidity and superconductivity in space, I also pointed to a possible superconductivity of nuclei-bosons (for example, α-particles) in the interior of white dwarfs and to superconductivity of protons which are present in a certain amount in neutron stars.

The possibility of the existence of superconductivity in some surface layer of the cold stars—white dwarfs was discussed in papers [166, 168]. The estimates give little hope. For example, for a density $\rho \sim 1$ g cm^{-3} the temperature is $T_c \sim 200$ K and, as the density increases, T_c falls rapidly.

Somewhat more interesting is the possibility of the superconductivity of metallic hydrogen in the depths of large planets—Jupiter and Saturn [168]. The estimates of the critical temperature T_c for metallic hydrogen, which are known from the literature, reach 100–300 K but the temperature in the depth of the planets is unknown. I am unacquainted with the present-day state of the problem but it seems to me that the existence of superconductivity in stars and large planets is hardly probable. The possibility of the appearance of superfluidity in the degenerate neutrino 'sea', whose existence at the early stages of cosmological evolution was discussed in some papers, was considered in note [167] (see also [168]). Such a possibility, as applied to neutrinos or some hypothetical particles now involved in the astrophysical arsenal, is currently of no particular interest but nevertheless it is reasonable to bear in mind.

7.7   High-temperature superconductivity

Beginning in 1964, I started investigating high-temperature superconductivity (HTSC) and from that time this problem remained, and remains, at the center of my attention although I was interested in many other things as well. My story about this work should, however, begin with quite a different question that concerns surface superconductivity. This question is as follows. Can two-dimensional superconductors in which the electrons (or holes) participating in superconductivity are concentrated near the boundary of, say, a metal or a dielectric with a vacuum, on the boundary between, e. g., twins (i. e. on the boundary of twinning), etc exist? It seems to me that surface superconductivity might be particularly well pronounced for electrons on surface levels which were first considered by I E Tamm as far back as 1932 [169]. The possibility of this particular superconductivity was discussed in paper [170]. The answer was affirmative—the Cooper pairing and the whole BCS scheme works in the two-dimensional case as well. The following possibility was also pointed out: electrons are located at volume-type levels but their attraction, which leads to superconductivity, takes place only near the body surface (or on the twinning boundary). Note that surface ordering, although absent in the volume, may certainly take place not only in the case of superconductivity, it is also possible, for example, for ferro- and antiferromagnetics, and ferroelectrics [171]. I subsequently saw experimental research testifying to the realistic character of such situations. But I did not follow the appearance of the corresponding literature and cannot, therefore, give any references. Besides, this is not the subject of the present paper. As to surface superconductivity, it was emphasized in 1967 that long-range superconducting order is impossible in two dimensions [172]. At the same time, as distinguished from the one-dimensional case, in two dimensions (the case of a surface) the fluctuations that destroy the order increase with the surface size L only logarithmically. Accordingly, even for surfaces of macroscopic size ($L \gg a$, where a is atomic size), the fluctuations may not be so large [173]. An even more important circumstance is that, in a two-dimensional system, there may occur a quasi-long-range order under which superfluidity and superconductivity are preserved. This is an extensive issue and we, therefore, restrict ourselves to mentioning paper [174] and the monograph [175 (chapter 1, section 5 and chapter 6, section 5)], where one can find the corresponding citations.[5*] Briefly speaking, superconductivity may well exist in two-dimensional systems. From an electrodynamic point of view, surface superconductors must behave as very thin superconducting films [176, 177]. In a certain sense, surface superconductivity is realized. For instance, superconductivity is observed in a $NbSe_2$ film with a thickness of only two atomic layers [178]. It would be more interesting to obtain surface superconductors on the Tamm (surface) levels [170]. It is obvious how interesting and probably important from the point of view of applications would be a dielectric possessing surface superconductivity. I am not, however, definitely sure that such a version may be thought of as radically different from a dielectric covered itself by a superthin superconducting film. But, after all, the difference does exist. The problem of surface superconductivity seems to be demanding and significant irrespective of the corresponding value of the critical temperature T_c.

The fates decreed, however, that surface superconductivity was to be associated with the problem of HTSC. To be more precise, the association appeared in my own work.

Before clarifying the matter, I shall make several remarks (henceforth, I shall sometimes use the text of my paper [179] which may prove to be unavailable to the reader).

For a full 65 years, the science of superconductivity was part of low-temperature physics, i. e. temperatures of liquid helium (and, in some cases, liquid hydrogen). Thus, for example, the critical temperature of the first known superconductor, mercury, discovered in 1911, is $T_c = 4.15$ K; and the critical temperature of lead, whose superconductivity was discovered in 1913, is $T_c = 7.2$ K. If I am not mistaken, higher T_c values were not achieved until 1930, although it was definitely understood that higher T_c were desirable. The next important step on this way was the synthesis of the compound Nb_3Sn with $T_c = 18.1$ K in 1954. Despite a great effort, it was not until 1973 that the compound Nb_3Ge with $T_c = 23.2$–24 K was synthesized. Subsequent attempts to raise T_c were unsuccessful until 1986, which saw the first indications (soon confirmed) of superconductivity in the La–Ba–Cu–O system with $T_c \sim 35$ K [180]. Finally, in early 1987, a truly HTSC $YBa_2Cu_3O_{7-x}$ with $T_c = 80$–90 K was created [181]. (This statement reflects my opinion that the term 'high-temperature' is appropriate only for superconductors with $T_c > T_{b,N_2} = 77.4$ K, where, obviously, T_{b,N_2} is the boiling nitrogen temperature at atmospheric pressure.)

The discovery of HTSCs became a sensation and gave rise to a real boom. One of the indicators of this boom is the number of publications. For example, in the period of 1989–91, about 15 000 papers devoted to HTSC appeared, i.e. on average, approximately 15 papers a day.[7]* For comparison, one of the reference books states that, in the 60 years from 1911 to 1970, about 7000 papers in total were devoted to superconductivity. Another indicator is the scale of conferences devoted to HTSC. Thus, at the conference M^{2}HTSC III in Kanazawa (Japan, July 1991) there were approximately 1500 presentations and the conference proceedings occupied four volumes with a total size of over 2700 pages (see [182]). Undoubtedly, such a scale of research is, to a large extent, explained by the high expectations for HTSC applications in technology. These expectations, by the way, from the very beginning appeared to me to be somewhat exaggerated and this was later confirmed in practice. But, of course, the potential importance of HTSC for technology, medicine (nuclear magnetic resonance tomograph), and physics itself leaves no doubts. Nevertheless, I still do not completely understand such a hyperactive reaction from the scientific community and the general public to the discovery of HTSC: it is some sort of social phenomenon.

Another phenomenon that may be attributed either to sociology or to psychology is the complete oblivion to which HTSC researchers, who began working successfully in 1986, consigned their predecessors. Indeed, the problem of HTSC was born not in 1986 but at least 22 years earlier— in its current form, this problem was first stated by Little in 1964 [183]. First of all, Little posed the question: Why was the critical temperature of the superconductors known at the time not so high? Secondly, he pointed out a possible way of raising T_c to the level of room temperature or even higher. Specifically, Little proposed replacing the electron–phonon interaction, responsible for superconductivity in the Bardeen, Cooper, and Schrieffer (BCS) model [18], by the interaction of conduction electrons with bound electrons or, in a different terminology which Little did not use, with excitons. In terms of the well-known BCS formula for the critical temperature

$$T_c = \theta \exp\left(-\frac{1}{\lambda_{\text{eff}}}\right) \qquad (7.75)$$

the meaning of the exciton mechanism is that the region of attraction between conduction electrons θ is set to be $\theta \sim \theta_{\text{ex}}$, where $k_B \theta_{\text{ex}}$ is the characteristic exciton energy. In contrast, for the electron–phonon mechanism of attraction in (7.75), we have $\theta \sim \theta_D$, where θ_D is the Debye temperature of the metal. Since the situation in which $\theta_{\text{ex}} \gg \theta_D$ is quite possible and even typical, it follows that for the same value of the effective dimensionless interaction parameter λ_{eff}, for the exciton mechanism T_c is $\theta_{\text{ex}}/\theta_D$ times higher than for phonons. Concretely, Little proposed to create an 'excitonic superconductor' on the basis of organic compounds by designing a long conducting (metallic) organic molecule (a 'spine') surrounded by side polarizers—other organic molecules [183].

It is not appropriate to go into details here. Let me just point out that Little's work did not remain unnoticed. Quite the opposite: it attracted a lot of attention. In particular, I also followed up Little's work by suggesting a somewhat different version: roughly speaking, replacing the quasi-one-dimensional conducting thread in Little's model with a quasi-two-dimensional structure ('sandwich'), i.e. with a conducting thin film placed between two polarizers (dielectric plates) [184]. More precisely, in paper [184], with a reference to the paper [170] on surface superconductivity, it was assumed that T_c may be raised with the help of some dielectric coverings of metallic surfaces. It was emphasized that quasi-two-dimensional structures are much more advantageous than quasi-one-dimensional structures [183] because of the considerably smaller role of fluctuations (this argument was worked out in [173]). Later on, I became engaged in earnest in the HTSC problem and concentrated on 'sandwiches', i.e. thin metallic films in dielectric and semiconducting 'coatings' and on layered superconducting compounds—these kind of 'files' of sandwiches [185–189, 175].

I should say that I write rather easily and, moreover, I even feel the necessity of expressing my thoughts in written form. As a result, during the 32 years in which I have been interested in the HTSC problem, I wrote many (probably, too many) papers on the subject, particularly popular papers. I do not think I need to refer to many of them here. Among the published works, special attention is deserved by the monograph [175]. This book was the outcome of the joint efforts undertaken by L N Bulaevskii, V L Ginzburg, D I Khomskii, D A Kirzhnits, Yu V Kopaev, E G Maksimov, and G F Zharkov (I E Tamm Department of Theoretical Physics of P N Lebedev Physical Institute of the USSR Academy of Sciences, Moscow) who had been 'attacking' the HTSC problem for several years. This monograph was published in Russian in 1977 and in an English translation in 1982 and was the first and, up to 1987, the only one devoted to this issue. In [175], a whole spectrum of possible ways of obtaining HTSC was considered.

I shall dwell on some of the results of our work.

A very important question is whether or not there are some limitations on admissible T_c values in metals, say, due to the requirement of crystal lattice stability. Such limitations are possible in principle and, moreover, in the 1972 paper [190], it was stated that it was the requirement of lattice stability that fully obstructs the possibility of the existence of HTSC. The point is that the dimensionless parameter of the interaction force λ_{eff} in the BCS formula (7.75) can be written in the form

$$\lambda_{\text{eff}} = \lambda - \mu^* = \lambda - \frac{\mu}{1 + \mu \ln(\theta_F/\theta)}. \tag{7.76}$$

Here λ and μ are, respectively, the dimensionless coupling constants for phonon or exciton attraction and Coulomb repulsion and $k_B \theta_F = E_F$ is the Fermi energy. At the same time, in the simplest approximation (homogeneity and isotropy of material, and weak coupling), we have

$$\mu - \lambda = \frac{4\pi e^2 N(0)}{q^2 \varepsilon(0, q)} \tag{7.77}$$

where $\varepsilon(\omega, q)$ is the longitudinal permittivity for the frequency ω and for the wavenumber q and the factor $1/q^2 \varepsilon(0, q)$ should be understood as a certain mean value in q; and $N(0)$ is the density of states on the Fermi boundary for a metal in the normal state. If, as was assumed in [190], the stability condition has the form

$$\varepsilon(0, q) > 0 \tag{7.78}$$

then, from equation (7.77), it follows that

$$\mu > \lambda. \tag{7.79}$$

This inequality and equation (7.76) imply that superconductivity (for which, certainly, $\lambda_{\text{eff}} > 0$) is generally possible only due to the difference between μ^* and μ, the T_c value being not large. It was, however, already known empirically that $\mu < 0.5$ and sometimes $\lambda > 1$ and, thus, that inequality (7.79) is violated. Apart from such and some other arguments already expressed in the early stages [188], it was later shown strictly (see [191, 175, 192] and the literature cited there) that the stability condition (7.78) is invalid and, in fact, the stability condition has the form (for $q \neq 0$)

$$\frac{1}{\varepsilon(0, q)} \leq 1 \tag{7.80}$$

i.e. is satisfied if one of the inequalities

$$\varepsilon(0, q) \geq 1 \qquad \varepsilon(0, q) < 0 \tag{7.81}$$

holds. It is interesting that the values $\varepsilon(0, q) < 0$ for large q, important in the theory of superconductivity, are realized in many metals [193, 194]. From the second inequality (7.81) and expression (7.77), it is obvious that the parameter λ may exceed μ. On the basis of this fact, our group came to the conclusion even before 1977 (I mean the Russian edition of the book [175]) that the general requirement of stability does not restrict T_c and it is quite possible, for example, that $T_c \lesssim 300$ K.

As has already been mentioned, the idea of the exciton mechanism is connected with the possibility of raising T_c by increasing the temperature θ in equation (7.75) which determines the energy range $k_B\theta$ where the electrons attract one another near the Fermi surface and, thus, form pairs. It is assumed that weak coupling takes place here, when

$$\lambda_{\text{eff}} \ll 1. \tag{7.82}$$

It is only under this condition that formula (7.75) and the BCS model are applicable. But the BCS theory is, on the whole, more extensive and admits consideration of the case of strong coupling [195], when

$$\lambda_{\text{eff}} \gtrsim 1. \tag{7.83}$$

Under conditions (7.83) the strong coupling formula (7.75) is, of course, already invalid although it is clear from it that the temperature T_c rises with increasing λ_{eff}. In the literature, a large number of expressions for T_c are proposed for the case of strong coupling (see [175, 192, 196, 197] and some references therein). The simplest of these expressions is as follows:

$$T_c = \theta \exp\left(-\frac{1+\lambda}{\lambda - \mu^*}\right). \tag{7.84}$$

Exactly as it should be under weak coupling conditions (7.82) or, more precisely, under the condition $\lambda \ll 1$, formula (7.84), of course, becomes (7.75). If, in equation (7.84), we set $\mu^* = 0.1$ then, for example, for $\lambda = 3$ the temperature is $T_c = 0.25\theta$. Therefore, for the value $\theta = \theta_D = 400$ K, which is readily admissible for the phonon mechanism, we already have $T_c = 100$ K. More accurate formulae also suggest that, for strong coupling (7.83), the phonon mechanism can already allow temperatures $T_c \sim 100$ K and even $T_c \sim 200$ K. But the analysis carried out in the book [175] and later showed that for 'conventional' superconductors with strong coupling, the temperature T_c is rather small. For example, for lead we have $\theta_D = 96$ K and, therefore, in spite of the high value $\lambda = 1.55$, the critical temperature is $T_c = 7.2$ K. For such a conclusion, i.e. that θ_D falls with increasing λ, there also exist theoretical arguments (see [175, chapter 4]). That was the reason why we (or, at least, I) did not hope for the creation of HTSCs at the expense of strong coupling but possessing the phonon mechanism. In any case, as I have already mentioned, in [175], a versatile and unprejudiced approach to the HTSC problem prevailed. Here I cite the last part of chapter 1 written by me for the book [175]:

> On the basis of general theoretical considerations, we believe at present that the most reasonable estimate is $T_c \lesssim 300$ K, this estimate being, of course, for materials and systems under more or

less normal conditions (equilibrium or quasi-equilibrium metallic systems in the absence of pressure or under relatively low pressures, etc.). In this case, if we exclude from consideration metallic hydrogen and, perhaps, organic metals, as well as semimetals in states near the region of electronic phase transitions, then it is suggested that we should use the exciton mechanism of attraction between the conduction electrons.

In this scheme, the most promising materials from the point of view of the possibility of raising T_c, are apparently layered compounds and dielectric–metal–dielectric sandwiches. However, the state of the theory, let alone the experiment, is still far from being such as to allow us to regard as closed other possible directions, in particular, the use of filamentary compounds. Furthermore, for the present state of the problem of high-temperature superconductivity, the most sound and fruitful approach will be one that is not preconceived, in which attempts are made to move forward in the most diverse directions.

The investigation of the problem of high-temperature superconductivity is entering into the second decade of its history (if we are talking about the conscious search for materials with $T_c \gtrsim 90$ K using exciton and other mechanisms). Supposedly, there begins at the same time a new phase of these investigations, which is characterized not only by greater scope and diversity but also by a significantly deeper understanding of the problems that arise. There is still no guarantee whatsoever that the efforts being made will lead to significant success but a number of new superconducting materials have already been produced and are being investigated. Therefore, it is in any case difficult to doubt that further investigations of the problem of high-temperature superconductivity will yield many interesting results for physics and technology, even if materials that will remain superconducting at room (or even liquid-nitrogen) temperatures will not be produced. However, as has been emphasized, this ultimate aim does not seem to us to have been discredited in any way. As may be inferred, the next decade will be crucial for the problem of high-temperature superconductivity.

This was written in 1976. Time passed but the multiple attempts to find a reliable and reproducible way for creating HTSC have been unsuccessful. As a result, after the burst of activity came a slackening which gave cause for me to characterize the situation in a popular paper [198] published in 1984 as follows:

It somehow happened that research into high-temperature super-

conductivity became unfashionable (there is good reason to speak
of fashion in this context since fashion sometimes plays a significant part in research work and in the scientific community). It
is hard to achieve anything by making admonitions. Typically it
is some obvious success (or reports of success, even if erroneous)
that can radically and rapidly reverse attitudes. When they sense
a 'rich strike', the former doubters, and even dedicated critics, are
capable of turning coat and become ardent supporters of the new
work. But this subject belongs to the psychology and sociology of
science and technology. In short, the search for high-temperature
superconductivity can readily lead to unexpected results and discoveries, especially since the predictions of the existing theory are
rather vague.

I did not expect, of course, that this 'prediction' would come true in
two years [180, 181]. It came true not only in the sense that HTSCs with
$T_c > T_{b, N_2} = 77.4$ K were obtained but also, so to say, in the social aspect:
as I have already mentioned, a real boom began and an 'HTSC psychosis'
started. One of the manifestations of the boom and psychosis was an
almost total oblivion to everything that had been done before 1986, as if
discussion of the HTSC problem had not begun 22 years before [183, 184].
I have already dwelt on this subject and in the papers [179, 196] and would
not like to return to it here. I will only note that J Bardeen, whom I always
respected, treated the HTSC problem with understanding both before 1986
and after it (see [199]).

The present situation in solid state theory and, in particular, the
theory of superconductivity does not allow us to calculate the temperature T_c or indicate, with sufficient accuracy and certainly, especially in
the case of compound materials, what particular compound should be investigated. Therefore, I am of the opinion that theoreticians could not
have given experimenters better and more reliable advice as to how and
where HTSC could be sought than was done in the book [175]. An exception is perhaps only an insufficient attention to the superconductivity
of the $BaPb_{1-x}Bi_xO_3$ (BPBO) oxide discovered in 1974. When $x = 0.25$,
for this oxide, we have $T_c = 13$ K which is a high value for a T_c when
it is estimated in a way similar to that used for conventional superconductors. In the related oxide $Ba_{0.6}K_{0.4}BiO_3$ (BKBO), superconductivity
with $T_c \sim 30$ K was discovered in 1988. Most importantly, the compound
$La_{2-x}Ba_xCuO_4$ (LBCO) in which superconductivity with $T_c \sim 30$–40 K
was discovered in 1986 [180] and is thought that the discovery of HTSC
belongs to the oxides. However even now, 10 years later, one cannot predict, even roughly, the values of T_c for a particular material and, moreover,
even the very mechanism of superconductivity in cuprates and, in particular, in the most thoroughly investigated cuprate $YBa_2Cu_3O_{7-x}$ (YBCO)

with $T_c \sim 90$ K is not yet clear.

It is inappropriate to dwell here extensively on the current state of the HTSC problem. I shall restrict myself to several remarks. At first glance, HTSC cuprates differ strongly from 'conventional' superconductors (see, for example, [53, 182, 200, 214]). This circumstance gave rise to the opinion that HTSC cuprates are something special—either the BCS theory is inapplicable to them or, in any case, a non-phonon mechanism of pairing acts in them. This tendency was very clearly expressed at the 1991 M^{2}HTSC III Conference [182].

Indeed, the phonon mechanism has no exclusive rights. In principle, the exciton (electronic) mechanism, the Schafroth mechanism (creation of pairs at $T > T_c$ with a subsequent Bose–Einstein condensation), the spin mechanism (pairing due to exchange of spin waves or, as it is sometimes called, due to spin fluctuations), and some other mechanisms (for some more details and references see, for example, [197, 214]) may all exist. Since I have always been a supporter of the exciton mechanism, I would be only glad if this very mechanism proves to act in HTSC. But there is not yet any grounded basis for such a statement. In the BKBO oxide and in doped fullerenes (fullerites) of K$_3$C$_{60}$ and Rb$_3$C$_{60}$ type (they all possess a cubic structure) with $T_c \sim 30$–40 K, the phonon mechanism obviously prevails. The same relates to superconductivity in MgB$_2$ (see article 6). The situation is more complicated with cuprate oxides which are highly anisotropic layered compounds. However, E G Maksimov, O V Dolgov, and their colleagues indicate, I believe, convincingly that the phonon mechanism may quite possibly also dominate in HTSC cuprates. In any case, omitting the important question of a 'pseudogap' [233], HTSC cuprates in the normal state differ from ordinary metals in only a quantitative respect. Formally, a standard electron–phonon interaction with a coupling constant $\lambda \approx 2$ accounts well for the high values $T_c \sim 100$–125 K as being due to the high Debye temperature $\theta_D \sim 600$ K (see [197, 201–205, 256] and the literature cited there).[14] The properties of the superconducting state of HTSC cuprates are a more complicated entity. To explain them, it is already insufficient to use a standard isotropic approximation in the model of a strong electron–phonon interaction. However, allowing for the anisotropy of the electron spectra and interelectron interaction, the electron–phonon interaction all the same may play a decisive role in the formation of a superconducting state. As has been shown [215, 216] (see also [206–208]), in the framework of multi-zone models allowing for standard electron–phonon and Coulomb interactions, one can obtain a strongly

---

[14] I find it necessary to note that the report [201] was, in fact, prepared by E G Maksimov alone. My name appeared in [201] only because there was a difficulty with including this report on the agenda and I had, by E G Maksimov's consent, to include my name which enabled him to participate in the 1994 M^{2}HTSC IV Conference. It is not a pleasure to speak about such morals and manners but this is a truth.

anisotropic superconducting gap including its sign reversal in the Brillouin zone, which imitates d-pairing. It is also possible that the electron–exciton interaction and peculiarities of the electron spectrum, which are almost insignificant for understanding the properties of the normal state, make their contribution to the formation of the superconducting state. I do not regard myself competent enough to think of such statements as proved. But it is beyond doubt that a general denial of the important role of the phonon mechanism of HTSC (in cuprates) typical of the recent past (see [182]) is already behind us [204, 256, 267, 268] (possibly I only hope so, see [263]).

Suppose, for the sake of argument, that in the already known HTSCs the exciton mechanism does not play any role. This is, of course, important and interesting but in no way discredits the very possibility of a manifestation of the exciton mechanism. As has already been mentioned, we are not aware of any evidence contradicting the action of the exciton mechanism. But it is actually not easy for the exciton mechanism to manifest itself. This will require some special conditions which are not yet clear (see, in particular, [205]).

The highest critical temperature fixed today (for $HgBa_2Ca_2Cu_3O_{8+x}$ under pressure) reaches 164 K. Such a value can be attained with the phonon mechanism. But if one succeeds in reaching a temperature $T_c >$ 200 K, the phonon mechanism will hardly be sufficient (when $\lambda = 2$, the temperature $T_c = 200$ K is obtained for $\theta_D \approx 1000$ K). As to the exciton mechanism, even room temperature is not a limit for T_c. A search for HTSCs with the highest possible critical temperatures is now being and will, of course, be undertaken. It seems to me, as before, that the most promising in this respect are layered compounds and dielectric–metal–dielectric 'sandwiches'.[15] It would be natural to use the atomic layer-by-layer synthesis here [209, 218, 247]. The role of a dielectric in such sandwiches can be played by organic compounds in particular. Still and all, the possibilities that may open on the way are virtually boundless. It is, therefore, especially reasonable to be guided by some qualitative consideration (see, for example, [175, chapter 1]).

For 22 long years (from 1964 to 1986), which, however, flew by very quickly, HTSC was a dream for me and to think of it was something like a gamble. Now it is an extensive field of research, tens of thousand papers are devoted to it, hundreds or even thousands of researchers are engaged in the study of one or another of its aspects. Much has already been done but much remains to do. Even the mechanism of superconductivity in HTSC cuprates is rather obscure, to say nothing of the myriad particular questions. I think that among these questions the first place belongs to the question of the maximum attainable value of the critical temperature T_c

---

[15] In addition to intuitive arguments [175, 186, 188, 189], there are also some concrete arguments [201, 205] in favor of such quasi-two-dimensional structures.

under not very exotic conditions, say, at atmospheric pressure and for a stable material. More concretely, one can pose a question concerning the possibility of creating superconductors with T_c values lying within the range of room temperatures (the problem of room temperature superconductivity (RTSC)). RTSC is, in principle, possible but there is no guarantee in this respect. The problem of RTSC has generally taken the place that had been occupied by HTSC before 1986–87. I am afraid that I do not see any possibility for myself to do something positive in this direction and it only remains to wait impatiently for coming events.

7.8   Concluding remarks

By 1943, when I began studying the theory of superconductivity, 32 years had already passed since the discovery of the phenomenon. None the less, at the microscopic level, superconductivity had not yet been understood and had actually been a 'white spot' in the theory of metals and, perhaps, in the whole physics of condensed media. The superfluidity of helium II had been discovered in its explicit form no more than 5 years before that time and its connection with superconductivity had only been outlined. The world was at terrible war and I myself hardly understand now why the enigmas of low-temperature physics seemed so tempting to me when I was cold and semi-starving in evacuation in Kazan. But it was so. Poor command of mathematics, an inability to concentrate on one particular task (I was simultaneously engaged in several problems), difficulties in exchange of scientific information, especially with experimenters, in the war and post-war years obstructed a rapid advance, and it was only in 1950 that something appeared completed (I mean the Ψ-theory of superconductivity). But this completeness is, of course, rather conditional because new questions and problems constantly arose.

At the same time, the character of studies in the field of low-temperature physics, as well as the whole of physics was changing radically. It is even hard to imagine now that it was only one laboratory that succeeded in obtaining liquid helium between 1908 and 1923. It is hard to imagine that applications of superconductivity in physics, to say nothing of technology, were fairly modest for three decades. And it was not until the 1960s that strong superconducting magnets were created and extensively used. At the present time, superconductivity finds numerous applications (see, for example, [71, 210]). Even the small book [211] intended for schoolchildren presents various applications of superconductivity, including giant superconducting magnets in tokamaks and tomographs. Creation of HTSCs (1986–87) gave rise to great expectations of the possibility of new applications of superconductivity. These expectations were partly exaggerated but nevertheless now, after 10 years, much has already been done in this direction, even in respect of electric power lines and strong magnets [212], not to mention some other applications [219]. I wrote in section 7.7 about the boom provoked by the creation of HTSC. Many thousands of papers and hundreds or even thousands of researchers—what a contrast with what was observed in, say, 1943 or as recently as 10 years ago!

In the light of the present state of the theory of superconductivity and superfluidity, much of what has been said in this paper is only of historical interest and, in other cases, is somewhere far from the forefront of the current research. At the same time, and this is very important, I have mentioned a large number of questions and problems which still remain unclear. This lack of clarity concerns the development of the Ψ-theory of superconductivity and its application to HTSC, the application of the Ψ-theory of superfluidity, the problem of surface (two-dimensional) superconductivity, the question of thermoeffects in superconductors (and especially their connection with heat transfer), the circulation effect in a non-uniformly heated vessel filled with a superfluid liquid and some other things, to say nothing of HTSC theory (see [256] for a review). The aim of the paper will have been attained if it at least helps to draw attention of both theoreticians and experimenters to these problems.[12*,14*]

Acknowledgments. Taking the opportunity, I express my gratitude to Yu S Barash, E G Maksimov, L P Pitaevskii, A A Sobyanin, and G F Zharkov for reading the manuscript and remarks.

Comments

1*. The paper was published in 1997 *Usp. Fiz. Nauk* **167** 429 [1997 *Phys. Usp.* **40** 407].

As distinct from all other papers of this collection, section 7.5, devoted to thermoelectric effects, has been changed compared to the text of the indicated publication. The matter is that after the paper had been published, I realized that some points were incorrect. I presented the corresponding modifications in a letter [213] to *Physics–Uspekhi*. It would of course be irrelevant to place section 7.5 here in its initial form and add the letter [213] to it. Instead, I have revised section 7.5 according to what was written in [213].

2*. In paper [220], the critical field for superheating was calculated to rather a high approximation (for $æ \ll 1$) with the result

$$\frac{H_{c\,1}}{H_{c\,m}} = 2^{-1/4} æ^{-1/2} \left(1 + \frac{15\sqrt{2}}{32} æ + O(æ^2) \right).$$

3*. The talented theoretician physicist Aleksandr Aleksandrovich Sobyanin died on 10 June, 1997 at the age of 54. Unfortunately, I am unaware of the fate of his last note mentioned in the footnote on p 219.

4*. For several years now (beginning from 1995), great attention has been shown in experimental studies of Bose–Einstein condensation (BEC) of rarefied gases at low temperatures. The theoretical analysis has mostly been based on the Gross–Pitaevskii theory (see reviews [221, 234, 237, 240]). The development of this theory, I believe, was significantly influenced by the Ψ-theory of superfluidity. It seems probable that the Ψ-theory of superfluidity, both in its original and generalized forms, may also be useful when applied to BEC in gases, particularly in the neighborhood of the λ-point. The thermomechanical circulation effect in superfluids [144, 159, 160] also can be interesting in BEC systems.

5*. In this connection, see the supplement to article 6 in the present volume.

6*. See also article 19 in the present volume.

7*. I have seen a statement in the literature that over 50 000 papers have been devoted to HTSC over 10 years.

8*. Paper [76] was elaborated in papers [230].

9*. According to [231], in superfluid ^{3}He the length $\xi(0) \sim 10^{-5}$ cm while, as we pointed in the text, for ^{4}He the length $\xi(0) \sim 10^{-8}$ cm. Clearly, in ^{3}He and in some other cases, the order parameter is not the scalar function Ψ. So one considers generalizations of the Ψ-theory with Ψ substituted by the corresponding order parameter.

10*. In the preprint [232], some scheme is elaborated that combines the generalized Ψ-theory of superfluidity with BEC theory. I do not see grounds for such a theory but, nevertheless, its analysis seems interesting. In paper [257], the Ψ-theory is somewhat generalized (by taking into account the Van der Waals forces) and compared with experiment. Unfortunately, the author's conclusions remain obscure to me.

11*. This presentation is based on the assumption that the Fermi-liquid model is applicable to cuprates (when they are considered). If, however, the Fermi-liquid notion is inapplicable to cuprates (and possibly to some other superconductors) (see [262] and references therein and [103, 104, 143] in the reference list to article 6), a special investigation (both theoretical and experimental) of thermoelectric phenomena in such materials will be necessary. With this fact in mind, it seems to me that the study of thermoelectric effects in superconductors acquires an additional interest.

12*. The problems I see in the field of superconductivity and super-fluidity are also discussed in my preprint [235]. They mostly coincide with the topics discussed in the present paper. Some new studies in the field of superconductivity are reported in article 6, where several publications from 2002–03 are cited.

13*. I have to state with regret that the Ψ-theory of superfluidity has not attracted any attention in the new publications known to me devoted to superfluidity in liquid helium and it is totally ignored in other cases. In fact, superfluidity in liquids is very rarely studied generally now, which can be understood, in particular, in connection with the enthusiasm for BEC in gases. Nevertheless, studies of liquid ^{4}He continue and, for example, I think it would be quite relevant to involve the Ψ-theory of superfluidity for analysis in papers [241, 242]. The same relates to studies of superfluidity in ^{3}He, in neutron stars, and in other cases.

14*. It should be borne in mind that the present paper was published in 1997. In presenting it in this book, we have added only a few notes and references to new literature, with the exception of section 7.5 (which is devoted to thermoelectric phenomena). Of course, this does not make the paper as it would have appeared should it be written anew in 2004. However, this concerns only the present state of the physics of superconductivity and not the history of its development, to which (though in the autobiographical aspect) this paper is devoted. So I hope the small changes to the original text, which somewhat violate its just proportions, prove to be justified.

I take the opportunity to make reference to several new papers (in addition to the already mentioned paper [247]) which have drawn my attention: [248]—superconductivity in thin films with tensions, [249, 250]—vortexes in alternating superconducting and ferromagnetic films, [252]—combinational (Raman) scattering of light in HTSC, [269]—λ-transition analysis in ^{4}He, [260, 261]—superconductivity in MgB_2.

References[16]

[1] Ginzburg V L 2001 *The Physics of a Lifetime: Reflections on the Problems and Personalities of 20th Century Physics* (Berlin: Springer)

[2] Ginzburg V L 1990 Notes of amateur astrophysicist *Annu. Rev. Astron. Astrophys.* **28** 1 [this paper is included in the present book (article 17)]

[3] 1988 *Reminiscences of Landau* (Moscow: Nauka)
     1989 *Landau. The Physicist and the Man* (Oxford: Pergamon Press)

[4] Landau L D 1941 *Zh. Eksp. Teor. Fiz.* **11** 592; 1941 *J. Phys. USSR* **5** 71

[5] Kapitza P L 1938 *Nature* **141** 74; 1941 *Zh. Eksp. Teor. Fiz.* **11** 581; 1941 *J. Phys. USSR* **4** 177; **5** 59

[6] Allen G F and Misener A D 1938 *Nature* **141** 75; 1939 *Proc. R. Soc. London* **172A** 467

[7] Kamerlingh Onnes H 1911 *Commun. Phys. Lab. Univ. Leiden* **124**c (this paper is included as an Appendix in a more readily available paper [9])

[8] Dahl P F 1992 *Superconductivity. Its Historical Roots and Development from Mercury to the Ceramic Oxides* (New York: American Institute of Physics)

---

[16] Papers written by the present author or those where he is a co-author are given with titles. This was done naturally with only the purpose of providing additional information because very little is said about some of these papers in the main text.

de Nobel J 1996 *Phys. Today* **49**(9) 40

[9] Ginzburg V L 1992 Research on superconductivity (brief history and outlook for the future) *Sverkhprovodimost: Fiz., Khim., Tekhnol.* **5**(1) 1 [1992 *Superconductivity: Phys., Chem., Technol.* **5** 1]

[10] Kamerlingh Onnes H 1911 *Commun. Phys. Lab. Univ. Leiden* **119**a; 1911 *Proc. R. Acad. Amsterdam* **13** 1093

[11] Keesom W H 1942 *Helium* (Amsterdam: Elsevier)

[12] London F 1954 *Superfluids.* vol 2 *Macroscopic Theory of Superfluid Helium* (New York: Wiley and Sons)

[13] Feynman R P 1972 *Statistical Mechanics* (Reading, MA: W A Benjamin)

[14] Ginzburg V L 1952 Present state of the theory of superconductivity. II. The microscopic theory *Usp. Fiz. Nauk* **48** 26 [Translation of the main part of the paper 1953 *Forschr. Phys.* **1** 101]

[15] Ogg R A Jr 1946 *Phys. Rev.* **69** 243; 1946 **70** 93

[16] Schafroth M R 1954 *Phys. Rev.* **96** 1149; 1442; 1955 **100** 463
Schafroth M R, Butler S T and Blatt J M 1957 *Helv. Phys. Acta* **30** 93

[17] Cooper L N 1956 *Phys. Rev.* **104** 1189

[18] Bardeen J, Cooper L N and Schrieffer J R 1957 *Phys. Rev.* **108** 1175

[19] Ginzburg V L 1944 Comments on the theory of superconductivity *Zh. Eksp. Teor. Fiz.* **14** 134

[20] Bardeen J 1956 in *Kälterphysik (Handbuch der Physik* **15**) ed S von Flügge (Berlin: Springer) p 274 [Translated into Russian 1959 in *Fizika Nizkikh Temperatur (Low-Temperature Physics)* ed A I Shal'nikov (Moscow: IL) p 679]

[21] Ginzburg V L 1944 On gyromagnetic and electron-inertia experiments with superconductors *Zh. Eksp. Teor. Fiz.* **14** 326

[22] Ginzburg V L 1944 On thermoelectric phenomena in superconductors *Zh. Eksp. Teor. Fiz.* **14** 177; 1944 *J. Phys. USSR* **8** 148

[23] Ginzburg V L 1943 Light scattering in helium II *Zh. Eksp. Teor. Fiz.* **13** 243; 1943 A brief report *J. Phys. USSR* **7** 305

[24] Ginzburg V L 1946 *Superconductivity* (Moscow–Leningrad: Izd. Akad. Nauk SSSR)

[25] Meissner W and Ochsenfeld R 1933 *Naturwissensch.* **21** 787

[26] Gorter C J and Casimir H 1934 *Physica* **1** 306; 1934 *Phys. Z.* **35** 963

[27] London F and London H 1935 *Proc. R. Soc. London* **149A** 71; 1935 *Physica* **2** 341

[28] Becker R, Heller G and Sauter F 1933 *Z. Phys.* **85** 772

[29] Ginzburg V L and Landau L D 1950 To the theory of superconductivity *Zh. Eksp. Teor. Fiz.* **20** 1064 [This paper is available in English in Landau L D 1965 *Collected Papers* (Oxford: Pergamon Press)]

[30] Lifshitz E M and Pitaevskii L P 1978, 1999 *Statisticheskaya Fizika* [Statistical Physics) Pt 2 *Teoriya Kondensirovannogo Sostoyaniya (Theory of the Condensed State)* (Moscow: Nauka) [Translated into English 1980 (Oxford: Pergamon Press)]

[31] Gorkov L P 1959 *Zh. Eksp. Teor. Fiz.* **36** 1918; **37** 1407 [1959 *Sov. Phys. JETP* **9** 1364; 1960 **10** 998]

[32] Ginzburg V L 1946 On the surface energy and the behaviour of small-sized superconductors *Zh. Eksp. Teor. Fiz.* **16** 87; 1945 *J. Phys. USSR* **9** 305

[33] Ginzburg V L 1950 Present state of the theory of superconductivity. Part 1. Macroscopic theory *Usp. Fiz. Nauk* **42** 169

[34] Landau L D and Lifshitz E M 1995 *Statisticheskaya Fizika (Statistical Physics)* Pt 1 (Moscow: Fizmatlit) ch XIV [Translated into English 1980 (Oxford: Pergamon Press)]

[35] Ginzburg V L 1949 The theory of ferroelectric phenomena *Usp. Fiz. Nauk* **38** 400

[36] Ginzburg V L 1955 On the theory of superconductivity *Nuovo Cimento* **2** 1234

[37] de Gennes P-G 1966 *Superconductivity of Metals and Alloys* (New York: W A Benjamin) [Translated into Russian 1968 (Moscow: Mir)]

[38] Saint-James D, Sarma G and Thomas E J 1969 *Type II Superconductivity* (Oxford: Pergamon Press) [Translated into Russian 1970 (Moscow: Mir)]

[39] Tilley D R and Tilley J 1986 *Superfluidity and Superconductivity* 2nd edn (Bristol: Adam Hilger)

[40] Schmidt V V 2000 *Vvedenie v Fiziku Sverkhprovodnikov (Introduction to the Physics of Superconductors)* 2nd edn (Moscow: MTsNMO) [Translated into English 1997 *The Physics of Superconductors: Introduction to Fundamental and Applications* (Berlin: Springer)]

[41] Abrikosov A A 1987 *Osnovy Teorii Metallov (Fundamental Principles of the Theory of Metals)* (Moscow: Fizmatlit) [Translated into English 1988 (New York: Elsevier)]

[42] Weinan E 1994 *Phys. Rev. B* **50** 1126; 1994 *Physica* **77D** 383

[43] Sakaguchi H 1995 *Prog. Theor. Phys.* **93** 491

[44] Bazhenov M V, Rabinovich M I and Fabrikant A L 1992 *Phys. Lett.* **163A** 87

Deissler R J and Brand H R 1994 *Phys. Rev. Lett.* **72** 478

Soto-Crespo J M, Akhmediev N N and Afanasjev V V 1995 *Optics Commun.* **118** 587

[45] Bethuel F, Brezis H and Helein F 1994 *Ginzburg–Landau Vortices* (Boston: Birkhauser)

[46] Kirzhnits D A 1978 *Usp. Fiz. Nauk* **125** 169 [1978 *Sov. Phys. Usp.* **21** 470]

[47] 1978 *Vitaly Lazarevich Ginzburg (Bibliography of Soviet Scientists* (Physics Series **21**) (Moscow: Nauka)

[48] Ginzburg V L 1947 On the non-linearity of electromagnetic processes in superconductors *J. Phys. USSR* **11** 93

[49] Ginzburg V L 1949 The theory of superfluidity and critical velocity in helium II *Dokl. Akad. Nauk SSSR* **69** 161

[50] Landau L D 1937 *Zh. Eksp. Teor. Fiz.* **7** 19; 627; 1937 *Phys. Z. Sowjetunion* **11** 26; 545

[51] Yang C N 1962 *Rev. Mod. Phys.* **34** 694

[52] Penrose O and Onsager L 1956 *Phys. Rev.* **104** 576

[53] Lynn J W (ed) 1990 *High Temperature Superconductivity* (Berlin: Springer)

[54] Ginzburg V L 1992 Theories of superconductivity (a few remarks) *Helv. Phys. Acta* **65** 173

[55] Ginzburg V L 1955 To the macroscopic theory of superconductivity *Zh. Eksp. Teor. Fiz.* **29** 748 [1956 *Sov. Phys. JETP* **2** 589]

[56] Boulter C J and Indeken J O 1996 *Phys. Rev. B* **54** 12407

Mishonov J M 1990 *J. Physique* **51** 447

[57] Ginzburg V L 1956 An experimental manifestation of instability of the normal phase in superconductors *Zh. Eksp. Teor. Fiz.* **31** 541 [1957 *Sov. Phys. JETP* **4** 594]

[58] Shoenberg D 1965 *Superconductivity* 3rd edn (Cambridge: Cambridge University Press) [Russian translation of the previous edition 1955 (Moscow: IL)]

[59] Abrikosov A A 1957 *Zh. Eksp. Teor. Fiz.* **32** 1442 [1957 *Sov. Phys. JETP* **5** 1174]; 1952 *Dokl. Akad. Nauk SSSR* **86** 489

[60] Ginzburg V L 1952 On the behaviour of superconducting films in a magnetic field *Dokl. Akad. Nauk SSSR* **83** 385

[61] Silin V P 1951 *Zh. Eksp. Teor. Fiz.* **21** 1330

[62] Ginzburg V L 1958 On the destruction and the onset of superconductivity in a magnetic field *Zh. Eksp. Teor. Fiz.* **34** 113 [1958 *Sov. Phys. JETP* **7** 78]

[63] Ginzburg V L 1958 Critical current for superconducting films *Dokl. Akad. Nauk SSSR* **118** 464

[64] Ginzburg V L 1951 On the behaviour of superconductors in a high-frequency field *Zh. Eksp. Teor. Fiz.* **21** 979

[65] Bardeen J 1954 *Phys. Rev.* **94** 554

[66] Ginzburg V L 1956 Some remarks concerning the macroscopic theory of superconductivity *Zh. Eksp. Teor. Fiz.* **30** 593 [1956 *Sov. Phys. JETP* **3** 621]

[67] Ginzburg V L 1956 To the macroscopic theory of superconductivity valid at all temperatures *Dokl. Akad. Nauk SSSR* **110** 358

[68] Ginzburg V L 1959 On comparison of the macroscopic theory of superconductivity with experimental data *Zh. Eksp. Teor. Fiz.* **36** 1930 [1959 *Sov. Phys. JETP* **36** 1372]

[69] Ginzburg V L 1963 Allowance for the effect of pressure in the theory of second-order phase transitions (as applied to the case of superconductivity) *Zh. Eksp. Teor. Fiz.* **44** 2104 [1963 *Sov. Phys. JETP* **17** 1415]

[70] London F 1950 *Superfluids.* vol 1 *Macroscopic Theory of Superconductivity* (New York: Wiley)

[71] Buckel W 1972 *Supraleitung* (Weinheim: Physik) [Translated into English 1991 *Superconductivity: Fundamentals and Applications* (Weinheim: VCH); translation into Russian 1975 (Moscow: Mir)]

[72] Ginzburg V L 1962 Magnetic flux quantization in a superconducting cylinder *Zh. Eksp. Teor. Fiz.* **42** 299 [1962 *Sov. Phys. JETP* **15** 207]

[73] Bardeen J 1961 *Phys. Rev. Lett.* **7** 162

[74] Keller J B and Zumino B 1961 *Phys. Rev. Lett.* **7** 164

[75] Ginzburg V L 1952 On account of anisotropy in the theory of superconductivity *Zh. Eksp. Teor. Fiz.* **23** 236

[76] Ginzburg V L 1956 Ferromagnetic superconductors *Zh. Eksp. Teor. Fiz.* **31** 202 [1957 *Sov. Phys. JETP* **4** 153]

[77] Bulaevskii L N 1987 in *Superconductivity, Superdiamagnetism, Superfluidity* ed V L Ginzburg (Moscow: Mir) p 69

[78] Laue M 1948 *Ann. Phys.* (Leipzig) **3** 31

[79] Laue M 1949 *Theorie der Supraleitung* (Berlin: Springer)

[80] Ginzburg V L 1952 Some questions of the theory of electric fluctuations *Usp. Fiz. Nauk* **46** 348

[81] Schmidt V V 1966 *Pis'ma Zh. Eksp. Teor. Fiz.* **3** 141 [1966 *JETP Lett.* **3** 89]

[82] Schmidt H 1968 *Z. Phys.* **216** 336

[83] Schmid A 1969 *Phys. Rev.* **180** 527

[84] Aslamazov L G and Larkin A I 1968 *Fiz. Tverd. Tela* (Leningrad) **10** 1104 [1968 *Sov. Phys. Solid State* **10** 875]

[85] Tinkham M 1996 *Introduction to Superconductivity* 2nd edn (New York: McGraw-Hill)

[86] Ginzburg V L 1960 Several remarks on second-order phase transitions and microscopic theory of ferroelectrics *Fiz. Tverd. Tela* (Leningrad) **2** 2031 [1961 *Sov. Phys. Solid State* **2** 1824]

[87] Levanyuk A P 1959 *Zh. Eksp. Teor. Fiz.* **36** 810 [1959 *Sov. Phys. JETP* **9** 571]

[88] Patashinskii A Z and Pokrovskii V L 1982 *Fluktuatsionnaya Teoriya Fazovykh Perekhodov* 2nd edn [*Fluctuation Theory of Phase Transitions*] (Moscow: Nauka) [Translation into English 1979 (Oxford: Pergamon Press)]

[89] Ginzburg V L, Levanyuk A P and Sobyanin A A 1983 Comments on the region of applicability of the Landau theory for structural phase transitions *Ferroelectrics* **73** 171

[90] Bulaevskii L N, Ginzburg V L and Sobyanin A A 1988 Macroscopic theory of superconductors with small coherence length *Zh. Eksp. Teor. Fiz.* **94** 356 [1989 *Sov. Phys. JETP* **68** 1499]; 1989 *Usp. Fiz. Nauk* **157** 539 [1989 *Sov. Phys. Usp.* **32** 1277]; 1988 *Physica* C **152** 378; 1988 **153–155** 1617

[91] Sobyanin A A and Stratonnikov A A 1988 *Physica* C **153–155** 1680

[92] Gorkov L P and Kopnin N B 1988 *Usp. Fiz. Nauk* **156** 117 [1988 *Sov. Phys. Usp.* **31** 850]

[93] Alexandrov A S and Mott N F 1994 *High Temperature Superconductors and Other Superfluids* (London: Taylor & Francis)

[94] Ginzburg V L 1987 On the Ψ-theory of high temperature superconductivity in *Proc. 18th Int. Conf. on Low Temperature Physics Kioto, Japan, LT-18*; 1987 *Japan J. Appl. Phys.* **26** (Suppl. 26-3) 2046

[95] Andryushin E A, Ginzburg V L and Silin A P 1993 On boundary conditions in the macroscopic theory of superconductivity *Usp. Fiz. Nauk* **163** 105 [1993 *Phys. Usp.* **36** 854]

[96] Ginzburg V L and Pitaevskii L P 1958 On the theory of superfluidity *Zh. Eksp. Teor. Fiz.* **34** 1240 [1958 *Sov. Phys. JETP* **7** 858]

[97] Cyrot M 1973 *Reps. Prog. Phys.* **36** 103

[98] Pitaevskii L P 1959 *Zh. Eksp. Teor. Fiz.* **37** 1794 [1960 *Sov. Phys. JETP* **37** 1267]

[99] Volovik G E and Gorkov L P 1985 *Zh. Eksp. Teor. Fiz.* **88** 1412 [1985 *Sov. Phys. JETP* **61** 843]

[100] Annett J F 1990 *Adv. Phys.* **39** 83; 1995 *Contemp. Phys.* **36** 423

[101] Sigrist M and Ueda K 1991 *Rev. Mod. Phys.* **63** 239

[102] Sauls J A 1994 *Adv. Phys.* **43** 113
Edelstein V M 1996 *J. Phys.: Condens. Matter* **8** 339

[103] Cox D L and Maple M B 1995 *Phys. Today* **48**(2) 32

[104] Barash Yu S, Galaktionov A V and Zaikin A D 1995 *Phys. Rev.* **52B** 665
Barash Yu S and Svidzinsky A A 1996 *Phys. Rev.* **53B** 15254
1996 *Phys. Rev. Lett.* **77** 4070

[105] Barash Yu S and Ginzburg V L 1975 *Usp. Fiz. Nauk* **116** 5; 1984 **143** 346
[1976 *Sov. Phys. Usp.* **18** 305; 1984 **27** 467]

[106] Ginzburg V L 1955 The surface energy associated with a tangential velocity
discontinuity in helium II *Zh. Eksp. Teor. Fiz.* **29** 254 [1956 *Sov. Phys.
JETP* **2** 170]

[107] Gamtsemlidze G A 1958 *Zh. Eksp. Teor. Fiz.* **34** 1434 [1958 *Sov. Phys.
JETP* **34** 992]

[108] Ginzburg V L and Sobyanin A A 1976 Superfluidity of helium II near the
λ-point *Usp. Fiz. Nauk* **120** 153 [1976 *Sov. Phys. Usp.* **19** 773]

[109] Ginzburg V L and Sobyanin A A 1982 On the theory of superfluidity of
helium II near the λ-point *J. Low Temp. Phys.* **49** 507

[110] Ginzburg V L and Sobyanin A A 1987 Superfluidity of helium II
near the λ-point in *Superconductivity, Superdiamagnetism, Superfluidity* ed V L Ginzburg (Moscow: Mir) p 242

[111] Sobyanin A A 1972 *Zh. Eksp. Teor. Fiz.* **63** 1780 [1973 *Sov. Phys. JETP*
36 941]

[112] Ginzburg V L and Sobyanin A A 1988 Superfluidity of helium II near the
λ-point *Usp. Fiz. Nauk* **154** 545 [1988 *Sov. Phys. Usp.* **31** 289]; 1987
Japan J. Appl. Phys. **26** (Suppl 26-3) 1785

[113] Goldner L S, Mulders N and Ahlers G 1992 *J. Low Temp. Phys.* **89** 131

[114] Ginzburg V L and Sobyanin A A 1972 Can liquid molecular hydrogen be
superfluid? *Pis'ma Zh. Eksp. Teor. Fiz.* **15** 343 [1972 *JETP Lett.* **15**
242]

[115] Collins G P 1996 *Phys. Today* **49**(8) 18

[116] Feynman R P 1955, in *Progress in Low Temperature Physics* vol 1
(ed C J Gorter) (Amsterdam: North-Holland) p 17

[117] Mamaladze Yu G 1967 *Zh. Eksp. Teor. Fiz.* **52** 729 [1967 *Sov. Phys. JETP*
25 479]; 1968 *Phys. Lett.* **A27** 322

[118] Dohm V and Haussmann R 1994 *Physica* B **197** 215

[119] Ginzburg V L and Sobyanin A A 1982 Structure of vortex filament in
helium II near the λ-point *Zh. Eksp. Teor. Fiz.* **82** 769 [1982 *Sov. Phys.
JETP* **55** 455]

[120] Pitaevskii L P 1958 *Zh. Eksp. Teor. Fiz.* **35** 408 [1959 *Sov. Phys. JETP*
35 282]

[121] Zurek W H 1996 *Nature* **382** 296

[122] Gasparini F M and Rhee I 1992 *Prog. Low Temp. Phys.* **13** 1

[123] Mikheev L V and Fisher M E 1993 *J. Low Temp. Phys.* **90** 119

[124] Zimmermann W 1996 *Contemp. Phys.* **37** 219

[125] Chan M, Mulders N and Reppy J 1996 *Phys. Today* **49**(8) 30

[126] Burton E F, Smith H G and Wilhelm J O 1940 *Phenomena at the Temperature of Liquid Helium* (New York: Reinhold)

[127] Donnelly R J 1995 *Phys. Today* **48**(7) 30

[128] Ginzburg V L and Zharkov G F 1978 Thermoelectric effects in superconductors *Usp. Fiz. Nauk* **125** 19 [1978 *Sov. Phys. Usp.* **21** 381]

[129] Ginzburg V L 1991 Thermoelectric effects in the superconducting state *Usp. Fiz. Nauk* **161**(2) 1 [1991 *Sov. Phys. Usp.* **34** 101]

[130] Geilikman B T and Kresin V Z 1972 *Kineticheskie i Nestatsionarnye Yavleniya v Sverkhprovodnikakh* [*Kinetic and Nonsteady-State Phenomena in Superconductors*] (Moscow: Nauka) [Translated into English 1974 (New York: J Wiley)]; 1958 *Zh. Eksp. Teor. Fiz.* **34** 1042 [1958 *Sov. Phys. JETP* **7** 721]

[131] Anselm A I 1962 *Introduction to the Theory of Superconductors* (Moscow–Leningrad: Gos. Izd. Fiz. Mat. Lit.) ch 8
Seeger K 1997 *Semiconductor Physics* (Heidelberg: Springer)

[132] Ginzburg V L 1989 Convective heat transfer and other thermoelectric effects in high-temperature superconductors *Pis'ma Zh. Eksp. Teor. Fiz.* **49** 50 [1989 *JETP Lett.* **49** 58]

[133] Kon L Z 1976 *Zh. Eksp. Teor. Fiz.* **70** 286 [1976 *Sov. Phys. JETP* **43** 149]; 1977 *Fiz. Tverd. Tela* (Leningrad) **19** 3695 [1977 *Sov. Phys. Solid State* **19** 2160]
Digor D F, Kon L Z and Moskalenko V A 1990 *Sverkhprovodimost': Fiz., Khim., Tekhnol.* **3** 2485 [1990 *Superconductivity: Phys., Chem., Technol.* **3** 1703]

[134] Arfi B *et al* 1988 *Phys. Rev. Lett.* **60** 2206; 1989 *Phys. Rev.* **39B** 8959
Hirschfeld P J 1988 *Phys. Rev.* **B37** 9331

[135] Jezowski A *et al* 1988 *Helv. Phys. Acta* **61** 438; 1989 *Phys. Lett.* **138A** 265

[136] Cohn J L *et al* 1992 *Phys. Rev.* **B45** 13144
Yu R C *et al* 1992 *Phys. Rev. Lett.* **69** 1431

[137] Cohn J L *et al* 1993 *Phys. Rev. Lett.* **71** 1657

[138] Hirschfeld P J and Putikka W O 1996 *Phys. Rev. Lett.* **77** 3909

[139] Ginzburg V L 1989 Thermoelectric effects in Superconductors *J. Supercond.* **2** 323; 1989 *Physica* C **162–164** 277; 1991 *Supercond. Sci. Technol.* **4** 1

[140] Ginzburg V L and Zharkov G F 1974 Thermoelectric effect in anisotropic superconductors *Pis'ma Zh. Eksp. Teor. Fiz.* **20** 658 [1974 *JETP Lett.* **20** 302]

[141] Selzer P M and Fairbank W M 1974 *Phys. Lett.* **A48** 279

[142] Gal'perin Yu M, Gurevich V L and Kozub V N 1974 *Zh. Eksp. Teor. Fiz.* **66** 1387 [1974 *Sov. Phys. JETP* **39** 680]

[143] Garland J C and van Harlingen D J 1974 *Phys. Lett.* **A47** 423

[144] Ginzburg V L, Zharkov G F and Sobyanin A A 1974 Thermoelectric phenomena in superconductors and thermomechanical circulation effect in a superfluid liquid *Pis'ma Zh. Eksp. Teor. Fiz.* **20** 223 [1974 *JETP Lett.* **20** 97]; Ginzburg V L, Sobyanin A A and Zharkov G F 1981 *Phys. Lett.* **A87** 107

[145] van Harlingen D J 1982 *Physica* **B109–110** 1710

[146] Gerasimov A M *et al* 1996 *Czechoslovak J. Phys.* **46** S 2 633 (LT21); 1995 *Sverkhprov.: Fiz., Khim., Tekhnol.* **8** 634; 1997 *J. Low Temp. Phys.* **106** 591

[147] Arutyunyan R M and Zharkov G F 1982 *Zh. Eksp. Teor. Fiz.* **83** 1115 [1982 *Sov. Phys. JETP* **56** 632]; 1983 *J. Low. Temp. Phys.* **52** 409; 1983 *Phys. Lett.* **A96** 480

[148] Ginzburg V L, Zharkov G F and Sobyanin A A 1982 Thermoelectric current

in a superconducting circuit *J. Low Temp. Phys.* **47** 427; 1984 **56** 195

[149] Ginzburg V L and Zharkov G F 1993 Thermoelectric effect in hollow superconducting cylinders *J. Low Temp. Phys.* **92** 25

[150] Ginzburg V L and Zharkov G F 1994 Thermoelectric effects in superconducting state *Physica* **235C–240C** 3129

[151] Arutyunyan R M, Ginzburg V L and Zharkov G F 1997 Vortices and thermoelectric effect in a hollow superconducting cylinder *Zh. Eksp. Teor. Fiz.* **111** 2175 [1997 *JETP* **84** 1186]; 1997 *Usp. Fiz. Nauk* **167** 457; [1997 *Phys. Usp.* **40** 435]

[152] Mattoo B A and Singh Y 1983 *Prog. Theor. Phys.* **70** 51

[153] Huebener R P, Ustinov A V and Kaplunenko V K 1990 *Phys. Rev.* **B42** 4831

[154] Ustinov A V, Hartmann M and Huebener R P 1990 *Europhys. Lett.* **13** 175

[155] Vinen W F 1971 *J. Phys. C: Solid State Phys.* **4** 1287; 1975 **8** 101

[156] Fabelinskii I L 1994 *Usp. Fiz. Nauk* **164** 897 [1994 *Phys. Usp.* **37** 821]

[157] Ginzburg V L and Sobyanin A A 1973 Use of second sound to investigate the inhomogeneous density distribution of the superfluid part of helium II near the λ-point *Pis'ma Zh. Eksp. Teor. Fiz.* **17** 698 [1973 *JETP Lett.* **17** 483]

[158] Ginzburg V L 1979 On the superfluid flow induced by crossed electric and magnetic fields *Fiz. Nizk. Temp.* **5** 299 [1979 *Sov. J. Low Temp. Phys.* **5**]

[159] Ginzburg V L and Sobyanin A A 1983 Circulation effect and quantum interference phenomena in a nonuniformly heated toroidal vessel with superfluid helium *Zh. Eksp. Teor. Fiz.* **85** 1606 [1983 *Sov. Phys. JETP* **58** 934]

[160] Gamtsemlidze G A and Mirzoeva M I 1980 *Zh. Eksp. Teor. Fiz.* **79** 921; 1983 **84** 1725 [1980 *Sov. Phys. JETP* **52** 468; 1983 **57** 1006]

[161] Ginzburg V L 1961 Second sound, the convective heat transfer mechanism, and exciton excitations in superconductors *Zh. Eksp. Teor. Fiz.* **41** 828 [1962 *Sov. Phys. JETP* **14** 594]

[162] Ginzburg V L, Gorbatsevich A A, Kopaev Yu V and Volkov B A 1984 On the problem of superdiamagnetism *Solid State Commun.* **50** 339

[163] Ginzburg V L 1979 Theory of superdiamagnetics *Pis'ma Zh. Eksp. Teor. Fiz.* **30** 345 [1979 *JETP Lett.* **30** 319]

[164] Gorbatsevich A A 1989 *Zh. Eksp. Teor. Fiz.* **95** 1467 [1989 *Sov. Phys. JETP* **68** 847]

[165] Ginzburg V L and Kirzhnits D A 1964 On superfluidity of neutron stars *Zh. Eksp. Teor. Fiz.* **47** 2006 [1965 *Sov. Phys. JETP* **20** 1346]

[166] Ginzburg V L and Kirzhnits D A 1968 Superconductivity in white dwarfs and pulsars *Nature* **220** 148

[167] Ginzburg V L and Zharkov G F 1967 Superfluidity of the cosmological neutrino sea *Pis'ma Zh. Eksp. Teor. Fiz.* **5** 275 [1967 *JETP Lett.* **5** 223]

[168] Ginzburg V L 1969 Superfluidity and superconductivity in the Universe *Usp. Fiz. Nauk* **97** 601 [1970 *Sov. Phys. Usp.* **12** 241]; 1971 *Physica* **55** 207

[169] Tamm I E 1932 *Phys. Z. Sowjetunion* **1** 733

[170] Ginzburg V L and Kirzhnits D A 1964 On the superconductivity of elec-

trons at the surface levels *Zh. Eksp. Teor. Fiz.* **46** 397 [1964 *Sov. Phys. JETP* **19** 269]

[171] Bulaevskii L N and Ginzburg V L 1964 On the possibility of the existence of surface ferromagnetism *Fiz. Met. Metalloved.* **17** 631 [1964 *Phys. Metals Metallography* **17**]

[172] Hohenberg P C 1967 *Phys. Rev.* **158** 383

[173] Ginzburg V L and Kirzhnits D A 1967 The question of high-temperature and surface superconductivity *Dokl. Akad. Nauk SSSR* **176** 553 [1968 *Sov. Phys. Dokl.* **12** 880]

[174] Ginzburg V L 1989 On two-dimensional superconductors *Phys. Scripta* **27** 76

[175] Ginzburg V L and Kirzhnits D A (eds) 1977 *Problema Vysokotemperaturnoi Sverkhprovodimosti* [*High-Temperature Superconductivity*] (Moscow: Nauka) [English translation 1982 (New York: Consultants Bureau)]

[176] Ginzburg V L 1984 On the electrodynamics of two-dimensional (surface) superconductors *Essays in Theoretical Physics (in Honour of D ter Haar)* ed W E Parry (Oxford: Pergamon Press) p 43

[177] Bulaevskii L N, Ginzburg V L and Zharkov G F 1983 Behavior of surface (two-dimensional) superconductors and of a very thin superconducting film in a magnetic field *Zh. Eksp. Teor. Fiz.* **85** 1707 [1983 *Sov. Phys. JETP* **58** 994]

[178] Frindt R F 1972 *Phys. Rev. Lett.* **28** 299

[179] Ginzburg V L 1996 Bill Little and high temperature superconductivity in *From High-Temperature Superconductivity to Microminiature Refrigeration* eds B Cabrera, H Gutfreund and V Kresin (New York: Plenum Press)

[180] Bednorz J G and Müller K A 1986 *Z. Phys.* **B64** 189

[181] Wu M K *et al* 1987 *Phys. Rev. Lett.* **58** 908

[182] 1991 *Proc. Int. Conf. on Materials and Mechanisms of Superconductivity, Kanazawa (Japan), July, 1991 High Temperature Superconductors III (Conf. M^2HTSC III); Physica* C **185**

[183] Little W A 1964 *Phys. Rev.* **A134** 1416
1965 *Sci. Am.* **212**(2) 21

[184] Ginzburg V L 1964 Concerning surface superconductivity *Zh. Eksp. Teor. Fiz.* **47** 2318 [1964 *Sov. Phys. JETP* **20** 1549]; 1964 *Phys. Lett.* **13** 101

[185] Ginzburg V L 1968 The problem of high-temperature superconductivity *Usp. Fiz. Nauk* **95** 91 [1968 *Contemp. Phys.* **9** 355]; 1970 **101** 185 [1971 *Sov. Phys. Usp.* **13** 335]

[186] Ginzburg V L 1968 The problem of high-temperature superconductivity *Contemp. Phys.* **9** 355

[187] Ginzburg V L 1971 Manifestation of the exciton mechanism in the case of granulated superconductors *Pis'ma Zh. Eksp. Teor. Fiz.* **14** 572 [1971 *JETP Lett.* **14** 396]

[188] Ginzburg V L 1972 The problem of high-temperature superconductivity *Annu. Rev. Material Sci.* **2** 663

[189] Ginzburg V L 1970 High temperature superconductivity *J. Polymer Sci.* C **29**(3) 133

[190] Cohen M L and Anderson P W 1972 *Superconductivity in d- and f-Band*

Metals (*American Institute of Physics Conf. Ser.* 4) ed D H Duglass (New York: American Institute of Physics) p 17

[191] Kirzhnits D A 1976 *Usp. Fiz. Nauk* **119** 357 [1976 *Sov. Phys. Usp.* **19** 530]

[192] Ginzburg V L 1992 Once again about high-temperature superconductivity *Contemp. Phys.* **33** 15

[193] Dolgov O V, Kirzhnits D A and Maksimov E G 1981 *Rev. Mod. Phys.* **53** 81

[194] Dolgov O V and Maksimov E G 1982 *Usp. Fiz. Nauk* **138** 95 [1982 *Sov. Phys. Usp.* **25** 688]

[195] Eliashberg G M 1960 *Zh. Eksp. Teor. Fiz.* **38** 966; **39** 1437 [1960 *Sov. Phys. JETP* **11** 696; 1961 **12** 1000]

[196] Ginzburg V L 1989 High-temperature superconductivity: some remarks *Prog. Low Temp. Phys.* **12** 1

[197] Ginzburg V L 1993 High-temperature superconductivity: its possible mechanisms *Physica* **C209** 1

[198] Ginzburg V L 1984 High-temperature superconductivity *Energiya* (9) 2

[199] Ginzburg V L 2001 John Bardeen and the theory of superconductivity (see Reference [1] p 451); 1986 *J. Supercond.* **4** 327

[200] Ginsberg D M (ed) 1989 *Physical Properties of High-Temperature Superconductors* I. vol 1 (Singapore: World Scientific) [Several other volumes of this series appeared later]

[201] Ginzburg V L and Maksimov E G 1994 Mechanisms and models of high temperature superconductors *Physica* **235C–240C** 193

[202] Shimada D *et al* 1995 *Phys. Rev.* **B51** 16495

[203] Maksimov E G 1995 *J. Supercond.* **8** 433

[204] Maksimov E G, Savrasov S U, Savrasov D U and Dolgov O V 1998 *Solid State Commun.* **106**(7) 409

[205] Ginzburg V L and Maksimov E G 1992 On possible mechanisms of high-temperature superconductivity (review) *Sverkhprovodimost': Fiz., Khim., Tekhnol.* **5** 1543 [1992 *Superconductivity: Phys. Chem. Technol.* **5** 1505]

[206] Abrahams E *et al* 1995 *Phys. Rev.* **B52** 1271

[207] Fehrenbacher R and Norman M R 1995 *Phys. Rev. Lett.* **74** 3884

[208] O'Donovan C and Carbotte J R 1995 *Physica* **252C** 87

[209] Bozovic I *et al* 1994 *J. Supercond.* **7** 187

[210] Cyrot M and Pavina D 1992 *Introduction to Superconductivity and High-T_c Materials* (Singapore: World Scientific)

[211] Ginzburg V L and Andryushin E A 1990 *Sverkhprovodimost' [Superconductivity]* (Moscow: Pedagogika) [English translation 1994 (Singapore: World Scientific)]

[212] Lubkin G R 1996 *Phys. Today* **49**(3) 48

[213] Ginzburg V L 1998 On heat transfer (thermal conductivity) and thermoelectric effect in superconducting state *Usp. Fiz. Nauk* **168** 363 [1998 *Phys. Usp.* **41** 307]

[214] Plakida N M 1996 *Vysokotemperaturnye Sverkhprovodniki [High-Temperature Superconductors]* (Moscow: Mezhdunarodnaya Programma Obrazovaniya [Translated into English 1995 *High-Temperature Superconductivity: Experiment and Theory* (Berlin: Springer)]

[215] Golubov A A *et al* 1994 *Physica* **235C–240C** 2383

[216] Combescot R and Leyronas X 1995 *Phys. Rev. Lett.* **75** 3732

[217] Maple M B 1995 *Physica* **215B** 110; 1995 *Physica* **215B** 127

[218] Bozovic I and Eckstein J N 1996 in *Physical Properties of High-Temperature Superconductors* vol 5 ed D M Ginsberg (Singapore: World Scientific); see also Reference [118] in the previous article 6 in this volume

[219] Ford P J and Saunders G A 1997 *Contemp. Phys.* **38** 63

[220] Dolgert A J *et al* 1996 *Phys. Rev.* **B53** 5650; 1997 **56** 2883

[221] Dalforo F *et al* 1999 *Rev. Mod. Phys.* **71** 463

[222] Klemens P G 1953 *Proc. Phys. Soc. London* **A66** 576; Klemens P G 1956 in *Handbuch der Physik* **14** ed S Flügge (Berlin: Springer) p 198 [Russian translation 1959 in *Fizika Nizkikh Temperatur* [*Low-Temperature Physics*] ed A I Shal'nikov (Moscow: IL)]

[223] Fedorov N K 1998 *Solid State Commun.* **106** 177

[224] Landau L D and Lifshitz E M 1992 *Electrodinamika Sploshnykh Sred* [*Electrodynamics of Continuous Media*] 27 (Moscow: Nauka) [Translated into English 1984 (Oxford: Pergamon Press)]

[225] Krishana K, Harris J M and Ong N P 1995 *Phys. Rev. Lett.* **75** 3529

[226] Mamin H J, Clarke J and van Harlingen D J 1983 *Phys. Rev. Lett.* **51** 1480

[227] Gulian A M and Zharkov G F 1978 in *Thermoelectricity in Metallic Conductors* ed F J Blatt and P A Schroeder (New York: Plenum Press) [The Russian text 1977 *Kratk. Soobshch. Fiz. FIAN* (11) 21]

[228] Gulian A M and Zharkov G F 1999 *Nonequilibrium Electrons and Phonons in Superconductors* (New York: Kluwer, Plenum)

[229] Zharkov G F 2002 *Zh. Eksp. Teor. Fiz.* **122** 600 [2002 *JETP* **95** 517]; 2003 *J. Low Temp. Phys.* **130** 45; 2004 *Usp. Phys. Nauk* **174** 1012 [2004 *Phys. Usp.* **47**]

[230] Zharkov G F 1958 *Zh. Eksp. Teor. Fiz.* **34** 412; 1959 **37** 1784 [1958 *Sov. Phys. JETP* **7** 278; 1959 **10** 1257]

[231] Simmonds R W *et al* 2001 *Phys. Rev. Lett.* **87** 035301

[232] Fliessbach T 2001 Effective Ginzburg–Landau model for superfluid ^{4}He, Preprint cond-mat/0106237

[233] Sadovskii M V 2001 *Usp. Fiz. Nauk* **171** 539 [2001 *Phys. Usp.* **44** 515] Loktev V M *et al* 2001 *Phys. Rep.* **349** 1

[234] Timmermans E 2001 *Contemp. Phys.* **42** 1

[235] Ginzburg V L 2001 Unthought, undone... Preprint 34 (Moscow: FIAN)

[236] Mo S, Hove J and Sudbo A 2002 *Phys. Rev.* **B65** 104501

[237] Kopnin N B 2002 *J. Low Temp. Phys.* **129** 219

[238] Kogan V G and Pokrovsky V L 2003 *Phys. Rev. Lett.* **90** 067004

[239] Kleinert H and Schakel A M 2003 *Phys. Rev. Lett.* **90** 097001

[240] Pitaevskii L and Stringari S 2003 *Bose–Einstein Condensation* (Oxford: Clarendon Press)

[241] Murphy D *et al* 2003 *Phys. Rev. Lett.* **90** 025301

[242] Chui T *et al* 2003 *Phys. Rev. Lett.* **90** 085301

[243] Lipavsky P *et al* 2001 *Phys. Rev.* **B65** 012507

[244] Ussishkin I, Sondhi S L and Huse D A 2002 *Phys. Rev. Lett.* **89** 287001

[245] Boaknin E *et al* 2003 *Phys. Rev. Lett.* **90** 117003

[246] Galperin Y M *et al* 2002 *Phys. Rev.* **B65** 064531

[247] Bozovic I *et al* 2003 *Nature* **422** 873

[248] Bozovic I *et al* 2002 *Phys. Rev. Lett.* **89** 107001

[249] Chen Y *et al* 2002 *Phys. Rev. Lett.* **89** 217001

[250] Carlson E W, Neto A H C and Campbell D K 2003 *Phys. Rev. Lett.* **90** 087001

[251] Gu J Y *et al* 2003 *Phys. Rev. Lett.* **90** 087001

[252] Misochko O V 2003 *Usp. Fiz. Nauk* **173** 385 [2003 *Phys. Usp.* **46** 373]

[253] Chu C P *et al* 2000 *J. Supercond.* **13** 679

[254] Shicll A B *et al* 2000 *J. Supercond.* **13** 687

[255] Schmideshoff *et al* 2000 *J. Supercond.* **13** 847

[256] Maksimov E G 2000 *Usp. Fiz. Nauk* **170** 1033 [2000 *Phys. Usp.* **43** 965]

[257] Mooney K P and Gasparini F M 2002 *J. Low Temp. Phys.* **126** 247

[258] Walker M B and Samokhin R V 2002 *Phys. Rev. Lett.* **88** 207001

[259] Cuoco M, Gentile P and Noce C 2003 *Phys. Rev. B* **68** 054521

[260] Nagamatsu J *et al.* 2001 *Nature* **410** 63

[261] Canfield P C and Crabtree G W 2003 *Phys. Today* **56**(3) 34

[262] Bel R *et al* 2004 *Phys. Rev. Lett.* **92** 177003

[263] Hwany J, Timusk T and Gu G D 2004 *Nature* **692** 714

[264] Ginzburg V L 2004 Nobel lecture *Rev. Mod. Phys.* **76**(3)

[265] Iida K *et al* 2004 *Prog. Theor. Phys. Suppl.* **153** 230

[266] Buzdin A I and Mel'nikov A S 2003 *Phys. Rev. B* **67** 020503(R)

[267] Gweon G-H *et al* 2004 *Nature* **430** 187

[268] Homes C C *et al* 2004 *Nature* **430** 539

[269] Gudstein D and Chatto A R 2003 *Am. J. Phys.* **79** 85

PART II

Article 8

A long, arduous, and varied life:

on the 100th anniversary of the birthday

of Igor Evgenevich Tamm[1*]

A centenary is a 'real' jubilee, although sometimes 'intermediate' ones—say, the 90th anniversary of a birthday—are celebrated. Lately I have read quite a few articles devoted to different jubilees, including centenaries. In some cases I knew the persons or had heard a lot about them. This is the law: the articles devoted to jubilees are one-sided, be they memoirs of a contemporary or, all the more, obituaries. The authors follow an ancient principle, following which one either speaks well about the deceased or does not speak at all—according to the Latin formula *"aut bene, aut nihil"*. Such an approach is understandable and, I think, inevitable. With the relatives, friends, and those who participated in the described events still being alive, who would write about the drawbacks, mistakes, and delusions of the person whose jubilee is being celebrated? Therefore, as a rule, only biographies or jubilee articles devoted to days long passed turn out to be fairly close to the truth.

I have made these remarks from thinking of the answer to the following question: what could I write about Igor Evgenevich Tamm if I were not bound by anything at all, say, by his children and grandchildren being alive? To my joy, I have concluded that I do not know anything about him that I would want to hide. Igor Evgenevich did not sign any malicious 'letters of scientists' or other 'letters of working people', never turned to self-reproach, and did not renounce his beloved executed brother or turn to intrigues. And all this during difficult times (when did Russia have easy times, though?).

Igor Evgenevich Tamm was born on 8 July, 1895, in Vladivostok but soon the family moved to Elizavetgrad in Ukraine. Here, in 1913, he graduated from the gymnasium and then studied at Edinburgh University in Scotland. After the beginning of the First World War, he changed to the Faculty of Physics and Mathematics at Moscow University and graduated from there in 1918. In his student years and later, Igor Evgenevich took an active part in the revolutionary movement. He was a menshevik-internationalist and was elected as a delegate to the First Congress of Soviets that took place in Petrograd in the summer of 1917. I remember Igor Evgenevich recalling that on some issue only he, from a multitude of menshevik delegates, raised his hand with a mandate in support of a bolshevik resolution and Lenin applauded him or expressed his satisfaction in some other way. To the end of his life, Tamm retained his revolutionary illusions and thought, like many others, that Leninism was merely distorted afterwards. How far we are now from those times, when we have learned about Lenin's atrocities, not to mention the bandit Stalin. Most probably, the life of Igor Evgenevich would have been even more difficult had he known the truth—he could console himself with proverbs like "you cannot make an omelette without breaking eggs" or suggesting that it was not Stalin but rather his apprentices, who were responsible for the bloody bacchanalia.

In 1919–20, Tamm lectured in southern Russia (in Simferopol and Odessa) where he met L I Mandelshtam. Igor Evgenevich considered Mandelshtam his teacher and kept close ties and friendship with him until Mandelshtam's death in 1944. From 1922, the scientific and pedagogical activities of Tamm were concentrated in Moscow. For many years, starting from 1930, he occupied the Theoretical Physics Chair at the Moscow State University (MSU) Physics Department. The ties with the MSU were broken only after the beginning of the war, in 1941, when the Physics Department was evacuated to Central Asia and Igor Evgenevich left, with the P N Lebedev Physical Institute of the Academy of Sciences of the USSR (FIAN), for Kazan. He had been working in this institute since 1934, since the Academy's transfer from Leningrad to Moscow, where he organized and headed the Theoretical Physics Department of FIAN. After the death of Igor Evgenevich (12 April, 1971), the department was named after him (it is now the I E Tamm Theoretical Physics Department of the P N Lebedev Physical Institute of the Russian Academy of Sciences).

Let us now turn to I E Tamm's scientific results. His first studies, made in collaboration with Mandelshtam, were devoted to the electrodynamics of anisotropic media. Let me mention that he published his first paper in 1924 [1, vol I, p 19], when he was already 29. By modern standards, this is quite late for a theoretical physicist and is basically explained by the turbulent times in which Igor Evgenevich started to work. At the same time, he never rushed to publish his results and suffered, in his own words, from 'agraphia'. I once dealt with such a situation when writing a paper on the theory of higher spins with him [2]. Work on the paper started in 1941 and continued, with interruptions, until 1945, when it was written but Igor Evgenevich refused to publish it. He had, however, his reasons, because we could not obtain important results.[1] I merely want to stress that Igor Evgenevich never hurried to publish his papers.

Electrodynamics in its various manifestations and applications always remained at the focus of Tamm's attention. In particular, he put a lot of effort into lecturing on it and in the only textbook written by him, *Foundations of the Theory of Electricity*, first published in 1929 (during the author's life eight editions of this wonderful book appeared and three have appeared since his death [3]).

This article is not aimed at describing the work of Igor Evgenevich in any detail. His results are reviewed in the introductory chapters in his collected works [1], collection of memoirs[2] [4, 5], and bibliographic reference book [6]. Here I would like to mention only those of outmost interest.

In condensed matter physics, he created a quantum theory for the photoeffect in metals. This work was done together with S P Shubin, whom Igor Evgenevich highly and deservedly esteemed. For example, in a letter to Shubin's relatives, he wrote: "Everybody having lived such a long, varying, and arduous life as mine is forming his own invisible Pantheon. A very special place in mine belongs to Semyon Petrovich. First, I always considered him to be the most talented one, not only among my students— and I was lucky with them—but among all our physicists of the same age" [7, p 347]. Alas, Shubin perished in exile in 1938 at the age of 30 and the location of his grave is unknown.

In 1930, Tamm published a rather detailed quantum theory of the scattering of light in crystals containing, in particular, a quantum explanation of the combinational (Raman) scattering of light discovered in 1928 in quartz and island spar by G S Landsberg and L I Mandelshtam (and simultaneously by Raman and Krishnan, so that the effect is most often named for Raman). In this work, Igor Evgenevich introduced quantized elastic sound waves which are nowadays called phonons. He was also the first one to introduce, in 1930, surface electron levels in crystals (these levels are now often named for Tamm).

In 1937, Igor Evgenevich developed (and published), together with I M Frank, classical theory of the Vavilov–Cherenkov effect discovered in the FIAN in 1934. He continued to study the radiation of uniformly moving charges in 1939 and developed a complete theory of this phenomenon in

[1] Let me note, in passing, that in the text of our paper in *Zh. Eksp. Teor. Fiz.* [2] there was the following footnote: "The paper will be published in English in 1947 in *J. Phys.* 11(3)". At this time, however, a campaign against servility towards everything foreign was launched and the wonderful *Journal of Physics of the USSR*, which had so much helped Soviet physics in getting international recognition, was hastily closed. Even the composed type was destroyed, so our paper never appeared in English.

[2] *The reminiscences of I E Tamm* was printed in two editions in 1981 and 1986. Despite all efforts, in neither of the two could we even include the name of his student, A D Sakharov, not to mention the possibility of printing his article. This is one of the reasons for preparing the third edition of *Reminiscences* that is now underway.

[1, vol I, p 77]. For an explanation of the physical origin of the Vavilov–Cherenkov effect, Tamm was awarded, in 1958, together with I M Frank and P A Cherenkov, the Nobel Prize in physics.

There is no doubt that the theory of Vavilov–Cherenkov phenomena presents an outstanding accomplishment but Igor Evgenevich himself did not consider it to be his best result. The work he regarded more highly was the that on the theory of beta-forces between the nucleons (1934, 1936). Igor Evgenevich put forward an idea that nuclear forces arise due to the simultaneous exchange of two particles, the electron and neutrino. The beta-forces turned out, in fact, to be insufficient to ensure nuclear stability. This idea provided, however, an impulse for the development and, one could say, a prototype for the theory of nuclear forces due to the exchange of mesons. Igor Evgenevich also worked on the problem of nuclear forces and related questions after that. In particular, in 1934, together with S A Altshuler, he came to the conclusion that the neutron possesses a magnetic moment. In 1953–55, Igor Evgenevich, together with his collaborators, studied photoproduction and the scattering of pions on nucleons in the non-relativistic (for nucleons) approximation. In this study, the possibility of the existence of nucleonic resonances having spin $\frac{3}{2}$ was considered. This approach was criticized at that time but later broad (having a small lifetime) nucleonic resonances found a firm place in nuclear physics.

Igor Evgenevich was always primarily interested in the fundamental problems of physics and, as a rule, worked at the forefront. His last passion was an attempt to introduce into relativistic field theory a certain fundamental length by considering curved momentum space. This hopefully would have removed the divergences (infinite expressions) in quantum field theory. This idea had also been considered earlier but Tamm introduced new moments in discussing the problem and obtained some preliminary results. In my opinion, this direction of research remains interesting even now, in spite of not being pursued. Who knows, however.... . The two last research papers by Igor Evgenevich [1, vol II, pp 218, 226] were, in fact, on curved momentum space (the second one, written together with V B Vologodsky, appeared after Igor Evgenevich's death). He was working on this problem while already seriously ill. In 1967, he fell ill with lateral amiotropic sclerosis and, during the last three years of his life, was 'chained' to a breathing machine.

In 1967, the USSR Academy of Sciences awarded Tamm with its highest annual distinction, the M V Lomonosov Great Golden Medal. It was already impossible for Igor Evgenevich to present the compulsory talk required by the Academy Charter to its General Assembly in 1968, so his lecture was read. This talk, titled "The Evolution of Quantum Theory", is, in my opinion, still of interest. It concludes the Tamm's collected works [1, vol II, p 48] and, in some way, crowns his whole life in science.

I have listed only the main articles by Tamm. Even within the chosen restrictions, it is necessary to mention the foundations of the theory of magnetic thermonuclear reactors, nowadays called tokamaks, developed along with A D Sakharov. Igor Evgenevich also authored a number of other articles classified as applied in his collected works. They should also be supplemented by various calculations he made at the 'Ob'ekt' [Unit] (in Arzamas-16) in relation to the development of the hydrogen bomb. However, I have never seen these calculations and cannot say anything about them.

Igor Evgenevich was a courageous person. It is probably sufficient to mention that he was a mountain-climber. I think, however, that his courage showed itself in a brighter and more non-trivial way during his terminal illness. As mentioned, with the help of a portable breathing machine, which made it possible to sit at his desk, he continued to work. Igor Evgenevich kept his sense of humor, played chess, and welcomed visitors, while at the same time making sad jokes, comparing himself to a pinned beetle.

He also possessed courage of a different sort—he fought fake science, in particular the Lysenko obscurantism. He made public statements about this at the General Session of the USSR Academy of Sciences during the Khrushchev era. One really had to have courage to do this—so many people, even those who were courageous in the physical sense, were afraid, in totalitarian times, to express their opinion in case it deviated from the one prescribed from above (one should recall that until Khrushchev's resignation, Lysenko enjoyed full support from the authorities).

Nowadays, when we are facing manifestations of religious and, more often, pseudoreligious feelings, it is appropriate to mention that Igor Evgenevich was a convinced and unreserved atheist. This also refers to all of his students whom I know. It seems unnecessary to say that such a person as Tamm had nothing to do with anti-Semitism but his disgust towards anti-Semites and other Black-Hundred members is worth recalling. I would also like to mention his politeness, which was in contrast to other examples known to me. Although addressing young people—students and postgraduate students—with their patronymic nowadays looks old-fashioned, it demonstrated that he genuinely belonged to the 'intelligentsia' with its inherent impossibility of being impolite. Igor Evgenevich could feel offended and blaze up but I am not aware of (and cannot imagine) his discontent turning into boorishness towards a colleague or, even less likely, an older person.

It is impossible to write about Igor Evgenevich without mentioning his charm, liveliness, and passion for mountain-climbing and hiking. This was mentioned by almost all of the 33 authors who contributed to the collection of memoirs *Reminiscences about I E Tamm: To the Centenary of I E Tamm's Birth*. A new supplementary edition is expected soon and, I hope, will be accessible.[2*] I should mention, however, that the first two editions constituted, cumulatively, 12 700 copies, so that they are present in many homes and libraries. It makes for fascinating reading.

I do not want to repeat something I have already written (in particular, in my article in [4, p 115]),[3*] so let me describe two stories told by Igor Evgenevich that I have not seen published.

Tamm's participation in the revolutionary movement was not limited to mere discussions. On occasion, he transported, during the Civil War in southern Russia, something forbidden—weapons or literature (I do not remember exactly what it was but most probably it was the latter) and was never caught by the corresponding security forces (again, I do not remember which forces exactly these were but the hunt was for bolsheviks, mensheviks, etc). So, here is a story. Igor Evgenevich traveled by train, did not have anything forbidden on him and, thus, felt completely relaxed. At the destination all the passengers were lined up and a security officer pointed to those who had to be searched (it was probably too cumbersome to search everybody). To Igor Evgenevich's surprise, he was among those chosen. When he was released, he asked the officer why he was selected to be searched. "You were nervous" was the reply. This episode seems to be of psychological interest to me. Evidently, when Tamm had something forbidden on him, he was focused, and did not strike one's eye. When he was not afraid of anything, his natural vivacity was not contained and created an impression of nervousness.

As he was a professor at MSU, Igor Evgenevich lived near the old university building in the former stables with a lavatory located in the yard. Because of a shortage of hotels or for some other reason, Tamm's friend, the great physicist P A M Dirac, stayed at Tamm's place during his visits to Moscow. Tamm was naturally ashamed of the condition of the lavatory and, during Dirac's second visit, expressed his regret with respect to this situation remaining unchanged. "You are mistaken"—was Dirac's answer—"The last time there was no light there, now it has appeared".

That Tamm lived in such difficult conditions and famous foreigners stayed at his place eventually had an effect and in the mid-1930s he got an apartment, quite a good one by the standards of those times, at Zemlyanoi Val. Apartments were rare things then and, in this house, they were distributed, if I remember correctly, by Molotov himself. Well-known people lived there, judging by the names I saw when coming upstairs to Tamm's apartment. I do not exclude the possibility that Tamm's international acclaim and his election as a Corresponding Member of the USSR Academy of Sciences in 1933 prevented his arrest. He was expecting it, though, and, as he said, had a backpack with everything necessary prepared. This hypothesis is, of course, not very convincing. People more prominent than Tamm were arrested and here was a former menshevik with an executed brother and, probably, with quite a few denunciations (accusing him of idealism, etc) registered against him. Luckily, somehow, this did not happen.

Another difficult period in Igor Evgenevich's life began, loosely speaking, in 1946. Elections to the Academy were announced. The Physics and Mathematics Branch had a corresponding vacancy and Tamm was considered to be an indisputable candidate for it. In those times, the Academy numbered many fewer members than now and the election procedure was, I would say, more rigorous. For example, there was a preliminary discussion of the candidates in the press, etc. I remember an article in one of the central newspapers discussing Tamm's candidacy. He was not elected, however.

In those times and later, the candidatures were discussed at the party section of the branch and approved by the CPSU Central Committee. It is known that a denunciation of Tamm had been written and Zhdanov 'himself' vetoed his election. The voting was secret and only full Academy members could vote. Characteristically, even with this secret ballot, the absence of the CPSU Central Committee recommendation was sufficient to preclude a candidate's election. Thus, 'political mistrust' of Tamm was expressed and this was quite important at that time. Undoubtedly, Tamm was hurt, annoyed, and, I think, insulted but he did not show it. Many others in such situations (and in less stressful ones, say just after not being elected) complained and even had heart problems.

During the war (1941–45) and right after it, Igor Evgenevich, like many theoretical physicists, found himself excluded from working on important projects. Although he performed calculations related to the demagnetization of ships, he was not engaged in anything really important. At the same time, already back in 1943, work on the so-called 'uranium project' aimed at constructing nuclear weapons had begun in the USSR (on 11 February, 1943, I V Kurchatov was appointed to lead the project). Igor Evgenevich was not only one of the best known Soviet physicists but also an expert in nuclear physics. Therefore, why he began working on the 'uranium project' only in 1947 is not clear. Even now, I do not know the reason for it.

One possible explanation is that Tamm's domain of expertise was far from the issues that were important during the first stages of the project (reactors, isotope separation, etc). The second possibility is a veto from the KGB (NKVD) because of his bad reputation with these organizations. The third one is related to Tamm's relations with Kurchatov. In 1943, when Kurchatov was appointed head of the atomic project, it was decided to boost his authority (this was perhaps the reasoning of officials) by electing him to Full Membership in the Academy. Some formalities were followed even at that time, so other candidatures for the same opening were allowed. This is how it happened that, besides Kurchatov, A I Alikhanov also participated in the elections. And, contrary to official wishes, he was elected. Usually, when telling this story, I ask the question: "What happened next?" This question tests the knowledge of the ways the Academy functioned. The answer reads: a phone call to the 'proper place' was made and immediately a new opening was 'given'. As a result, in September 1943, Kurchatov was elected as a Full Member of the Academy as well.[6*] Such a practice, unlawful and impertinent in its essence, where officials, including those of the Academy, could arbitrarily provide places in the Academy, sometimes even preventing unwanted candidates from being elected, existed until recently.

The reason for mentioning these elections in the present article is that at the Academy Assembly, Tamm supported Alikhanov, i.e. against Kurchatov. I think that in this case Tamm was completely wrong—not because Kurchatov was supported by the officials and the place was 'given' to him. But by speaking against Kurchatov, Tamm merely demonstrated his independence. For reasons completely unrelated to the 'uranium project' and official wishes, I am convinced, however, that it was Kurchatov rather than Alikhanov who deserved to be elected at that time. This opinion is based on studying the scientific achievements of the candidates. It suffices to say that before Kurchatov began working in nuclear physics, he had done classical studies in the physics of ferroelectrics. In the domain of nuclear physics, he, within a short period of time, had also obtained valuable results, in uranium fission first and foremost. All this is, however, not important here. What is essential is that Kurchatov had reasons to feel offended by Tamm. So, my third explanation as to why Kurchatov did not take Tamm on board from the very beginning of the 'uranium project' is based on the fact that he felt offended. I have no grounds to insist that this assumption being right, though. Undoubtedly, Kurchatov had a large personality and was, perhaps, above such things. For whatever reasons, it was only in 1947 that Tamm finally started working on the 'uranium project' and, more specifically, on issues related to the development of the thermonuclear bomb (see comment 4 to article 18 in the present book).

Here I will allow myself a small digression on Kurchatov's work. His role in constructing the nuclear bomb is well known [8]. In my opinion, he deserves great credit for inviting Tamm to participate in these activities, despite objective and subjective hindrances. Furthermore, as far as I know, it was Kurchatov who saved Soviet physics from 'lysenkization' in 1949. The point is that, using the notorious VASKhNIL session in 1949 as an example, an All-Union Congress of Physicists had been prepared. The idea of its organizers was to 'restore order' in physics, uproot idealism, etc. It is unfortunate that verbatim reports of the meetings of the organizing committee of this Congress during the three months of its preparation have not been published. The fragments from them that have appeared in *Priroda* [9] speak for themselves. A stench coming from these 'fighters against idealism and cosmopolitism' can be clearly felt. The Congress was scheduled to open on 21 March, 1949 but was cancelled at the last moment. There exist grounds to think that it was cancelled by Stalin after Beria reported to him a statement by Kurchatov that, without the theory of relativity and quantum mechanics, the bomb could not be constructed. This was precisely a critical moment: the preparations for the first Soviet nuclear bomb test were underway (this test took place on 29 August, 1949).

Let me finally mention Kurchatov's role in declassifying and promoting research on controlled thermonuclear fission. This work was started in 1950 at the initiative of Sakharov and Tamm. This research was top secret (code 'top secret, special file'). This secrecy was, to a certain extent, understandable, because during the first phases of research it seemed that the goal was near and it would be possible to build a controlled thermonuclear reactor in the near future. I always thought that the most attractive aspect of all this was the possibility of obtaining a practically unlimited source of energy. I N Golovin, who kindly read the first version of this article, told me, however, that at that time nobody thought about this aspect of the project. It was thought that the reactor would be used as a source of tritium and neutrons. Tritium and neutrons are formed in reactions of synthesis in the course of collisions of deuterium nuclei (it was planned to fill the reactor chamber with deuterium).[4*] I was also working on the problem of controllable thermonuclear synthesis and in 1950–51 wrote a number of reports on this issue. It was good to learn from the book by Golovin [8] that Kurchatov and others had read them. I mention this because either at the end of 1951 or at the beginning of 1952, the 'First Department' terminated my access to my own classified working materials on thermonuclear synthesis (reports, workbooks), i. e. my permit to work on this topic was effectively withdrawn—so high was its value.[3] With time it became clear, however, that experimental realization of controllable thermonuclear synthesis is extremely hard to achieve. I am sure, nevertheless, that because of inertia this work would still have been kept secret for years. Luckily, Kurchatov was bold and resolute and ensured that the thermonuclear synthesis problem was declassified. He did it in a dramatic way by giving a lecture on this topic at the British Centre for Atomic Research in Harwell on 25 April, 1956 [11].

It is now time to return to the main idea of this article. In 1947, Kurchatov tasked Tamm with the hydrogen bomb development project. According to Sakharov, this happened in 1948 [12]. It is possible that he was right but I still think this took place in the second half of 1947. Before that, the possibility of constructing a hydrogen bomb had already been analyzed by Ya B Zeldovich and his group—with relatively pessimistic conclusions. This explains, perhaps, why the work on the thermonuclear bomb was very secret but still not top secret. This is why Igor Evgenevich could ensure that I joined the project (at that time I was a Dr Phys Math Sci and a Deputy Director of the Theoretical Physics Department headed by Tamm

---

[3] When the work on thermonuclear synthesis was declassified I published, in a some sort of revenge, extracts from these reports [10], see also comment 5.

but my wife was in exile). That this was quite, so to say, non-trivial, is clear from the fact that Tamm could not do the same for E L Feinberg (before their marriage Feinberg's wife had lived for some time in the USA). At the next stage, when Tamm and Sakharov left for Yu B Hariton's Arzamas-16 in 1950, I was already prevented from following them and remained in Moscow heading, so to say, a support group. We were still working 'behind the guarded door' but our work was not directly related to the weapon, just in performing calculations coordinated with Tamm and Sakharov during their visits to Moscow. This was the previously mentioned beginning of my work on the thermonuclear problem. Let me note that I think I was quite lucky with this. On the one hand, I participated in an extremely secret project and was thus protected from attacks (at that time I was accused with cosmopolitism, my professorship was not approved, I was expelled from FIAN's Scientific Council, etc.). On the other, I could work on interesting things like thermonuclear synthesis and completely open issues like superconductivity, and so on. There is no need to describe this period of Igor Evgenevich's life in more detail: one can read about it in Sakharov's *Memoirs* [12]. The third edition of recollections about Tamm, which is currently under preparation, will include new material on this topic.[2*] Let me only mention that the first Soviet hydrogen bomb was successfully tested on 12 August, 1953.

After Stalin's death (5 March, 1953) and the execution of Beria that followed shortly after, there began a new, less terrible epoch. It was remembered, in particular, that the previous elections to the USSR Academy of Sciences had taken place in 1946, so new ones were scheduled for October 1953. Of course, I did not participate in these elections directly. I remember only having to forge Tamm's signature on the corresponding documents (he was not in Moscow). At this time, his election to Full Member of the Academy was smooth. At the session of the Physics and Mathematics Branch of the Academy, there was only one vote against him. As far as I know from Igor Evgenevich's words, it was Kurchatov who had the process of elections in this Branch under his control. From Tamm's group, Sakharov became a Full Member of the Academy and I became a Corresponding Member.

At the very end of 1953 or at the beginning of 1954, many participants in the 'bomb project' received state awards. Tamm received a Golden Star of the Hero of Socialist Labor and the Stalin Prize of the first degree (converted into the State Prize several years later). Besides that, he was given a dacha, an apartment, and a large sum of money as a premium. So, Igor Evgenevich became a *persona grata*. Soon after this he could return to Moscow, although his links to the 'Ob'ekt' were still not cut. This happened, probably, after Igor Evgenevich began travelling abroad, which he did for the first time (after the war) in 1956, although this was a visit to the DDR. In 1956, Tamm was awarded the Nobel Prize in physics and, naturally, visited Stockholm. In the same year, he actively participated in the Second World Conference on the Peaceful Use of Atomic Energy. Later, he visited the USA, England, Poland, India, China, France, and Japan. Igor Evgenevich was fluent in languages and liked visiting new places and having discussions with colleagues. It is clear that he enjoyed having this possibility to travel abroad. By the way, at that time many participants in the atomic project (and not only them) either could not travel abroad at all or faced significant difficulties in this respect. It is surely common knowledge that before 1987–88, travelling abroad was a big problem.

In the USSR, Tamm already enjoyed recognition before the war but in the 1950s it became, so to say, official and worldwide. He was elected to a number of foreign Academies of Sciences (as a foreign member). These were the Polish Academy of Sciences, the Swedish Royal Academy of Sciences, the American National Academy of Arts and Sciences, the German Academy of Naturalists "Leopoldina", and the New York Academy of Sciences. Let me specially mention his election to the prestigious National Academy of Sciences of the USA (1961).

There is no doubt that Igor Evgenevich was an ambitious person but these were 'healthy' ambitions. Lacking these it is hardly possible to be successful in science. What I mean here is the aspiration to write a good paper and to get recognition for it. As for shallow ambitions (vanity), Igor Evgenevich was very modest in this respect. Even the Nobel Prize, not to mention the state awards, did not excite strong feeling in him. As far as I remember, he never mentioned his election to foreign academies (a real honor indeed), so it is only now that I have learned about it from the Academy reference book.

As Igor Evgenevich wrote in 1953 in the previously cited letter, he had had a 'long, varied and arduous life'. It is comforting to realize that after 1953 this life continued in favorable conditions for a number of years. He was acclaimed widely, was wealthy, traveled a lot, and, which was of great importance for him, was engaged in active and successful research in physics. During these years he was socially active as well and took great interest in molecular biology and fought for its development in our country. I remember how many people participated in 1965 in the celebration of Igor Evgenevich's 70th birthday in the Conference Hall of FIAN. This was a demonstration of love and respect towards him. He was on his best form, making jokes. He had only two years left before a grave illness caught him...

It is appropriate to mention that the Academy of Sciences (in particular, its then President M V Keldysh) did everything possible to take care of Tamm. To the end of his days, he stayed at home (in his Moscow apartment or at his dacha), medical personnel were on duty by his bed, and a breathing machine was available at all times. I have already written that it is during this illness that Tamm's courage and passion for science were more visible than ever, allowing him to continue to work in the most difficult circumstances.

Igor Evgenevich Tamm left behind first-class results in physics and the memory of a respected, courageous, and noble person.

Comments

1*. This article was published in 1995 *Vestn. Ross. Akad. Nauk* **65**(6) 520 [1995 *Herald Russ. Acad. Sci.* **65** 231]. The article includes several photographs omitted in the present edition.

2*. This edition has now appeared: 1995 *Reminiscences about I E Tamm: To the Centenary of I E Tamm's Birth* ed E L Feinberg (Moscow: IzdAT) 3rd edn (Moscow: Nauka).

3*. This article was also published in my book 1995 *On Physics and Astrophysics* (Moscow: Byuro Kvantum) p 350 [2001 *The Physics of a Lifetime: Reflections on the Problems and Personalities of 20th Century Physics* (Berlin: Springer) p 351].

4*. This is not quite correct. Most probably it was planned to produce tritium by bombarding ^{6}Li with neutrons (see article 18 of the present book).

5*. Regarding this issue, see also the comment 11 to article 18.

6*. Because I did not attend the Session myself, I cannot guarantee that my account is exact. I have heard, in particular, a version of this story, according to which the second opening had already been available before Alikhanov was elected. That Tamm supported Alikhanov and not Kurchatov is beyond any doubt.

References

[1] Tamm I E 1975 *Collected Scientific Papers* 2 vols (Moscow: Nauka)
[2] Ginzburg V L 1947 I E Tamm *Zh. Eksp. Teor. Fiz.* **17** 227
[3] Tamm I E 2003 *Fundamentals of the Theory of Electricity* 11th edn (Moscow: Fizmatlit)
[4] 1981 *Reminiscences about I E Tamm* (Moscow: Nauka) 1st edn (2nd edn 1986)
[5] 1972 *Problems of Theoretical Physics: In Memory of I E Tamm* (Moscow: Nauka)
[6] 1974 *Igor Evgenevich Tamm (1985–1971)* (*Biographic Materials on Scientists in the USSR, Physics Series* **16**) 2nd edn (Moscow: Nauka)
[7] Shubin S P 1991 *Izbrannye Trudy po Teoreticheskoi Fizike: Ocherk Zhizni, Vospominaniya, Stat'i* (*Selected Papers on Theoretical Physics. Life Essay. Memoirs. Papers*) Eds S V Vonsovskii and M I Katsnelson (Sverdlovsk: Izd. UrO AN SSSR)

[8] Golovin I N 1978 *I V Kurchatov* (Moscow: Atomizdat)

[9] Sonin A S 1990 *Priroda* **3** 57; **5** 93; Sonin A S 1994 *Idealism in Physics. Essay on One Ideological Campaign* (Moscow: Fizmatlit)

[10] Ginzburg V L 1962 *Trudy FIAN SSSR* **18** 55

[11] Kurchatov I V 1956 *Usp. Fiz. Nauk* **59** 603

[12] Sakharov A D 1990 *Memoirs* (New York: Chekhov Publ.); 1996 (Moscow: Human Rights); Magazine versions 1990 *Znamya* **10–12**; 1991 **1–5**

Article 9

'Singing' electrons[1*]

In his Nobel lecture on 11 December, 1958 in Stockholm Igor Evgenevich Tamm said:

> We perceive the Mach waves radiated by a projectile as its familiar hissing or roaring. That is why, having understood the quite similar mechanism of the Vavilov–Cherenkov radiation of light by fast electrons, we have nicknamed it 'the singing electrons'. I should perhaps explain that we in the USSR use the name 'Vavilov–Cherenkov radiation' instead of just 'Cherenkov radiation' in order to emphasize the decisive role of the late Prof S Vavilov in the discovery of this radiation. [1]

Along with a 'singing electron', the term 'shining electrons' gained some currency but both terms are, in fact, only rarely used, so one usually speaks of the Vavilov–Cherenkov radiation—or, in the West, Cherenkov radiation. This can be explained primarily by the fact that, in the article in which the effect was first described, only Cherenkov's name appeared [2].[2*]

Vavilov's own publication in the same journal issue [3] was only concerned with the nature of the newly-discovered glow (which has to do with Compton electrons, Vavilov argued, not with gamma rays). But it was Vavilov who had suggested the idea of the experiment and who actually guided it. A historical review by Frank [4] and especially a note by Dobrotin, Feinberg and Fock [5] give Vavilov full credit for the discovery—further indicating that only the name Vavilov–Cherenkov effect can be used properly.[3*]

Discovered in 1934 [2, 3], i.e. 60 years ago, the Vavilov–Cherenkov (VC) effect has been explained theoretically in 1937 by I E Tamm and I M Frank [6].[1] They did what was, in fact, the most important thing to

---

[1] The order of the authors' family names in the Russian publication [6] is due to the letter T preceding F in the Russian alphabet. In the Latin alphabet, the order is reversed,

do at the time: they established the nature of the phenomenon (incidentally not at all a trivial achievement) and obtained the formulae

$$\cos\theta = \frac{c}{nv} \tag{9.1}$$

$$W = \frac{q^2 l}{c^2} \int\limits_{vn/c \geq 1} \omega\left(1 - \frac{c^2}{n^2(w)v^2}\right)d\omega \tag{9.2}$$

which determine the direction of the radiation (θ is the angle between the particle velocity v and the wavevector of the light) and its intensity (W is the energy which a particle with charge q moving with a velocity v in a transparent medium with refractive index $n(\omega)$ loses to VC radiation over a distance l). Of course, you cannot ask for much more of a short *Doklady* paper [6] than that and I E Tamm and I M Frank, in fact, did not stop there and continued with their study. Interestingly, both the style and character of their research were quite different for the two. For Frank, VC radiation and the radiation from uniformly moving sources in general (the subject which, apart from the VC effect, also includes the Doppler effect and the transition radiation) were, in fact, his life's work (along with neutron physics, though). The reader can refer to Frank's monograph [4] for a summary of his research on the VC effect.

For I E Tamm, in contrast, VC studies were yet another episode in his career and did not, in fact, take much of his time. Following the pioneering paper [6] co-authored with Frank, a number of points that still remained unclear were clarified by Tamm in his own study [7]—at the end of this, Tamm writes that he had extensive discussions with L I Mandelstam and answered his questions when preparing the paper. In particular, Tamm considers radiation from a charge whose uniform motion is limited to a certain time interval. It is assumed, specifically, that the charge which is at rest up to time $-t_0$ is instantaneously accelerated to a velocity v and then is instantaneously stopped at time t_0. The radiation field in this case is a combination of radiations emitted, due to acceleration, at times $\pm t_0$ plus (provided $v > c/n(\omega)$) the VC radiation field. Thus, it proves possible to find the relative contributions from both radiation types depending on t_0 and the radiation frequency ω. Further, unlike the usual approach, the VC radiation was calculated not in a laboratory frame but in one where the radiator (electron) is at rest. Finally, the charge's field itself was found and a number of useful observations were made. Besides reference [7], Tamm's other publications on the subject include only reference [8], in part a review, which is co-authored by Frank and the Nobel lecture already mentioned [1].

which is reflected in the way the relevant reference is given in Tamm's English language publication [7]. Also Frank and I are in the same situation, with me preceding Frank in Russian publications and following him in English ones.

At first sight, it may seem that as far as its theoretical aspects are concerned, the subject of VC radiation was largely exhausted in Tamm and Frank's work. This was by no means the case. As is usually the case with phenomena that are rich in physics, new questions emerged, new applications in various areas of physics appeared, etc. To overview the 60-year history of VC research and its applications is beyond the scope of the present note but there are some points that are worth mentioning here.

First, the detection of VC radiation can serve as a way to register the passage of a charged particle through a detector (this device is often referred to as the Cherenkov counter). In this way, one also determines the direction of the particle's motion because θ in equation (9.1) is the angle its velocity makes with the wavevector of the emitted wave. Second, given the refractive index $n(\omega)$ of the counter medium, measuring the angle θ determines the particle velocity v. By simultaneously measuring the particle momentum $p = mv/\sqrt{1 - (v^2/c^2)}$ (from the deflection in a magnetic field, for example), the mass of the particle m and its energy $E = mc^2/\sqrt{1 - (v^2/c^2)}$ are also determined. Third, the quantity q in the radiation intensity formula (9.2) is the charge of the particle, so that measuring W gives q. If the particle charge is $q = eZ$, where e is the electron charge, then measuring q yields Z. It is these features which, in principle, determine the operation of the Cherenkov counters—devices which are currently widely used in high-energy physics [9–11]. In particular, VC radiation is used in studying the extensive air showers produced by energetic enough gamma photons (with $E_\gamma > 10^{11}$–10^{12} eV in practical terms; see [10]). The use of the VC effect in high-energy physics (in particular, in gamma-ray and neutrino studies) is a big chapter in physics, with intensive research being done and numerous facilities running or being projected [10, 11].

Putting aside the applied aspects, a number of important theoretical problems have been clarified since Tamm and Frank's studies. These include the VC effect in anisotropic media (crystals) [12, 17, 18], the quantum theory of the VC effect [13, 17], VC radiation in channels and slits [14, 17], the inclusion of absorption [15, 16], radiation from the magnetic monopole and various kinds of electric and magnetic multipoles [4, 13, 17], and the simultaneous emission of VC radiation and transition radiation [19]. References [4, 17–19] are review literature.

Clearly, today, after 60 years of work, a kind of saturation has been achieved in both theoretical and experimental research into the 'singing electrons'. But this, of course, is not an absolute saturation—nor will it ever be. Life moves on[2*] and charged particles keep 'singing' in numerous laboratories and in the atmosphere—reminding us again and again of the classical work of Igor Evgenevich Tamm, whose birth centenary we are celebrating today.

Comments

1*. This note was published in the journal *Priroda* [*Nature*] dedicated to the 100th birthday of I E Tamm (no 7 1995 p 105).

2*. This is not exactly so. To my knowledge, neither the publication [2] nor S I Vavilov's paper [3] which immediately followed it produced a noticeable reaction. However, after it became clear that a new phenomenon was, in fact, reported and after the theory of the effect had been constructed, a paper by P A Cherenkov alone was sent by S I Vavilov to *Nature* and—after being rejected—to the *Physical Review* [20]. See article 2 of this collection for more details.

3*. See also article 2 in this book and a note by V L Ginzburg (2002 *Usp. Fiz. Nauk* **172**(3) 373 [2002 *Phys. Usp.* **45**(3) 341]).

References

[1] Tamm I E 1959 *Usp. Fiz. Nauk* **68** 387 [1960 *Science* **131** 206]; 1975 *Sobranie Nauchnykh Trudov* [*Collected Works*] vol 1 (Moscow: Nauka) p 121

[2] Cherenkov P A 1934 *Dokl. Akad. Nauk SSSR* **2** 451 [1934 *C. R. Acad. Sci. URSS* **2** 451]

[3] Vavilov S I 1934 *Dokl. Akad. Nauk SSSR* **2** 457 [1934 *C. R. Acad. Sci. URSS* **2** 457]

[4] Frank I M 1988 *Izluchenie Vavilova–Cherenkova. Voprosy Teorii* [*The Vavilov–Cherenkov Radiation. Theoretical Problems*] (Moscow: Nauka)

[5] Dobrotin N A, Feinberg E L and Fock M V 1991 *Priroda* **11** 58

[6] Tamm I E and Frank I M 1937 *Dokl. Akad. Nauk SSSR* **14** 107 [1937 *C. R. Acad. Sci. URSS* **14** 107]

[7] Tamm I E 1939 *J. Phys. USSR* **1** 439

[8] Tamm I E and Frank I M 1944 *Trudy FIAN SSSR* **2** 63
Tamm I E 1975 *Sobranie Nauchnykh Trudov* [*Collected Works*] vol 1 (Moscow: Nauka) p 113

[9] Zrelov V P 1968 *Izluchenie Vavilova–Cherenkova i ego Primenenie v Fizike Vysokikh Energiy* [*The Vavilov–Cherenkov Radiation and Its Applications in High-Energy Physics*] (Moscow: Atomizdat)

[10] 1990 *Cherenkovskie Detektory i ikh Primenenie v Nauke i Tekhnike* [*Cherenkov Detectors and Their Application in Science and Technology*] (Moscow: Nauka)

[11] 1994 *CERN Courier* **34**(1) 22

[12] Ginzburg V L 1940 *Zh. Eksp. Teor. Fiz.* **10** 608; 1940 *J. Phys. USSR* **3** 101

[13] Ginzburg V L 1940 *Zh. Eksp. Teor. Fiz.* **10** 589; 1940 *J. Phys. USSR* **2** 441

[14] Ginzburg V L and Frank I M 1947 *Dokl. Akad. Nauk SSSR* **56** 689 [1947 *C. R. Acad. Sci. URSS* **56** 689]

[15] Fermi E 1940 *Phys. Rev.* **57** 485

[16] Kirzhnits D A 1989 *Nekotorye Problemy Sovremennoi Yadernoi Fiziki* [*Some Problems of Modern Nuclear Physics*] (Moscow: Nauka)

[17] Ginzburg V L 1987 *Teoreticheskaya Fizika i Astrofizika* [*Theoretical Physics*

and *Astrophysics*] (Moscow: Nauka) [1989 *Application of Electrodynamics in Theoretical Physics and Astrophysics* (New York: Gordon and Breach)]

[18] Bolotovskii B M 1957 *Usp. Fiz. Nauk* **62** 201; 1961 *Usp. Fiz. Nauk* **75** 295 [1962 *Sov. Phys. Usp.* **4** 781]

[19] Ginzburg V L and Tsytovich V N 1984 *Perekhodnoe Izluchenie i Perekhodnoe Rasseyanie* [*Transition Radiation and Transition Scattering*] (Moscow: Nauka) [Complete translation into English 1990 (Bristol, New York: Adam Hilger)]

[20] Cherenkov P A 1937 *Phys. Rev.* **52** 378

Article 10

Lev Davidovich Landau:
the physicist and the man[1]*

10.1 Introduction

Lev Davidovich Landau was only 54 (short of two weeks) when, on 7 January 1962, he met with a car accident. Landau lived for six more years (he died on 1 April, 1968) but, in those years, he was a very sick man and was totally unable to work. Thus, as a physicist he left us about 30 years ago and, hence, today only those older than 45 or so remember him as a physicist and especially his famous seminars. For the younger generation of physicists (and they are the majority), Landau and everything related to him have become a legend, a living legend so to say, since many of his papers and especially his books are still widely used. For this reason (and several others), the interest in his life as that of a remarkable physicist and an outstanding personality remains. The best way to satisfy this interest is to read the book *Landau: The Physicist and the Man*.[1] However, not all is covered in this book and stories about Landau are still very popular. For instance, more than 200 people gathered to listen to my recent report about Landau's life as a physicist (I presented this report on 9 September, 1992, at my weekly theoretical seminar held at FIAN, the P N Lebedev Physical Institute of the Russian Academy of Sciences). It was on the basis of this report the magazine *Priroda* [*Nature*] suggested I write the present paper.

I do this with mixed feelings. First, there are sure to be repetitions of what has already been said or published. Second, experience has shown that in my reminiscences about the deceased I inevitably write about myself, too. On the one hand, this is understandable but, on the other, it can

---

[1] The English translation of this book was published in 1989 by Pergamon Press, Oxford; the Russian edition was published in 1988 by Nauka, Moscow, and was titled *Vospominaniya o L D Landau* [*Reminiscences about L D Landau*]. I will refer to this book in the text as *Reminiscences*.[2]*

annoy the reader. Unfortunately, there is nothing I can do about this and would only like to assure the reader that I do this not with the purpose of reminding him or her about myself—simply, most events are remembered better if they primarily involve oneself. Third, I do not think that reminiscences about a person should present that person as an icon, which is often the case. Even a great person is only a human being and cannot always be flawless and right. My recollections about Landau are filled with respect and kind feelings but I do not forget about imperfections in his character and his errors. By the way, these errors concerning physics were always interesting and instructive. Finally, the material in this article is rather fragmentary but no one should expect a full picture from a magazine article.

10.2   Biography

Lev Davidovich Landau was born on 22 January, 1908 in Baku. His father was a petroleum engineer and his mother a physician, who, at one time, was engaged in scientific work in physiology. Lev Landau had a sister Sophia, who was somewhat older than he. I do not think that young Lev could be called a child prodigy but his talents revealed themselves very early in life. He graduated from school at the age of 13, studied calculus on his own, and later used to say that he hardly remembered himself not being able to differentiate and integrate. For a year after school he attended a college but then, in 1922, he enrolled at Baku University in the Physics and Mathematics Department. In 1924, Landau transferred to the Physics Department of Leningrad University. It must be noted that at that time Leningrad was the main center of Soviet physics. There Landau first made the acquaintance of theoretical physics, which was going through a turbulent period (suffice it to say that quantum mechanics was created in the period from 1924 to 1927). He worked very strenuously, which was characteristic of him in all his future endeavors. The incredible beauty of theoretical physics and physics as a whole always amazed him. According to his closest friend and pupil, Evgenii Mikhailovich Lifshitz, Landau used to say that the study of quantum mechanics 'not only gave him delight in appreciating the true glamour of science but also an acute realization of the power of the human genius, whose greatest triumph is that man is capable of understanding things beyond the pale of his imagination' (see *Reminiscences*, p 10). In 1927, Landau graduated from the university but even earlier, in 1926, at the age of 18 he published his first scientific paper. Incidentally, Landau liked to measure the age of physicists according to the time since their first publications. For instance, he used to say that he was 13 years my senior, since my first paper was published in 1939, though biologically he was only approximately nine years my senior.

In 1927, Landau started work at the Leningrad Physicotechnical Institute, and in 1932 he moved to Kharkov, where he created and then became head of the Theoretical Division of the newly organized Ukrainian Physicotechnical Institute. But even during the Leningrad period Landau traveled abroad and for one and a half years worked in Denmark, Great Britain, and Switzerland. To him the most important part of his trip was his stay in Copenhagen at the Institute of Theoretical Physics headed by Niels Bohr. Lev Davidovich always considered himself as a disciple of Bohr, while I never heard of anybody else having a strong influence on him. Landau visited Copenhagen twice more, in 1933 and 1934, during his Kharkov period, which ended in 1937, when he transferred to the Institute of Physical Problems in Moscow. The importance of these trips abroad can hardly be overestimated. They allowed Landau to converse with all the leading figures in theoretical physics of those times. He found himself at the hottest possible spot, so to say. True, Landau used to complain (half-jokingly) that he was born too late and was not old enough at the time when quantum mechanics was created. Therefore, as he used to say, all the important work had already been done and all the beautiful girls had already married. And indeed, Landau had time only to do one important investigation concerning the diamagnetism of electron gas (1930), which is believed to be a classical work in non-relativistic quantum mechanics.

In Kharkov, Landau began his teaching carrier, established his own school of theoretical physics, and did valuable scientific research. In the same period, he created what is known as his theoretical minimum and began to write, together with his pupils, his famous course in theoretical physics, and also other books. At this point, it would be proper to dwell on the role of Landau as a teacher, an outstanding teacher. However, the scope of the present article does not allow for this. The interested reader can turn to the *Reminiscences*; here I will only make a general remark. The importance and place of Landau in the physics of the 20th century are determined by a combination of three factors: his scientific achievements, his unique mastering of all theoretic physics, and, finally, his mission as a teacher. It is the product, so to say, of these three factors that is especially large and characteristic of such a phenomenal person as Landau.

The generally peaceful flow of Landau's life was terminated by his arrest on 27 April, 1938. Landau spent almost exactly a year in prison (he was released on 28 April, 1939: I discuss the events during this period in the following section). Fortunately, he was able to recuperate rapidly and he returned to his research, at the center of which was the problem of the properties of liquid helium. From then on, up to the already mentioned car disaster on 7 January, 1962, Landau continued his active work, including the period he spent in Kazan during the war.

In 1946, Landau was elected a Full Member (Academician) of the

USSR Academy of Sciences (he was then 38), thus skipping the usually inevitable period of being a Corresponding Member. Of course, this election was due to his outstanding achievements. But also credit must be given to the Soviet physicists of the older generation, since apparently many of them were very irritated by the keen criticism and somewhat peculiar manners of Landau. It really is, as life shows, very difficult to crush such feelings.[3*,7*] There were also other numerous outward manifestations of the recognition of his contributions—he was awarded a number of orders and medals and was elected to many foreign academies of sciences. Here I only specifically mention the fact that, in 1962, he was awarded the Nobel Prize in physics 'for his pioneering theories for condensed matter, especially liquid helium'.

10.3   Arrest and prison

The nameplate that Landau hung on the door of his office at the Ukrainian Physicotechnical Institute read 'L LANDAU. BEWARE, HE BITES!' Of course, this was a manifestation of his sense of humor but, as they say, many a true word is spoken in jest. Landau was, especially in his youth, ardent and incisive in discussion. He did not pay too much attention to the form of his remarks. There is this case that comes to mind in this connection. After a report made by a respectable professor, Landau picked the professor's arguments to pieces. The professor was very offended and upset. When Landau was told about this, he replied: 'I don't understand why he took offence. I did not call him an idiot, I just said his theory was idiotic'. Behind the harshness of his remarks, there was usually no malevolence but not everyone could understand and appreciate this. His pupils adored him and many other physicists knew how important Landau's work was and realized how greatly his criticism benefited the scientific community. But there were also physicists, leave alone non-physicists, that were put off by Landau's manners and even felt apprehensive, if not to say hostile, toward him owing to his brilliance and the ease with which he grasped complex physical problems. Understandably, Landau had many personal enemies, given the 'witch-hunts' for saboteurs in the Soviet Union in the 1930s. In the Ukrainian Physicotechnical Institute, as almost everywhere in the country, the atmosphere became worse and worse, and clouds gathered over Landau's head, and even the transfer to Moscow in the spring of 1937 did not help. Some of his friends in Kharkov were arrested: among these, I will mention the outstanding experimental physicist Lev Vasil'evich Shubnikov, who perished in the NKVD (a forerunner of the KGB) dungeons (in 1937 he was shot). In this situation, it would be natural to connect Landau's arrest with denunciations and false testimonies beaten out of the employees of the Physicotechnical Institute who had been arrested earlier. I must confess that, for a long time, I believed that this was the case. Incidentally, I was often a witness to the following scene. A former inmate of the Gulag is asked for what he or she was sentenced to a labor camp or prison term. And the indignant person answers: "What do you mean for what? Of course, for nothing. Somebody informed against me." As it turned out, however, in Landau's case, the situation was not that simple.

In 1991, an article titled 'Lev Landau: A Year in Prison' appeared in the Soviet press,[2] which was followed by two articles by G E Gorelik,[3] who gained access to the KGB archives.

It appeared that Landau was arrested on the basis of entirely absurd information about sabotage at the Ukrainian Physicotechnical Institute. But he was also accused on another count: participation in the writing (already in Moscow) of a leaflet in which Stalin and the Stalinist regime were sharply criticized. The text in the leaflet was such that in those times this alone could mean the death penalty. I often heard from Landau about him being a Marxist in his early years and that only prison removed the scales from his eyes. Hence, it was difficult for me to believe that Landau understood the essence of Stalinism much earlier in the same way as we understand it today. But the leaflet is proof of this. In this connection, one would think that the leaflet was a provocation staged by the NKVD. However, the investigation conducted by Gorelik[3] attests to the authenticity of the leaflet and that Landau could have seen it before his arrest.[4*] Anyhow, fortunately for Landau, the NKVD did not pay much attention to the leaflet. When compared with the much more serious (false) accusation of sabotage at the Physicotechnical Institute in Kharkov and other charges, the leaflet business was not considered convincing enough. But even without the leaflet charge, Landau was certain to be sentenced, if not to death than to a long term in prison or labor camp. And actually that would mean death since he was always in delicate health. He also used to say (and to me personally) that he was already close to death in prison, since he could not eat the gruel that was apparently the staple food. The person who saved Landau was P L Kapitza, who wrote letters to Stalin, Molotov, and Beria defending Landau. These letters can be found in the Appendices to the *Reminiscences*. There one can also find a similar letter by Niels Bohr to Stalin. As a result, Landau was freed on the grounds of the personal guarantee of P L Kapitza. Landau's case was closed and the charges were dismissed only on 23 July, 1990 (see the previously cited article in the *Bulletin of the CPSU*). Kapitza, of course, should be given full credit for his role in freeing Landau but that did not give him the right to treat Landau so rudely as he did, of which I was witness. I once asked Landau how he could stand such rudeness and he answered in the following way: "Kapitza transferred me from the negative state to the positive, so

---

[2] 1991 *Izvestiya TsK KPSS* [*Bulletin of the Central Committee of the CPSU*] no 3 134
[3] 1991 *Priroda* [*Nature*] no 11 93; 1992 *Svobodnaya mysl'* [*Free Thought*] no 1 45. See also Gorelik G 1997 *Sci. Am.* **277**(2) 52

I feel powerless to contradict him". Generally, Landau often said that in prison he 'became a Christian', i. e. as I understand it, will not fight with his boss, etc. Fortunately, prison did not break Landau as a physicist.

In prison, it was impossible to write (no paper, pens, or pencils were allowed). Hence, Landau did theoretical physics by memory. In particular, he excelled in deriving certain hydrodynamic equations. They proved to have already been derived earlier by other physicists but, nevertheless, his mental work in prison helped Landau and E M Lifshitz later in writing the volume of their Course of Theoretical Physics devoted to fluid mechanics.

The KGB did not leave Landau alone after prison. His telephones were tapped and secret agents wrote their reports. All this became clear from the recently published[4] KGB top-secret report to the Central Committee of the CPSU, dated 20 December, 1957. It is a very interesting document characterizing Landau's views which, in many cases, are very modern. But what impressed me most was that, according to this document, the denunciations were written also by people close to Landau. Here is one quote: "A person very close to Landau, when asked of his/her opinion about a proposed visit of Landau abroad, said the following:" and then comes the advice, which was most likely taken, not to let Landau leave the country. And another quote: "A report from one of the agents very close to Landau states that Landau believes democracy will succeed in the Soviet Union only after the class of bureaucracy is overthrown". And who are these people that were 'very close' to Landau? Why the public should not know about them? We have the example of when the agents among the upper hierarchy of the Russian Orthodox Church were named.[5] I believe the same must be done with respect to Landau.[5*]

10.4   Could Landau have accomplished more? Was he conservative?

Landau's talent was so bright and he commanded the techniques of theoretical physics so skillfully, that one is inclined to think he could have done a lot more and solved still more difficult problems. Once, in a conversation with him, I expressed my feeling that he could achieve more than he had. He replied immediately and very definitely: "No, this is wrong—I have done everything I could". I believe he was right, since he generally

---

[4] 'The KGB was lucky. They "listened" to Landau' (the newspaper *Komsomol'skaya pravda*, 8 August, 1992). More material was published in the journal *Istoricheskii arkhiv* [*Historical Archive*] no 3, 151 (1993). My name is mentioned in one episode there too, but I do not recall the event and, to my mind, I was mistaken for somebody else (fortunately, there was nothing discreditable in that event, since otherwise I would have to prove the impossible).

[5] See, for example, M Pozdnyev's paper in the magazine *Stolitsa* no 36, 1 (1992); earlier two papers on the same topic were published in the weekly magazine *Ogonek*.

worked very hard and tried to solve very difficult problems. For example, he spent much effort trying to build a theory of second-order phase transitions beyond the self-consistent field approximation. Landau once told me that no other problem had taken so much of his strength as this one, with which he still had not made much progress. He also often asserted that he was not an inventor, that he had invented nothing. As far as the invention of devices is concerned, it is true that he had no talents as a designer, no traits in his character that are needed to make inventions. The sober mind of a highly educated physics theorist and analyst works in a way that is somewhat orthogonal to the search-in-the-dark, trial-and-error methods of the inventor. But Landau was very inventive when it came to solving complicated problems and searching for new methods.

Landau's highly critical attitude, his practice of pronouncing many ideas, especially glimpses of ideas, 'pathological', stemmed from his sober mind and his comprehension and profound knowledge of physics. Besides, Landau did not really care about presenting his remarks and opinions disingenuously. He was often very frank and direct. This often made him look opinionated and unwilling to adopt new ideas. But that would have been a misimpression, because Landau very often agreed, maybe not immediately, with disputable hypotheses and, in general, with new trends. So I do not support the opinion about Landau's conservatism that I sometimes hear. It is certainly difficult to weigh the degree of conservatism on a precision balance. It is also difficult to decide where the boundary lies between true conservatism and 'healthy' conservatism—by the latter I mean, for example, the recognition that the old order should not be broken without good reason. Landau's last published paper "On the fundamental problems" is to me vivid proof that he was not a conservative. It appeared in 1960 in a collection in commemoration of Wolfgang Pauli.[6] In this paper Landau expressed the opinion that the "Hamiltonian method for strong interactions is dead and must be buried". So Landau was ready to face the breakup of fundamental theory, although the Hamiltonian method was later found to be far from exhausted and now forms the basis of quantum chromodynamics.

10.5 How Landau made mistakes. Did he impede the creation of 'great theories'?

He is lifeless that is faultless—this well-known maxim is completely valid. Less trivial, however, is another maxim: the mistakes of an outstanding

[6] Landau L D 1960 in *Theoretical Physics in the Twentieth Century* eds M Fierz and V F Weisskopf (New York: Interscience) p 245; see also Landau L D 1956 *Collected Papers* vol 2 (Moscow: Nauka) p 421 [1965 *Collected Papers* (London: Pergamon Press) p 800].

person can be instructive. And this, undoubtedly, is true in relation to Landau.

I will start with an example that is interesting in many respects. In the only paper I wrote with Landau,[7] published in 1950, we constructed a phenomenological, or macroscopic, theory of superconductivity. (Perhaps it would be better to refer to this theory as quasi-macroscopic, for I think Landau preferred that term.) The crucial point in our paper is the equation for a certain order parameter Ψ, which we called the effective wavefunction of superconducting electrons. The term in this equation that depends on the vector potential A takes the form

$$\frac{1}{2m^*}\left(-i\hbar\nabla - \frac{e^*}{c}A\right)^2\Psi$$

which is obviously very similar to the corresponding term in the Schrödinger equation for a particle of charge e^* and mass m^*. The mass m^* can be chosen arbitrarily because the quantity $|\Psi|^2$ cannot be measured directly. At present, it is most often assumed (and this is convenient) that $m^* = 2m$, where m is the free-electron mass. But what is the meaning of the charge e^*? Since Landau and I were dealing with a phenomenological theory, it seemed to me from the start that e^* was some effective charge that could well be different from the free-electron charge e. But Landau rejected this idea and, in our paper, there is a statement, typical of Landau, that 'there is no reason to believe that e^* differs from the electron charge'. At the time we wrote that paper, I had no specific reasons for insisting on the introduction of an effective charge $e^* \neq e$ and I do not recall any detailed discussion on this topic. Several years later, however, I came to the conclusion (I arrived at it from a comparison with experimental data) that the introduction of an effective charge $e^* = (2 - 3)e$ made the agreement between theory and experiment much better. Our theory involves a dimensionless parameter æ, with

$$æ = \frac{2(e^*)^2}{\hbar^2 c^2}H_{cm}^2\delta_0^4$$

where H_{cm} is a critical magnetic field and δ_0 is the penetration depth of the external magnetic field into the superconductor. These two quantities, H_{cm} and δ_0, can be measured directly (in our paper we were concerned with what are now called type I superconductors, for which $æ < 1/\sqrt{2}$). The parameter æ can also be determined directly from the data on the surface energy between the superconducting and normal phases and the limiting field for supercooling. Thus, knowing H_{cm}, δ_0, and æ, one can determine the value e^* (for more details see articles 6 and 7 in the present collection). Naturally, I informed Landau of my result but he once again

[7] Ginzburg V L and Landau L D 1950 *Zh. Eksp. Teor. Fiz.* **50** 1064; reprinted in Landau L D 1956 *Collected Papers* vol 2 p 126 [English edition 1965 p 546].

objected to the introduction of the effective charge $e^* \neq e$ explicitly. He argued—he probably had the argument worked out even at the time we wrote the paper—that the effective charge, like the effective mass, might depend on pressure, temperature, metal composition, and other factors. This meant that e^* could also depend on the coordinates (by virtue, say, of sample inhomogeneity or the dependence of temperature on the coordinates). But such behavior would break gauge invariance. I tried somehow to avoid this difficulty but did not succeed and it was clear that Landau's thoughts were fairly set. So I wrote a paper (1955 *Zh. Eksp. Teor. Fiz.* **29** 748 [1956 *Sov. Phys. JETP* **2** 589) in which I pointed out the possibility of substantially improving agreement between theory and experiment by introducing the charge $e^* = (2-3)e$, and I also mentioned in it, with Landau's permission and of course with a reference to him, Landau's arguments against an effective charge different from the free-electron charge. As is well known, the microscopic theory of superconductivity of Bardeen, Cooper, and Schrieffer (the BCS theory), in which the charge $e^* = 2e$ arises due to pair formation, appeared soon after. Even now I feel sorry and, to some extent, ashamed to think that I did not consider the possibility of pair formation, since Landau's objection to introducing an effective charge no longer arises if the effective charge is universal and equals $2e$ irrespective of temperature, composition, etc. But neither Landau nor anyone else thought of the possibility of introducing a universal charge $e^* \neq e$. There are reasons why the idea of pairing, which seems trivial now, was not at all obvious at that time and I will come back to them later.

Now I would like to turn to another topic. It has often been said (I have heard it myself) that Landau's sharp criticism of the works of others prevented them from creating something great or at least from obtaining and/or publishing exceedingly important results. Indeed, Landau did criticize others, hotly and not always politely, but that was his style and those who knew him were convinced that even his sharpest expressions usually did not imply any personal hostility. As for Landau preventing others from carrying on their work or from publishing papers—that was totally out the question, at least in the cases I know. The story of the effective charge is not atypical: although Landau was resolutely opposed to the introduction of the effective charge, not only did he not put obstacles in the way of my publishing my paper but, as I have already said, he allowed me to present our respective arguments. I must note, however, that I did not work in the Department of Theoretical Physics headed by Landau, I was not his pupil, and, unfortunately, did not take his theoretical minimum but I think that this factor never played an important role.

There is one more aspect that I would like to touch on here. Important scientific achievements and discoveries do not appear from nothing. More than one person, and sometimes even many people, may think of the same thing—at times, they may even attain the same objective. However, if someone leaves something out or does not understand a problem to the bottom, what is to blame for that? Firstly, chance plays an exceedingly large role here. Of course, I do not mean that the theory of relativity or, more generally, great and profound ideas arise merely by random occurrence. But when we speak of some particular effect, phenomenon, or theorem, there may be an infinitude of reasons why one or another physicist did not think it through to the bottom, did not 'hit' it and appreciate and publish the results. Secondly, an author's own underestimation of the results he obtained (or was on the verge of obtaining) is the best evidence that he made the discovery accidentally or that he did not really make it at all but only claimed later that he had. (When this happens, as it does sometimes, it is in some cases not a deliberate deception but rather a not uncommon psychological effect.) Let me relate a story in this connection: Physicist A mentioned to Physicist B that he had derived the Schrödinger equation before Schrödinger but had not published this result because he had not thought it to be of sufficient importance. To this B replied: 'I advise you not to tell anybody else about this, for it is no shame not to derive the Schrödinger equation, but to obtain such a wonderful result and fail to appreciate its importance really is a shame'. I think I heard this story from Landau and believe it was he who played the role of Physicist B. At any rate, Landau was of precisely the same opinion as B. In short, Landau might have failed to comprehend or support some obscure idea brought to him for his opinion, or he might have criticized it, but it would simply be foolish to make him responsible for the fact that afterwards, in someone else's hands, that idea proved very fruitful. I would also like to note that, on the whole, Landau was very tolerant. (Some exceptions cannot change this conclusion because most people have some things about which they are anomalously sensitive.) In particular, Landau was rather liberal regarding publications. He objected to publications only of those papers that contained no new results but presented, say, another way of obtaining known results. Generally, Landau spoke with contempt and irritation about 'giving new grounds' (from the German word 'Neubegrundung', which Landau liked to use in this connection). I am not sure, however, that he interfered actively with the publication of even such papers. Of course his approval was not required for publication of most papers as long as the paper did not contain any blatant errors. He did not like scientists being too casual in publishing papers and he, obviously, did not publish or rush into print everything he did.

It is in my opinion that the question of whether to be casual or restrained about what one allows oneself to publish cannot be decided unanimously, for the answer depends on the author's taste and style. I think that Landau's restraint was, to some degree, determined by, among other things, the fact that he did not like to write: it is common knowledge that even papers of which he was the sole author were usually written by someone else. It also seems to me that Landau was affected by the thought that a physicist of his rank should not publish trifles.

The two volumes in Russian of Landau's *Collected Papers* contain 98 papers. (The English edition is somewhat different but not in any essential way.) Seventeen papers and notes are not reprinted in the *Collected Papers* but are listed in it. Of these 17 publications, some are short reports on the papers included in the collection. Those papers that Landau himself regarded as erroneous are also not included in the collection. Among these is a paper, published in 1933, that attempts to explain superconductivity on the basis of the spontaneous current hypothesis. This paper may seem erroneous but, in fact, it contains a very interesting idea.[8] I turned to the paper in 1978, almost 45 years after it was published. I was glad I knew it existed. Returning to the question of Landau's publications, I would like to note that the number of papers that Landau published could have been much larger. As I have already said, Landau did not publish everything he did; and he obviously did not in the least try to publish papers just to lengthen his publication list. Also many results reported in papers by other authors, in fact, belong partly to Landau. I am not saying that these were instances of plagiarism but rather that without Landau's inestimable advice and criticism, some of the papers would have been of much less value. Sometimes Landau simply renounced co-authorship. In 1943, for example, I was working on the problem of the field acting in a plasma (since applications involved the Earth's ionosphere, the plasma was assumed rarefied). Some authors at that time believed that the effective field E_{eff} was equal to the mean macroscopic electric field E, while others supposed that E_{eff} did not equal E but, for instance, $E_{\text{eff}} = E + (4\pi/3)P = [(\varepsilon + 2)/3]E$, where $P = [(\varepsilon - 1)/4\pi]E$ is the polarization, and ε is the dielectric constant. To sort out the right answer was not an easy task, given the state of plasma theory at that time. I got confused, did not know how to achieve consistency in my analysis of the problem, and asked Landau for advice. He was convinced from the start that $E_{\text{eff}} = E$ but he did not think that this was obvious and could go without proof. With Landau's help, I proved the equality of the two fields and then wrote a paper reporting the proof and brought it to Landau, whom I had included as my co-author. But he did not accept the offer and the pa-

---

[8] Landau considered in that paper a phase transition into a state with a non-zero spontaneous current density j. This breaks the gauge invariance, which must be maintained. Instead of j, however, one can take for the order parameter the density of the toroidal dipole moment T, which possesses the same transformation properties as the current density. This order parameter leads us to a new type of magnet, toroidal magnets (see 1984 *Solid State Commun.* **50** 339).

per appeared under my sole authorship. In the paper I acknowledged Landau for 'a detailed discussion of the problem' and for instructive advice on how to take close collisions into account. In an article published in the *Reminiscences*, I give another example of Landau renouncing coauthorship. But I would not like to dwell here on this case, which is unpleasant for me, although it is less trivial than the one narrated earlier. I do not know exactly why Landau renounced authorship when he did: I suppose that, in these cases, Landau simply did not consider the results interesting or valuable enough. By the way, some who sought Landau's advice or help could not appreciate it (which was indeed not always easy to do), and for this reason published the results under their name only. I know of results, some of then popular in the literature, that in fairness should bear Landau's name and not the names of his colleagues only.

I now return to the theory of superconductivity and superfluidity. It was no accident (or so it seems to me) that Landau did not surmise the existence of pair formation following my suggestion that the effective charge of our phenomenological theory might be twice the free-electron charge. Landau had long been of the opinion that Bose statistics and Bose–Einstein condensation had nothing in common with the superfluidity of He II. He based his assertion on the fact that an ideal Bose gas is not a superfluid. Besides, it seemed to Landau that one did not need to invoke the Bose statistics of ^{4}He atoms to explain the superfluidity of He II. But it is essential for the superfluidity of He II that atoms of ^{4}He obey the Bose statistics (as far as I know, this was first clearly shown by Richard Feynman in 1953). Unfortunately, I do not remember how Landau reacted to the fact that liquid ^{3}He, first obtained in 1949, is not superfluid down to very low temperatures 0.1 K (this clearly shows the role of statistics, since ^{3}He atoms, in contrast to ^{4}He atoms, obey the Fermi statistics). In any case, before the formulation in 1957 of the BCS theory of superconductivity, the idea of electron pairing was alien to Landau, as to very many other people.

I had in mind other examples of how Landau's objections were usually interesting and instructive, even if it was later found that they were of limited importance. However, because of a lack of space, I will only note that, for some time, Landau believed that plasmons in solids cannot be 'good' quasi-particles because their damping time must be small, i.e. of the same order as the period of oscillations. This view reflected one of Landau's favorite theses, namely that electrons cannot form an almost ideal Fermi gas in a normal (non-superconducting) metal 'because nobody has yet repealed Coulomb's law'. It was, of course, none other than Landau who explained all this later in the theory of the Fermi liquid (1956). As for plasmons, they do exist in simple metals, i.e. their damping is relatively small.

10.6 Landau's attitude toward general relativity

Now I would like to say some words concerning Landau's attitude toward general relativity. He categorically denied the possibility of introducing the Λ term, as well as of somehow changing or generalizing general relativity theory, even when such changes did not violate the good agreement between the theory and the known observational data. This was the reflection of Landau's great admiration of general relativity theory, which he called 'the most beautiful of the existing physical theories'. Though I fully share his admiration, I could not understand why it should rule out the existence of the Λ term. I do not remember that Landau ever put forward any physical arguments against the Λ term but, at that time, there were no realistic arguments for its existence either and Einstein himself believed, I think, that his introduction of this term in 1917 had been a mistake. Pauli, too, expressed a negative attitude toward the use of the Λ term. It is now common knowledge that introducing the Λ term is equivalent to using the equation of state $p = -\varepsilon$ and the term is now the subject of much discussion in the theory of the early universe (see article 1 in the present collection).

What do I wish to illustrate by this discussion? Landau, like his great older contemporaries Einstein and Pauli, attached much importance to logical simplicity and beauty in a fundamental theory. He realized that emphasizing logic and beauty is inevitable and necessary when considering problems for which there are not enough experimental data and for which the theoretical possibilities are numerous. This attitude was, above all, a manifestation of Landau's pragmatism, since in those days research in general relativity and especially in relativistic cosmology was not yet in a position to justify the introduction of the Λ term or, the more so, to extend in some way the general theory of relativity.

10.7 My first encounter with Landau

I would now like to tell the story of how I got acquainted with Landau or, to be exact, how I met him on scientific grounds, so to say (I saw Landau before this encounter and maybe even had been introduced to him by somebody). For a short time starting from the end of 1939 (but mainly in 1940), two groups of physicists, one headed by Lev Landau at the Institute of Physical Problems (IPP) of the USSR Academy of Sciences, and the other headed by Igor Tamm at the FIAN (the P N Lebedev Physical Institute of the Russian Academy of Sciences), met for joint seminars, held alternatively at the two institutes. I belonged to Tamm's group. Both groups were very small: of Landau's coworkers at the time I now remember only Evgenii Lifshitz. I recall, in particular, two joint meetings. At one of these, held at the IPP, Landau reported on the theory of superfluidity and Tamm proposed the term 'roton'. At the other meeting, at the FIAN,

Tamm started discussing one of my first papers devoted to the quantum theory of Vavilov–Cherenkov (VC) radiation. I had shown that the conditions for the emission of this radiation follow when one applies the laws of conservation of energy and momentum to a particle radiating a photon of energy $\hbar\omega$ and momentum $\hbar\omega n/c$, where $n(\omega)$ is the refractive index of the medium.[6*] I, of course, had also calculated the intensity of the VC radiation but we did not come up with the intensity at that seminar. Landau immediately expressed skepticism about my results, saying that they were of no interest since the effect is classical and there was no point in treating it quantum mechanically. He was right in some sense, because quantum corrections in the problem of VC radiation are of order $\hbar\omega/mc^2$ (m is the particle mass) and are, therefore, small in the optical region. But very often a new interpretation, approach, or conclusion turns out to be fruitful. This was the case in my work on VC radiation: the quantum approach and the use of the conservation laws appeared to yield new results in, for example, studies of the Doppler effect in a dispersive medium.[6*]

I have dwelt on this experience with Landau, first, to demonstrate once again his pragmatism and his dislike for 'Neubegrundung'. Second, this example is a striking illustration of the role of tastes and feeling in science. I literally love problems dealing with VC radiation and, generally, with radiation emitted by uniformly moving sources. Landau was quite indifferent to this range of questions and did not consider the VC effect beautiful. This incident is not the only illustration of this predilection: I remember telling Landau about a paper—I think it was the paper by D Bohm and D Pines (or E P Gross)—in which the plasma (longitudinal) wave damping discovered by Landau was interpreted as the inverse VC effect. But Landau remained quite indifferent to this interpretation.

10.8   Being self-critical: Landau and Feynman

I have already discussed the 'accusations' that Landau's bitter criticism might have been an obstacle in someone's way. The claims that Landau was a conservative and that he 'considered himself the most clever man' are groundless as well. I am not inclined to idealize Landau and to pass in silence over his weak points (when this is appropriate). Landau himself was not beyond criticizing even those he regarded as great and he was critical of himself. The latter trait showed up in many ways. For example, Landau considered himself inferior 'in class' to a number of other physicists and his contemporaries. Here it must be explained that Landau classified physicists according to the significance of their contribution to science. This classification was based on a logarithmic scale of five: thus, a first-class physicist supposedly accomplished 10 times as much as a second-class physicist, and so on ('pathological types' were ranked in the fifth class). On this scale of 20th century physicists, only Einstein occupied the highest class of 0.5, while Bohr, Heisenberg, Schrödinger, Dirac, and certain others were ranked in the first class. Landau modestly ranked himself for a long time in class 2.5 and it was only comparatively late in his life that he promoted himself to the second class (and still later, I think, to class 1.5). De Broglie was also ranked in class 1. This suggestion met with some skepticism but Landau was firm—the scientific achievement of de Broglie, although it was not supported by his later work, is indeed enormous (here we are speaking of waves of matter, of course). He also placed R Feynman (who was 10 years younger than Landau) in class 1. In 1962, I met Feynman at a conference in Poland. Feynman was concerned about Landau's health and inquired after him. (The two had never met.) I mentioned in our conversation how highly Landau thought of Feynman's results and that he ranked them higher than his own. Feynman was somewhat embarrassed by this and resolutely declared that Landau was wrong in so judging him. Such comparisons are not my main point, however. Even Landau mentioned his 'classification' less often as the years passed and he started to take a more sober view of classifying people. By the way, of all the physicists I have known personally, nobody resembled Landau more than Feynman. This resemblance extends to many things, scientific style, aspects of their personalities and behavior, and interest in pedagogical ideas. Even among famous physicists, there is a diversity of talents. Niels Bohr and Landau, for instance, were on opposite extremes. Landau and Feynman I see as having had very similar talents. The similarity seems to me to be genetic. The differences, which one could attribute to their different surroundings and different upbringing, are certainly also very large. What a pity that these two remarkable physicists never met. It really pains me to think of this 'product' of our past.

10.9 Concluding remarks

In this article, I realize, I could promote an understanding of Landau's style and his scientific image only to a very small degree. But I find consolidation in the thought that it is extremely difficult to provide real insight into the nature of this remarkable physicist.

I would, however, like to make one final remark. Few are those to whom I have returned in my memory as often as I have to Landau, even though he left us many years ago. Some of my colleagues have expressed the same feelings. I cannot explain this fact only by my friendly feelings toward him and by his tragic and bitter demise. Another fact might provide a clue to this phenomenon: Landau was a unique physicist and teacher of physicists.

So my attitude toward him is inseparably linked with our attitude toward physics, which is so dear to all of us.

Comments

1*. The paper was published in the magazine *Priroda [Nature]* no 2, 92 (1993) under the title "A unique physicist and teacher of physicists" and the title of the present article "Lev Davidovich Landau. The physicist and the man" served as the subtitle. The original article also contained some photographs. I also wrote two similar articles for foreign publications, one for *Physics Today* (May 1989, p 54) and the other for the collection 1989 *Advances of Theoretical Physics* ed A H Luther (Oxford: Pergamon Press) p 1.

2*. My two papers about Landau published in the *Reminiscences* have also been published in my book *O Fizike i Astrofizike [On Physics and Astrophysics]* (Moscow: Byuro Kvantum, 1995) pp 364 and 368 [English translation 2001 *The Physics of a Lifetime* (Berlin: Springer) pp 367 and 371]. In the same book, on p 442, the reader can find my article "*The Course* (in memory of L D Landau and E M Lifshitz)". This paper is excluded from the English translation of my book for lack of space.

3*. Moreover, as specified in the following section of the present paper, Landau's case in the KGB was closed only many years after his death (see footnote 4 on page 272). Therefore, his election to Full Member of the USSR Academy of Sciences in 1946 appears to me especially non-trivial. I think that the relevant circumstances deserve the attention of historians of science (see comment 7).

4*. On the whole, the picture is still unclear to me. More about the political evolution of Landau, the authenticity of the leaflet or its provocative nature, and other aspects of the whole business can be found in the previously cited article by G E Gorelik. The text of the leaflet is partially given in the article about Landau in Feinberg E L 1999 *Epokha i Lichnost [The Epoch and the Individual]* (Moscow: Nauka; 2nd edn 2003 (Moscow: Fizmatlit)). This article of Feinberg is generally important for understanding what person Landau was.

5*. Materials regarding Landau's biography continue to appear, so that the last words have yet not been said. Of course, it would be most interesting to see the unpublished documents from the KGB that refer to the results of covet surveillance of Landau's activities. Unfortunately, there is little hope that we will ever find out anything about this, since in Russia (in contrast to Germany and the Czech Republic, I believe) such information is unavailable. Of the new sources of information about Landau, I will mention the book by Landau's widow Cora Landau-Drobantseva *Akademik Landau. Kak My Zhili [Academician Landau. How We Lived]*

(Moscow: Zakharov, AST, 1999) and notes by doctor Kirill S Simonyan published after his death: Simonyan K "The secret of Landau: Reminiscences about L D Landau", *Okna* (Israel) of 2, 9, and 15 April, 1998 (*Okna* is a supplement to the newspaper *Vesti*). About the widow's book, I do not even want to write;[9] all my acquaintances and I have a very low opinion about it. I would only like to note that everything she has to say about E M Lifshitz is shear nonsense and, I would say more, slander (see in this connection my answer to the question of the editor of the journal *Prepodavanie Fiziki v Vysshei Shkole* [*Teaching Physics in Institutions of Higher Learning*] no 18, 24, 2000). Simonyan's evidence is very important. He states (and I tend to believe him) that, to a large extent, the death of Landau was due to mistakes made by the doctors. The pains (in particular, the pains in the abdomen) of which Landau constantly complained were not phantom pains: what was needed was an operation (the autopsy proved this point quite clearly). But the doctors who treated Landau did not want to risk an operation. What was also new to me was that, as Simonyan states, shortly before his death Landau's mental faculties had been restored to a considerable degree. In any case, Landau was able to read newspapers and books (at the time many of us did not think this was possible). It is quite possible that the real pains hampered the recovery of Landau or, in any case, did not give him the possibility of living an almost normal life. What a tragedy! I cannot stop thinking about his fate.

I would also like to mention the article by B Gorobets titled "A secret agent close to Academician Landau" and published in the newspaper *Nezavisimaya gazeta* in the science supplement no 9 of 19 July, 2000. In this article, the author hypothesized about the person (or persons) who could have informed the KGB about Landau's activities.

6*. See article 2 in the present collection.

7*. Only now, after so many years, have I understood, or simply found out, that what I wrote in the text was incorrect. Incidentally, in comment 3 I justly noted that all is not clear with the election of Landau to Full Member of the USSR Academy of Sciences. I never knew anything about and was not really interested in, the special assignments that Landau did in the first stages of the atomic project. It turned out that I V Kurchatov, at least from 1945 and possibly even earlier, was very eager to get Landau to work on the atomic bomb. Suffice it to say that in their recently published article of historical nature titled "On the development of the first Soviet atomic bomb" (2001 *Usp. Fiz. Nauk* **171** 79 [2001 *Phys. Usp.* **44** 71]), G A Goncharov and L D Ryabev devote an entire section (section

---

[9] In connection with that book I wrote a note titled "Once more about Lev Davidovich Landau and something more", which was not intended for publication. It was written on the spur of the moment and rather carelessly. What is more, I do not want to touch on this subject any more. However, several people have this note and sometimes in the future it may prove to be useful for Landau's biography.

11) to the attempts of Kurchatov to get Landau to do calculations for the atomic project. And, in 1946, these attempts succeeded, After the first explosion on 29 August, 1949, Landau was among the people who were awarded the Order of Lenin. Thus, I now have no doubts that Landau became a Full Member of the USSR Academy of Sciences on 30 November, 1946 as a result of the actions taken by Kurchatov, who ironed out the terms with the Central Committee of the CPSU. As I wrote in the article about I E Tamm (article 8 in the present collection), he was out-voted in the same voting session, although among physicists, as far as I known, he was the first candidate for Academician. But the Central Committee of the CPSU did not 'recommend' the election of Tamm and this proved sufficient, notwithstanding the secret ballot. Incidentally, I think that Kurchatov's job (of getting Landau elected and, hence, working on the atomic project) was made easier by the fact that Ya P Terletskii, sent on a spy mission to Niels Bohr in 1945 (see the comment 1 in article 18 in the present collection), reported to Beria about the high opinion that Bohr had about Landau (see the journal *Voprosy Istorii Estestvoznaniya i Tekhniki* [*Problems of the History of Natural Sciences and Technology*] no 2, 21, 1994).

In view of what has been said, it is clear that my interpretation in this article of the reasons for electing Landau a Full Member of the Academy are not only wrong but are also proof of my naivety and, if you wish, a misinterpretation of the situation in the country and in the USSR Academy of Sciences, in particular. Of course, in this new edition, I could have deleted the corresponding place from the text but, as I said before, I will not cover up my mistakes in that manner.

Article 11

Physics, the '*Course*', and life:
on the 80th anniversary of Evgenii Mikhailovich Lifshitz[1*]

On 21 February, 1995, Evgenii Mikhailovich Lifshitz would have been 80 years old. Most regretfully, this is the tenth year that he is not among us: he died on 29 October, 1985.

The death of Evgenii Lifshitz was tragic. I will never forget the evening when we were all phoning each other trying to find out the outcome of the operation. And the answer was: 'The doctors are trying but the heart does not want to "start up".' This was an operation on the open heart. In the West such operations are common and the percentage of failures is low. It is said that today in Russia the percentage of failures of such operations is low, too, but ten years ago in the USSR they were fairly rare and possibly not everything had been worked out. Be this as it may, the heart did not 'start up' and Evgenii Mikhailovich Lifshitz died. This was even more tragic than it might have seemed since he had great potential, both physically and spiritually, and could have done a lot more—he had plans, worked to the very end, and his thoughts were clear.

Evgenii Lifshitz achieved a lot in his life and one could even say he lived, on the whole, a happy life. But, of course, life in the 20th century, especially in the Soviet Union, was not very smooth. The first 18 years of Evgenii's life were quite happy, as far as I can judge. His father was a well-known doctor and the children (Evgenii and his younger brother Ilya, who also became a well-known physicist) received a good education at home, studied foreign languages, and Evgenii even went abroad, which was really a rare thing in those times. Evgenii studied in school only two years (he started school in the sixth grade), then studied for two years at a chemical college, and went in 1931 to the Physics and Mechanics Department of the Kharkov Mechanics and Machine Building Institute (KMMBI). It was in the Institute that his outstanding abilities manifested themselves: he graduated in 1933, only two years after starting his studies, having completed the examinations and presenting a diploma thesis, which was accepted. And he was only 18 years old! In the same year (1933), he became a graduate student at the Ukrainian Physicotechnical Institute (UPTI) under Lev Landau, who had just started his work in Kharkov and this is certainly proof that Lifshitz was well prepared for graduate work. The next year, in 1934, he completed the course and took the PhD examinations. In his autobiography, Evgenii M Lifshitz does not mention when he got his degree but there is a document from which it follows that, by 1936, he was already a Candidate of Physical and Mathematical Sciences (PhD). His first paper (in co-authorship with Landau) was published in 1934 and was devoted to the generation of electrons and positrons in the collisions of two particles. This involved the use of quantum electrodynamics and Dirac's theory. Lifshitz developed the ideas in this work by himself in his next paper, published in 1935. Thus, he was a mature physics theoretician by the age of 20. The theoreticians I know who graduated from Moscow State University (including myself) lagged behind, so to say, by four to five years.

But not everything went smoothly in those 20 years. Young Evgenii's achievements did not please someone and attempts were made to impede his advancement. This is clear from the following document, which was supplied by Zinaida Ivanovna Gorobets, E Lifshitz's widow (here I would like to take the opportunity to thank her for this and for her useful remarks):

Resolution of the Secretariat of the VUSPS (25 June, 1933):[1]

'Re: Comrade Lifshitz, student of the Kharkov

Mechanics and Machine Building Institute'

Regarding the special academic progress of Comrade Lifshitz, who, according to the professorate, showed outstanding abilities in the area of theoretical physics (at the age of 18 he completed, in two years, the institute course and graduated as a physics engineer), the decision of the Trade-Union Bureau of the Physics and Mechanics Department aimed at undeserved discredit and attempts not to let Comrade Lifshitz do scientific research must be considered incorrect and harmful. In this way the Bureau has followed the path of gross distortions of the directives issued by the Central Committee of the Party on stimulating students that excel in their academic studies. Despite the repeated instructions of the Central Bureau of the Student Section of VUSPS to reconsider and repeal the politically incorrect and harmful decision of

---

[1] The Russian abbreviation of the All-Ukrainian Trade Unions Council.

the Trade-Union Bureau of the Physics and Mechanics Department backed up by the Trade-Union Committee of the Mechanics and Machine Building Institute, the latter not only did not repeal this decision but even insisted on it.

The Secretariat of VUSPS decrees that

1. Comrade Lifshitz certainly deserves to be recommended for scientific work as a person who graduated very successfully from the Institute and demonstrated considerable abilities in theoretical physics.

The Central Bureau of the Student Section of VUSPS is to repeal the decision of Trade-Union Bureau of the Physics and Mechanics Department of the Institute and to verify compliance with the given decree.

2. For distorting the Decree of the Party and Government concerning the training of highly qualified specialists and research workers, the operational section of the Trade-Union Bureau of the Physics and Mechanics Department of the Institute is to be strongly reprimanded.

The operational section of the Trade-Union Committee of KMMBI, which backed the decision of the Trade-Union Bureau of the Physics and Mechanics Department of the Institute, is to be instructed about the inadmissibility of such mistakes in its future work. The given decree of the Secretariat should be implemented and thoroughly studied at student meetings.

3. The local trade-union organizations at institutions of learning must record every case when a student successfully graduates from the institution and use it as an example for other students.

The present decree is to be published in the press.

For the Secretary of VUSPS   *Zlatopol'skii*

Here I have presented this document in full since it is also interesting in that it gives the reader a feeling of the times. But the decree certainly makes clear what a hard time Evgenii Lifshitz had before he was able to overcome the hostile attitude displayed toward him by the local trade-union officials, both at the department level and at the institute level.

Evgenii's carefree life in the theoretical department of the UPTI ended in 1937. The 'dark' forces using demagogy under the pretext of closer links with industry tried to 'destroy' the departments at the UPTI that conducted basic research. In particular, Lev Landau was considered fair game in this hunt, so he had to leave Kharkov in a hurry—he moved to Moscow and started his work at the Institute of Physical Problems. Possibly, this saved his life, since several talented physicists from the UPTI, among them such an outstanding physicist as L V Shubnikov, were arrested. In the same year (1937), Shubnikov and some of his colleagues were shot. This appalling fact was kept secret for many years and has only lately been revealed. Officially it was said that Shubnikov and others were detained in prisons and/or camps without the right to correspond with their loved ones. For instance, in the well-known handbook written by Yu A Khramov in 1983 *Fiziki* [*Physicists*] (Moscow: Nauka), it is stated that Shubnikov died in 1945 (without giving the exact date of his death).

It is easy to imagine how awful Evgenii Lifshitz felt after his closest friend and teacher Lev Landau left for Moscow and after many of his colleagues were arrested (at the end of April 1938 Landau was also arrested in Moscow). I do not know the details but, one way or another, Lifshitz had to leave the UPTI. For a short time, he taught at various institutions of higher learning, first in Kharkov and then in Moscow. In 1938, he actually fled to the Crimea and hid there for three months. Fortunately, the 'organy' (the NKVD, a forerunner of the KGB) ether simply forgot about him or it was simply luck that saved him. Exactly a year later, Landau was released from prison and returned to the Institute of Physical Problems. Lifshitz returned to Moscow in September 1939 and immediately started work at the same institute. There he stayed to the end.

All these years Evgenii Mikhailovich worked very hard, not only doing research but also writing the famous *Course of Theoretical Physics*. For example, the first edition of *Statistical Physics* (now volume 5 in the *Course*) was written in 1937. But we should not balance his achievements in research against his achievements in writing the volumes of the *Course*: the *Course* contains so much mew material, at least in methods, that such contraposition seems artificial.

In 1992 Lifshitz's collected papers were published in Great Britain[2] (unfortunately, the situation in Russia with scientific publishing is such that there is simply no money to publish his collected papers in Russian).[3] Altogether there are 48 papers in this collection. This might seem a modest number but if we add the 10 volumes of the *Course* (altogether about 5300 pages), we can see how much was achieved by him. Of course, the number of papers and pages is not really important.

I have already mentioned the first two papers by Evgenii Lifshitz. These were followed by a fundamental paper, written together with Landau, devoted to the dynamics of magnetic moments in ferromagnets (1935). Then Evgenii Mikhailovich did research in the theory of the photo-electromagnetic force in semiconductors (1936), the kinetic equation for electrons

[2] *Perspectives in Theoretical Physics: The Collected Papers of E M Lifshitz* (Oxford: Pergamon Press, 1992).
[3] The Russian edition of this book (*Sobranie Nauchnykh Trudov E M Lifshitza* [*The Collected Papers of E M Lifshitz*]) was finally published in 2004 by Fizmatlit, Moscow.

in a magnetic field (1937), the theory of collisions of deuterons and heavy nuclei (1938 and 1939), the theory of second-order phase transitions (1942), phase transitions in monomolecular films (1944), the theory of molecular (van der Waals) forces in condensed media (1954 and later), the problem of gravitational stability in the expanding universe (1946), and, finally, beginning in 1961 and up to 1984 (the years when the respective papers were published), he published (in co-authorship with other researchers) a number of papers devoted to relativistic cosmology. Of course, this list is incomplete but I will make only two more remarks here.[4] In 1944, Evgenii Mikhailovich published a paper in which he showed that second sound in superfluid helium can be generated by a periodically heated solid. And this was the way in which the second sound was actually discovered later. Finally, in a note written by Landau and Lifshitz and published in 1955 (*Dokl. Akad. Nauk SSSR* **100** 699), the two assumed that the apparent entrainment of the superfluid part of a liquid by a rotating cylindrical vessel can be explained by the formation of coaxial cylindrical surfaces of discontinuity of the superfluid velocity. However, it soon became evident that what formed were not discontinuity surfaces but vortex filaments, since energetically such a situation is preferable. For this reason, the work in question was assumed to be an error and, for example, was not included in Landau's collected papers. But recently it was found that, under certain conditions, layered structures are realized in the superfluid phase of ^{3}He. Thus, this work (no 24 in the English version of Lifshitz's collected papers), which was assumed to be an error, came in handy, so to say.

The works of E M Lifshitz, those already mentioned and the others, are of unquestionable merit and some have become part of the most valuable work done in theoretical physics. But one cannot think of them as being unique, since there is a lot of other good research done in theoretical physics. Nonetheless, the *Course of Theoretical Physics* compiled by Landau and Lifshitz is indeed unique—there is no other course of such magnitude and excellence. It has been translated in full into six languages and separate volumes have been published in ten more languages.[5]

In the setting of Landau's brilliance, Lifshitz remained in the shadows.

---

[4] More details about his scientific work can be found in his obituary published in 1986 *Usp. Fiz. Nauk* **148** 549 [1986 *Sov. Phys. Usp.* **29** 294] and the collected papers mentioned in footnote 2 on page 288.

[5] Earlier I wrote an article devoted to the *Course*. It was published in *Nauka i Zhizn'* [*Science and Life*] no 3, 86 (1986) and in my book 1995 *O Fizike i Astrofizike* [*On Physics and Astrophysics*] (Moscow: Byuro Kvantum) p 442; the same article was published in the previous edition of this book (Moscow: Nauka, 1992). The article *The Course* is one of the appendices of the book *Vospominaniya o L D Landau* [*Reminiscences about L D Landau*] (Moscow: Nauka, 1989). The book has been translated into English (*Landau: The Physicist and the Man* [Oxford: Pergamon Press, 1989]).

True understanding of the role that Evgenii Mikhailovich played in writing the *Course of Theoretical Physics* came only after a tragic quirk of fate (indeed, a paradox of life). On 7 January, 1962, Lev Davidovich Landau met with a car accident and, after that, was totally unable to work. At that time the *Course* was still not finished—there were still three volumes out of ten to be written, not to mention the republication with additions of other volumes. I must confess that I thought at the time (and not only I) that the *Course* would never be finished. But Evgenii Mikhailovich thought differently. He spent many years after this working on and finished the *Course* (in collaboration with L P Pitaevsky and, in relation to volume 4 devoted to quantum electrodynamics, with additional help from V B Berestetsky).

The *Course of Theoretical Physics* is a memorial to E M Lifshitz. After Lifshitz's untimely death, L P Pitaevsky continued republishing the *Course* and this work is yet to be completed. With the deterioration in publishing fundamental science in Russia, the absolutely necessary republishing of the *Course* has been put into jeopardy. I am convinced that the duty of the Russian Academy of Sciences (RAS) and, specifically, the Institute of Physical Problems of the RAS is to do everything possible to make certain that the *Course* continues to be used and, in this way, ensure the further development and the very existence of theoretical physics in Russia.[2*]

Evgenii Lifshitz loved music and poetry and liked to travel—he was not a pedant totally immersed in science, to the exclusion of everything else. And yet the most important thing in his life was physics, his work in physics, connected mainly, in the last period of his life, with his work on the *Course*. He was constantly thinking of how to improve it and made notes in special notebooks of defects in presentation, errors, and thoughts concerning the *Course* that could be used in future work. Each time I noticed something in the *Course* that required (in my opinion) elaboration, I phoned Evgenii Mikhailovich and I think others did the same. He never parted with his notebooks, even on vacation and in the hospital. When I visited him in the hospital shortly before his death, we also talked, as usual, about the *Course*.

Such an attitude toward work, with such total dedication to it, is the keystone of success and manifestation of profound professionalism.

I must also note the irreconcilable attitude of Evgenii Mikhailovich toward pseudoscience, his high standards in science, and his enormous contribution to editing *Zhurnal 'Eksperimental'noi i Teoreticheskoi Fiziki* [*Journal of Experimental and Theoretical Physics*]. He was an honest and highly scrupulous man.[3*]

Evgenii Mikhailovich Lifshitz did not live long enough to accomplish all he hoped to but the memory of him and his work will last forever.

Comments

1*. On Lifshitz's birthday (21 February, 1995), when he would have been 80 years old, a meeting in his memory was held at the Institute of Physical Problems of the Russian Academy of Sciences. The proceedings of this meeting and some other material devoted to E M Lifshitz were published in the journal *Priroda* [*Nature*] no 11, p 86 (1995). The present article is my report at that meeting. The text in *Priroda* was slightly edited and, what is more important, photographs and bibliographic citations of Lifshitz's papers were added.

2*. L P Pitaevsky is continuing his work on the *Course* and it is being republished in full by Fizmatlit (Moscow).

3*. As noted in comment 5 to article 10 devoted to L D Landau, I consider the attacks on E M Lifshitz to be found in the book by Landau's widow slanderous (see the journal *Prepodavanie Fiziki v Vysshei Shkole* [*Teaching Physics in Institutions of Higher Learning*] no 18, 25 2000). Issue 24 of *Prepodavanie Fiziki v Vysshei Shkole* [*Teaching Physics in Institutions of Higher Learning*] for 2002 was devoted completely to E M Lifshitz. In particular, it contains the full list of Lifshitz's works and a number of biographical notes and reviews.

Article 12

In memory of David Abramovich Kirzhnits[1*]

Very soon after David Abramovich Kirzhnits's death on 4 May, 1998, the idea of a tribute collection of memoirs and David Abramovich's selected works emerged. It will probably take two years or so for this collection to appear, so I took a decision to write these notes right now: I am 10 years older that DA, today (in 1998) I am already 82, and it is better not to postpone things for the uncertain future.

DA was a distinguished physicist. This simple phrase said, and having in mind that all notions and characteristics devalue over time, some points need to be immediately clarified. As the Charter of the Academy of Sciences of the Russian Federation (formerly, of the USSR) specifies: scientists who 'have enriched science with distinguished scientific achievements' (Clause 10), are eligible for the title of the Academy's Corresponding Member. Academicians (the Full Members of the Academy), however, are supposed to have enriched science with 'achievements of paramount scientific significance'. So, by calling DA a distinguished physicist, we only say that he was deservedly elected a Corresponding Member in 1987. The fact, however, is that DA was a prominent scientist on an international scale of values, the author of some excellent studies, and one of the highest-ranking theoretical physicists in the USSR and Russia. It is not worth ranging and ranking, though. Turning to the substance of the matter, the material of the future collection demonstrates quite clearly both the rich content and high level of DA's work. I shall here mention only three of those research areas in which DA worked—and these are the ones to which he most significantly contributed.

First, it is research into the equations of state and structure of dense matter, with application to stars and artificially compressed objects. Second, DA was able to express the conditions for the stability of a substance in terms of its dielectric permittivity. It had been widely accepted that for, say, a metal to be stable, its static dielectric permittivity must be positive.

This requirement is not correct, as DA showed and stability is also possible if this property has a negative value (see section 7.7 in article 7 of this book). Indeed, it is precisely this situation which occurs in some materials. All this is of special relevance to high-temperature superconductivity studies (HTSC). Third, DA was, to my knowledge, the first to consider phase transitions in a vacuum, with application to the cosmology of the early Universe (or, more precisely, to phase transitions close to spacetime singularities under conditions of superhigh temperatures). These are, to repeat, what I consider the most important areas of DA's research. But it should be noted that he performed many other studies, some of them very interesting. A broadly educated theoretical physicist, he was in full control of the mathematical apparatus and, by and large, worked at a very high professional level.

As far as I remember, I first met DA in the early 1950s in the city of Gorky, where I held a part-time professorship at Gorky State University and where he worked in industry and came to see me on some business one day. Late in 1954, DA was employed in FIAN's Theoretical Department (now the I E Tamm Theoretical Physics Department), with me a deputy head at the time; later, after I E Tamm's death, I headed the department from 1971 to 1988 and I still keep working there today. So DA and I worked alongside one another for a full 44 years. We co-authored a number of papers, and he participated actively in my Wednesday theoretical seminars. On 16 October, 1996, a regular 1512nd session of the seminar was held to mark DA's 70th birthday (he was born on 13 October, 1926). DA attended the seminar and took the floor at the end of it.

If memory does not let me down (mine is rather poor, I would say, and now especially so), this was the last time I saw DA alive. He was already gravely ill and spoke with difficulty. Still, we spoke on the phone on and off, both before and after that seminar, and one such conversation I remember well. With the 1996 presidential elections then just around the corner, I wrote an article which I titled "Communists are the main threat for Russia and the rest of the world". My idea was, as the article's title suggests, to warn against the communist (bolshevik) danger at this new stage of our history. As of now, the article has been published in full in my book *About Science, Myself, and Others* (Moscow: Fizmatlit, 1997),[2*] at the end of it but at the time of the story, before the elections, I only managed a limited publication (see comment 1 to article 26 for more details). Well, so I sent the manuscript to DA and he read it and called me, and his sharply negative reaction was quite a surprise to me. What he told me was something like this: "My advice is, hide this in your desk and, don't publish it ever by any means" and 'don't even show it to anybody', I believe he added. Sure enough, I was somewhat shocked and perhaps even hurt. What was the matter, I still ask myself today and I have no clear answer. But I think that DA might have suspected that I was insincere when writing this before the

20th CPSU Congress in 1956 I had not understood the nature of the Soviet system and bolshevism and the role of Stalin. He might have thought, in other words, that I was acting against my conscience trying—consciously or, at best, subconsciously—to justify my behaviour or, more precisely, my stupidity. That millions of people were in a sense blind in Soviet times is a very intriguing issue and so I have decided to write in somewhat more detail on this here. The reason is that I gained a better insight into this theme very recently from DA's autobiographical essay "Where Chelyabinsk 40 will arise..." included in the D A Kirzhnits memorial collection.[5*]

This essay gives a clear picture of how hard DA's childhood was, both in general and, it is safe to say, in political terms. His father was arrested when DA was 11. His mother, an educated woman, apparently knew well the how and why of such things, as did those in their surroundings, many of whom were themselves victims of lawlessness. Commonplace though it may sound, the dictum 'the social being determines consciousness' may have much truth in it after all. So, by and large, it is quite understandable why DA realized very early in his life the great gulf between communist mottos and the way things actually were in the USSR. However, for many— including myself—life evolved in an entirely different way. My father, an old engineer, never meddled with politics; and my mother, who died when I was four years old, and her sister, who tried to replace her, were also uninterested in politics. None of our relatives, nor even a close acquaintance, was subject to persecution. As for school and newspapers—the only sources of information I had—it was all beating drums, glorifying the revolution, and eulogizing the 'great leader'. These, of course, were not suitable conditions for anyone to gain insight and, as for me personally, I still think I was especially slow-witted in this respect and quite insensitive to social matters. Nor could any of my university classmates boast of understanding the true nature of these phenomena, not even Semen Zakharovich Belenkii, whom we considered the most intelligent of us (the reason I mention him is that he worked in our Theoretical Department up to his untimely death in 1956). What is even more amazing, my illusions did not leave me and I kept wandering in darkness—or at least in a fog—even after my marriage in 1946 to N I Ermakova (Ginzburg), who had spent a year in the Gulag before being exiled to Gorky (formally, even Gorky was off-limits to her, a village of Bor across the river Volga being her official place of residence). My wife provided me with, shall I say, first-hand information of what the KGB was about but, even so, I was unable to grasp the nature of the bolshevist dictatorship and the role of Stalin. It is, however, out of place to dwell on this, and I refer the reader to my article "Communists are the main threat..." mentioned earlier for a more detailed account of the life of my wife and myself. My point was only to give an idea of how difficult it was, in the environment of total hypocrisy and lies, to see things correctly even if for some people they were, by force of circumstances, crystal clear.

In short, I understand why DA reacted to my article the way in which he did. However, I think that his own rich and bitter experience somehow prevented him from understanding how unfathomable the ignorance of people of a different background might be. As I have a special interest in this issue, I have asked many people for their opinion on this matter and, by and large, I saw my view supported. The following vivid example impressed me most.

One of our summer cottage neighbours is Aleksandr Nikolaevich Yakovlev, M S Gorbachev's comrade-in-arms and former member of the CPSU Central Committee Politbureau. He and I talk with each other now and then and sometimes exchange literature, occasionally of our own production. Yakovlev's hard-line anti-bolshevist position is well known to everyone from his political statements, articles, and books. But is he sincere? This question has long puzzled me and I have come ultimately to the conclusion that he is and that what he says is truthful. Of course, this is an intuitive statement, as E L Feinberg would put it, i.e. one which I cannot prove. Well, the other day the course of our conversation somehow led AN to give us his impressions of Khrushchev's famous 'classified' report to the 20th CPSU Congress. Yakovlev, then an instructor at the ideological Central Committee department, in charge of schools, television, etc, had not been a delegate but had been given an invitation ticket and so had attended the congress. There was dead silence in the hall, he recalls, people were stupefied, they would not look into each other's eyes, and left the hall silent and, in fact, shocked—AN himself included. By the way, some time after that our conversation, I happened, quite by chance, to see AN repeat the story in the 1956 edition of 'The Old Apartment' TV show. (This is the reason, incidentally, why I did not ask his permission for my placing the story here.)[3*]

Of course, all this is not new and there are many people who testify to the same. Just remember hosts of foreign communists and 'sympathizers' who, in spite of the savage trials of the 1930s in the USSR, kept believing in Stalin and, more than that, kept spying for him at the risk of their life. I admit that very many people, including myself, somehow turned a blind eye to the truth and tried to avoid understanding it[1] because, with the full realization of that horrendous reality, life itself would perhaps seem not worthwhile living. Today, when I reflect back, compassion for David

---

[1] The important aspect of life in a totalitarian society is fear—fear of persecution of one kind or another, from public disgrace and expulsion from the Party (or from, say, the Union of Writers), to arrest and even execution. This fear humiliated people and had a shattering psychological impact on them. D A Granin's recent essay *Fear* (St Petersburg: Informatsionnyi Tsentr Blitz, 1997) explores this theme in considerable depth. I conducted an analysis of the fear issue with respect to myself but it would be out of place to dwell on this here. I think, the problem of fear played a major role in DA's life. But I don't have enough information on that and am not, in fact, entitled to touch on such a delicate theme as far as another person is involved.

Abramovich, for his bitter life as a child, and for all the hardships he endured up to 1954 fills me.

Of course, young people can—and did—find ways to make up for negative emotional experiences. Religious people may find support in their faith. For us atheists, the acquisition of scientific knowledge and its application to research are the primary source of positive emotions. Denied access to so many things—material well-being, freedom of speech, and freedom to live, to work and to have a rest where and the way in which you wish—physicists (and other scientists) of my generation primarily derived their strength from their work, from scientific studies—as did physicists of DA's and successive generations. I remember in this connection what a prominent American physicist wrote after attending, in about 1956, one of the first international conferences held in the USSR after many years of interruption. The physicist was amazed at the enthusiasm of Soviet physicists, their love for science, and it appears at their professional level as well. "But then", he explained, "there is nothing more for them to enjoy".[4*]

While this 'diagnosis' is, by and large, correct, it is greatly exaggerated. I, for one, did not feel unfortunate or especially deprived of anything. True, I was extraordinarily lucky personally. But how many people were not in those bitter Soviet times! M Bronshtein, S Shubin, L Shubnikov, and many many other very talented physicists were shot by a firing squad or murdered in Stalinist penal camps for no known reason. Not to mention the repression of the survivors, or those young men and women who were unable to find employment and had to invoke and demand their right to emigrate. Today, as my life comes to an end, all the tragedy of our Soviet past is especially clear to me. There was only a short period of time following the break with the bolshevist system when we were filled with hopes. Well, today, it seems the hope ebbs away. Today life is hard for Russian physicists, as indeed for their fellow scientists in other fields. Still, it is my conviction that it would be incorrect and indeed mean to belittle the significance of the changes that have taken place. Most important, we have become free people. It is a great pity that illness prevented David Abramovich from availing himself fully of this freedom.

But it is comfortable to think that for 44 years, i. e. for most of his life, DA worked in the theoretical department of FIAN, free to do what he liked, highly regarded and respected by those around. It is even more comfortable that David Abramovich Kirzhnits was very successful in his work and made an invaluable contribution to physics.

Comments

1*. This note was written in September 1998 for the D A Kirzhnits memorial volume. See comment 6.

2*. This paper is also published in the present edition (article 26).

3*. The story is also included in A N Yakovlev's book *Omut Pamyati* [*The Whirl of Memory*], first published in 2000 by Vagrius (Moscow) and then republished.

4*. This is discussed in somewhat more detail in the next (13) article of this collection.

5*. This paper is also published in the November 2001 issue of the journal *Priroda* [*Nature*] (p 70).

6*. In 2001 and 2002, D A Kirzhnits' two-volume *Studies on Theoretical Physics, and Memoirs* were published by Fizmatlit (Moscow), with the first volume featuring the present note. Added to the note was a postscriptum, written back in 1998, which I reproduce here.

Three people close to me were unanimous in being dissatisfied with the present note. The main point of their criticism is that I say disproportionately little about David Abramovich and, instead, too much about myself and an issue of only indirect relevance to DA. This would, of course, be absolutely right if an obituary or even a note for an ordinary collection of memoirs were at issue. However, my idea of the D A Kirzhnits memorial collection is different: I would recommend that it also include some works and unpublished notes by DA himself, in particular his autobiographic article "Where Chelyabinsk-40 will arise...". In such a collection, I believe that apart from the memoirs in their 'pure form' also some materials only indirectly related to DA—the so-called *à propos* commentary—are appropriate. In my understanding, publications of this type are generally more informative and interesting than collections of memoirs usually are. With this in mind, I left the article intact. Probably not all will find this justifiable. But, with already so many years of life experience, I have learned that, however favourably you accept criticism—which is very useful sometimes—still do not follow somebody's advice if this runs counter to your own well thought-out opinion.

Article 13

About Efim Fradkin[1]*

For many years, I have not been occupied with the problems that are the subject of the present conference. But it was organized in the memory of Efim Fradkin and, therefore, I decided to give a talk. The point is that I have known Efim (Fima, as we used to call him) longer than anyone present. I shall describe later how and why I met Fima for the first time but now I shall begin with his biography.[2]*

Efim Samoilovich Fradkin was born on 30 November, 1924, in the provincial Belorussian town of Shchedrin located within the so-called 'Jewish pale'. Not everyone in this audience, especially the foreigners, will know what this means. In tsarist Russia, i.e. before 1917, Jews had the right to reside only within certain limited territories. Exceptions were only made for christened Jews, rich merchants, and so on, and the Fradkins belonged to none of those groups. Theirs was a poor family with many children. Their life was hard and their father, a former rabbi, was subjected to repression and died in prison. Fortunately, no racial limitations or, simply speaking, state-encouraged anti-Semitism existed in the USSR in the 1930s and Fima could enter Minsk University in 1940. He studied there for only a year before the Soviet Union entered the Second World War, i.e. before 22 June, 1941. Fima managed to leave Belorussia before it was occupied but his mother, two sisters, and a younger brother were killed by the Nazis. Of all the family, only Fima and his elder brother, who was in the army, survived. Fima was evacuated for some time, to Bashkiria, where he worked as a school teacher and, at the beginning of 1942, voluntarily joined the army as a common soldier. He was badly wounded near Stalingrad and, after leaving hospital, he was sent to an artillery school. Then he took part in combat again but this time as an officer. He was rewarded for his services in battles. Along with his service in the army, Efim studied by correspondence at the Lvov University from 1945. But only after demobilization in 1946, could he study normally and he graduated from the university in 1948. He even wrote two diploma theses. One of them, which unfortunately was not published, was devoted to the effects of an electric field upon some transitions in atoms. In his second diploma thesis, Efim considered the behavior of a relativistic particle of spin $\frac{5}{2}$. He chose this subject himself after he read in a library my paper analyzing spin $\frac{3}{2}$ [4]. He wanted to extend my consideration to the case of a higher spin (this paper was later published [5]).

In Lvov, there were apparently no specialists in relativistic quantum theory and that is why Fima decided to move to Moscow in 1947, being entitled to do this as an ex-serviceman. I do not remember exactly but we may have exchanged letters before that. However, I remember well how Fima appeared in FIAN (the P N Lebedev Physical Institute of the USSR Academy of Sciences) in order to speak to me. We met in the old FIAN building in Miusskaya Square, now occupied by the M V Keldysh Institute of Applied Mathematics. I had a tiny office, some walled-off cubbyhole. And there I saw a slender short youth dressed in a greatcoat. I gave him the only chair in the room and sat down myself on the table. Fima told me later that he had been amazed: he thought he would meet a dignified and pompous professor, for in Lvov, which had been a Polish town before 1939, there probably remained such a professorate. I was then 31 years old and was neither dignified nor pompous. None of us at the Theoretical Department of FIAN were dignified or pompous, and the founder of our department Igor Tamm was no exception, although he was already 52 at that time. The atmosphere at the department was friendly and democratic. A detached view would be more exact, whereas I have worked at the department for more than 60 years (since 1940). But I will permit myself to express the opinion that our department is not typical, and during all these 60 years I had only one serious conflict caused by the dismissal of one research workers. What is typical of our department is respect for youth and the impossibility of putting one's name on another person's paper. In particular, Fima was my postgraduate student and we frequently discussed various issues but we have no joint publications.

I E Tamm and I appraised Fima's abilities and recommended that he should enter the postgraduate course at FIAN. But this was not at all easy because state anti-Semitism had already come to reign in this country. Fima was accepted to the postgraduate course with great difficulty in 1948 and only, I think, because he was a war veteran and had been wounded. It seems to me, by the way, that Fima first appeared at our department in late 1947. As far as I understand, Fima was happy (he told me about it himself [1]), for after so many years of a very hard life he finally found himself in the right place. And he 'responded' with selfless work: he obviously believed, and not without reason, that much time had been lost. Efim's capacity for work, his devotion to science and work attracted attention, although none of us were idle. In addition, Fima was a bachelor and it can be said that all his effort was directed to work. He was first of all interested in fundamental questions and it was not accidental that he had set himself to study spin theory even before he came to FIAN. The young research workers and postgraduate students of the department, including Andrei Sakharov (he was three years senior to Fima, having entered the postgraduate course of the department in 1945, and defended his candidate's dissertation in November 1947), were good company for him.

Unfortunately, or maybe fortunately, because this played a positive role in his fate, Efim was unable to give all his strengths to the solution of fundamental problems for several years. The point was that in 1948 or 1949 he was included in the group headed by I E Tamm and admitted to secret work (the content of this work became known only in 1990(!) after the death of A D Sakharov: the work was aimed at creating a hydrogen bomb). For several years, E Fradkin was engaged in a number of problems, namely transport processes in hot plasmas [6], hydrodynamics [7], and the theory of turbulent mixing [8] (these studies were published with a delay after their content was declassified). But, as I have already mentioned, in his heart Efim longed for another kind of problem and published not only the previously mentioned paper about spin $\frac{5}{2}$ [5] but also a paper about the reaction of radiation in the classical theory of electron [9]. The main thing is that he also found time to follow the current literature on elementary particle physics, as this field was then called. He was also interested in quantum statistics. Beginning with 1953 or 1954, Efim could give almost all his strength to investigate in these particular fields (quantum field theory and quantum statistics). At that time he also began a 'new life' in another respect—in 1955, Fima got married. This was a very happy marriage. The whole department was present at his wedding, which I described in paper [1] and I would not like to repeat myself here.

From 1955 up to his death (he passed away a year ago, on 25 May, 1999), Efim, a person of studious habits, was completely absorbed in his work. The only obstacle in his way was his poor health, which had been badly affected by his hard childhood and youth and the wound. In 1955 or so, I practically stopped working on the previously mentioned problems, which were Efim's prime concern. That is why it would be out of place if I dwelt here on the results of his work on quantum field theory and quantum statistics, the more so as this has been covered in the collection of papers [2], in the obituary [3], and will certainly be reflected by the present conference.

I would like, however, to make two more remarks.

Here is what A D Sakharov wrote in his *Memoirs* [10, p 108]:

Of all our company, Fradkin was the only one to have reached the level of a highly professional 'forefront' theoretical physicist, of which we all had dreamed. He has great achievements in almost all basic directions of quantum field theory (the Green function method in renormalization theory, functional integration, gauge fields, unified theories of strong, weak, and electromagnetic interactions, the general theory of quantization of systems with constraints, supergravity, string theory, etc.). He was the first to discover the 'Moscow Zero' independently of Landau and Pomeranchuk. Many of the results obtained by Fradkin are classical. Fradkin has no equal in methodical questions.

I think that this is a just appraisal. And, incidentally, during Sakharov's exile in Gorky, Efim visited him several times and helped him in all possible ways.

Efim Fradkin was a brilliant representative of a whole generation of Soviet physicists, who were involved in science research with great enthusiasm. Meanwhile, the financial conditions of our life were rather bad according to American and European standards. In the Stalin period, particularly during the Cold War, only the chosen few could go abroad. Since 1947, the persecution of so-called cosmopolitans began, the remarkable *Journal of Physics USSR* was no longer issued, and our Russian periodicals were not translated into English. I do not even mention the complete lack of freedom of speech under the totalitarian regime. But we worked, I repeat, with great enthusiasm to the amazement of some of our foreign colleagues. It seems to me that it was in 1956 that a large group of such highly qualified theoretical physicists, of whom Sakharov wrote, came to the USSR for the first time after many years. F Dyson was among them. After he returned home, in one of his papers, he specially commented on what I have said about (the enthusiasm of Soviet colleagues) and explained it as follows: "They have nothing else" (I quote from memory). In other words, 'everything has gone to science' and, in such a way, they can forget about their hard life. This is a profound remark (here Efim is a vivid specimen) and for a long time I believed it to be quite correct.[3*] But now I no longer consider such an explanation to be exhaustive.

Indeed, after the fall of the villainous bolshevik Lenin–Stalin regime in Russia, we now have freedom of speech and freedom of migration. Research workers, as all citizens, can go abroad practically unlimitedly and meet their colleagues all over the word or correspond with them through either ordinary or electronic mail. Our main journals are translated into English. Of course, there are still many enthusiasts, who give all their strengths to science. But the tone, the general spirit is now quite different. A lot of young people leave science (say, for business), others go abroad or work reluctantly and do not attend seminars regularly. Elderly people often think that 'everything was better' in the days of their youth. But I am sure that it is not this effect that explains my diagnosis. In my opinion, the explanation is basically as follows: the social status of physicists in Russia has changed. In the USSR, physicists and representatives of some other professions were, so to say, the salt of the earth. To be a physicist was prestigious. And, in addition, the salary of research workers was nearly the highest in the country, apart from those of higher party and Soviet functionaries. Now the conditions of science in Russia are very hard in any respect. There is not enough money for equipment and literature and the salaries are very low not only according to international standards but also compared to all types of clerks and secretaries in banks and firms even in Russia. At the same time, many rich people have appeared, sometimes simply rogues, who earn incomparably more than any first-class physicist. I do not think that our postgraduate students and candidates of science (approximately the PhD level) are worse financially than they were in the 1950s, to say nothing of the 1930s and 1940s. But they are beggars compared to the so-called 'new Russians', all sorts of swindlers. This cannot but have its effect. But I am still not inclined to exaggerate and hope that Russia and, in particular, physicists in Russia will raise their heads in the near future. However, the former students and colleagues of Efim Fradkin do not hang their heads even today and, in many respects, have adopted his anxious attitude and devotion to science. I believe that the present conference is one of the proofs of this. I hope the conference will be successful and I wish you this success.

Comments

1*. A talk at the Conference 'Quantization, Gauge Theory and Strings' (dedicated to the memory of Professor Efim Fradkin) held on 5 June, 2000, published in 2001 in *Usp. Fiz. Nauk* **171** 874 [2001 *Phys. Usp.* **44** 831], as well as in the Proceedings of the mentioned conference (Moscow: Scientific World, 2001, v 1, p 5).

2*. I shall partly repeat here what I wrote in paper [1] published in the collection [2] devoted to E S Fradkin's 60th birthday. See also his obituary [3].

3*. It is this particular paper by F Dyson that is mentioned in my paper [11] dedicated to the memory of D A Kirzhnits.

References

[1] Ginzburg V L 1987 The paper in collection [2], vol 2, p 15
[2] Batalin I A, Isham C J and Vilkovisky G A (eds) 1987 *Quantum Field Theory and Quantum Statistics. Essays in Honour of the Sixtieth Birthday of E S Fradkin* (Bristol: Adam Hilger)

[3] Batalin I A *et al* 1999 "Pamyati Efima Samoilovicha Fradkina" (To the Memory of Efim Samoilovich Fradkin) *Usp. Fiz. Nauk* **169** 1281 [1999 *Phys. Usp.* **42** 1175]

[4] Ginzburg V L 1942 *Zh. Eksp. Teor. Fiz.* **12** 425

[5] Fradkin E S 1950 *Zh. Eksp. Teor. Fiz.* **20** 27

[6] Fradkin E S 1957 *Zh. Eksp. Teor. Fiz.* **32** 1176 [1957 *Sov. Phys. JETP* **5** 1004]

[7] Fradkin E S 1965 *Tr. Fiz. Inst. Akad. Nauk SSSR* **29** 250; 257

[8] Belen'kii S Z and Fradkin E S 1965 *Tr. Fiz. Inst. Akad. Nauk SSSR* **29** 207

[9] Fradkin E S 1950 *Zh. Eksp. Teor. Fiz.* **20** 211

[10] Sakharov A D 1996 *Vospominaniya* [*Memoirs*] (Moscow: Prava Cheloveka) [English version 1992 (New York: Vintage Books)]

[11] Ginzburg V L in *Sbornik Statei Pamyati D A Kirzhnitsa* [*Collection of Papers in the Memory of D A Kirzhnits*] (Moscow: Nauka) (this is the preceding article 12 in this volume)

Article 14

In memory of Vitya Shvartsman[1]*

Talented, brilliant, gifted, well-educated, sympathetic—these are the epithets that come to my mind first when I remember Viktor Pavlovich (for me, simply Vitya) Shvartsman. I never worked with him, nor did we met very much, but to talk to him, to have a chat with him for a while always was a pleasant prospect for me. I enjoyed speaking or arguing with him—and people like this are rare in one's life. As, in fact, people of Vitya's calibre and type are. Here both these aspects seem important to me, calibre being in a sense a measure of a person's talent and gift and type rather referring to personality, character traits, behaviour. I met people of outstanding abilities but who were very reserved, non-communicative, and kind of closed to those around—not exactly persons to chat with, for all their intellectual depth. Vitya was a happy blend of depth and brilliancy, I would say. I saw him last at a conference on the large-scale structure of the Universe, at Lake Balaton in Hungary, in July 1987 I think it was—just weeks before the tragic end of his young life. My wife, Vitya and I went for a walk one day and I remember I asked him a lot about his life. I asked him, I don't know why, whether he had married and his answer was no, he had not, but he was not lonely either. His work, which he appreciated very much (or to which he devoted so much effort, rather), did not live up to his expectations—being tied to a telescope high in the mountains was not exactly his idea of what happy life was like. Still, he did not seem to be in a blue mood and did not complain. He did not ever complain, for that matter—not to me, at any rate. The reason I am writing this is that his untimely death gives an idea of how Vitya must have suffered sometimes. He was ill with depression, as far as I know—a condition which affects a large number, up to tens of percent, of people. When the disease is in its gentle form, one speaks of being in 'bad mood', down-hearted, etc. For some, this is not even a disease at all. In severe cases, however, depression causes what can safely be called agonies and often makes the patient commit or attempt suicide. From what I have heard, this disease is due to certain substances being secreted into the blood, and is currently amenable to drug treatment. I am not a specialist and it would be out of place to delve deeper into the subject but I feel sorry for Vitya and for all those who are missing him. I feel like giving a warning that being in 'bad mood' for too long may be a symptom of a disorder, that this is by no means a trifle and, at a certain stage, consulting a doctor becomes a real necessity— which does not at all mean being a hypochondriac and wasting your time. Vitya's relatives will forgive me if I am being irrelevant but I only know that he visited a doctor—but did he really undergo a treatment, I am not so sure. It is a known fact that more than occasionally the tragic end of depression or the agonies it inflicts (which is also tragedy) come about as a result of false shame, shyness, of the fear of betraying oneself—feelings which cause one to disguise oneself. However, skepticism about the power of medicine may be a factor or the lack of attention and understanding on the part of those around. We see a cheerful and seemingly happy man— but in actual fact there is a gnawing in his heart and, if he says he feels bad, people tend to answer in a come-on-stop-it manner: you are totally OK, there is nothing wrong with you, and such like. My point here is that people like Vitya Shvartsman need protection, that we should give them a helping hand whenever possible. Was it possible to help Vitya? I do not know but surely this is the question which all of us who knew him should try to answer—at least to draw a lesson for the future.

Comments

1*. Viktorii Favlovich Shvartsman was born on 22 July, 1945, lived only 42 years, and took his own life on 27 August, 1987. We lost a talented, brilliant, and sympathetic person. There is no point in repeating here what I wrote in my note. There are some reasons why I find appropriate that this note—not very substantive, admittedly—be reprinted here. First, it is the desire once again to pay tribute to Vitya's memory. Second, it gives me the opportunity to draw attention to a Vitya Shvartsman memorial collection *In Search of Unity*. Unfortunately, this 404-page book is not easy to find, the more so since the publisher, year of publication (1995 perhaps) or the circulation are not indicated—it is not available in libraries, I am afraid. The collection features memories (in particular, the present article, p 15, and an excerpt from my letter to Vitya, p 26) and also includes a selection of V F Shvartsman's papers, excerpts from his diaries, and some of his poems—materials which clearly show what a heavy loss Vitya Shvartsman's untimely death is.

Article 15

In memory of Vadim Sidur[1*]

I would like to say something of interest here—or at least something of substance—but surely this is not that simple in this case if you are neither an art critic nor an art historian. Indeed, as far as appreciating sculptures and pictures is concerned, my scale of values is just 'terrific—great—not bad—just disgusting'—surely not enough to speak of Dima Sidur. Sure, I think very highly of Dima's work but there were some things I did not like that much. This hurt Dima's feelings and he ascribed this to my being undeveloped in the arts. He was right perhaps: it was, to a large extent, he who formed my taste but he moved on and on—whereas I lagged behind. 'Mutations' or 'coffin art'—all that was and still is beyond me.

Nor is there much I could say about Dima as a man: just epithets such as kind, clever, and the like. So I decided instead to say some words about the hardships Dima went through. There is a hidden motive behind my decision, though—as I shall make clear at the end of my speech.

Now Dima's hardships and ordeals were many and I shall only touch on those of them in which I myself was personally involved.

In 1971, the prominent physicist Igor Evgenevich Tamm died. We, those people close to him, were faced with the problem of building a funerary monument for him and I turned to Dima for help. Dima offered one of his sculptures and it is this one which was ultimately placed on I E Tamm's grave at the Novodevich'e cemetery and which all those present here know—from photographs, at least. That was the time, it is recalled, when Sidur's works were neither exhibited nor recognized and erecting cemetery monuments was a major channel for a sculptor to present at least some of his works to the public.

The memorial to E S Varga, set up in Novodevich'e cemetery prior to Tamm's memorial, as far as I know, is considered to be one of the most famous and valuable sculptures in the cemetery.

So, the late Tamm's friends and relatives decided for Sidur's project and that seemed to be it, one might say. Far from it. The Novodevich'e cemetery was, shall I say, a 'party and government' cemetery, so nothing could be erected there unless an 'arts council' at the Ministry of Culture gave its approval. I happened to attend a session of the council as a representative of the 'client'—it was in 1972—and the memory of this still causes me disgust and indignation. Poor Dima, he had to listen to the malicious ravings of the 'Sculptor Laureates' without being able to say what exactly he thought of them. Some people defended the project, though. All that seemed so totally irrational to me, so much so that I cannot even say what the subject of the discussion exactly was. Of course, I gave vent to my exasperation—only to make things worse, I am afraid. The voting proved to be indecisive, the for and against votes being equal in number. Voices were heard to the effect that as the chairman voted in favor, the total outcome should also be considered in favor of—is not it funny—the 'defendant' (or was it 'injured party', I do not remember which term exactly was used). The monument was erected after all—although there was much yet to overcome—but sure all of us and Dima in particular—were indeed an 'injured party'. The client was no less than the USSR Academy of Sciences, Sidur was already a big name in sculpture, and there was, in fact, nothing that shocking about the monument—and still it was entirely at the mercy of a gang of ... well ... a group of officials, whether the public would or would not see it.

'Don't give them an inch'—such was the motto of the country where Sidur—and many many others—lived all their working life.

Let me give another example.

In 1967 Sidur created a portrait of Einstein, possibly the artist's most known sculpture. I think most highly of it, and this is the case when my opinion may perhaps carry some weight: for me, Einstein (along with Leo Tolstoy) is the greatest of the great and I have seen, I think, most of the photographs of him, including those of his sculptures—and occasionally the sculptures themselves. Sidur's Einstein was made from alabaster, and to translate it into bronze was not possible for Sidur and, indeed, no one in the country could erect the sculpture at the time. That is why Dima agreed to sell it abroad after some people from the West visited his studio and were impressed by the work. To make a long story short, official letters came from the USA asking for permission to buy the sculpture. Absurd though this might have seemed to foreigners (and, fortunately, to us—now) but those were times when an artist was all but rightless and it was generally impossible for him to sell his work abroad. So, the struggle for permission to send the sculpture to the USA was on the agenda and I wrote a letter to P N Demichev, a CPSU Central Committee secretary, then, to the best of my recollection, 'in charge of culture'! It was later that the guy became a minister of culture and so a better known person. But then, in the early 1970s, I remember it took me some effort to learn his first and second names. So, I sent him a letter with the necessary documents from the

USA attached. He vouchsafed me no answer, sending all the papers to the Ministry of Culture instead, from where, in turn, a lady 'specialist' was dispatched to Sidur's studio. She took some photographs from there and reported 'to her superiors'. The result was that a deputy minister called me (I think his family name was Popov but as to his initials, this I don't know) and said that "he wouldn't even hang Sidur's pictures or their reproductions in his home". I got angry, naturally enough, and told him rather sharply that my opinion was quite the opposite and that Sidur's were the only pictures hanging in my home (which they were)—but that this was not the point at all—the point was that some people like the sculpture and want to buy it and what about allowing this. I don't remember what else was said between us but, anyway, the sculpture was not sold. For Sidur, even though he certainly was in need of money, what really mattered was to cast the sculpture in bronze and, above all, to exhibit it. So later he accepted a plan (which was worked out in fact without me, so I don't remember the specifics) that he should donate the sculpture to the USSR Academy of Sciences, which in turn presents it under the same terms to the Fermi Laboratory (in the USA, near Chicago). The plan worked and, in 1975, the Einstein arrived in the USA, was cast in bronze, and erected. I saw the sculpture there in 1987, when I was allowed to visit the USA and the Fermilab in particular (for the first time in 18 years, by the way: not only artists were oppressed and discriminated in those times). Unfortunately, the sculpture looks less than impressive there. It stands in a corner, on a wooden pedestal, with no inscription on it if I remember correctly. Which is not exactly what Dima would like, because his Einstein's head is double-sided and must rotate and so it would be natural for the sculpture to be in full view of the public. Unfortunately, the head of the laboratory (of this huge institute I could say) was absent at the time and, anyway, I did not feel like 'throwing my weight around' as they say. There are things people must understand for themselves. Fortunately, this sculpture won fame anyway: it was featured twice in the *CERN Courier*, and can be seen established in four research institutes in Germany.

Finally, the last example. Immediately after Sidur's funeral, attempts were made by MOSKh (the Moscow Division of the Union of Painters) to lay its hands on his studio—a damp basement where he had worked for as long as 30 years. The rental of 9 roubles a months (true, with a disabled soldier's discount) is the telltale indication of the basement's value—but there were about 500 sculptures there, for which no other premises were available. So an uphill struggle to save the studio began. First, on 22 September, 1986, a letter signed by 16 writers, physicists, etc was sent to MOSKh—to no avail. Then, with the help of L Pol'skaya, of *Literaturnaya Gazeta [Literary Newspaper]*, I published an article in the 29 April, 1987, issue of her newspaper. The article was titled 'Sculptures Which We Don't See (about the fate of Vadim Sidur's works)' and its closing words were:

"Please do consider this an open letter to the Soviet Foundation of Culture and to the USSR Ministry of Culture: do prevent the destruction of Vadim Sidur's studio, don't let this act of vandalism take place". Whether it was the prospect of gaining a reputation as vandals or whether the changes that took place in the country in general played their role—or maybe both—but the MOSKh ultimately gave up. The studio was saved.

There are three reasons why I have recalled these three—shall I say—episodes. First, one now reads and hears all the time that perestroika has failed, that we have been on the wrong track for six years, and things like that. My understanding is that only people deprived of memory and some other abilities—a sense of justice, for example—can see things this way. I think only the blind cannot see how dramatically things have changed since 1985 and especially since 1987. Today we speak and write what we like, we are free—or almost free—to go abroad, and there is no longer any—or almost no—threat of a nuclear war. Of course, the situation is very bad economically but does this mean that we can run down all around us and call for strikes? I hope my opinion on that is quite clear from what I have said. Second, I wish to stress that however hard Dima's life was, he did not became embitter, nor became a destructionist in a sense. He showed a soft wisdom—a quality which so many people lack. He and I argued sometimes but we understood one another. I think that our ideas on the present situation would be similar and, even if not, we certainly would not fall foul of one another—recognizing that everybody has the right to have his/her opinion and should not be afraid to express it. Today, it happens more than often that people are ready to hate and all but crucify anyone who disagrees. There was nothing like this in the character and attitudes of Vadim Sidur of blessed memory.

Comment

1*. Vadim Abramovich Sidur (28 June, 1924 – 26 June, 1986) was a prominent sculptor and painter and he also wrote prose and poems. Under the Soviet régime, Sidur was out of favour and his life was hard. Today, fortunately, his work has received a well-deserved recognition—not only abroad but in Russia as well. Since 1989, the Sidur Museum functions in Moscow (Novogireevskaya ul., 37; tel.: (095) 918 51 81). Close to the museum, Sidur's monument to the Muscovites who fell in the Afghan War is erected.

The present note is the text of the speech I gave at the Sidur commemoration meeting at the Moscow House of Cinema on 27 March, 1991.

Article 16

About Sergei Ivanovich Vavilov[1*]

My speech is nothing more than several remarks and rather odd ones at that. I only hope that these remarks will not be without interest and to some extent complement the in-depth analysis inherent in E L Feinberg's articles [1] and his speech to this audience.

Sergei Ivanovich Vavilov was the Director of the P N Lebedev Physical Institute (FIAN) from its inception in 1932 up to his decease in 1951. As for me, I have been working at this institute since 1940 but, in actuality, I have been related to the FIAN for a longer time (approximately since 1938). The Institute was small at that time (about 200 staff members) and the Director's activities were generally transparent to all staff members. Furthermore, some issues with which I was concerned interested Sergei Ivanovich, though I recall only one scientific discussion with him. One day, which was before the war, Sergei Ivanovich asked me to what extent the acceleration of a Vavilov–Cherenkov radiation source could be neglected— for such an acceleration was seemingly inevitable due to the radiation loss. I gave the correct answer: the acceleration is generally insignificant, which follows from calculations neglecting the acceleration. However, later I gave some more thought to the issue and grasped the heart of the matter without any computations. First, it is possible to compensate for the source (then, as well as today, an electron was usually referred to) acceleration, say, by an external electric field. Second, if the source mass is large enough, the effect of radiation, which is responsible for the variation of the source velocity, can always be treated as being arbitrarily small. In other words, the source velocity can be quite legitimately considered to be given and, in particular, constant. I also note that Sergei Ivanovich was, to some extent, familiar with my work in other fields as well, for he communicated about ten of my papers to *Dokl. Akad. Nauk SSSR* from 1940 to 1946.

I note this incidentally, for I am not going to discuss physics today. I would like to touch upon other aspects and appraisals of Sergei Ivanovich's activity.

310

Among these there are found some distinctly negative appraisals concerning both Sergei Ivanovich's scientific level and accomplishments in physics and the political attitude he assumed. There exists an opinion that we may either speak highly of those who are gone or not speak at all (*de mortuis aut bene aut nihil*). This viewpoint is acceptable when we are dealing with an epitaph or even with an obituary. But after a long lapse of time, a different formula appears to be solely correct to me: one must either say nothing or speak the truth. I, therefore, believe that the negative appraisals of Sergei Ivanovich should be asked and answered. What are these judgements?

True, not only was Sergei Ivanovich never subjected to any repression but he also became the President of the USSR Academy of Sciences, whereas his elder brother Nikolai Ivanovich Vavilov was arrested in 1940 and perished in prison on 26 January 1943. This gave some cause to accuse Sergei Ivanovich of betraying his brother. In particular, A I Solzhenitsyn's *The Gulag Archipelago* [2] reads as follows: "Academician Sergei Ivanovich Vavilov came, after the massacre of his great brother, to be the servile President of the Academy of Sciences. (The joker with a big moustache devised this with a jeer, trying the human heart.)"

Furthermore, Sergei Ivanovich was President in the most macabre time of Stalin. And more than once he had to say things which are just monstrous from today's viewpoint and which concerned Stalin 'the coryphaeus of all sciences', the unscientific (so-called Michurin's) biology, etc. Naturally, such behavior may be condemned by those who were not familiar with Sergei Ivanovich, his status, and the general situation. I am convinced, however, that all such accusations against S I Vavilov are absolutely inconsistent. E L Feinberg writes about it convincingly and in sufficient detail [1]. I will only note here that three years ago I wrote a letter to A I Solzhenitsyn to inform him (with the enclosure of the corresponding materials) that his appraisal of S I Vavilov was wrong. Aleksandr Isaevich rang me up in response to express his satisfaction with the elucidation of the truth. I hope that the incorrect remark concerning Sergei Ivanovich will be missing from subsequent editions of *The Gulag Archipelago*.

By the way, apart from documents and dates, the attitude of Nikolai Ivanovich's sons to Sergei Ivanovich for me is a convincing argument as regards the relations between the Vavilov brothers. In particular, Yurii, N I Vavilov's younger son, speaks warmly of 'uncle Serezha', just as he would speak of his father. Sergei Ivanovich tried his best to take care of Yu N Vavilov and his mother—his brother's widow. As far as I was able to find out, Oleg, the elder son of N I Vavilov, who perished tragically in 1946, treated Sergei Ivanovich in precisely the same way (there are grounds to suspect that O N Vavilov was killed in revenge for his openly stated indignation at the destruction of his father). So, everything is absolutely clear about Sergei Ivanovich's alleged betrayal of his brother.

Concerning Sergei Ivanovich's activity as the President of the USSR Academy of Sciences, elected in the July of 1945, the following remark is in order. The election of the President of the Academy of Sciences by the General Meeting at that time was purely formal in character. The president was appointed by Stalin. In this particular case, his choice was, I believe, the best from purely business-like considerations (the case in point is a physicist and a good administrator as well). True, it is quite probable that a part was played by inherently Stalin's foul and insidious wish to appoint to a high post the brother of N I Vavilov, whom he had destroyed. Could Sergei Ivanovich decline the appointment? To my knowledge, declining Stalin's proposal was mortally dangerous in those times. What is more, immediately after the victorious completion of the war, in society at large there were widespread hopes for a slackening of the dictatorship and a certain democratization of the regime. Finally, Sergei Ivanovich realized that certain of the other possible presidential candidates would not do the good for USSR science he himself was able to do (see [1]). Therefore, I am convinced that there are no grounds to reproach Sergei Ivanovich for giving his consent to becoming president. It is to be regretted that the expectations for the slackening of the dictatorship and for taking the path of civilized development were not borne out. The Cold War with the outer world began and the arbitrary rule in the realms of culture and science continued as in the pre-war fashion. It would suffice to recollect the defamation of the great Akhmatova and Zoshchenko, the Lysenkohood (Lysenko husbandry), and the defamation of so-called cosmopolitans. In these terrible conditions, Sergei Ivanovich did, as far as I know, all he could to cushion the blows and save the science. In doing this, he had to act against his conscience and agree to disgusting compromises. That was very hard. Hence, the heart attacks and his untimely decease in 1951 on the eve of his 60th birthday.

To summarize, I believe that the reproaches cast upon Sergei Ivanovich—all that I know—are absolutely groundless and we should feel only a deep gratitude for his work as the President of the Academy of Sciences. I will note, by the way, that I have, naturally, no respect for those who allow themselves to hurl stones at Sergei Ivanovich. Instead of hurling reproaches at other people, one should first of all look at oneself. To exemplify this, it would suffice to recall the letter condemning A D Sakharov signed by USSR Academy Members in 1975. It was signed by 72 people and only five Academicians refused to do so (the Presidium of the USSR Academy of Sciences reported their names to the Central Committee of the Communist Party of the Soviet Union; see [3, p 430]). And this occurred in the time of Brezhnev, when the perils of arrest beatings, and shooting was quite low.

Regrettably, this unfair attitude towards Sergei Ivanovich extended to science too and was held by some physicists. The most vivid example is P L Kapitza's letter (1936) to Rutherford, which was conveyed to Rutherford personally by P L Kapitza's spouse and not sent by post. Clearly, this was a private letter but it survived in Rutherford's archive and was published abroad [4]. This letter was later published in Russian as well [5]. I think that publishing such private letters before the lapse of many years (say, 50 years, as is customary in many cases) after the death of their author is generally incorrect. However, once this has happened, there is no escape from noting that I consider this letter to be disgraceful. I will not dwell upon the letter, the more so as I touched upon it earlier (see [6, p 399]). Furthermore, and this is more important, S P Kapitza informed me that his father had radically changed his opinion of Sergei Ivanovich by the end of his (SI's) life and probably regretted writing this letter. Another unfair appraisal of S I Vavilov as a physicist, and not only as a physicist, is found in S E Frish's 'memoirs' [7]. Here, as in some other cases, the role of Sergei Ivanovich in the discovery of the Vavilov–Cherenkov effect is assessed quite incorrectly. To those familiar with the history of the discovery of this effect, the decisive role played by Sergei Ivanovich in this brilliant accomplishment is quite evident. He proposed the topic and the method of investigation and realized at the decisive moment that the case in point is not luminescence. All this is addressed in greater detail in the books by I M Frank [8] and E L Feinberg [1]. One way or the other, the solely correct name for the beautiful effect of emission by uniformly moving charges should bear the name of Vavilov. The term 'Cherenkov effect' adopted in the West and partly in our country is absolutely unfair. This name probably stems from the fact that initially (1934) separate papers by Cherenkov [9] and Vavilov [10] appeared. After the nature of the effect was elucidated by Frank and Tamm [11], a manuscript was sent abroad in 1937 about the discovery with Cherenkov alone as the author [12]. The paper was sent by S I Vavilov and it remains unknown why he did not include his name on the author list, as he had good reason for doing so. I have made some guesses about it but I will not cite them here because they are merely guesses. The Vavilov–Cherenkov effect was then perceived as an unexpected and not nearly obvious phenomenon. This becomes clear from the fact that the manuscript of [12] was initially sent to *Nature* but was rejected. This manuscript was later sent to and appeared in *Physical Review* but was initially not understood, either. The latter follows from the fact that the Vavilov–Cherenkov experiments were repeated and confirmed with the use of an electron beam [13] but Collins and Reiling never understood the nature of the phenomenon—they believed they were dealing with bremsstrahlung.

Sergei Ivanovich did not know, of course, of the previously mentioned letter written by P L Kapitza but he was aware of Kapitza's strong disapproval of him and may be of some insults. I write about this to mention Sergei Ivanovich's characteristic conduct and judgement, which was reported by B P Zakharchenya [14]. When Kapitza was in disgrace and worked in his country cottage at Nikolina Gora (in the 'hut-laboratory', as they used to say at that time), he applied to the Supply Department of the USSR Academy of Sciences with a request to provide him with some materials and simple instruments but was bluntly refused. In response to P L Kapitza's complaint, not only did the Academy President Sergei Ivanovich Vavilov put the rude fellows in their place and order the delivery of the requisite equipment to Nikolina Gora but he also came personally to Kapitza presumably with an apology. Next I cite [14, p 40]: 'Knowing about the strained relations between Kapitza and Vavilov, N A Tolstoi, who may be referred to as Sergei Ivanovich's student, asked him: "Why did you make this fine gesture? Didn't he criticize you severely in the past?" Vavilov's reply was: "A noble act and politeness in this case are the just revenge of an intelligent person".' As previously mentioned, P L Kapitza finally realized who Sergei Ivanovich was.

Next I would like to emphasize the following: not only was Sergei Ivanovich an outstanding physicist and science administrator but also a connoisseur of the history of physics and culture in general. This is a topic in its own right, which is discussed, among other topics, by E L Feinberg [1]. I will only note that S I Vavilov translated Newton's *Opticks* from the Latin, is the author of several popular articles and the book *The Eye and the Sun*, and was the Editor-in-Chief of the *Big Soviet Encyclopedia* and several other publications. He was, as they say, 'a Renaissance man'.

I will enlarge on Sergei Ivanovich's contribution to Newtonology. The 300th anniversary of Newton's birth (4 January 1643 according to the new style) fell on the most difficult (and at the same time the turning-point) period of the Second World War. It is, therefore, reasonable that the Newtonian anniversary celebrations were rather modest in character and, in Newton's native land, to my knowledge, no new books devoted to him made an appearance. But this is what is striking: thanks to S I Vavilov's care, as many as five books were published in the USSR in connection with this jubilee! Standing out among them is the biography *Isaac Newton* written by S I Vavilov [15]. For a small volume, this book is rich and deep in content and is perfectly written. There is no escape from mentioning the conditions in which Sergei Ivanovich wrote the book and prepared its second edition. The foreword to the first edition, which was published in early 1943, was dated November 1942. The foreword to the second edition (published in 1945) bore the date December 1944. At that time S I Vavilov lived primarily in Ioshkar Ola, for that was the site of the State Optical Institute (GOI) of which he was in charge. But he spent part of the time in Kazan, since he retained the post of the Director of FIAN, which had been evacuated to Kazan:

the hardest time full of hardships and intensive labor. What is more, Sergei Ivanovich worried about people who were dear to him but who were detached from him. Clearly, the work on Newton's biography was done in his 'spare time', a work into which he put his heart and soul. The feelings and ideas of Sergei Ivanovich were undoubtedly embodied in this book and especially in the forewords to its first and second editions. I cannot reread these forewords without emotion. This may be a manifestation of the fact that I remember that time and was in Kazan myself together with FIAN. I believe that those who know of the war-time only by books would not remain indifferent to these forewords, either. Here, for example, is an extract from the foreword to the first edition:

> Devoting the primary effort today to assist our heroic Red Army, the USSR Academy of Sciences cannot pass by the remarkable date of the 300th anniversary of the birth of one of the greatest creators in culture—Isaac Newton. The Academy of Sciences set up a special committee to commemorate Newton's jubilee. This biography was written following the proposal made by the committee.

And here is a fragment from the foreword to the second edition:

> The second edition of Newton's biography is being prepared in the days when the war is undoubtedly approaching its victorious completion. The peoples of Europe liberated by the Red Army and the Allied Forces from the dull and ferocious yoke of the 'race of masters' are regaining access to live culture and freedom. In such a time, many may be encouraged and inspired by the story of the life and work of the 'decoration of the human race'.

I do not pretend to a deep knowledge of the entire history of physics. However, I am rather well familiar with precisely Newton's activity, for in 1987 I wrote a long article on the occasion of the 300th anniversary of the fundamental *Philosophiae Naturalis Principia Mathematica* [*Mathematical Principles of Natural Philosophy*] by Isaac Newton [16]. I, therefore, believe to be able to expertly assess Newton's biography written by Sergei Ivanovich [15] and, as stated earlier, the appraisal is very high. In this connection, Viktor Vavilov, the late son of S I Vavilov, and I myself undertook a new edition of the book *Isaac Newton* in 1989, which was provided with my foreword and an additional article (the latter is close to that of [16]).

Sergei Ivanovich Vavilov has left a deep trace on physics and the history of the development of Russian science. I am glad that I still had the opportunity to give tribute to his blessed memory during today's ceremonial meeting.

Comment

1*. My talk, published here, was given, among other talks, at the grand meeting of the Presidium of Russian Academy of Sciences (RAS), Bureau of the Division of General Physics and Astronomy of RAS, and the Scientific Council of the P N Lebedev Physical Institute of RAS devoted to the 110th anniversary of Sergei Ivanovich Vavilov (24 March, 1891 – 25 January, 1951) and held on 28 March, 2001. This and all the other talks were published in 2001 *Usp. Fiz. Nauk* **171** 1070 [2001 *Phys. Usp.* **44** 1017].

References

[1] Feinberg E L 1999 *Epokha i Lichnost'. Fiziki. Ocherki i Vospominaniya* [*Epoch and Personality. Physicists. Essays and Memoirs*] (Moscow: Nauka). Two papers and a short article in this book are devoted to S I Vavilov (pp 137–75)

[2] Solzhenitsyn A I 1990 *Arkhipelag Gulag* vol 2, part 4, ch 3 [*The Gulag Archipelago*] (Moscow: Tsentr 'Novyi Mir') p 260 [Translated into English 1991–1992 (New York: WestviewPress)]

[3] Gorelik G 2000 *Andrei Sakharov: Nauka i Svoboda* [*Andrei Sakharov: Science and Freedom*] (Moscow–Izhevsk: RKhD)

[4] Badash L 1985 *Kapitza, Rutherford, and the Kremlin* (New Haven, CT: Yale University Press)

[5] Kapitza P L 1989 *Pis'ma o Nauke: 1930–1980* [*Letters about Science: 1930–1980*] (compiler P E Rubinin) (Moscow: Moskovskii Rabochii)

[6] Ginzburg V L 1995 *O Fizike i Astrofizike* [*Physics and Astrophysics: a Selection of Key Problems*] (Moscow: Byuro Kvantum) [Translated into English 2001 *The Physics of a Lifetime: Reflections of the Problems and Personalities of 20th Century Physics* (Berlin: Springer)]

[7] Frish S E 1992 *Skvoz' Prizmu Vremeni* [*Through the Prism of Time*] (Moscow: Izd. Politicheskoi Literatury)

[8] Frank I M 1988 *Izluchenie Vavilova-Cherenkova. Voprosy Teorii* [*Vavilov-Cherenkov Radiation. Theoretical Problems*] (Moscow: Nauka)

[9] Cherenkov P A 1934 *Dokl. Akad. Nauk SSSR* **2** 451 [1934 *C. R. Acad. Sci. URSS* **2** 451]

[10] Vavilov S I 1934 *Dokl. Akad. Nauk SSSR* **2** 457 [1934 *C. R. Acad. Sci. URSS* **2** 457]

[11] Tamm I E and Frank I M 1937 *Dokl. Akad. Nauk SSSR* **14** 107 [1937 *C. R. Acad. Sci. URSS* **14** 107]

[12] Cerenkov P A 1937 *Phys. Rev.* **52** 378

[13] Collins G B and Reiling V G 1938 *Phys. Rev.* **54** 499

[14] Zakharchenya B P 1995 "Nepovtorimyi Nikita Alekseevich" [The Inimitable Nikita Alekseevich] *Avrora* (11) 29

[15] Vavilov S I 1989 *Isaac Newton (1643-1727)* 4th edn (Moscow: Nauka) [Translated into English 1951 (Berlin: Akademie)]

[16] Ginzburg V L 1987 *Usp. Fiz. Nauk* **151** 119 [1987 *Sov. Phys. Usp.* **30** 42].

This article was also inserted in V L Ginzburg's book *O Fizike i Astrofizike* (see [6, p 258])

Article 17

Notes of an amateur astrophysicist[1*]

17.1 Introduction

I was surprised to receive the kind invitation to write the prefatory chapter for this volume of the *Annual Review of Astronomy and Astrophysics* (1990), in that I do not consider myself a real astronomer. Rather, I am an amateur astrophysicist and, in any case, am somewhere on the periphery of the astronomical community. For this and other reasons, I did not think that one day I would be asked to write such a paper. On receiving the invitation, however, I immediately decided to make an attempt because I have always liked the idea of publishing in review volumes autobiographical papers or other non-standard reviews including autobiographical elements.

The content of scientific knowledge does not, of course, depend on who has established some facts or how they have carried out observations and measurements. However, researchers are human beings with their own tastes, methods, passions, and fates, and it is much more difficult to understand a person than to decipher the spectrum of a star. At the same time, isn't it a natural temptation to desire insight into the inward life of a 'comrade-in-battle' whom you meet at conferences and/or on journal pages? This problem is solved to some extent by obituaries and, particularly, other posthumous publications but for a number of reasons they can in no way substitute for autobiographical narration.

Here I shall try to outline my astrophysical attempts or, more precisely, the results of these attempts (which has partially been done already in [1, 2]). But I shall also dwell on my biography, which I hope will be of interest to my colleagues in the West and help them better understand the living conditions in the Soviet Union, especially in those now already remote years when my generation of physicists and astronomers was brought up.

17.2   Autobiography I

I was born on 4 October, 1916, in Moscow, where I have lived all my life
(with the exception of two war years that I spent in Kazan). My father
was an engineer engaged in water purification and held several patents.
He married for the first time in 1914 at the age of 51. My mother, a
physician, was then 28. I was the only child in the family. My mother died
of typhoid fever in 1920 and I have almost no memories of her. After she
died, her younger sister came to live with us and was as good a mother to
me as she could be. It was a very hard time—the First World War and
then the Civil War. Moscow became the capital and was, in general, in
a privileged position but all the same there was little food and diseases
raged. My memory is generally poor or, at any rate, has a high threshold.
Above this threshold I recall one scene that I witnessed not far from our
home in the center of Moscow in about 1920: a cart carrying coffins, a
carter dragging himself along, and arms and legs sticking out of the coffins.
Another reminiscence, not so ghastly but still typical, also comes to mind:
we managed to buy fresh meat somewhere but it appeared to be dog meat;
and under normal conditions, dog meat was never used as food in Russia.
Nevertheless, my family suffered much less from what are called practical
difficulties, I think, than did most people in the country at that time—we
had a roof over our heads, Moscow was not occupied by troops, and there
was no real hunger. What I had in excess was loneliness. It was aggravated
by the fact that I did not go to elementary school and went only to the
fourth form at age 11. I do not remember why this was so. The schools,
like almost everything in the country at that time, were being subjected to
all kinds of reorganization and my parents probably thought it better for
me to get my education at home. I believe this was legal at that time. (I do
not know of such cases now, except for sick children.) It was undoubtedly a
mistake, since when I finally did going to school, I found that it was not so
bad. My school was a former gymnasium where many of the old teachers
worked. But bad luck is bad luck. When in 1931 I finished the seventh
year at school, it was decided, somewhere by somebody, that seven years
were quite enough, and thus the remainder of the secondary schooling was
liquidated. (Secondary school has been different at different times: some
time later, they changed their minds and returned to the ten-year school,
as it is today.) My total school education, however, amounted to only four
years.

After seven years of school, one was supposed to enter a factory-plant
school, where one was simultaneously to advance in education and to train
for worker's qualifications. But I did not take this route and instead became a laboratory assistant at an X-ray diffraction laboratory at one of
the high schools. There I mainly communicated with two other laboratory
workers who were three years my senior and took a great interest in physics

and invention. (By the way, both my fellow lab workers later became good physicists.[2*]) I did not learn much but was imbued with something more important than knowledge—enthusiasm and an interest in work. In 1933, for the first time in many years, one could become a student at a university through open competitions at the entrance exams and I decided to attempt to enter the physical faculty of Moscow State University. It took me three months to 'cover' the three-year course corresponding to the eighth, ninth, and tenth years of school. I passed the entrance exams but was not admitted—preference was given to those with better biographical particulars (i. e. social origins and occupations of parents). But there was no special discrimination—the results of my exams were not brilliant. I decided not to wait until the next year, left my work, entered a correspondence course at the University, again studied almost without assistance, and finally, in 1934, at the age of 18, started attending lectures and studying normally 'like other students'.[3*]

17.3   Amateur astrophysicist

Why all these details? My aim is only to warn people against this route because I am deeply convinced that one should not follow my example if given the choice. Of course, the school curriculum can be mastered in much less than ten years but for this one generally must pay a high price. In my case, this price was Russian grammar and, to some extent, my ability to write in this fairly rich language. If this does not become your profession, mastering a language requires exercise, exercise, and again exercise. I hated cramming and did not do it without pressure. The same also applies to mathematics. To understand something, to solve a dozen or so problems, and that is all. But at school, I would have solved ten, if not a hundred, times more of such problems, which would have provided the necessary automatism. All this is obviously so clear that further explanations are not needed here and, in any case, I have already written about this in more detail elsewhere [3]. I shall give a less trivial example. First-year students of the direct department (i. e. not including students taking correspondence courses) were taught astronomy and my school-mates recalled it with pleasure. However, I somehow contrived to become a second-year student without passing exams in astronomy or in chemistry. For this reason and because of my lack of corresponding school knowledge, even on becoming a professor of physics I remained illiterate in astronomy and chemistry. In both cases, I am sure, a retribution followed. As far as chemistry is concerned, I am still paying the price even now, because since 1964 I have been engaged in research on high-temperature superconductivity, which is currently the center of gravity of my work (see, e. g., [4]) and this research appears to be closely connected with some chemical problems of which I am ignorant. As for astronomy, I shall begin with a funny thing. Many times, especially when I took interest in the 'new astronomy' with its quasars, pulsars, etc, I have spoken at lectures or to my friends about various astronomical discoveries, about the radio, X-ray, and gamma-ray 'sky'. But the ordinary stellar sky is unfamiliar to me and, when asked what star or constellation it is, I must honestly say that I do not know. And if I have called this situation funny rather than shameful, it is because I do not consider myself to be a professional astronomer. I am somewhat ashamed nonetheless but such is my life. When in 1946, at the age of 30, I wrote my first paper in astronomy, I had already authored many papers in physics and an even greater number were waiting their turn. I had neither the spare time nor extra strength and the life was hard. How under such conditions could I learn the map of the stellar sky, remember it, and get used to it?

This may seem strange but the facts show that a lack of elementary knowledge in one or another field—I mean now astronomy and physics—is not yet an obstacle for obtaining interesting and important results in these fields. Examples are especially numerous among mathematicians engaged in solving (and very successfully) various physical problems in spite of their ignorance of physics as a whole, to say nothing of numerous details. Analogously, quite a number of physicists, myself included, have written papers in astronomy that are of interest and were published in the astronomical literature without any allowances in spite of the very poor general astronomical background of the authors.

This is my personal opinion and, although I have not had a chance to discuss it with anyone, it would be of interest to know what other people think. This is the first reason why I have entered upon this discourse here. The second is my desire to explain why I consider myself to be an amateur astrophysicist, as can be seen from the title of this article. Finally, the third reason, which is the most interesting to me, is as follows. I wonder how it would influence my work in astronomy if I had a 'normal' astronomical education, i.e. the same as any astronomer who chose this profession as a university student, if not earlier. Unfortunately, it is extremely difficult, indeed almost impossible, to answer such a question. Alas, one cannot start one's life anew. The best of all imaginable ways of obtaining the answer would be to trace the life of identical twins developing under different conditions. But given the absence of not only a twin brother but even analogous examples of other people, I can only make assumptions. I do not, of course, mean a fairly trivial general assertion on the benefit of being well educated and well informed. I refer to quite concrete hypotheses, results, and even discoveries for which I have asked myself whether I could have been the author of some of them. Sometimes the answer was negative, whereas in certain cases I am sure I would have immediately given a correct answer if asked or, if acquainted with the corresponding astronomical material, I would have asked myself this or that question. But can a man who has never heard of neutron stars ask why they can rotate rapidly, possess a huge magnetic field, and be superfluid in some part of them? In short, and this is of course common knowledge, once a question is formulated, the work is sometimes half done. And in order to formulate a question concerning astronomical objects or effects, one must be acquainted with them.

Overall, I have long come to the conclusion that it was a mistake that when I was first entering the field of astronomy, I restricted myself to only the material directly connected with what I was working on (at first it was the Sun; see [5]). I should have eliminated my astronomical illiteracy in spite of all the obstacles that I have already mentioned. I was only 30 years old then and, of course, I could, if I had realized this, have postponed some $(n + 1)$st paper in physics and studied a textbook in astronomy instead. Today, as in the past, some physicists and engineers who are far from astronomy make attempts, say, to measure cosmic gamma-ray emission or to detect gravitational waves. It seems to them that they may remain unacquainted with astronomy as a whole. They are wrong. I know, of course, that such advice is usually not taken but, all the same, I try to convince people that to extend their astronomical horizons and to increase their knowledge is right even within an exclusively pragmatic approach.

17.4   Autobiography II

I shall, however, return to my biography. From 1934 to 1938, I studied quite conscientiously at the physical faculty of Moscow State University. This was a flourishing period for the physical faculty (one that ended with the beginning of the Second World War and the evacuation from Moscow). My sympathy was from the very beginning with L I Mandelstam and his school. (Representatives of this school were I E Tamm, G S Landsberg, and others, including A A Andronov, although by that time he had already moved to Gorky and I made his acquaintance later). The name of L I Mandelstam is little known in the West, although he made great scientific achievements in optics and radio physics. (Suffice it to say that Mandelstam and Landsberg simultaneously and, of course, absolutely independently of C V Raman discovered the effect named after Raman; in the USSR, however, we prefer the term 'combinational light scattering'.) L I Mandelstam (1879–1944) delivered elective courses on various topics that had a wide audience. These lectures were published.[4*] These lectures, as well as subsequent discussions and occasional seminars, were a perfect school of true physics.

In the third or fourth year (the whole course consisted of five years, after which a student obtained a degree), I had to make a precise choice of speciality. The Department of Theoretical Physics was headed by I E Tamm and seemed most attractive to me. But my mathematical aptitude is rather modest, while in theoretical physics mathematics is considered to be not simply important (this is undoubted) but very essential. Therefore, I chose experimental optics (the department was headed by Landsberg) and started working under the guidance of S M Levi. We tried to study the spectrum of canal rays.

In hindsight it is now clear that our experimental potentialities were not adequate to this difficult task. Nonetheless, I received my degree. More interesting were my contacts with Levi. He was a Jew from Lithuania who had worked for a long time in Berlin, if I am not mistaken, in R Landenburg's laboratory. The advent of fascism to power impelled him to leave for Moscow. Jumping ahead, I shall mention that later (in 1937 or 1938) Levi was dismissed from the physical faculty but, fortunately, not imprisoned. He could move abroad and found himself in the USA. In the 1960s I visited the USA three times (1965, 1967, and 1969) and tried to find Levi. Miss Helen Dukas, the former secretary of Einstein, and some other people tried to help me but all in vain. Later, when back in Moscow, I learned Levi's address when I complained to an acquaintance of mine of the bad luck in my attempts to find Levi in the USA. It turned out that this man had, for a long time, exchanged New Year's greetings with Levi. Such tangles are, of course, not surprising. It is more interesting that as far back as the 1930s Levi (and obviously many others) clearly realized the sense and the possible role of induced emission. Levi told me straightforwardly: "Create an overpopulation at higher atomic levels and you will obtain an amplifier; the whole trouble is that it is difficult to create a substantial overpopulation of levels."

As is now known, it is not so difficult to create an overpopulation and, more importantly, by using mirrors the optical pathlength can be extended so that we obtain a laser. Why lasers were not created as far back as the 1920s, I do not understand. Much becomes obvious in hindsight. Perhaps there were some obstacles or simply nobody thought of using mirrors. This idea did not occur to me either and here I definitely cannot explain this by my lack of knowledge. But the grains dropped by Levi have produced shoots. (I do not mention the general influence upon me of this pleasant and educated man.) As mentioned later, at the beginning of the war I was engaged in work on radio-wave propagation and, in particular, paid attention to the decisive role of induced emission for the propagation of radio waves in some ranges of the Earth's atmosphere [6], The point is that O_2 molecules possess magnetic moment and, therefore, in the Earth's magnetic field the O_2 molecule levels (the lower electron levels) split. But the difference between the magnetic sublevels is $\hbar\omega \sim e\hbar H/mc \sim 10^{-19}$ erg. Therefore, at a temperature $T \sim 10$–300 K, the energy is $kT \gg \hbar\omega$ and the sublevels are filled almost equally. As a result, waves with frequencies corresponding to the energy difference between sublevels propagate by means of a continuous alternation of absorption of a wave (radio photon) from a lower sublevel and an induced emission of the same photon from the upper sublevel. If the induced emission is ignored in this situation, the resulting absorption will be $kT/\hbar\omega \sim 10^5$ times stronger than the actual absorption. It is now, of course, a well-known fact that it is necessary in astrophysics that allowance be made for induced emission. But in application to the Earth's atmosphere, my paper [6] was pioneering to the best of my knowledge. However, radio-wave propagation in the ionosphere and atmosphere has long been beyond the scope of my interests, and I shall not dare estimate the role of the paper [6]. This role, I think, was negligible because the resulting effect is small and the paper [6] itself most likely remained unnoticed. In 1942, people had other troubles.

On graduating from the university in 1938, I became a postgraduate student[1] but not everything went smoothly. The war was coming and deferments for postgraduates were rescinded. I was called up and remember receiving a document in which I was called an 'espirant' (because, I think, the word 'aspirant'—a postgraduate student—was associated with the then-popular language Esperanto). The physical faculty still managed to grant deferments to its postgraduate students but in 1938 this happened for the last time. I do not doubt that had I been then inducted into the army. I would not be writing this paper now. Only a few of my university fellows who entered the army survived the war. The situation on the whole, however, was more complicated. Three more times I quite accidentally failed to get in the army: two of the three times I tried to volunteer. All of this, as well as much else that I mention only casually, is not without interest but I am afraid that I have already paid too much attention to my biography and further details would be out of place. (Some of these details, however, will become clear from what follows.)

While awaiting the summons (or, more precisely, the call-up 'to present myself with things'), I did not rush to the black-walled windowless room where I conducted measurements. Instead, I tried somehow to explain the strange angular dependence of the canal-ray radiation that we were attempting to find. (The indications of it found in the old literature were erroneous, as I am now sure.) It occurred to me that the electromagnetic field of a moving charge may play the same role as a photon flux and, in particular, induce radiation. This idea is erroneous, since a charge field is not equivalent to a free field. But in those times the situation was less clear. At any rate, when in the fall of 1938 (if I am not mistaken, this historical event in my life happened on 13 September) I asked I E Tamm a corresponding question, he got interested, advised me to look through the literature, and wished me well. I soon found out that the difference

---

[1] This is a three-year period intended for preparation of a candidate dissertation, which generally corresponds to a PhD dissertation in the West.

between the methods used in classical and quantum electrodynamics gave rise to misunderstandings and obscurities which I partially managed to clarify and that it was not a complicated mathematical technique that was needed but the understanding of the formulation of the problem. This is how I became a theoretician and left experimentation for life. Obtaining these (maybe modest) results on my own, I realized that I could fruitfully work in theoretical physics. This had a great effect on me and my new life began. In this respect, I am much obliged to I E Tamm and his attitude toward people. I do not write about I E Tamm in more detail here, since the interested reader may refer to my reminiscences about him [7]. Here I only emphasize how important a friendly atmosphere and mere kindness are for many (although not all) beginners (for more details, see [7]). Neither do I dwell on the content of my first three papers devoted to quantum electrodynamics and published in 1939. The corresponding references can be found in my book [8] and paper [9], where the subject itself[2] is also considered, of course.

In 1934, S I Vavilov and P A Cherenkov discovered the phenomenon that is now known as the Vavilov–Cherenkov effect or radiation. (True, in the West, it is most often referred to as Cherenkov radiation but many Soviet authors (I among them) who know well the history of this discovery use only the term Vavilov–Cherenkov (VC) radiation.)[5*] The classical theory of the effect was formulated in 1937 by I M Frank and I M Tamm, who showed that a charge uniformly moving in a medium at a velocity $v > c/n$ (where n is the refractive index of the medium) faster than that of light must radiate. This happened already within my memory and it was, thus, very natural that while studying quantum electrodynamics, I constructed in 1940 the quantum theory of VC radiation as well as considered this effect in an anisotropic medium. Since then, the radiation of uniformly moving sources (including, besides the VC effect, the transition radiation and the Doppler effect) has become one of my favorite fields. Yes, I like all of these effects, they seem beautiful to me. Of course, the definition of 'beautiful' is always subjective. For example, L D Landau (1908–68) was almost absolutely indifferent to the previously mentioned effects. By the way, I regard L D Landau, as well as I E Tamm (1895–1971), to be one of my teachers. Like Tamm, Landau was an outstanding personality (see the book of reminiscences about Landau [10], which includes my papers).[6*] Therefore, it may not be out of place to narrate, as an illustration of what has just been said, how we first met (and, in fact, got acquainted) on scientific grounds. At one of the joint seminars (for more details, see [10, 11]), Tamm mentioned the quantum theory of VC radiation formulated by me,

---

[2] Here and in what follows, I try to refer to more recent and accessible literature, especially in what does not concern astronomy. For greater details about my papers, especially those noted earlier, see the article in *O Fizike i Astrofizike* [*On Physics and Astrophysics*] [2001 *The Physics of a Lifetime* (Berlin: Springer)] [7].

which Landau immediately commented on the effect that this is absolutely unnecessary because the effect is classical. As was usual in such cases, Landau's criticism had serious grounds—the quantum corrections in the theory of VC radiation are of the order of $\hbar\omega/mc^2$ (ω is the radiation frequency and m is the mass of a radiating particle) and are small because at high frequencies ω, the refractive index $n(\omega)$ tends to unity and VC radiation disappears. However, the quantum theory of VC radiation is, in fact, not of narrow methodical interest only, since it also substantially clarifies the situation when applied to the Doppler effect in a medium (for a discussion of this and practically all my papers devoted to the theory of uniformly moving sources, see [8, 11]).[5*]

The papers mentioned here were written within about a year and a half and were the basis for my candidate dissertation, defended at Moscow State University in May 1940. A candidate dissertation and the corresponding candidate degree in the USSR correspond to the PhD in the West. Of course, the level of candidate dissertations is sometimes very low but in physics and at good institutes and universities, the level of candidate dissertations seems to me to be rather high. We also have another degree—the Doctor of Sciences (DSc). In this case, and again in the proper places, a doctoral thesis is usually defended by an author of dozens of papers, quite an independent research worker not younger than 30 to 40. The Doctor's degree gives one the right to hold a professor's post, although there are few professors in our country without a Doctor's degree. At that time (1940), the USSR Academy of Sciences had special vacancies for those who were to prepare a Doctor's dissertation within three years. There was one such vacancy in the Department of Theoretical Physics at the P N Lebedev Physical Institute of the USSR Academy of Sciences (FIAN), which I filled on 1 September, 1940. Ever since, I have been working in this department, which was founded by Tamm in 1934. (Until mid-1941, he had also been head of the Theoretical Department at the Physical Faculty of Moscow State University.) Before entering FIAN, I had already been engaged in the theory of particles with higher spin states. This work continued for many years (partially in collaboration with I E Tamm and also with young colleagues). I shall restrict myself to mentioning only my most recent papers [12, 13] that touch upon this problem.

As is well known, Germany invaded the USSR on 22 June, 1941. Several cities had already been bombed for many hours when, at about midday, Molotov announced over the radio that war had broken out. I clearly remember listening to his speech with my two-year-old daughter on my lap. I also remember Stalin's speech on 3 July, when this dictator, who had covered the country with blood and tears, called his audience 'brothers and sisters', I presume for the first and last time. I was a 'private untrained' and, therefore, was not mobilized. But for people like me and for all those who were not mobilized, a people's volunteer corps was organized. Soon a great many of these people were killed or captured near Moscow. Of course, I immediately joined the volunteer corps and spent the whole day in a school building, where we were 'formed', and in the evening dismissed with an order to come 'with things' at the first call. But soon there came a decision to evacuate the Academy of Sciences and, if I am not mistaken, exactly one month after the war broke out, my elderly father (he was 78 years old), aunt, wife, and I left for Kazan. (My only daughter had been evacuated with her grandmother some time before.) Before our departure, Moscow underwent only one air attack by the Germans. When we heard the sound of the alert, my wife and I happened to be near a metro station, where we had to spend that night. In the morning, I saw only traces of the attack. I shall note that in Kazan, I had to join the 'labor front'—we dug trenches not far from the town, which fortunately were not used. I had no deferment ('reserve') because, in those times, it was intended only for people of higher standing. But having had no military training I was of no value to the army and, in any event, I was not called up. While waiting for the summons, I wished to finish my Doctor's dissertation as soon as possible, and, in May 1942, I defended my thesis on the theory of higher-spin particles. I have no reason for considering my thesis to be weak and, indeed, it was highly estimated. Under peaceful conditions, however, I would not have hurried to get my Doctor's degree. But in the situation described, and taking into account the fact that I changed my subject of research (see below), my haste was justified.

By the way, the defense of a thesis did not have any influence, as far as I know, upon one's being called to military service, at least in those extremely hard times—by that time, the Germans had reached the Volga. Furthermore, I was invited to volunteer to become a paratrooper and I tried to do so but was rejected because of the state of my health under an anecdotal set of circumstances. In Kazan, our institute occupied a floor in the wing of the Kazan University building and there were no more than 100–200 people in the institute. We were very often sent to fulfill different kinds of jobs, of which I remember unloading a barge of logs on the Volga. Everybody worked, including Tamm, Landsberg, and others who then seemed to me to be very old people although they were no more than 50. As for me, I carried big logs 'on a crate'—this is a setup in which straps are put on like a rucksack and on one's back a 'ground' is then fixed on which a log or something else is put. Such a contrivance was widely used in Russia for carrying bricks, etc. I was not at all mighty (180 cm tall and weighing about 60 kg at that time). Using this contrivance, however, I carried rather heavy short logs that two men could hardly put on the 'ground'. But the load was apparently too heavy and one time I found blood in my mouth—a small vessel must have broken. Tuberculosis was suspected and I was sent to the dispensary, where they found some 'petrificated foci', after which I was registered. Therefore, I was regarded as unfit for the landing force. I am not going to play the hypocrite and say that I regret this. But at that time, I felt that it was better to die in a battle than to find myself under German occupation and then in a death camp.

Life in Kazan was not easy, of course. My family lived in a small room in the university hostel where the temperature in winter fell below zero. My father could not endure such a life and died in mid-1942. The Academy was supplied with some food and thus we did not actually starve but we did feel hungry all the time. I remember a canteen where waitresses served pea porridge for a corresponding recompense. People came in with a tin stuck into a briefcase, filled the tin with porridge or soup, and then carried away the briefcase as if it were filled with papers.

At the end of 1943, the members of FIAN began returning to Moscow as victory approached. Later I narrate another important chapter of my biography so now I shall only note that, on returning to Moscow, the theoretical department started growing. For several years, I was Tamm's deputy and, after his death in 1971, I had to become head of the department. I say 'had to become', for I did not want this position but in the Academy it was necessary "for the welfare of the department staff" that its head be a man of position. And in 1971 there were only two Full Members of the Academy (Academicians) in the department: A D Sakharov and myself. Sakharov had already become a dissident and was not suitable as a chief, so I had to be head of the department for 17 years. A new rule has recently been introduced obliging all members of the Academy beyond the age of 70 to resign their administrative posts with the right to become a counselor (adviser) without a decrease in salary. Of course, to be head of the department that now numbers nearly 60 workers, including three full and four Corresponding Members of the USSR Academy of Sciences (the same number as in the rest of our great institute), required time and strength. But I do not complain, for our department is known for its friendly terms and very kind-hearted atmosphere, and during the lifetime of the department (54 years!) we have not had a single serious conflict.

Since 1943, I have been engaged in the theory of superconductivity and have written a number of papers in this field: the most well-known among them [14] (in collaboration with L D Landau) was published in 1950. I have also been concerned with the theory of ferroelectrics, superfluidity, and many other things, including astrophysics, which will be a special subject here. But all this was a little later. In spite of all the warnings, the coming war was not openly regarded as inevitable in the USSR. Therefore, to the best of my knowledge, no direct preparation for the war was conducted by the Academy. In any event, when the war broke out on 22 June, 1941, physicists were, in general, out of it. Suffice it to recall that even I V Kurchatov set himself the problem of the demagnetization of ships against magnetic mines and it was only at the end of 1942 that he was appointed to head the nuclear program. As has already been mentioned, I was occupied in an absolutely abstract analysis of equations for higher-spin particles. As the war began, we all started searching for applications of our abilities closer to practice, if not to defense. Accidentally I was given advice to concern myself with radio-wave propagation in the ionosphere, which was supposedly an important problem for defense. I followed this advice and spent much time on this subject.

17.5   Lateral associations

There seems to be a special type of person with 'lateral associations'. I belong to this type, in that I am constantly developing various associations that require comment.

I have already mentioned the title 'Academician', the use of which in the West has always surprised me and aroused an unpleasant feeling. I first experienced this feeling many years ago, when in the list of participants of some congress I came across the names of Professor P Dirac, Professor N Bohr, and Academician X. Why a special title for a Soviet scientist? Does it not suffice to be a Professor, especially in the same row with Bohr and Dirac?

In the USSR Academy of Sciences, there exist two 'degrees' (titles): Full Member (or Academician) and Corresponding Member. Unfortunately, both these titles are often used in the literature. I think that this is the influence of the pre-revolutionary habit not to forget to mention the title of count or duke, if not grand duke. The German influence, whereby a colleague is thanked as, say, Herr Geheimrat Professor Doctor X, was also strong. The progressive tendency in our country is now to omit titles, say, in acknowledgements (acknowledge X, Y, or Z but not Professor, Doctor, or Academician X, Y, or Z). In the West, as far as I know, such a style has already become conventional. In any case, I am a resolute opponent to the use of the Academician title in the English literature. However, one cannot totally abandon titles in the USSR, because otherwise some authority will not answer the telephone, among other things.

Since I have already touched upon the subject, it seems reasonable to say some more about the USSR Academy of Sciences. (We have numerous other academies as well.) This academy numbers about 800–900 members, of which about one-third are Academicians and the rest are Corresponding Members. Such a structure was created before the revolution of 1917. Merely for the title, an Academician now receives 500 roubles a month (without any duties) and a Corresponding Member receives half this sum. These are rather large sums if one takes into account that my salary as head of a department or counselor is also 500 roubles, while a young candidate of sciences gets 150–200 roubles a month. Members of the Academy have some other privileges as well; for instance, an Academician may call a car with a driver for several hours a day. This may greatly surprise some readers but one should remember that the salary of 500 roubles a month is often much less than the salary of a man of equal qualifications in the West. As for the use of a car, some time ago the members of the Academy were fewer in number than now, the majority of them were elderly people (the average age of Academicians in 1985 was still 69.9), and there were very few private cars.

I am convinced that, in principle, the only correct thing to do is to abolish all material privileges for members of the Academy. But, in our society, this is evidently a long way off.

Academic privileges have, naturally, a very pernicious effect upon the elections: not only scientists but also designers (engineers), high-ranking officials, etc, try to join the Academy. Despite this, the USSR Academy of Sciences, especially in the exact sciences, is on the whole a focus of the most qualified people of this country. Besides, the Academy does not act exclusively on the basis of arbitrariness but is still guided by a charter that provides for a competitive ballot and not 'one place – one candidate'-type elections. However, until recently the charter has not formally but in essence been repeatedly violated. (For example, special vacancies were allotted for candidates welcome to the authorities.) But under the conditions of the deep changes now taking place in my country, I hope that the charter will be made more precise and, most importantly, observed. The Academy and all of Soviet science have, however, problems that are no less important. One of them is the necessity to fight against fantastic bureaucracy. It would be out of place to dwell on this subject here, the more so as I can refer the interested reader to my paper [15] that elucidates this problem (for those who wish to have insight into the state of Soviet society in mid-1988, I advise to get acquainted with the whole of the volume containing [15]).[7*]

Another 'association' arises in connection with the fact, mentioned at the end of the preceding item, that I am the author of a large number of works and a still larger number of papers. By 'work', I mean some scientific results in the form of a paper or papers that I have registered in a special list. Long ago, when I was defending my theses, by some formal requirement I had to start such a list and I have since continued it voluntarily because it is convenient for references. The list now numbers 325 items and one item often includes several papers devoted to one problem. A number of notes and reviews, popular scientific papers, etc, are not included nor will the present paper figure in the list, although it was more difficult for me to write this one than a whole number of scientific papers. The total number of my publications probably approaches 1000 and the number of my 'works' is also large. What does this prodigality signify? Is it good or bad? I have had to face such or similar questions and opinions in this connection are very diverse. For example, I have been reproached for publishing papers very easily. There also exists an opinion that authors who publish many papers do so for the sake of glory, for increasing the number of references to their papers, etc.

It is quite obvious that a large number of publications or even 'works' cannot in itself be put to one's credit. One significant work may, of course, be more valuable than 1000 weak papers. So in estimating the contribution of one or another author, the decisive role is undoubtedly played by the content of his publications. All other conditions being equal, the number of publications characterizes basically the style of the activities and tastes of the author. Personally, I write rather easily if, of course, I have an idea of what I am going to write about. Moreover, I do not feel satisfied until the paper is written; of course, dissatisfaction often remains afterward but, nonetheless, the very process of writing is, at least for me (and I think for many people), an important element of the work itself. With some exceptions, a paper is written not as a duty but as a result of some interest. And if the author him/herself is interested, why not share this with others and send a paper or a note to press? As for those who write little, there exist different cases. Some merely have nothing to write about; others find it difficult to write or believe that they have already reached such a level that publication of a not very important result, a popular paper, etc, will add nothing to their scientific reputation and may even arouse ironical smiles. I have long been well aware of the fact that many of my publications only give rise to criticism. But I have no fear for the 'clamours of Boeotians', for I do not want to be led by snobs. Sometimes there are mistakes and I regret having sent a paper to press but idlers alone never make mistakes. Finally, tastes and opinions are diverse (a very typical example is given in [3]). The author must heed his/her inner voice and not try to please everybody. If a paper is not interesting to someone, s/he can choose not to read it. This, by the way, consoles me with respect to this paper, because I realize that some may not like it.

The value of the works and their quality is an important question in scientific activities. For young people who are only beginning their scientific life, it is even of vital importance but, irrespective of age and position, one wants to hear a just opinion. An unambiguous and reliable way of judging the merits of works probably does not exist. In the West, one common way of estimating papers is based on the citation index. This method is seldom used in the USSR and is, in general, unpopular but I do not think that there is enough reason for such a negative attitude. The citation index may, of course, absolutely distort the picture. For instance, an erroneous but sensational paper may give rise to a lot of references. Another source of errors is that only the first author's name is taken into account.[3] Finally,

---

[3] By the way, I, along with the majority of my colleagues in the USSR, use only an

papers published in Russian are not as frequently cited as those appearing in English in well-known journals. But information from the citation index is, nonetheless, interesting and significant. True, I do not look in the index. (The library of our institute does not have it, and if I am not mistaken it is available only in two places in the whole of Moscow.) But once in a Russian publication, I came across a listing [16] of the 249 most frequently cited scientists in the world, based on the overall number of references to their papers in 1961–75, i.e. within 15 years. In this list, L D Landau was in the second place (18 888 references) and five other Soviet authors were mentioned (four physicists and one chemist), of which I had the largest number of citations (6834 references or an average of 456 references a year); overall, I was in 66th place. In subsequent years, the number of references to my papers has become somewhat smaller, which is natural. (In recent years the papers that I have published have been less in number and maybe also in quality; I note, however, that some of my old papers, especially [14], are now most frequently cited in the text without any mention in the list of references.)[24]* Unfortunately, I do not know which place is taken by references to my astronomical papers (including cosmic-ray astrophysics).

In spite of all the limitations of conclusions that can be drawn from the citation index, it would nonetheless be of interest to have such information with respect to astronomical themes. (True, it is not always easy to decide to what branch of science this or that paper should be ascribed.)

17.6   Autobiography III

In 1985, the Soviet Union entered a new epoch, one of the most important and distinctive features of which is *glasnost* (openness). We now (especially beginning with 1987) read books and articles in magazines and newspapers whose publication was unthinkable some time ago. Possessing, to say nothing of spreading, such literature would lead to arrest and many years imprisonment. *Glasnost* also allows me to dwell here on the part of my biography which I could not write or speak on publicly before. The ignorance of some foreign colleagues with respect to life in the Soviet Union has often amazed us, especially as quite a lot of materials have already been published fairly sufficient to remove such 'illiteracy' for anyone who can read. But not everybody is fond of reading and, for astronomers, a concrete example concerning an 'amateur' astrophysicist but still one of

alphabetical ordering of names for my publications. (I have written only two or three papers in which this order was more or less accidentally violated.) One should bear in mind, however, that the Russian alphabet differs from the Latin one. In my case, it is particularly important that the letter Γ (G) is the fourth in the Russian alphabet and the third letter of the Russian alphabet (B) corresponds to V or W. Also, the Russian Φ (F) is toward the end of the alphabet. Therefore, when a paper is translated from Russian into English, the alphabetical ordering often appears to be violated.

their colleagues may be of some additional interest.

When the war ended in 1945, physicists in the USSR were already held in great esteem. The existing institutes grew rapidly and new scientific institutions and high schools appeared. As far back as the 1930s, at Gorky University there was a strong group of physicists and mathematicians, the most outstanding of which was A A Andronov. This group decided to organize a specialized radio physical faculty. Since they were short of their own experts in this field, they invited three Professors from Moscow (from FIAN) who were expected to come to Gorky from time to time to deliver lectures. I was one of these three and was put in charge of organizing and heading the department of radio-wave propagation because, as I have already mentioned, I had been occupied with precisely this problem since the beginning of the war and had already published a number of papers. I remember, at the end of 1945, arriving in Gorky by train (it is a night's journey) in far from comfortable conditions and hiring a fellow to carry my suitcase by sledge (we were walking) from the station to the central part of the town, where the university and the hotel were located. At the beginning, if I am not mistaken, there was only one colleague (M M Kobrin) in 'my department' and one strange student, who later left the university. But soon there appeared capable students, followed by postgraduates. Many of those who graduated from the department have long since obtained their Doctor's degree and the department continues to exist. That I headed this department under hard postwar conditions now seems to me to have been a venturesome enterprize. But I was 29 then and, as a doctor of sciences, had the right to be at the head of a department. I was willing to teach youngsters but in Moscow it was difficult to realize this ambition. I think I would have left Gorky in a couple of years, as did my colleagues from Moscow, had fate not willed things differently. I met there my present (second) wife Nina Ermakova and, in 1946, we got married.

I would not, of course, even mention here my family life if it were not for some special circumstances—namely, that Nina was, in fact, in exile in Gorky and had no right to move to Moscow. In short, the story (which has even been mentioned in some publications) is this. Her father was arrested in 1938 and died in Saratov's prison in 1942. (He was in the same prison as the well-known geneticist N I Vavilov, who died there during the same period.) A student of the mathematical faculty of Moscow State University, Nina was arrested in July 1944 and along with her fellows (some of them, as she herself was, were the children of persecuted parents) accused of plotting an attempt on Stalin's life—Stalin was supposedly to be shot from the window of her flat in the Arbat.[8*] But the 'scriptwriters' from the KGB had never taken the trouble before the arrest to check everything and only later was it established that the windows of the room where Nina lived with her mother did not face the Arbat. For all the nonsense of the accusations brought by KGB, the investigators somehow tried not to make easily refutable assertions. In any event, the accusation of terror was remitted and there remained 'only' the accusation of counter-revolutionary group anti-Soviet activities (Clauses 58.10 and 58.11 of the then Criminal Code). She spent nine months in prison and in March 1945, without any trial and by the decision of the so-called special assembly, was sentenced to three years' confinement in a camp. This was a very 'short' term for a person convicted under Clause 58. Maybe it was for this reason that when an amnesty was announced on the occasion of the end of the war, even prisoners charged under Clause 58 were given amnesty. (The majority of those charged under Clause 58 had longer terms and were not included in the amnesty.) Thus, in September 1945 Nina was released but without the right of residing in a number of large cities.[4] She had an aunt in Gorky and, therefore, she chose Gorky but, nonetheless, she could get registered only on the other bank of the River Volga in the village of Bor. Despite this, she managed (and this was non-trivial—kind people helped her) to enter the Polytechnical Institute in Gorky, from which she graduated in 1947. Until 1949, she lived illegally in my room but, at the end of 1949, she was registered in Gorky by A A Andronov's application. (This happened after a large accident on the River Volga on 29 October, 1949, in which a ship carrying people from Gorky to Bor was wrecked: out of about 250 passengers, Nina was among some dozen or so who survived.) I, naturally, handed in an application each year (it could not be done more frequently) with a request to allow my wife to move to Moscow but all in vain. It was only in 1953, after Stalin's death, that a new amnesty was announced and Nina could leave for Moscow at last. In 1956 she as well as all her fellows from the 'anti-Soviet group' were completely rehabilitated—i. e. the accusation was recognized as fully groundless. To characterize the rehabilitation process, I will say that an inspector came to Nina's mother's place to draw up a statement, in the presence of witnesses, that the windows of the room did not face the Arbat.

In addition to the reasons that I have already mentioned, I narrate this story to explain why I taught students and conducted research work in Gorky for so long. I had students, postgraduates, and colleagues there and much research was conducted. Therefore, after 1953 I continued, although more and more seldom, to visit Gorky. I headed the department, I believe, until 1961 and the last two times (in 1980 and, if my memory does not fail me, at the end of 1983)[9*,25*] I went to Gorky mainly to see Sakharov, who, by an irony of fate (at least I feel so) was also exiled to Gorky.

In 1942 I became a candidate member of the CPSU (Communist Party of the Soviet Union) and, in 1944, a CPSU member. The war was in full swing then and today, when we know what Stalin's true face was, it is

---

[4] In order to live in a given place, Soviet citizens must be registered there by the police and get visas in their passport. Amnestied people were given passports with limitations for visas.

even hard to believe that many millions of people, including myself, were absolutely blind about him. Even when I learned the 'criminal' story of my wife and her fellows, I did not think yet that everything came from the 'great leader'. My eyes were opened wide only after N S Khrushchev's report on 25 February, 1956, at the 20th Congress of the CPSU.[10*]

If I continued writing my biography in further detail, it would, of course, overstep the admissible page limits of the present paper. Therefore, I shall restrict myself only to a few more remarks. By an unwritten rule, CPSU members were not, of course, supposed to marry 'counter-revolutionaries'. In addition, on 4 October, 1947, in the *Literaturnaya Gazeta* an article appeared in which, on the initiative of a certain physicist D Ivanenko (he wrote denunciations of many others as well, including Tamm), I was accused of 'servility' and 'cosmopolitism', i.e. admiration for bourgeois science. This gave me much trouble and still more was to be faced. I think it would have cost me my head if it were not for I V Kurchatov, who invited I E Tamm in 1947 to take part in nuclear research.[11*] Tamm, in turn, enlisted me, A D Sakharov, and a number of other workers in our department. Soon I made an important proposal; another was made by Sakharov. Unfortunately, even now that the USSR and the USA exchange military observers and the Secretary of Defense of the USA has recently examined the new Soviet armaments, these 40-year-old works still remain classified.[12*] Therefore, so as not to get into new troubles (I have had enough of them in my life), I shall only say the following. Tamm, Sakharov, and some others left for far-off lands sometime in 1948.[11*] As one that aroused suspicion (the wife, you know, was in exile), I remained in Moscow but a sentry was posted at the door of my office. I knew no real secrets and was not at all interested in them but, in 1950, on the initiative of A D Sakharov and I E Tamm, research work in the field of controlled thermonuclear synthesis was begun. At the first stage, this was, of course, a physics problem and I was also engaged in it and obtained some results. By 1952 (or the late 1951), this work was considered so important and secret that I was altogether removed from it.[26*] Maybe it was for this reason that after this problem was declassified (this was due to Kurchatov, who made his well-known report on the thermonuclear problem in England in 1956), I decided to publish my old reports [17] in 1962.

As is well known, in 1952 and early 1953, the situation in my country became still more heated. Stalin went absolutely mad and the vivid evidence of this was the notorious 'case of the physicians'. New monstrous persecutions came. Fortunately, on 5 March, 1953, 'the greatest man of genius of all times' was finally gathered to his fathers. Many rapid changes followed in the country, the 'case of the physicians' was disavowed, and they were released, the next amnesty was announced (I have already mentioned it), and Beria—one of the closest bloody assistants of Stalin—was shot. It was noted that the USSR Academy of Sciences had not called elections since 1946 and, late in 1953, the elections were finally held. I was then elected a Corresponding Member. At the end of the year (or in the early 1954), I was awarded the order of Lenin (the highest civil order in this country) and the Stalin Prize of first degree (later this prize was renamed the State Prize, and three degrees were abolished and reduced to only one) for the previously mentioned 'secret' works. In general, by 1954, from a half-disgraced man, I had become a person. Since then, my life has been generally normal. From 1962 to 1970, I could even frequently go abroad, sometimes with my wife (this was a rare thing for Soviet citizens but an exception was made for Full Members of the Academy; I was elected to be an Academician in 1966). Since 1971, true, I have gotten into new troubles but these were not a threat to my life and, therefore, I shall say only a few words about them. First, I was almost never allowed to go abroad. For instance, I could not deliver my Darwin Lecture myself. (It was delivered at a meeting of the Royal Astronomical Society from a text that I had submitted beforehand [18].) Only after several indignant letters sent by me 'to the very top' did I manage to go to several international conferences but each time with such nervous strain that I would not wish it on my enemy. Only since 1987 have I been going abroad without extreme difficulties, although, as is quite a usual thing with us, with a great many bureaucratic impediments (see [15]). Second, complications arose owing to the fact that in 1969, A D Sakharov again started working in our department. I refused to sign any letters condemning his activities and, generally, the attitude toward him was quite loyal in the department, even when in 1980 he was exiled to Gorky. On our initiative, he remained a worker in our department and, with permission from the administrators, his colleagues from the department went to see him there.[9*] When, at the very end of 1986, Sakharov could finally return to Moscow, he appeared in the department the very day of his arrival and was heartily greeted. He is, of course, working in our department now. Being head of the department, I had certainly to deal with 'the Sakharov case'. I think that I have done my best and I have no reason to reproach myself but, on the whole, it is not up to me to judge.

In general, as can be seen from the aforesaid, it was only in 1987 that I got out (is it forever?) of 'external' difficulties. In return, by virtue of what is called the 'law of conservation', my so to speak 'internal' difficulties increased. The point is that beginning at the age of about 65, it has become more and more difficult to work fruitfully, although I try to overcome this obstacle (see [3]).

Concluding this section, it only remain to answer one, obviously natural question: what was the influence of my severe trials upon my scientific activities? A simple answer suggests itself—under better conditions I would have managed to do more. But, frankly speaking, I am not sure of that. During all of my 'scientific life' (since 1938), I could almost without interruption be engaged in anything I wanted and not worry about earning my bread, even though it was not always with butter. This was number one. Number two, and this is also typical of many of my colleagues, science and scientific activities occupied a predominant place in our lives: we simultaneously work, play, and rest (and even a narcotic). I think that had I lived under better conditions, I would have probably been happier and have rested and seen more. But the integral of my scientific activity, if I may say so, would not most probably be larger than it is.

17.7   Radio astronomy[5]

As has already been mentioned, in the middle of 1941, I turned to the study of radio-wave propagation in the ionosphere and generally to plasma physics. This activity is reflected in the monograph [19] as well as in [8]. It was, in fact, the ionosphere that was the starting point for my astronomical and, more precisely, radio-astronomical studies.

The well-known Soviet physicists and radio specialists L I Mandelstam and N D Papaleksi contemplated the problem of Moon radiolocation (radar) long before the Second World War. In 1944, stimulated by the progress in radiolocation, N D Papaleksi returned to this idea; and he also considered, naturally, radiolocation of the planets and the Sun. In this connection, at the end of 1945 or the beginning of 1946, he asked me to clarify the conditions of radio-wave reflection from the Sun. Certainly, it was essentially a typical ionosphere problem and I had all the formulae at hand. The results of the calculations did not seem to be very optimistic, since for a large set of parameters, such as electron concentration and temperature in the corona and the chromosphere (which then remained unknown in many respects), radio waves should be strongly absorbed in the corona or the chromosphere, and not reach the level of reflection. (The reflection due to inhomogeneities was not considered and the 'point' $n(\omega) = \sqrt{1 - \omega_\mathrm{p}^2/\omega^2} = 0$ was, roughly speaking, taken as the level of reflection.) But this was immediately followed by a more interesting conclusion: The source of solar radio emission must not be the photosphere but rather the chromosphere and, for longer waves, the corona. At the same time, the corona was already assumed to be heated up to hundreds of thousands or even a million degrees. Thus, even under equilibrium conditions, i. e. in the absence of perturbations and sporadic processes, the temperature of emission from the solar corona for waves longer than about a metre must reach approximately a million degrees at a photospheric tempera-

---

[5] In this and the following sections, I have used material from and, sometimes, even the exact texts of my papers [1, 2]. Since these papers are easily accessible, I refer those (probably very few) interested in the details to [1] and [2]. A number of references omitted here can also be found in these papers.[13*]

ture $T_{\mathrm{ph}} \approx 6000$ K. All this was presented in my first astronomical paper [5]. In the same year, analogous conclusions were published by Martyn [20] and Shklovsky [21]. (I can only say that my paper [5] was submitted for publication on 27 March, 1946, whereas the dates of submission of the other two papers are not indicated; they appeared, respectively, in *Nature* on 2 November, 1946 and in the November–December 1946 volume of *Astronomicheskii Zhurnal* [*Astronomy Journal*].[14*] In the calculations, I used absolutely clear and reliable formulae known from the theory of ionospheric wave propagation [19]. Martyn did not present the formulae [20] but he evidently acted in the same way. As for Shklovsky, he believed [21] that the absorption due to 'free–free' transitions and the absorption due to electron–proton collisions should be taken into account separately and then summed up, whereas, in reality, these represent one and the same mechanism [19]. However, this circumstance was of no importance, since the parameters of the corona were not exactly known at that time.

In the paper by Pawsey, published immediately after Martyn's paper [20], the existence of thermal radio emission of the corona at $T \sim 10^6$ K was confirmed in which such emission was shown to play the role of a lower limit that is reached as soon as the sporadic component of solar radio emission becomes sufficiently weak.

A weak point of radio astronomy of that period was a low angular resolution that prevented investigations of the Sun even for regions of sizes of arcminutes—it is difficult to believe this now that the angular resolution of radio interferometers far surpasses that of the best optical telescopes. In this connection, N D Papaleksi suggested measuring solar radio emission during the total solar eclipse of 20 May, 1947, with the help of a 1.5 m wavelength antenna which installed on board a ship and had a wide directivity pattern of several degrees. These measurements were successful (see the literature cited in the book containing [1]) and appeared to be the first of their kind. Whereas the intensity of optical emission from the Sun decreases during a total eclipse by several orders of magnitude, at a wavelength of 1.5 m the intensity during the eclipse decreased by no more than by 60%. Thus, metre-wave radio emission was proven to come from the corona which remains uncovered by the Moon even during a total optical eclipse.

I took part in the Brazil expedition of the USSR Academy of Sciences, on board the ship "Griboyedov", in which radio observations of the Sun were conducted. It seems that I was included on the staff of the expedition in recognition of my work in the field of radio astronomy, which was then only beginning in the USSR. I did not take part, however, in the measurements themselves—they were carried out on board the ship while the main part of the expedition made its way inside Brazil to take optical measurements, which were, unfortunately, unsuccessful because of bad weather. This main part of the expedition also included a small ionospheric group headed by Ya L Al'pert. (I was in this group but the weather could not, certainly, prevent ionospheric measurements.)

The previously mentioned activities drew me further into radio astronomy and I became for a while almost a professional radio astronomer. (I tried to acquaint myself with all the available material, methods of measurement, etc.) As a result, I wrote two reviews of radio astronomy [22, 23], which were possibly the first in the world literature. (I cannot, however, vouch for this.)[6] Now that 40 years have passed, it is difficult for me to judge the value of these papers and I do not feel like analyzing them in more detail. I shall only mention the suggestion that I made in [23] that the radio-wave diffraction on the Moon edge should be used to increase the angular resolution of details on the Sun during eclipses. This question I considered in further detail in my paper with G G Getmantsev (see the reference in [1]), where we were already thinking more of discrete sources of cosmic radio emission than of the Sun. The method of diffraction of radio emission on the Moon edge has since been widely used and, therefore, I have nothing to add here except the following. Already in [23], I mentioned the possibility of enhancing the angular resolution still more by observing a source located on the line that joins the Moon center with the point of observation. (Here we are evidently dealing with the Arago–Poisson light spot.)[7] The non-spherical shape of the Moon and the necessity of having a source on or very near to the indicated line strongly hamper such observations, of course, and I am unaware of any attempts to exploit this method. But maybe one should, nonetheless, analyze such a possibility in more detail, not only for the Moon but also for the planets and their satellites and artificial screens (both flat and spherical). Perhaps this has already been done and I do not know about it, for I have not looked through the corresponding literature.

If I tried to dwell on my subsequent papers in so much detail, it would take too much space. My work in astrophysics has been rather sporadic and chaotic and the portion closest to radio astronomy may somewhat conditionally be divided into three main trends:

---

[6] How far I still remained from astronomy as a whole, in spite of what has been said, can be seen from my note [24] in the volume dedicated to the 80th birthday of J Oort. There, I recount that on the way back from Brazil, the participants of our expedition were lucky enough to visit Leiden, although quite accidentally. And while there, instead of taking the opportunity of meeting J Oort and generally taking part in the discussion of astronomical problems, I rushed to the Kamerlingh Onnes Cryogenic Laboratory because I was then most interested in low-temperature physics.

[7] As is known, as an objection to the wave theory of light, Poisson pointed out a consequence of this theory that seemed to him quite absurd: On the axis of the geometrical shadow of a round opaque screen a light spot must be observed. (The source is considered to be point-like and is positioned behind the screen on the axis perpendicular to its plane.) The experiment conducted by Arago immediately after this confirmed the existence of the light spot. For an opaque sphere as screen, the conditions for observing a central spot are facilitated compared with those for a flat screen.

(1) Twinkling of radio sources in and beyond the ionosphere, oscillations of the intensity of solar emission, the use of polarization measurements, and satellite measurements.

(2) The theory of sporadic emission from the Sun. V V Zheleznyakov and I began studying this range of questions in 1958 [25]. A number of other papers followed but it would be irrelevant to refer to them here, since the corresponding results (with references) are fully given in the book by Zheleznyakov [26].

(3) The theory of cosmic synchrotron radio emission, and its connection with the problems of the cosmic-ray origin and with high-energy astrophysics as a whole. This was the subject of my main astrophysical interest and I still work in this field, although not so much as before. From the point of view of historical information that I can offer, these problems also play the most important role. Therefore, they are discussed in the section to follow (for more details, see [1, 2]).

I shall now dare to mention some more of my papers close to radio astronomy. The origin of radio galaxies and, specifically, the question of the energy source of their radio emission were not immediately clear. As an example, one can mention the hypothesis of colliding galaxies, which proved incapable of explaining the mechanism of energy output in radio galaxies. Another suggested idea was that of a sharp increase in the number of supernova flares in radio galaxies, an idea that has always been rather groundless. Therefore, in my opinion, the note [27] was not without value that the required energy output and cosmic-ray acceleration in radio galaxies are, in principle, easily provided by gravitational energy—in particular, in the process of star formation. I pointed out that "it seems more attractive to associate the galactic flares not with supernova flares but with another large-scale mechanism, for example, the gravitational instability of a galaxy or of its central part, schematically outlined above". This trend of thought continued to a certain extent in later papers devoted to quasars (see [28] and the literature cited therein, as well as [29]).

The discovery of pulsars gave rise to the temptation to clarify the mechanism of their radio emission. V V Zheleznyakov and I (and partially V V Zaytsev) published several articles on this subject, the last one where I participated being the review [30]. But the problem turned out to be much more complicated than it seemed at first. (A problem happens sometimes to be particularly sophisticated when this is not suspected in advance.) For this reason, I decided long ago to leave this field and I am sure I was right to do so. Only representatives of the younger generation (or even generations) were able to gain a real understanding (though not to the bottom) of this very interesting but essentially complicated and many-sided range of questions—the mechanisms of pulsar radiation and the whole theory of pulsar magnetospheres (see [31]).

17.8   Synchrotron radio emission, cosmic-ray astrophysics, and gamma-ray astronomy

I shall now proceed to cosmic-ray astrophysics and gamma-ray astronomy, which are very important areas of astronomic research, in which I have taken part. My activity in these fields started with the theory of synchrotron radio emission.

It is reasonable to comment first on the terminology, which had not yet become conventional. Cosmic rays are now usually understood as charged particles (protons, other nuclei, electrons, etc) of cosmic origin that possess high energy (say, kinetic energy $E_k > 100$ MeV but this limit is rather conditional). The whole range of questions connected with the origin and role of cosmic rays in space may be called cosmic-ray astrophysics, although it is frequently referred to as 'the problem of cosmic-ray origin'. However, the origin (in the literal sense of the word) of cosmic rays is only part of cosmic-ray astrophysics. The term 'high-energy astrophysics' is also exploited and includes, besides cosmic-ray astrophysics, gamma-ray astronomy, X-ray astronomy, and high-energy neutrino astrophysics. Many authors, I among them, have written much on these subjects. I refer to the reviews [32–34] and particularly to the Proceedings of the International Cosmic-Ray Conferences (ICRC); the 20th ICRC was held in 1987 [35]. (Paper [36] is my introductory talk at this conference.[15*]) Therefore, I shall only briefly comment on the early stages in the development of the theory of cosmic synchrotron radiation and the appearance of cosmic-ray astrophysics (for more details, see [1, 2]).

Approximately in 1947–49, it became quite clear (as far as I reckon) that comparatively long-wave, non-solar cosmic radio emission (including, in particular, the very first radio-astronomical measurements made by K Jansky at a wavelength of about 15 m) possesses an effective temperature T_{eff} exceeding 10^4 K. Hence, it was impossible to interpret such radio emission as thermal radiation from interstellar gas, since this gas, generally speaking, has a temperature $T \lesssim 10^4$ K and, in any case, radiation with $T \gg 10^4$ K cannot be explained by thermal radiation of gas. Thus, one had to assume the existence of a non-thermal source analogous, for example, to sources of non-thermal (sporadic) solar radio emission. This was how the 'radio-star hypothesis' arose in a quite natural way. According to this idea, some stars are anomalously powerful radio sources responsible for the non-thermal, cosmic radio emission with its continuous spectrum and diffuse directional distribution [37–39]. The radio-star hypothesis, however, faced many difficulties, mainly involving assumptions (sometimes arbitrary and unrealistic) concerning the hypothetical radio stars. Moreover, an alternative soon appeared and with time became stronger and stronger—the synchrotron hypothesis of the origin of non-thermal radio emission, which in the end proved to be valid.

The competition or struggle between these two hypotheses took several years. Although synchrotron radiation had already been known and understood from the physical point of view for many years [40], and in the 1940s was especially widely discussed in the physical literature in connection with the analysis of synchrotrons, it was only in 1950 that the first papers [41, 42] appeared in which the synchrotron mechanism was considered as related to cosmic radio emission: radiation from radio stars was discussed in [41] and that from interstellar space in [42]. It seems rather strange to me that both these papers appeared in a journal of physics, and, besides, in the form of short letters. In any case, as far as I know, they did not attract the attention of astronomers. I, on the contrary, believed at once that the synchrotron mechanism was responsible for non-thermal cosmic radio emission. I do not attribute this to any keen insight on my part but rather to the previously mentioned fact that I was closer to physics and far from classical astronomy. In this situation the synchrotron mechanism seemed clear and realistic, whereas hypothetical, strange radio stars remained purely speculative. I immediately verified the calculations [41, 42] and, if I am not mistaken, did not add anything essential (I have not now compared all the expressions and estimates from [41, 42], and my own paper [43], submitted for publication on 31 October, 1950, because it does not seem essential here, the more so I have never claimed priority). But the fact remains that my paper [43] was the first and, for some time, the only one that responded to the proposals in [41, 42] to use the synchrotron mechanism in astronomy. The reaction of astronomers was probably quite the opposite, i. e. the synchrotron mechanism seemed mysterious and speculative, whereas radio stars, although posing riddles, were more acceptable, for what kinds of stars cannot exist? In this respect, I S Shklovsky was not an exception. He not only developed the radio-star hypothesis [39] but also positively denied the synchrotron hypothesis ("which for a number of reasons does not seem to us to be acceptable" [39]). This quotation from Shklovsky is presented in more detail in [2]. I dare make this remark here only because it is just Shklovsky's paper [39] that has been repeatedly cited in the world literature as nearly the first and principal one discussing the application of the synchrotron hypothesis. A similar error is spread concerning the question of exploiting polarization of synchrotron radiation as a criterion for establishing the validity of the synchrotron origin of cosmic radiation (for more details, see [1, 2]). After the appearance of a very important paper by Pikel'ner [44], who emphasized that the interstellar magnetic field exists in the entire galactic volume, Shklovsky [45] realized that his earlier objection [39] to the efficiency of the synchrotron mechanism had been groundless. For reasons clear from the section to follow, I do not here dwell in more detail on Shklovsky's paper [45] and his subsequent papers (see [1, 2]).

Judging from the Proceedings of the 1955 Manchester IAU Symposium on Radio Astronomy [46], which included a paper on the 'radio-star' hypothesis while my paper sent to the symposium was not even published, 'astronomical public opinion' in 1955 was still on the side of the radio-star model. However, by the next 1958 Paris Symposium, the synchrotron mechanism was unconditionally accepted as the dominant production mechanism of nonthermal cosmic radio emission. (I do not mean, of course, the radiation coming from the solar atmosphere and generally from relatively dense regions.) This time my report "Radio astronomy and the origin of cosmic rays" was included in the Proceedings of the Symposium [47], although I myself was not able to participate.

The establishment of connection between radio astronomy and cosmic rays has led, as a matter of fact, to the appearance of a new field of research in astronomy, namely cosmic-ray astrophysics and then also high-energy astrophysics. The point that although cosmic rays were recognized earlier (prior to 1950–53) to be cosmic objects, they were investigated only on the Earth (in the atmosphere and on its boundaries) exclusively for the purposes of high-energy physics. As is well known, the cosmic-ray results have led to exceedingly important discoveries in physics (positron e^+, $\mu^{\pm}$-leptons, $\pi^{\pm}$-mesons, and some other particles were discovered; see the collection cited in [2]). But since cosmic rays are highly isotropic, studying them near the Earth is similar to making a spectral analysis of the light of all stars taken together. It is clear that under such conditions, without additional information about celestial bodies, astronomy could not have developed. The reception of synchrotron radio emission from cosmic rays (more precisely, from their electron component) has sharply changed this situation. It has become clear that cosmic rays are a universal phenomenon—they are present in interstellar space, in supernova remnants, and in galaxies. (Radio galaxies and then quasars have been discovered.) Radio-astronomical data, together with the information on primary cosmic rays near the Earth and with the available astronomical concepts, have promoted further advances. I note that as far back as 1934, Baade and Zwicky [48] associated cosmic-ray generation and neutron star formation with supernova flares. In 1949, in considering cosmic rays as a gas of charged particles, Fermi [49] discussed a possible mechanism for their acceleration. These ideas, together with radio-astronomical data, were the basis for constructing the galactic model of the origin of the main part of cosmic rays observed near the Earth. Since 1953, I have supported only the galactic model with halo. The halo problem aroused many discussions (see, e. g., [50]), which seem to me to result from misunderstanding (see [50, 51]). It is impossible to write here about this and many other problems of cosmic-ray astrophysics in further detail, and I once again refer the reader to the literature (see [32–36, 51]).[15*] True, it is necessary to lay special emphasis on the principal achievement of recent years or, more precisely, of the last two decades. I mean gamma-ray astronomy and, more specifically, the reception of gamma-ray emission due to π^0-meson decay, with the energy of photons $E_\gamma > 30 - 50$ MeV. π^0-mesons are produced, in turn, during collisions of protons and nuclei, which make up nearly 99% (by number of particles) of cosmic rays, with gas nuclei in the interstellar medium. Therefore, identification of gamma-ray photons generated by π^0-meson decay provides information on the proton-nuclear component of cosmic rays far from the Earth. In this respect, gamma-ray astronomy plays the same role as radio astronomy plays in the study of the electron (or, more precisely, electron–positron) component of cosmic rays. Thus, only through the combination of radio and gamma-ray astronomical methods can we, in principle, obtain rather complete information on cosmic rays far from the Earth. Unfortunately, gamma-ray astronomy is not yet sufficiently developed. The reason is that it is necessary to use special satellites (gamma-ray observatories) to detect photons in the range 30 MeV$< E_\gamma <$1–5 GeV but not a single gamma-ray observatory has been in operation since the very successful satellite COS-B finished its work in 1982.[16*]

The problems to be solved by gamma-ray astronomy are fairly diverse and cover the whole gamma-ray range from $E_\gamma > 10^5$ eV to $E_\gamma > 10^{14}$ eV (gamma-ray bursts, gamma-ray lines, diffuse gamma-ray emission, radiation from discrete sources, in particular, gamma-ray emission from SN 1987 A [52]). Here, too, I restrict myself only to references [34–36, 53].[15*]

The proton-nuclear component of cosmic rays generates neutrinos as well and the potentialities of high-energy neutrino astrophysics are very large. (Since I have participated in a single work [54] in this field, I shall only refer to it and the review contained in [34].) The prospects for the development of cosmic-ray astrophysics and the whole of high-energy astrophysics have been repeatedly discussed, by me among others (most recently in [36]).[15*]

Summarizing, I can say that within the memory of my generation, astronomy has undergone a profound transformation—from optical to all-wave; in the course of this process, cosmic-ray astrophysics has been added, and in the near future it will also include high-energy neutrino astrophysics. Along with physics, I was lucky to plunge rather early into astronomical problems as well.

17.9   A few words about priority

Questions of priority have long played an important role in the life of scientific community. Suffice it to recall how much time and strength the great Newton spent on priority questions (see, e. g., [55]) even in the period when he had already obtained general recognition. His was, of course, another era but in not so remote times and even now the priority passions have flared and do flare up, although people have learned to hide them. At the same time, a great increase in the number of scientific publications and various conferences has created additional difficulties.

On 3 August 1987, I gave an introductory talk at the 20th International Cosmic Ray Conference (ICRC 20, Moscow)—in the talk itself I did not mention the names of the authors for the majority of the papers upon which I touched. That is why I gave some explanations that were absent in the published version [36]. For this reason and for purposes of further discussion, I shall briefly dwell on these explanations here.

First, the gleam of names in a talk (and, say, on overhead transparencies) hinders discussion of the material itself in that it diverts attention from the content. Second, papers and authors now are so numerous that it is impossible to mention everybody, while selection often brings displeasure and offense. In this connection I showed a transparency with the following two phrases: "Priority questions are a dirty business. Priority mania or supersensitivity is a disease."

I then advised that not much importance should be attached to the absence of references. In most cases, this omission is explained not by evil intent but rather by the abundance of literature, by ignorance, by the desire to refer only to the most recent review, etc. It is only a negligible minority of authors who avoid citations deliberately and act from ill-intentioned motives. Such people and such actions, with some exceptions, do not deserve attention. As I understood, my transparency and comments were taken exactly as I expected, i.e. as advice given half-jokingly, inspired by my experiences of many years in the fields of physics and astrophysics. By the way, two weeks after (on 21 August, 1987), I made a remark in the same spirit at the 18th International Conference on Low-Temperature Physics but in quite different conditions and atmosphere, due to which some people misunderstood me (see [4]). But, of course, my way of thinking is exactly as I have tried to explain. My point is not that I am somehow indifferent to the question of priority. Yes, I do notice when my papers are cited and when not but I never lay claims to authors who do not refer to my works. Besides feeling that such claims (which are, unfortunately, not rare) do not seem to be very decent, I think that, in most cases, the reasons for which a citation is sometimes omitted are not ill intentioned.

I decided to touch upon this question of priority for two reasons.[17*] First, it seems to me that it deserves attention because some people are deeply agitated about it and get upset. Therefore, one should not pretend that such a problem does not exist. Second, from time to time I happen to be, to this or that extent, involved in priority disputes or some collisions. I may state that I have never tried to prove or defend my priority. In all cases my indignation was provoked by the tactless behavior of my opponents. It is, of course, up to the reader to decide whether or not I am sincere—a man is a sophisticated creature and not so seldom may play the hypocrite

or distort, even if unconsciously (subconsciously?), the truth. But I also
have the right to express my opinion, which I have done here. I would
not do this, however, if I did not think it necessary to somehow explain
here my relations with I S Shklovsky (1916–85). It has been well known
in my country and possibly abroad that, since the end of 1967, we were
on bad terms, did not speak to one another or greet each other. I do
not know about the West but in our country such a fantastic form of
relations is sometimes encountered. By the way, it does not necessarily
show some special enmity. In this case it simply so happened and neither
of us wanted afterward to break the ice of estrangement. In the opinion
of some colleagues, the break-off of my relations with Shklovsky was due
to my priority claims. In my view, this is absolutely wrong. Fortunately,
I need not prove this assertion here, for it is quite clear from papers [1, 2]
written while I S Shklovsky was alive.

Of course, he made use of my papers and took into account my results
but I did the same with respect to his papers. True, we had different
manners of citation but this question is of secondary importance and is
not, therefore, the point here.

We were of the same age and studied almost simultaneously at Moscow
State University but we made a closer acquaintance only during the Soviet expedition to Brazil, aimed at observing the total solar eclipse of 20
May, 1947, in which we both participated. From then our relations, if not
very close, were quite friendly up to 1966. By the way, all those 'claims'
that I could, in principle, lay to Shklovsky refer to the pre-1966 period
(see references in [1, 2]). But I did not pay attention to his tactlessness
and attached no importance to it. But from 1966 on, beginning with the
IAU conference on radio astronomy in Nordwijk, Shklovsky was ill dis-
posed toward me for some reason, of which I am unaware and can only
guess. This had no consequences until the fall of 1967, when Shklovsky
published his paper [56] in which he associated powerful X-ray radiation from a number of sources with binary systems, in which one of the
components is a neutron star onto which plasma accretes. Meanwhile,
all of this had, as a matter of fact, been said at a round-table discussion held during the Nordwijk conference the year before [50]. G Burbidge, who was the chairman of this discussion, reported the results at
the conference (see [50, p 463]). I was present at the discussion and so
was Shklovsky, who, as far as I remember, did not utter a word. In
his paper [56], he did not refer to this discussion, which was published
some time later [50]. The history of the interpretation of powerful X-
ray radiation has been described in more detail by G Burbidge [57], who
also narrated what I have said. I spoke at the round-table discussion
and after Burbidge's report (see [50]) as well but I have not published
anything on this subject and have no special priority claims. But even
before that, Ya B Zeldovich, together with I D Novikov, had published

a paper [58] containing the idea of accretion onto a neutron star. According to Zeldovich, Shklovsky knew about it but made no reference to it in [56]. Ya B Zeldovich was generally very sensitive to priority questions and, with respect to the paper [56], he was indignant and told me about this. I, in turn, told him about the Nordwijk round-table discussion. Eventually, we wrote (of course, with both our signatures) a corresponding letter to Shklovsky, and I added there the examples presented in [1, 2]. I have made no attempts now to find the copy of this letter among my papers. The letter was not distributed—it was sent to Shklovsky only.[18*] However, his close colleagues read this letter. Ya B Zeldovich and I were led by the naive idea that Shklovsky's eyes would be opened but nothing of the sort happened. He wrote an answer but I was not shown it because I left for England for three months and when I returned, Shklovsky was either ill or fell ill soon thereafter and, thus, his friends and coworkers decided to 'damp down' this matter. After that I did not communicate with Shklovsky but he remained on diplomatic terms with Zeldovich. I think that in sending this letter, Ya B Zeldovich and I made a mistake, the more so as it did not change in Shklovsky's position any way. In particular, in his paper written in 1982 on the occasion of the 20th anniversary of X-ray astronomy, and even in spite of the paper by Burbidge [57], Shklovsky insisted on the priority of his paper [56].[19*]

Shklovsky was a very capable man and had a number of actual achievements in the field of astrophysics. As regards his unfortunate statements concerning radio astronomy and X-ray astronomy, they were harmful, first of all, to himself—I will not try to understand his motives. In his last years, Shklovsky wrote autobiographical short stories. At the end of 1987 I was given a volume of these short stories, entitled *Echelon*. I found this book very interesting and hope it will someday be published.[20*] Influenced by Shklovsky's short stories, I wrote a rather long letter (addressed to L S Marochnik) in which I narrated the story of my relations with Shklovsky and the cause of our quarrel. This letter was read by Shklovsky's former co-workers from the Institute for Space Research, and I asked one of them (B V Komberg) to keep this letter and, if possible, other documents[21*,27*] as well. Historians of astronomy (or psychology?) may one day be interested in these documents.

It only remains to answer the following question: Was this worth writing? Shklovsky is unfortunately not with us now and so my position is rather delicate—he cannot object to me and how can I avoid suspicions of ignobly settling a score with him posthumously? I understand this and hesitated at first but I still decided not to delete this section. First, all my remarks are based on already published materials, which are cited in [1, 2] and here. Thus, those interested may check everything. Second, some unpublished material (including the letter—Shklovsky's reply—that

I have not read) is with his former co-workers, and I will only be too glad if they now publish everything that they may care to, including their own comments. Finally, as I write this paper (in the fall 1988), I am already 72 years old and I am not at all sure I shall live to see it published. Under such conditions, it seems to me simply preposterous to write not with the goal of clarifying the truth but from some other considerations. I hope nobody will doubt that.

17.10   A few concluding remarks

In this paper I have not intended even to enumerate all my papers that refer to astronomy. The main references, however, were made, directly or indirectly, when I referred the reader to papers [1, 2] and others. In view of the fact that this paper is to be published in the *Annual Review of Astronomy and Astrophysics*, it does not seem pertinent, however, to refer separately to my papers [59, 60], which were published in this series and devoted to the theory of synchrotron radiation. I shall also mention papers [61, 62] on X-ray and gamma-ray astronomy. In [63] I discussed the problem of superfluidity and superconductivity in astrophysics.

Besides these, I shall only dwell on the papers connected with the general theory of relativity (GTR). GTR is, of course, a physical theory but it is so closely connected with astronomy that it can be referred to as astrophysics as well. This, in particular, is what I did in my popular book [64] (see also my book 2001 *The Physics of a Lifetime: Reflections on the Problems and Personalities of 20th Century Physics* (Berlin: Springer), mentioned in [3, 7, 24]).

In general, I did not work hard on GTR and do not belong to the so-called relativists. I shall mention the study of the collapse of a magnetic star [29, 65] which led to the conclusion that the magnetic moment of the star tends to zero as the radius of its surface approaches the Schwarzschild radius. This was the first indication of an effect that was later described by the phrase 'a black hole has no hair'.[22*] How large the magnetic field of such stars can be was pointed out in [29], several years prior to the discovery of pulsars and, therefore, neutron stars.

The application of the equivalence principle, the fundamental principle of GTR, in taking into account the quantum effects and, specifically, zero-point fluctuations of different fields is discussed in [66]. Finally, several of my papers were devoted to a discussion of the experimental verification of GTR (my last review on this subject is [67]; see also [64]) and to the limits of its applicability (I refer only to [68] and [69]). The latter problem is associated with the question of limitations on certain physical concepts when applied to astronomy, the question that has attracted attention of some astronomers. My Darwin lecture [18] was devoted to this subject. Today I am of the same opinion as that expressed in this lecture, namely that one can give no definite guarantees but it is most probable and realistic that if some deviations from the known physical laws and, specifically, from a quite definite classical gravitation theory can be expected, they may only happen at very early stages of cosmological evolution. (I mean the observable, expanding Universe, which, from the point of view of some inflationary cosmological models, may appear to be part of a larger or even infinite system; see the review [70].) True, I am not going to touch upon the regions of black holes that are deep below the horizon (i.e. for a Schwarzschild black hole for $r \ll r_g = 2GM/c^2$) nor upon mini-black holes, cosmic strings, etc. One of the possible ways in which GTR might be violated is connected with the existence (purely hypothetical) of a fundamental length $l_f \gg l_g = \sqrt{G\hbar/c^3} \sim 10^{-33}$ cm (for more details, see [69]).[23*] At later stages of cosmological evolution (for example, for the red shift parameter $z < 10^3$), there is absolutely no reason to expect that the well-known laws of macroscopic physics do not apply. Of course, this is, on the whole, a large and complicated problem and it is impossible to consider it here more thoroughly (some thoughts concerning this problem were mentioned in article 1, which opens the present collection).

In concluding, I would like to ask the question: have I contributed anything essential to astronomy? This is undoubtedly up to other people to decide but the author himself also has the right to have and normally has his own opinion, even if it is erroneous. For instance, I believe that I have made a noticeable contribution to the theory of superconductivity but as to astrophysics, I do not have any definite opinion. I have done some things, of course but how important are they and what role have they played? It would be interesting to have the answer but, unfortunately, I do not know a single work that considers in detail and assesses the history of 'new astronomy', i.e. the development of astronomy over the last 40–50 years. On the contrary, I have come across rather many papers and even booklets or books containing errors concerning this history. Absurd assertions, so to say, adapted by repetition, roam from one such paper or book to another. It is surprising how credulous some authors who elucidate history are not to consult the original literature and check their assertions. I hope that the situation will be clarified with time and believe that the prefatory chapters published in the *Annual Reviews* will make their contribution to this process. I hope the present paper is not an exception, although I have not tried (with some exceptions; see [1, 2]) to take on the role of historian, i.e. to compare, verify, and assess the results and assertions of various authors. Furthermore, against my own expectations and original aim, this paper appeared to be devoted not so much to astrophysics as to autobiographical and other remarks. But maybe this is not so bad.

Comments

1*. For many years in the well-known *Annual Review of Astronomy and Astrophysics* as the first or Prefatory chapter, has been in the form of an autobiography of an astronomer or astrophysicist together with a photograph selected by the editorial board. In 1988, I received an invitation to write such an article, which I did. This article (1990 *Annu. Rev. Astron. Astrophys.* **28** 1) has never before been published in Russian, since it was written with the Western scientific community in mind. The article is given without practically any alterations.

2*. These are my colleagues, Veniamin Aronovich Tsukerman and Lev Vladimirovich Al'tshuler. They, among other things, participated very actively in the Soviet atomic and hydrogen bomb program at Arzamas-16. See the book *Lyudi i Vzryvy [People and Explosions]* by V A Tsukerman and Z M Azarkh published in 1994 in Arzamas-16.

3*. More information about this period and the years that followed can be found in article 19 in the present collection and also [3] in this article.

4*. See L I Mandelshtam 1948–50 *Polnoe Sobranie Trudov* [Complete Collection of Works] vols 1–5 (Moscow: USSR Academy of Sciences). I do not know whether the lectures have been translated into English. The original article does not contain this note.

5*. See articles 2 and 9 in the present collection.

6*. See also article 10 in the present collection.

7*. The reader should bear in mind that this article was written at the end of 1988. I have not changed anything in it since, if one does that, where does it lead? I spoke on different topics of academic life at the General Meetings of the Russian Academy of Sciences (RAS) in 1996, 2000, and 2002 (RAS is the successor of the USSR Academy of Sciences) and at sessions of the Presidium of RAS. The texts of my speeches have been published in *Vestn. Ross. Akad. Nauk [Herald Russ. Acad. Sci.]* (see 1997 **67** 223 [1997 **67** 149]; 1999 **69** 702 [1999 **69** 271]; 2000 **70** 903 [2000 **70** 417]; 2001 **71** 702 [2001 **71** 354]; and 2002 **72** 879, 890 [2002 **72** 468, 480]). Moreover, some of the articles have been published in the newspaper *Poisk*, which incidentally is a publication sponsored by RAS (see 1999 *Poisk* no 52, 2000 no 5, 2001 no 48, and 2002 nos 17, 45, 46). I would also like to point out the articles in *Literaturnaya Gazeta [Literary Newspaper]* for 16 October, 2002, 26 February and 9 July, 2003.

8*. One of the few still alive at the time this article was written, Valerii Frid (who died on 7 September, 1998), was also involved in the case with my wife and described the whole case in his book *58 1/2. Zapiski Lagernogo Pridurka [58 1/2. Notes of an Easy-Job Convict in the Gulag]* (Moscow: Rusanov, 1996).

9*. I visited A D Sakharov in Gorky (Nizhnii Novgorod) on 11 April,

1980 and 22 December, 1983. For more details about my relations with Sakharov, see my article "The Sakharov Phenomenon" published, among other places, in the collection *O Fizike i Astrofizike* [*On Physics and Astrophysics*] [7] [2001 *The Physics of a Lifetime* (Berlin: Springer) p 471]. The article was also published in the collection 1996 *On Mezhdu Nami Zhil... Vospominaniya o Sakharove* [*He Lived among Us. Memories of Sakharov*] (Moscow: Praktika).

10*. For more details, see articles 26 and 12 in the present collection.

11*. Here and in what follows my memory may have failed me. In his recollections, A D Sakharov writes (Sakharov A D 1990 *Vospominaniya* [Memoirs] (New York: Chekhov); 1996 (Moscow: Prava Cheloveka)) that we started the 'bomb business' not in 1947 but in 1948. Tamm and Sakharov left for Arzamas-16 (known as the 'Ob'ekt') not in 1948 but in 1950.

12*. Since the death of A D Sakharov (14 December, 1989), the history of the Soviet hydrogen bomb in its first version has finally been published. Today there are many published materials on this topic. The respective references can be found in the comments to article 18 in the present collection.

13*. The papers [1, 2] have never before been published in Russian. They are included in the present collections as articles 4 and 3, respectively.

14*. Regarding the time when the paper [21] was written, see comment 2 to article 4 in the present collection.

15*. In 1996 I published paper [71], which was a brief review of the problem of the origin of cosmic rays in its development up to 1995 (in it I also included the data on the 24th International Cosmic Ray Conferences, which took place in 1995). There is also a brief review [72] and a review devoted to gamma-ray astronomy [73].

16*. In 1991, the Compton Gamma-Ray Observatory (CGRO) was launched (see [71]). It was a great success.

17*. On the question of priority, see also my book *O Fizike i Astrofizike* [*On Physics and Astrophysics*] [7], p 195 [2001 *The Physics of a Lifetime* (Berlin: Springer) p 232].

18*. I have finally found the letter, which is an outline of a 'Letter to the Editorial Board of Astronomicheskii Zhurnal' entitled 'Concerning certain works and claims to priority of I S Shklovsky'. But, as I have said, this letter was not sent to the editorial board, instead we passed it on to two persons that were close to I S Shklovsky, S B Pikelner and N S Kardashev, with a sort of covering letter, in which we expressed our hope that there would be no need to send the letter in question to the editorial board since Shklovsky would comment in the press about the infirmity of some of his claims to priority. Ten years ago this covering letter was published by Zeldovich's daughter M Ya Ovchinnikova in a collection of articles in memory of Ya B Zeldovich 1993 (*Zhakomyi Neznakomyi Zel'dovich* [*The

Well-known Unknown Zeldovich] (Moscow: Nauka) p 88.

19*. In the recently published collection by I S Shklovsky 1996 *Razum, Zhizn', Vselennaya* [*Intelligence, Life, and the Universe*] (Moscow: Yanus), I discovered (p 68), to my great surprise, something similar in relation to the prediction of polarization of the radiation from the Crab Nebula (for more details see comment 3 to article 3 in the present collection).

20*. Some of these short stories have since been published (I S Shklovsky 1991 *Eshelon* [*Echelon*] (Moscow: Novosti)).

21*. Since then I have also passed on to B V Komberg the copy of the outline of the letter to the editorial board of *Astronomicheskii Zhurnal* mentioned in comment 18 (written by me and Ya B Zel'dovich) and my remarks not intended for publication entitled 'Why Iosif Shklovsky and I were Not on Speaking Terms for 18 Years and Some Other Recollections'. See also comment 4 to article 3 in the present collection.

22*. This sentence was not in the first version of the article but it appeared due to the persistent advice of K S Thorne, who read the article prior to publication and made a number of remarks. Naturally, in the English text of the article, I thanked Thorne for this.

23*. I have also written about this in my book *O Fizike i Astrofizike* [*On Physics and Astrophysics*] [7] on pp 75, 104, 117, and 118 [2001 *The Physics of a Lifetime* (Berlin: Springer) pp 69, 96, 108, and 109], and also in article 1 of the present collection.

24*. A list that exists in the Internet (http://www.scientific.ru) contains data on the complete number of references to the works of scientists in the USSR and Russia that appeared after 1985, i. e. approximately for 15 years (the data for 2001 are incomplete). There are 5907 references to my works in that list, i. e. about 400 references per annum. I believe that from 1976 to 1985 there were about 4000 references to my works. Thus, altogether from 1961 to the present, there should be about 17 000 references to my works. My first papers were published in 1939, and I really do not know the number of citations of my works before 1961. In the new list published in Russia in the first half of 2003 (Scientific.ru.Forum), there are 8962 references to my works published after 1985 (up to December 2002). There it is also mentioned that, according to ISI 1945, there are 19 519 references to my works after 1945.

25*. The last time I was in Nizhnii Novgorod was at the end of January 2001, where I attended the celebration of the 70th birthday of my former student, postgraduate worker, and co-author V V Zheleznyakov, today an Academician of the RAS. There I delivered a lecture for a broad scientific audience on the problems of physics and astrophysics (its contents are reflected in article 1 in the present collection) and also made a report of a historical nature at the Institute of Applied Physics of the RAS. Incidentally, in 1998, I was given the Degree of Honorary Doctor by the Nizhnii

Novgorod University.

26*. See comment 11 to article 18 in the present collection.

27*. The letter to L S Marochnik (true, with some inaccuracies) has recently been published as an Appendix to the book by T K Breus entitled *Institut* [*Institute*] (Moscow: KoNT, 2002). A copy of my letter to T K Breus containing some commentaries in connection with this publication has been passed on to B V Komberg.

References

[1] Ginzburg V L 1984 in *The Early Years of Radio Astronomy* ed W T Sullivan (Cambridge: Cambridge University Press) p 289 (article 4 in the present collection)

[2] Ginzburg V L 1985 in *Early History of Cosmic Ray Studies: Personal Reminiscences with Old Photographs* eds Y Sekido and H Elliot (Dordrecht: Reidel) p 411 (article 3 in the present collection)

[3] Ginzburg V L 1986 *Priroda* (10) 80. A more complete text has been published in Ginzburg V L 1995 *O Fizike i Astrofizike* [*On Physics and Astrophysics*] (Moscow: Byuro Kvantum) p 288 [Translation into English 2001 *The Physics of a Lifetime* (Berlin: Springer) p 285]

[4] Ginzburg V L 1989 *Prog. Low-Temp. Phys.* **12** 1

[5] Ginzburg V L 1946 *Dokl. Akad. Nauk SSSR* **52** 491; 1946 *C. R. Acad. Sci. URSS* **52** 487

[6] Ginzburg V L 1942 *Dokl. Akad. Nauk SSSR* **35** 302; 1942 *C. R. Acad. Sci. URSS* **35** 270

[7] Ginzburg V L 1995 in: *Vospominaniya o I E Tamme* [*Reminiscences about I E Tamm*] (Moscow); see also Ginzburg V L 1995 *O Fizike i Astrofizike* [*On Physics and Astrophysics*] (Moscow: Byuro Kvantum) [Translation into English 2001 *The Physics of a Lifetime* (Berlin: Springer)].

[8] Ginzburg V L 1979 *Theoretical Physics and Astrophysics* (Oxford: Pergamon Press). The 3rd augmented Russian edition of this book, *Teoreticheskaya Fizika i Astrofizika* [*Theoretical Physics and Astrophysics*] was published in 1987 (Moscow: Nauka) [Translation into English 1989 *Application of Electrodynamics in Theoretical Physics and Astrophysics* (New York: Gordon and Breach)]

[9] Ginzburg V L 1983 *Usp. Fiz. Nauk* **140** 687 [1983 *Sov. Phys. Usp.* **26** 713]

[10] Ginzburg V L 1988 in *Reminiscences about L D Landau* (Oxford: Pergamon Press); see also article 10 in the present collection

[11] Ginzburg V L 1986 in *The Lessons of Quantum Theory* (Amsterdam: Elsevier) p 113. Unfortunately, in that publication my report was reduced considerably by the editorial board (for lack of space, I believe). The report has been published in Russian in full 1986 *Trudy (Proc.) P N Lebedev Physical Institute* **176** 3 [Translated into English in the US by Nova Science, New York, 1988) (see also article 2 in the present collection)

[12] Ginzburg V L and Man'ko V I 1976 *Fiz. Elem. Chastits At. Yadra* **7** 3 [1976 *Sov. J. Part. Nucl.* **7** 1]

[13] Ginzburg V L 1987 in *Quantum Field Theory and Quantum Statistics* (Bristol: Adam Hilger, 1987) vol 2 pp 15–33

[14] Ginzburg V L and Landau L D 1950 *Zh. Eksp. Teor. Fiz.* **20** 1064; reprinted in Landau L D 1969 *Collected Papers* vol 2 (Moscow: Nauka) [English edition 1965 (London: Pergamon Press) p 546]

[15] Ginzburg V L 1988 in *Inogo ne Dano* [*No Other Way*] (Moscow: Progress) p 135

[16] Garfield E 1977 *Current Contents* **49** 5

[17] Ginzburg V L 1962 *Trudy (Proc.) P N Lebedev Physical Institute* **18** 55

[18] Ginzburg V L 1975 *Quart. J. R. Astron. Soc.* **16** 265; 1995 *O Fizike i Astrofizike* [*On Physics and Astrophysics*] (Moscow: Byuro Kvantum) p 204 [Translation into English 2001 *The Physics of a Lifetime* (Berlin: Springer) p 241]

[19] Ginzburg V L 1970 *Propagation of Electromagnetic Waves in Plasmas* (Oxford: Pergamon Press) [Russian edition 1967 (Moscow: Nauka)]

[20] Martyn D F 1946 *Nature* **158** 632

[21] Shklovsky I S 1946 *Astronom. Zh.* **23** 333

[22] Ginzburg V L 1947 *Usp. Fiz. Nauk* **32** 26

[23] Ginzburg V L 1948 *Usp. Fiz. Nauk* **34** 13

[24] Ginzburg V L 1980 in *Oort and the Universe* (Dordrecht: Reidel); 1995 *O Fizike i Astrofizike* [*On Physics and Astrophysics*] (Moscow: Byuro Kvantum) p 452 [Translation into English 2001 *The Physics of a Lifetime* (Berlin: Springer) p 457]

[25] Ginzburg V L and Zheleznyakov V V 1958 *Astronom. Zh.* **35** 694 [1958 *Sov. Astron. J.* **2** 653]

[26] Zheleznyakov V V 1970 *Radio Emission of the Sun and Planets* (Oxford: Pergamon Press) [Russian edition: 1964 (Moscow: Nauka)]

[27] Ginzburg V L 1961 *Astronom. Zh.* **38** 380

[28] Ginzburg V L and Ozernoy L M 1977 *Astrophys. Space Sci.* **50** 23

[29] Ginzburg V L 1964 *Dokl. Akad. Nauk SSSR* **156** 43 [1964 *C. R. Acad. Sci. URSS* **9** 329]

[30] Ginzburg V L and Zheleznyakov V V 1975 *Annu. Rev. Astron. Astrophys.* **13** 511

[31] Beskin V S, Gurevich A V, and Istomin Ya N 1986 *Usp. Fiz. Nauk* **150** 257 [1986 *Sov. Phys. Usp.* **29** 946]; 1988 *Astrophys. Space Sci.* **146** 205; 1993 *Physics of Pulsar Magnetosphere* (Cambridge: Cambridge University Press)

[32] Ginzburg V L and Syrovatskii S I 1964 *The Origin of Cosmic Rays* (Oxford: Pergamon Press) (see article 4 in the present collection)

[33] Ginzburg V L and Ptuskin V S 1976 *Rev. Mod. Phys.* **48** 161; 675

[34] Ginzburg V L (ed) 1984 *Astrofizika Kosmicheskikh Luchei* [*Astrophysics of Cosmic Rays*] (Moscow: Nauka) (2nd edn 1990) [Translation into English 1990 (Amsterdam: North-Holland)]

[35] 1987 *Proc. 20th Int. Cosmic Ray Conf., Moscow, USSR, 2–15 August 1987* eds V A Kozyarivsky *et al* (Moscow: Nauka)

[36] Ginzburg V L 1987, see [35] **7** 7; 1988 *Usp. Fiz. Nauk* **155** 185 [1988 *Sov. Phys. Usp.* **31** 491]

[37] Ryle M 1949 *Proc. Phys. Soc.* **A62** 491

[38] Unsold A 1949 *Z. Astrophys.* **26** 176

[39] Shklovsky I S 1952 *Astronom. Zh.* **29** 418

[40] Schott G A 1912 *Electromagnetic Radiation and the Mechanical Reactions Arising from It* (Cambridge: Cambridge University Press)

[41] Alfven H A and Herlofson N 1950 *Phys. Rev.* **78** 616

[42] Kipenheuer K O 1950 *Phys. Rev.* **79** 738

[43] Ginzburg V L 1951 *Dokl. Akad. Nauk SSSR* **76** 377

[44] Pikel'ner S B 1953 *Dokl. Akad. Nauk SSSR* **88** 229

[45] Shklovsky I S 1953 *Astron. Zh.* **30** 15

[46] van de Hulst H C (ed) 1957 *Radio Astronomy* (IAU Symp. No. 4) (Cambridge: Cambridge University Press)

[47] Bracewell R N (ed) 1959 *Paris Symposium on Radio Astronomy* (IAU Symp. No. 9) (Stanford, CA: Stanford University Press) [Translated into Russian 1961 (Moscow: IL)]

[48] Baade W and Zwicky F 1934 *Phys. Rev.* **46** 76; 1934 *Natl Acad. Sci. USA* **20** 259

[49] Fermi E 1949 *Phys. Rev.* **75** 1169

[50] van Woerden H (ed) 1967 *Radio Astronomy and the Galactic System* (IAU Symp. No. 31) (New York: Academic Press)

[51] Ginzburg V L 1989 *Essays on Particles and Fields: MGK Menon Festschrift* eds R R Daniel and B V Sreekantan (Bangalore: Indian Academy of Sciences)

[52] Berezinsky V S and Ginzburg V L 1987 *Nature* **329** 807

[53] Dogiel V A and Ginzburg V L 1989 *Space Sci. Rev.* **49** 311

[54] Berezinsky V S and Ginzburg V L 1981 *Mon. Not. R. Astron. Soc.* **194** 3

[55] Westfall R S 1980 *Never at Rest: A Biography of Isaac Newton* (Cambridge: Cambridge University Press)

[56] Shklovsky I S 1967 *Astrophys. J. Lett.* **148** 1

[57] Burbidge G 1972 *Comm. Astrophys. Space Phys.* 4(4) 105

[58] Novikov I D and Zeldovich Ya B 1966 *Nuovo Cimento Suppl.* **4** 810

[59] Ginzburg V L and Syrovatsky S I 1965 *Annu. Rev. Astron. Astrophys.* **3** 297

[60] Ginzburg V L and Syrovatsky S I 1969 *Annu. Rev. Astron. Astrophys.* **7** 375

[61] Ginzburg V L and Syrovatsky S I 1965 *Space Sci. Rev.* **4** 267

[62] Ginzburg V L 1966 *Usp. Fiz. Nauk* **89** 541 [1967 *Sov. Phys. Usp.* **9** 543]

[63] Ginzburg V L 1969 *Usp. Phys. Nauk* **97** 601 [1970 *Sov. Phys. Usp.* **12** 241]; 1971 *Physica* **55** 207

[64] Ginzburg V L 1985 *Physics and Astrophysics (A Selection of Key Problems)* (Oxford: Pergamon Press)

[65] Ginzburg V L and Ozernoi L M 1964 *Zh. Eksp. Teor. Fiz.* **47** 1030 [1965 *Sov. Phys. JETP* **20** 689]

[66] Ginzburg V L and Frolov V P 1987 *Usp. Fiz. Nauk* **153** 633 [1987 *Sov. Phys. Usp.* **30** 1073]

[67] Ginzburg V L 1979 *Usp. Fiz. Nauk* **128** 435 [1979 *Sov. Phys. Usp.* **22** 514]

[68] Ginzburg V L, Kirzhnits D A and Lyubushin A A 1971 *Zh. Eksp. Teor. Fiz.* **60** 451 [1971 *Sov. Phys. JETP* **33** 242]

[69] Ginzburg V L, Mukhanov V F and Frolov V P 1988 *Zh. Eksp. Teor. Fiz.* **94**(4) 1 [1988 *Sov. Phys. JETP* **67** 649]

[70] Barrow J D 1988 *Q. J. R. Astron. Soc.* **29** 101

[71] Ginzburg V L 1996 *Usp. Fiz. Nauk* **166** 169 [1996 *Phys. Usp.* **39** 155]

[72] Ginzburg V L 1993 *Usp. Fiz. Nauk* **163**(7) 45 [1993 *Phys. Usp.* **36** 587]

[73] Ginzburg V L and Dogel' V A 1989 *Usp. Fiz. Nauk* **158** 3 [1989 *Sov. Phys. Usp.* **32** 385]; 1989 *Space Sci. Rev.* **49** 311

Article 18

The business of bygone days: reminiscences of my participation in the 'atomic project'[1*]

The *Moscow News* no 41 (8 October, 1989) published an interview with I N Golovin about the first Soviet atomic bomb, its making and testing (on 29 August, 1949). Simultaneously, the *Argumenty i Fakty* no 40, 41 (7 13 and 14–20 October, 1989) published the article "The atomic bomb mystery" also containing some information about this subject.

In what was told by I N Golovin, I was especially impressed by the episode dealing with I V Kurchatov's letter addressed to Niels Bohr. I had never heard anything of the kind—this matter is extremely surprising and unclear to me. That is why on 9 October I called I N Golovin but, unfortunately, he could not say hardly anything besides what had already been said in the interview. To tell the truth, though I am all for openness, I doubt that mentioning the 'correspondence' of Kurchatov with Bohr was admissible without revealing the content of the letter (and this content is unknown to I N Golovin). Now this question must be fully examined (otherwise a shadow might be cast over the memory of Bohr and of Kurchatov as well).[2*]

In the conversation with I N Golovin, naturally, the question was raised of the next stage of the 'atomic project'—of creating and testing a hydrogen bomb. Though I took part in the work at this stage, it was only several days ago, from the book *I V Kurchatov* by T M Chernoshchekova and V Ya Frenkel' (Moscow: Prosveshchenie, 1989) that I learnt that the testing of the first thermonuclear (hydrogen) bomb took place in the USSR on 12 August, 1953.

I learned from I N Golovin that even he until recently had not known about my role in creating the hydrogen bomb. Still earlier I had had a chance to make sure that this fact was unknown to G N Flerov and, at one time, even to B P Konstantinov, who had guided the creation of the unit for division of lithium isotopes (though, as will be clear later, it had been done for carrying out my proposal).

It is clear that now, at last, the ban preventing the events of 40 years ago from being known will be lifted and publications about the hydrogen bomb will appear.[3*] I N Golovin said (so I understood it) that he was collecting material on this topic and asked me to write what I remembered.

So I am writing. I must say that I have a bad memory and my participation in the 'project' was peculiar and limited. But then I am only going to relate what I remember, without any claims to being complete and so forth. Besides, I am not going to circulate this text. I will only give it to I N Golovin—so we agreed.

In 1947 I V Kurchatov enlisted I E Tamm to the work on the project to create the hydrogen bomb, i. e. an explosion based on reactions between hydrogen isotopes. Digressing, I will note that I have always considered this act of I V Kurchatov to be a manifestation of his high moral qualities. Whether it was really so, I am not absolutely sure, as I do not know anything except the following. In 1943, after I V Kurchatov was made the head of the 'atomic project', someone somewhere decided, as it was customary with us (I hope it will never be customary in the future), 'to consolidate' I V Kurchatov's position by electing him an Academician. A corresponding seat was 'allotted' and, in 1943, in Kazan, the election took place. But A I Alikhanov was elected to this place.[12*] Those who know the ethical conduct at the Academy will at once answer the question that I asked many times: what did the Academy's authorities do after 'the instructions from on high' to elect I V Kurchatov were not carried out? The answer is: the right person called another right person and 10 minutes later a second seat was 'allotted', after which I V Kurchatov was elected. I am writing about this here because I E Tamm, then a Corresponding Member of the Academy, took an active part in this business and openly agitated for A I Alikhanov's election. Of course, he was absolutely wrong. Even at that time, as I am sure, the achievements of I V Kurchatov were incomparably higher than those of A I Alikhanov. Under such circumstances, I V Kurchatov had every reason to feel offended by I E Tamm. I do not know if he really felt offended but the fact is that he enlisted I E Tamm to the work, though I E Tamm was far from being in favor and there was no dire necessity for him. It is this fact that I consider to be a manifestation of I V Kurchatov's high moral qualities, not every person would be capable behaving like that.[4*]

At that time I had been a DSc for 5 years and I E Tamm's deputy at the theoretical department of P N Lebedev Physical Institute. Naturally, I E Tamm enlisted me to the work, as well as a number of other workers of the department: S Z Belen'kii (it seems that he was then working for his doctorate under IE's guidance), his postgraduates A D Sakharov and Yu A Romanov, my postgraduate E S Fradkin and also, it seems, someone else (I remember that afterwards, probably in 1950, this group was joined by V Ya Fainberg, who came to work at the P N Lebedev Institute.)

We started by exploring 'combustion' in liquid deuterium. The reactions here were

$$d + d \rightarrow {}^3He + n + 3.27 \text{ MeV}$$
$$d + d \rightarrow t + p + 4.03 \text{ MeV}$$
(18.1)

here d and t are the nuclei of deuterium and tritium, and n and p are a neutron and a proton.

I do not remember whether the reaction with tritium

$$d + t \rightarrow {}^4He + n + 17.6 \text{ MeV}$$
(18.2)

was considered from the start, maybe it was not, since tritium is radioactive and practically non-existent in nature.

All this can (and should) be made more precise by looking into the reports (mine, among others), which can be probably found at the Ministry of Medium Machinery Construction.[5*] I only remember that some time later (it might have been the beginning of 1949) I came across an article dealing with reaction (18.2) in the *Physical Review*. The article informed me that this reaction cross section was far bigger than for reactions (18.1). I remember that I was surprised that the Americans published this article. They must have not yet been aware of tritium's possible role. Anyway, for the d + t mixture, the 'combustion' (i. e. explosion, propagation of the shock wave) was going slowly and the possibility of creating a hydrogen bomb looked rather doubtful even with the d + t mixture. Besides, as has already been said, using tritium did not look very realizable.

It was here that it occurred to me that we could use the reaction

$${}^6Li + n \rightarrow t + {}^4He + 4.6 \text{ MeV}.$$
(18.3)

Lithium-6 (${}^6Li$) is present in natural lithium in the quantity of several percent (the main isotope being ${}^7Li$) but extracting it would not necessarily be a great problem (as ${}^{235}U$ had already been separated from ${}^{238}U$). As I learned, many years later, from B P Konstantinov, that it was he who headed the team who constructed a plant for extracting ${}^6Li$. For 'igniting' reaction (18.2) at the very beginning, it is necessary, of course, to have a certain quantity of tritium. So this is probably done (this is how it was the American press is even now shouting in connection with the temporary close of the plants producing tritium). But, after that, tritium can be regenerated by way of using reaction (18.3).[6*]

The Americans could not have thought at first of using reaction (18.3) or, in any case, they must have underestimated this opportunity. The first testing of the American hydrogen bomb or, to be more exact, some prototype of it, is known to have been carried out with the use of a cumbersome ground unit without ^{6}Li. When, on 12 August, 1953, the first Soviet hydrogen bomb already containing ^{6}Li was blown up, Americans found ^{6}Li in the atmosphere and were astounded, I read somewhere. So, *post factum*, many years after 1953, I understood that my suggestion had been of great importance for our 'atomic project', where I had been removed from closer cooperation (see later). However, using ^{6}Li is not enough in itself—even in the pure mixture d + t, if it is a question of a liquid or a solid under normal pressure, the reaction is too slowly for the explosion to emerge. So, the mixture should be compressed. For solving this problem at that time (I do not remember whether it was before my proposal to use ^{6}Li or some time later), A D Sakharov came up with an idea of how to achieve compression. Let AD write about that himself, I can only say that quite recently I asked AD if his idea had been used. He answered in the negative.[7*] But then (at the very end of 1947 or the beginning of 1948)[8*] both suggestions (A D Sakharov's and mine) 'went to the top'. The result was the following: it was decided that all this was realizable and Tamm's group was told to transfer to Yu B Khariton's facility. (This was a closed site where secret work was carried out, termed the 'Ob'ekt' (Unit) by the group of physicists involved.) But, in fact, those who left were I E Tamm, A D Sakharov, and Yu A Romanov. C Z Belen'kii was also recorded as having left. But he was already seriously ill (he died in 1956, at the age of only 40). However, the order had been given and signed probably by the great leader himself. So for several years Belen'kii was recorded as working at the 'Ob'ekt', where, in fact, he had never been. But he lived in Moscow—nobody (probably not even I V Kurchatov, to say nothing of I E Tamm) was able to do anything else.[1]

As for me, I was not sent to the 'Ob'ekt' and remained in Moscow as head of a small 'supporting group', so to say. We were given assignments, Tamm and Sakharov came from time to time. C Z Belen'kii and E S Fradkin, working in Moscow, made rather valuable calculations: it seems that they were connected with the IPM research institute, which is now named after M V Keldysh.

The reason why I was not sent to the 'Ob'ekt' must have been my political unreliability.[9*] Though in 1942 I was accepted for candidate membership of the CPSU and, in 1944, was admitted to the party, in 1946 I married

---

[1] In this connection I even wrote 'a poem' (it seems that I did it only once in my life):

Our esquire is bursting to go into action,

Where the thunder is roaring,

Where the tam-tam is beating.

But Motherland decided he must stay,

As from Moscow he should not be away.

So, instead of exciting life, he remained with his wife.

To be understood I will add that Belen'kii's friends called him esquire.

Nina Ivanovna Ermakova (since 1947 her surname has been Ginzburg), who had been subjected to repression in 1944 and in 1945 was living in Gorky, in fact, in exile. NI's father, an engineer and a rather old member of the party, had been arrested in 1938 and died in Saratov in prison, in 1942 (he was there with N I Vavilov, who died in the same place at the same time). Nina, a student of the Mechanics and Mathematics Department of the Moscow State University (MGU), was arrested together with a whole group of young people on the charge of preparing an attempt on Stalin's life. It is a rather well-known case, already described in the literature, though not exactly (see 1978 *Pamyat'* (*Istoricheskii Sbornik*) [*Memory* (*Historical Collection*)] issue 1 (New York) p 219). According to the KGB's script, it was planned to shoot Stalin out of the window of the apartment on Arbat Street where Nina lived. But the KGB 'scriptwriters' had not bothered to check everything before the arrest and only found out later that the windows of the room where Nina lived with her mother did not face Arbat. With all the absurdity of the charges fabricated by the KGB, the investigators tried, to a certain extent, to avoid assertions which could have been easily refuted. Anyway, the charge of terrorism was withdrawn from Nina but what remained was 'only' the charge of counter-revolutionary group anti-Soviet activities (clause 58.10 and 58.11 of the then Penal Code). In March 1945, having spent about nine months in prison, she was sentenced, without any trial, by the decision of the so-called 'special conference', to three years' imprisonment in a camp. With regard to clause 58, such a 'small' term very rarely was given. Probably because of that when the amnesty was announced at the end of the war, even those imprisoned under clause 58 but with only a three-year term fell under the amnesty (while the overwhelming majority of those 'convicted' under clause 58 were not subject to the amnesty). So, in September 1945, Nina was released but without the right to live in a number of major cities. Her aunt lived in Gorky, so she chose Gorky but she was able to get official registration only in the village of Bor on the other bank of the Volga. However, she managed to enter the Polytechnic Institute in Gorky (this was quite unusual but some good people helped her), which she graduated from in 1947. Until 1949 she lived illegally in the room given to me in Gorky but at the end of 1949 she was registered in Gorky itself owing to A A Andronov's petitioning (it happened after the serious accident on the Volga on 29 October, 1949—a launch taking people from Gorky to Bor capsized: out of some 250 people who were on board only 13, it seems, were saved, Nina among them). Naturally, every year (i. e. as often as was possible) I put in applications for a permit for my wife to move to Moscow but every time I was refused. Only in 1953, when Stalin's death was followed by a new amnesty, was Nina able to move to Moscow. In 1956 she, like all her fellow-sufferers from the 'anti-Soviet group', was completely rehabilitated, i. e. all charges against her were declared groundless. To characterize the process of rehabilitation, I can tell that an investigator came to the apartment of Nina's mother and in the presence of witnesses composed a document confirming that the windows of the room really did not face Arbat.

But I have digressed, thinking that this story is also interesting, though now such stories come as no surprise. So, being considered politically unreliable, I was not sent to the 'Ob'ekt'. I was only too glad, because after Tamm and Sakharov had left, I had little classified work and could do something for my private satisfaction. It is not that I neglected my duties but calculations and mathematical physics are not my cup of tea, and it was these things that we had to do then (as I have already written, in our Moscow group this was successfully done mainly by S Z Belen'kii and E S Fradkin). That is why I was happy when, in 1950, research in the field of controlled thermonuclear fusion began on the initiative of A D Sakharov and I E Tamm. At the first stage, it was certainly a physical problem. I, too, went in for it and did something. Then this work was considered so important and secret that about the year 1952 (or at the end of 1951) I was removed from it. It might have been for this reason that after this work was declassified (owing to I V Kurchatov, who made, in 1956, in England, his well-known report on the thermonuclear problem), I published, in 1962, some old reports of mine (1962 *Trudy FIAN* [*Works of the P N Lebedev Physical Institute*] **18** 55–104.[11]*

After being removed from the work on thermonuclear fusion, I did not do anything classified work at all, as far as I remember, and nobody demanded anything of me. Unfortunately, in 1955, I was sent to Yu B Khariton as a member of a commission of experts headed by I E Tamm. But since at that time I was not at all interested in all that, I do not remember anything of what was shown to us. After that I did absolutely nothing secret but my work was still recorded as 'classified' until 1987—a further 32 years (and maybe I still am 'classified'?). In this connection I was repeatedly refused permission to go abroad, which was explained by 'the objections of the Ministry of Medium Machinery Construction'. As I have already said, this ordeal continued up to 1987, though I wrote letters to Brezhnev, Zimyanin, and Ryabov, to say nothing of letters to the Presidium of the USSR Academy of Sciences.

Of course, I was not alone in suffering from this idiotic secrecy. However, the references to secrecy were nothing but a pretext for not letting people go abroad and probably for taking revenge for some imaginary transgressions which figured in the reports of those who informed on me (I know that there were many such reports, as I have a glib tongue and quite a few enemies).

But that is another matter. While, on the whole, I must say that taking part in the 'atomic project' turned out to be rather positive for me. Firstly, it may have saved my life. The thing is that from 1947 onwards, the press started to persecute me as a cosmopolitan. It started with an article published in the *Literaturnaya Gazeta* [*Literary Newspaper*] on 4 October, 1947 (by the way, my birthday). In this article signed by Academician Nemchinov, mentioning me was inspired by the not unknown D D Ivanenko (this I knew absolutely for sure).[10*] On the same day (here is one more coincidence), the Plenum of the VAK (the High Assessment Commission), on the initiative of the same D D Ivanenko, did not confirm my promotion to the rank of full professor. After that they began to refer to me wherever and whenever possible, accusing me of cosmopolitanism and servility to the West. I was also to figure as one of the victims at the All-Union Conference on Physics, which was planned for March, 1949 but which was cancelled at the last moment (as I heard, this was also owing to I V Kurchatov). I think that as the husband of a woman subjected to repression, a cosmopolitan, and a Jew, at that period I would have been arrested and would have probably perished together with my wife. But, as was said, exactly in 1947 my secret work started and thereby I got a 'safe conduct' (I think that in 1953 it would have no longer worked but here the 'era of Stalin' finished, at last).

Secondly, at the very end of 1953 I was rewarded, and even rather generously—given the Order of Lenin and the Stalin Prize of first degree. Someone told me that I even got a double sum of this prize, 20 000 roubles (if compared with the contemporary prices, it was 200 000 then). However, as I remember, A P Aleksandrov at the same time got the prize of 10 000 but was also given the title of Hero of Socialist Labor. I am sure that I was not given this title only for the reason clear from what has been said. But thanks at least for what was given. Moreover, in 1953, I was elected a Corresponding Member of the USSR Academy of Sciences. Undoubtedly, it was also one of the consequences of the mentioned classified work, though I was the author of many other works (and rather good works in a number of cases, I dare say). But who would have elected me had it not been for the approval of I V Kurchatov.

That is, as a matter of fact, all I can tell. I hope that the story of the creation of our hydrogen bomb will be published, as concealing all that now is simply absurd (once again see comment 3).

Comments

1*. This article is actually the letter (of 15 October, 1989) which I sent to I N Golovin. It goes without saying that this letter has not been previously published.

2*. Everything mentioned here can be read in the first edition of the book (1997). In accordance with my self-accepted rule—not to change what has already been published but only comment on the text—I am doing so in this case as well. Here it is all the more justified as in such a way it is possible to demonstrate the level of secrecy which existed in the USSR. Indeed, I N Golovin was one of I V Kurchatov's deputies but, even 40 years after the first explosion of an atomic bomb, he did not know the details of Kurchatov's contact with Bohr. The corresponding publications appeared only in 1994, i. e. after the collapse of the USSR. I will not give any references, as they can be found in the easily accessible article by P E Rubinin (1997 *Usp. Fiz. Nauk* **167** 101 [1997 *Phys. Usp.* **40** 95]). From this article, and also from the subsequent article by the same author (1997 *Usp. Fiz. Nauk* **167** 1349 [1997 *Phys. Usp.* **40** 1249]), the picture is already clear enough. In 1945, "under the guise of looking for the equipment from Soviet research institutions taken away by the Germans", a group of KGB workers was sent to Denmark "to get into contact with Bohr and obtaining from him information on the problem of the atomic bomb"—this is a quotation from L P Beriya's letter addressed to I V Stalin on finishing the spy 'operation'. On 14 and 16 November, 1945, the KGB agent, physicist Ya P Terletskii, who was a member of the mentioned group, talked with Bohr and asked him 22 questions, afterwards reporting to his superiors about the answers. Terletskii was received by Bohr because he gave him a letter from P L Kapitza, which said, in particular: "This letter will be given to you by the young Russian physicist, Terletskii. As a young and capable professor of the MGU, he will explain to you himself the purpose of his trip abroad. With him you will be able to send me your answer..." (the quotations are taken from the previously mentioned article by P E Rubinin). According to P E Rubinin, Bohr told Terletskii what 'one could learn having read attentively "Smith's account" just published in the USA'. He gave a rotary duplicator copy of this report to Terletskii.

What should we think of P L Kapitza's previously mentioned letter, as well as of his earlier letter to Bohr, of 28 October, 1943, with the suggestion that he should come to the USSR? This is not a simple question. P E Rubinin's opinion is clear from his articles. A somewhat different view on the situation is given in the book: Pais A 1991 *Niels Bohr's Times* (Oxford: Clarendon Press). I do not have a clear picture, on the whole but I wonder why P L Kapitza, who was detained in the USSR in 1934, thought that the same might not happen to Bohr.

3*. What has been said here rather well characterizes the degree of secrecy around the work at the 'atomic project' in our country. Even in his book *Vospominaniya [Reminiscences]*, the preface to which was written, probably, in 1989, A D Sakharov does not reveal the essence of the ideas used in creating the hydrogen bomb (he calls his suggestion the first idea, and mine, the second idea; it is about this second idea, using ^{6}Li, that I am writing here). After Sakharov's death (on 14 December, 1989) the first and second ideas were declassified at last (this was done in the articles by Yu A Romanov and V I Ritus, published in the journal *Priroda [Nature]* no 8, 10, 20, 1990). What followed was an inexhaustible flow of publications dealing with the history of creating atomic (nuclear) and thermonuclear weapons in the USSR. However, with regard to the hydrogen bomb, this was done with enough detail only in 1996, in the following articles: Khariton Yu B, Adamskii V B and Smirnov Yu N 1996 *Usp. Fiz. Nauk* **166** 201; Goncharov G A 1996 *Usp. Fiz. Nauk* **166** 1095 [1996 *Phys. Usp.* **39** 185; 1996 *Phys. Usp.* **39** 1033].

From these articles, especially from the article by G A Goncharov, I have learned a lot of new things, to say nothing of remembering the content of my reports, in former times strictly classified.

4*. In the article on the occasion of I E Tamm's 100th birth anniversary (see article 8 in this collection), I am discussing the question of why I E Tamm was not enlisted to the work at the 'atomic project' for rather a long time. In that it happened at last (in 1947 or 1948), the beneficial role of I V Kurchatov is beyond doubt. I think, however, that the main obstacle for engaging I E Tamm in this extremely secret work was objections from the KGB (see article 8 in this collection).

5*. See the articles mentioned in comment 3.

6*. Strictly speaking, I do not know whether having some amount of tritium in a bomb from the start is really necessary. It is a question of kinetics and the efficiency of the 'product'. In principle, it is possible, of course, not to introduce the 'igniting' tritium but to expect that it will be produced a result of the reaction (18.3), due to the neutrons of the fission, from the atomic (plutonium or uranium) charge, initiating a thermonuclear explosion.

7*. Now I do not remember this conversation in detail. I am somewhat surprised by Sakharov's answer, as nowadays it is known that in the first two Soviet hydrogen bombs the compression of light elements was achieved exactly by using Sakharov's 'first idea', i.e. applying the so-called 'layer structure' ('sloika'). Later the 'layer structure' was substituted with a system with compression by radiation (this is the third idea mentioned by Sakharov in his book of reminiscences). Apparently, when answering me (see the text), Sakharov meant contemporary hydrogen bombs.

8*. For some reason I remember it as the end of 1947 but I must be mistaken. In any case, our suggestions 'went to the top' only in 1948 and Tamm and Sakharov went to the 'Ob'ekt' (Arzamas-16) only in 1950.

9*. Here again I am touching upon the subject already described in my book 1995 *O Fizike i Astrofizike* [*On Physics and Astrophysics*] (Moscow: Byuro Kvantum) [2001 *The Physics of a Lifetime* (Berlin: Springer)] and especially in article 26 of this collection. I apologize for the repetitions but each time something different is written and, in general, I decided not to abridge anything, as in abridgement something will be lost or may be distorted.

10*. By chance, at the home of some acquaintances, I met the author of the article in the *Literaturnaya Gazeta* (at first it was called, it seems, "To Court Honor" but it was published with the heading "Against Servility"). This author (his surname is Shneiderman, I do not remember the initials) was working at Timiryazev Agricultural Academy or had some connection with it. So he had been given the assignment to write an article stigmatizing the opponents of Lysenko, in particular the renowned biologist A R Zhebrak. Also working at the Agricultural Academy at that time was D D Ivanenko. Having learned about the article being prepared, he decided to take advantage of the opportunity. Specifically, he persuaded Shneiderman, so as to make the article sound more generalized and weighty, not to limit himself to examples from the field of biology (or, to be more exact, to criticize not only those who were against Lysenko) but also to come down on the 'servile' physicist Ginzburg, who would not acknowledge the achievements of 'the truly patriotic scientist' Ivanenko and so forth.[2] However, in the *Literaturnaya Gazeta*, they did not want to publish the article with the surname Shneiderman and, I do not know how, got Academician V S Nemchinov, the then Rector of the Agricultural Academy, to sign it. As far as I know, V S Nemchinov was a qualified person and not altogether a partisan of Lysenko, he was just exposed to tripping up. How bitter it is to remember this damned bolshevist past of ours!

11*. It was not until 50 years (!) had passed that I learned something about the organization of research in the field of controlled thermonuclear synthesis in the USSR, in which I took part in 1950–51. The corresponding historical data are given in several articles published in 2001 *Usp. Fiz. Nauk* **171** 877–908 [2001 *Phys. Usp.* **44** 835]. Here I will only dwell on my work in the field of thermonuclear reaction, which I probably took up in August or September, 1950. It happened quite informally, it was just that I E Tamm or A D Sakharov, during one of their flying visits from the 'Ob'ekt' to Moscow, told me about the problem. I was glad to join in this business, since, as I have already said in this text, at that time I had little other 'closed' work. I wrote several 'thermonuclear' accounts, they 'went' somewhere. And so, as I have now learned (see the mentioned publication in *Usp. Fiz. Nauk*, p 896 [2001 *Phys. Usp.* **44** 854]), I V Kurchatov, I N Golovin and A D Sakharov in their letter of 11 January, 1951, addressed to L P Beriya, suggested, among other things, that in developing corresponding research 'a working contact should be established' with 'the worker of Lebedev Physical Institute, Comrade Ginzburg V L, who has conducted important theoretical research into the nuclear magnetic reactor'. In response L P Beriya commissioned a number of things, pointing out

---

[2] A recently published article (Sonin A S 2004 *Voprosy Istorii Estestvoznaniya i Tekhniki* (Problems of the History of Natural Science and Engineering) 1 18) gives the verbatim account of D D Ivanenko's speech (1947) to the plenum of the Supreme Certification Commission. In this speech Ivanenko demanded that I should not be conferred the title of Professor, accusing me of 'hushing up the achievements of compatriot physicists and advocating foreign works'. Naturally, by 'compatriot physicists' were meant Ivanenko and his co-authors.

that "Taking into account the particular secrecy of working out the new type of reactor, a careful selection of people and measures of due secrecy of works should be ensured". Obviously, I did not stand up to this 'careful selection of people' and was removed from this work (officially started by the Decree of the USSR Council of Ministers, approved by Stalin on 5 May, 1951). Therefore, it was probably as early as in May 1951, or maybe a little bit earlier, that they stopped to show me my own calculations on the thermonuclear reaction.

12*. See, however, comment 7 to article 8 of this collection.

Article 19

How I became a theoretical physicist and, in general, about myself... [1*]

10 October, 1979. I have been confined to a hospital bed for 35 days (now I am beginning to get up). More precisely, this is the third hospital, the one in which I was operated on (19 September) and I had been hoping to go home in a couple of days. But my temperature jumped (nobody knows and probably will never know why). And so I will have to stay in the hospital for five more days at the least. I am bored, I read, of course, and look through Xerox copies of papers but attempts to devise or do something in physics are unsuccessful. The times when I used to undertake brainstorming with some effect, which I hope to outline later, have supposedly gone never to return. Yes, physics is the game of the young and I was 63 six days ago. My mind, however, is clear (or seems to be). By no means am I going to surrender and I am planning to work in the years to come. I believe, Rubakin-senior (has there been a writer with this name?) was right to remark: "the only bad thing about old age is that it also passes" (with the reservation, however, that there are no illnesses, which prevent one from working and bring about suffering). In any event, I do not feel like working and the only useful pursuit (?) that occurs to me is to write something like a memoir. Not 'in general' but on a narrow topic. And not for publication, just for no particular reason. However, it may be that some day will be published it but this does not matter. Or I would let my close friends read it.

I have already related several times the story of how I became a theoretical physicist to my students and postgraduates—certainly I do not write in order to speak of myself once again. It is simply that I am concerned with many aspects of the question of choosing the path and profession of physicist and this question is undoubtedly vital and topical for young people! In any case, they have always listened with interest. The role of chance, of a friendly hand... all this cannot be formalized and, for this reason alone, is interesting.

To complete the picture, I will tell the story in a roundabout way. My school years coincided with maybe the most unfortunate period in the history of Soviet secondary education. Of the old school (the gymnasia, etc) only some buildings and individual teachers remained. Chaos reigned over the rest. In 1931, when I graduated from a seven-year school, everything terminated with it—nine-year schools, etc had been abolished. After school, one was supposed to enter an FZU (a factory training school) and later may be enter a rabfak (a worker's preparatory school). The seven-year school system was itself on its last legs. History was not taught at all—there was 'social science' and, for example, in the last (seventh) form only comrade Stalin's report to the 15th (?) Congress of the VKP(b) (Communist Party of the Soviet Union [bolsheviks]) was studied. 'Social' work flourished. I remember a shortish lad (his surname was Rybin) head of the 'uchkom' (educational committee): he later was dismissed to be criticized for 'rybinhood'. However, there were several old and supposedly skilled teachers (Tsetsiliya Leonardovna taught us all subjects (?) except for mathematics which was taught by Nikolai Pavlovich). For some reason, my parents (more precisely, my father and my aunt) had sent me to the fourth form and so I studied at school for only four years. I do not know the present-day school program and it may well be that my schooling was not too bad. But the absence of a proper 'educational' atmosphere, in the family in particular, had the effect that I am under the impression that I gained little from school. Nevertheless, an interest in physics emerged even in those years and a steady one, though I myself do not know why. I was fond of O D Khvolson's book *Fizika Nashikh Dnei* [*The Physics of Our Days*], which I read even at school or immediately after graduation, it seems to me. All in all, I never hesitated about going in for physics since those school days (they taught us physics, even with demonstrations but I can recall neither the teacher nor the textbooks).

I should not have gone into these dim recollections—they can hardly give anything. On graduation from school I did not enter an FZU and, I remember, would feel very uncomfortable and lonely. Then I was somehow fixed up in a job as a laboratory assistant in the Bubnov Moscow Evening Machine-Building Institute. Later (when Bubnov was supposedly imprisoned), Bubnov was replaced with Lepse. Later, it seems to me, Lepse also 'disappeared' but I am not sure (alas, it is even possible that I have quoted the names of Bubnov and Lepse in the wrong order). Initially I was 'in training' in A A Bochvar's laboratory of the Institute of Non-Ferrous Metallurgy (we still remember and shake hands with one another—since 1932—though I worked there for only half a year). Then I found myself in the X-ray Laboratory of the Lepse Institute. The chiefs were E F Bakhmet'ev and N K Kozhina (for some time Ya P Silisskii was also there). The major power was Venya Tsukerman. Leva Al'tshuler was also there, though I do not remember what list he was on. The three of us were on friendly terms and worked together. Of course, I ranked third, the lads were three years older and knew essentially more. Ven'ka called us the 3Vs: Venya, Vitya, and Vladimirovich Leva or 'vsegda vperedi vsekh' (always ahead of all).[1] But this is a special topic which wanders away. The work in the Laboratory was of benefit to me but it taught me resourcefulness (following Venya's example) and experimental skills rather than physics. In physics, to say nothing of mathematics, I made no significant progress. The year 1933 saw the first 'free' (i.e. 'competitive' rather than by assignment) enrollment to the Moscow State University (MGU), and I decided to enter the Department of Physics. It is either that I left the Laboratory or somehow otherwise but, for three months, I was perseveringly preparing myself with two, it seems to me, tutors, i.e. visited them to take lessons. And so during these three months I went formally through the eighth, ninth, and tenth grades. But the program was actually shorter and, say, mathematical analysis was not required at all—I remember becoming proficient in it on my own a year later using the textbook by Grenville and Luzin. I remember I knew only some 'Bezout theorem' in addition to the program in mathematics: at present, I do not remember what it is about and it may be that all schoolboys nowadays know it. From a formal viewpoint, I was prepared but I am convinced that the lack of good, regular schooling had an adverse effect on me. While a schoolboy (a good one, which I would have been despite having no better than a modest ability for mathematics) solves, say, 100 or 1000 problems on trigonometry, logarithmics, etc, the number I solved was 10 or 100 times smaller. The same was true of arithmetic for I entered the fourth form. And this is always said to me: I perform calculations badly, slowly, with effort, automatism is lacking. I have always feared and disliked calculations. Of course, behind it is the absence of an ability for mathematics (fortunately, not total but in comparison with the corresponding abilities of the overwhelming majority of my fellow-theorists). But this is precisely the reason why the lack of training had such a pronounced an effect.

Of course, the lack of regular schooling adversely affected other aspects, too (I did not sense it only in physics). At the age of about 30, I read for the first time A Gertzen's *Byloe i Dumy* [*The Past and Meditation*] and many other works of literature (however, I am not sure that this is a drawback). Of greater significance is 'the Russian language'. When I was in my second year at the MGU, all of us took dictation and I made eight mistakes (I wrote 'koleblyashchiisya' instead of 'koleblyushchiisya' and made seven more punctuation mistakes) to be marked 'unsatisfactory'. This was common, it seems to me, to half the students and for some time we would attend Russian lessons. Even now I write with mistakes and seek somehow to 'shift the blame' onto a typist. However, making grammatical mistakes

---

[1] See comment 2 in article 17 of this book.

is not as significant as the ability to write—the mastery of style and language. My language is somewhat poor and my phrases are frequently not quite literate. In this connection I recall my conversation with G S Gorelik. He had the ability to write well and, to my question "What helps you write so well?", he replied with a question: "How many times a week did you write compositions at school?". I answered something like "once a week or once in two weeks, I do not remember". GS remarked that he had studied in Switzerland and wrote compositions every day. The same topic is touched upon in Irina Erburg's (I Erenburg's daughter) book entitled *Latinskii Kvartal* [*The Latin Block*], if my memory does not fail me (I read it many many years ago). All in all, no educational institution can make one into a very good writer, physicist, or mathematician, unless s/he exhibits the corresponding aptitude. However, first, inclinations alone would not suffice. How many talented people never 'realize' their potentialities and what role was played by the shortcomings in their education? Second, a good background, training, etc are supposedly able to make a worthy professional out of a person of average abilities, who would otherwise be a drudge, become a failure, find no satisfaction in work, etc. On the whole, it is all clear. I write wherever I am led by my pen; and this topic has been touched upon because I have often pondered over the question as to what losses I have 'incurred' due to unfavorable conditions at school. Of course it is impossible to give a clear answer. On the one hand, as I believe, I was extremely lucky as regards the 'realization' of my modest abilities (see later). But, on the other hand, what would have been if I had studied in a good ten-year school, to say nothing of 'professional' family support (there was none)?

Here, I would like to touch upon yet another 'favorite' topic which I ponder over quite often. Take, for example, a sportsman who covered, say, a 100-m distance in 9.9 seconds to become an Olympic champion and a sprinter who did it in 10.2 seconds and coming fourth missed even the bronze medal (the figures are, of course, arbitrary). Here, random circumstances might have played their part: how he had slept, whom he had slept with, what had eaten, how he had pushed off the shoe, etc. Fortunately, in science this is not the case: the lot of the fourth is much better, he makes his contribution, writes good papers (with the understanding that the first writes very good ones). But the role of chance and of good luck may still be critical. This is not so for titans like Einstein, for the 'safety margin' and the outstripping of others are too large. The talents of Maxwell, Bohr, Planck, Pauli, Fermi, Heisenberg, and Dirac were scarcely dependent on the fluctuations of good luck, accidental idea, etc. De Broglie and even Schrödinger were, it seems to me, a different matter, to say nothing of numerous Nobel Laureates. M von Laue was a well-qualified physicist but it is stated that the idea of X-ray diffraction in crystals was a 'beer idea' (Bieridee). The Braggs, Roentgen, Zeeman, Stark, Lenard, Josephson,

Penzias and Wilson, Hewish and Ryle, Cherenkov, Basov, and Prokhorov, as well as three-quarters of the entire list were largely strokes of luck rather than 'divine' revelations. I only want to emphasize that chances of success depend both on a lucky strike and a variety of other factors, which include health, a timely read article or book, activity, ambition (as a stimulus), and perhaps many others—an interesting topic.

It is time, however, to revert to my attempt to enter the Department of Physics of the MGU in 1933. I took an examination in physics with Yu Kushnir, who was red-haired and likeable. He proved to be (it turned out, of course, many years later) a friend of M A Markov (regrettably, he is gone). He examined me rather strictly and yet benevolently. He gave me a 'good' mark, if my memory does not fail me. I got an 'excellent' mark for something else and the remaining marks were 'good' and 'satisfactory'. On the whole, I passed my examinations so so and not through the lecturers' fault. At the exams there was not a slightest trace of discrimination. It showed, however, in the fact that I was not admitted, though people with even worse marks were. Among them was Sema Belen'kii (likeable, small, and dark-haired). But his total mark was only about one point less and his personal particulars probably had some advantages (he must have been a comsomol member or his parents' 'indices' were better). Anyway, it did not look like a crying injustice. It was that I simply was not admitted and nothing else (by the by, Misha Galanin was and his brother Liks was not, though he was quite good).

Before going further, one more digression. When I was not admitted, Venya decided to undertake something, namely to go to the rector of the MGU. Venya was 20 and certainly not solid enough for a matter like that. That is why spectacles were found, simple glass was put into them (at that time Venya did not need glasses, though his eyes were already failing him but in a different way—he saw badly in the dark). So here I am sitting in Alexandrovskii Garden and waiting for Venya who has gone to the rector (who was the rector then? Was it Butyagin?). Venya told the rector about 'our young capable worker', i.e. about me, the rector must have mumbled something and that was the end of it. I remember that while sitting in Alexandrovskii Garden I was watching some strange couple (such are the whims of memory—it was 46 years ago, I have forgotten a lot of things since then but this couple I remember). He—probably some mathematician—was rather odd-looking and tall. She was also somewhat unusual, also tall, weary of something, and looking as if she were not a country woman but something different. The impression remained, there is no point in analyzing it.

The worst thing about my not being admitted was that it once more hit my sore spot. I decided not to wait (like Liks) for the next year and entered the correspondence courses of the Physics Department. I studied by myself and, in addition, there were some lectures for external students.

But afterwards (I entered the second year), having compared my studies and those of my comrades, I understood that I had lost quite a lot in mathematics. Reading Grenville and Luzin could not completely substitute for lectures and seminars and, above all, the competitive spirit—the first-year students were carried away with the theory of sets and so forth. Besides, I somehow cheated and did not take exams in astronomy and chemistry and did not study these subjects at all, which I feel even now. It may seem funny but I do mean it. Since 1945 I have been working in astronomy and some people have long considered me an astronomer (even formally, I have long been a member of the International Astronomical Union; and was elected, in 1970, a foreign member of the Royal Astronomical Society and also, in 1975, I gave the Darwin lecture of this society). But at the same time I am a complete ignoramus in elementary school astronomy—I do not know the constellations, I know celestial coordinates badly and so forth. Of course, learning all that is not so difficult. But it is exactly 'learning' that is my weak point—I cannot learn anything uninteresting, here my will is not strong enough (which it is in some other cases). That is why I still have some gaps in my school knowledge. Disgracefully, I also do not know foreign languages, though, thank God, I have somehow mastered English (but I only can speak, though with mistakes, and make reports, while I am almost unable to write on my own without someone checking it). I am writing all that because I have firmly come to the conclusion that a person needs quite a lot to do real work and achieve success and satisfaction. Not knowing languages is, as a matter of fact, a disgrace, to say nothing of the harm to the business. The Europeans do not have such a problem. Any Dutch physicist knows English well and probably also knows German and French. Having a facility for languages, one could master a language even without studying it at school—having started from childhood and so forth. But what if a person does not have linguistic abilities? These are specific abilities indeed. I, for example, am absolutely unable to remember poetry and, in general, am not able to learn anything by heart (as, for instance, a report). Whereas in a natural way, in childhood, in the school years I would probably have been able to cope with all that. But enough of this subject, though its repetition is not accidental here. All my life I have felt regret that I do not know languages, that I could know more about this and about that. However, when your work is in progress and there are so many interesting things in it, will you learn verbs or the names of constellations? I, for one, have never been capable of doing that.

One more reminiscence (perhaps the only one left) of the external learning. At the head of it was Ul'pi—a big gray-haired man. Once he was reading a lecture to the students of our year or addressing them with some speech. Among other things, he was quoting Marx (leading to the peaks of science is not a broad road but a steep mountain trail, and so on and so forth). I was charmed. But it somehow happened that I attended Ul'pi's lecture (speech) for another stream. And he was saying the same! He was sounding quite natural. But that left an indelible impression on me— since then I have been afraid to repeat witticisms, quotations and so forth, though sometimes I have to do it and, to a certain extent, it is inevitable (of course, there is a way out—not to deliver one and the same report twice and not to be carried away too much with your own eloquence).

When entering the second year, I was sent to a military hospital in order to decide in which group I should be enrolled, a military or a civil one (in the latter groups there was no military training, whereas the former trained military officers). I was examined by an elderly doctor (he seemed elderly to me but he might well have been below 40). I was very skinny, with the height of 180 cm, my weight was hardly 60 kg. The doctor poked his finger at my throat, pronounced the word 'struma' (it is some swelling in the thyroid gland) and directed me into the civil group. The struma has not made itself felt up until now, while all my life would have been different if I had been enrolled into a military group (suffice it to say that the majority of our students from these groups were killed in the war). Such is the role of chance and, in my life, there were many things like that. Thinking about it, I even feel some superstitious fear. In any case, I understand why a person can be superstitious: even nowadays a man, his life, and fate are like a small boat in the rough sea, always in danger of capsizing. By the by, I have always been superstitious but have always been ashamed of it. However, here in the hospital, I have a certain support. When, answering the question of doctors or nurses, I say that I am feeling well, they say "Touch wood"!

So, in 1934, I became, at last, a second-year student in the MGU's Physics Department. I studied conscientiously and had the necessary minimum of abilities for getting only excellent marks. In all that time I probably did not get a single mark, in any subject, which was not excellent. But with the existing standards and system (in contrast to the system of the Physico-Technical Institute where a student can early show not only his/her ability to perceive but also to do something on his/her own) getting only excellent marks did not mean very much. This in itself could not be the sign of something promising. In our group there were five people (of about 20–25) who certainly stood out: V V Vladimirskii, M D Galanin, S Z Belen'kii, L M Levin, and me (V Struminskii was also definitely capable but already at that time he was somewhat 'on a different plane'). Out of these five, Vladimirskii was certainly the first—both in mathematics and in physics and 'in general'. Out of the remaining four, I was probably the weakest in mathematics but because of my striving to catch up it was not very noticeable and in other subjects I was not worse. Thus, we reached the fourth year, or maybe the end of the third year, when we were to choose our specialty. It was an excruciating process. Sema chose theoretical physics, seemingly without any doubts, just because he was not good with his hands and simply could not become an experimenter. As for me, I was handy enough and already had the experience of working in a laboratory. So, I (like Leva Levin and Misha Galanin, I do not remember anything about Vladimirskii) took up optics. It was not accidental. The chair of optics was headed by G S Landsberg, and we felt that it was one of the best (if not the very best) parts of the Physics Department (there was also a connection with L I Mandelshtam and others). The scientist who became my supervisor was Saul Maksimovich Levi, a Jew from Lithuania, who had worked in Germany (in the laboratory of R Ladenburg) and emigrated to the USSR in order to save himself from fascism. It seems that I am digressing again but should I set S M Levi aside? Then, more than 40 years ago, he was in his late thirties. He was a nice and cultured person whom at that time I was not yet able to appreciate to the full. But something I already did understand, I felt it instinctively, in spite of my lack of general education, lack of understanding, and even blindness. Hinting that 'everything is not so simple', Levi engendered in me some doubts of a general (philosophical) character. But the main thing was that he explained induced emission and so forth (it was studied at Ladenburg's laboratory) to me. He had a quite clear understanding of the possibility of building a quantum amplifier—by creating an overpopulation in the upper level. It was the latter that seemed difficult. In fact, an aberration was taking place there. Overpopulation for some mixture had been created, as it seems, as early as at the beginning of the 1920's but it was not noticed, not understood. As there was no concept of feedback, it would have been difficult to gain a noticeable coefficient of amplification. That is why Ladenburg and his school worked with dispersion (the negative dispersion) but, as it turned out, it seems, fairly recently, they had made mistakes in something. The idea of the amplifier, as S E Frish once told me, was understood by many as early as the 1920s (it seems that he mentioned Rozhdestvenskii and Ehrenfest). Anyway, Levi explained everything to me absolutely clearly in 1936 or 1937. And this turned out to be very important for me (see later).

By the by, in the 1960s, the Bureau of the General Physics and Astronomy Division of the USSR Academy of Sciences confirmed the award of the S I Vavilov Gold Medal (it seems that it was this award) to V A Fabrikant. I E Tamm, as the chairman of the commission, was speaking about Fabrikant's services in the matter of inventing the quantum amplifier and so forth. I objected, saying that I was in favor of awarding the medal to Fabrikant for his works but for the reasons mentioned earlier I do not consider the question of the quantum amplifier to be dealt with correctly by I E and the commission (I do not remember the exact wording). Tamm was displeased. Fortunately, by secret voting, the medal was given to Fabrikant unanimously, so nobody could suspect me of anything. But if someone had merely abstained, it would have been considered to be me, without any chance for me to prove anything.

In 1937 or 1938, S M Levi was dismissed from teaching (he came from Germany!) but, fortunately, he and his wife managed to leave for the USA, instead of going to Siberia or Oswiecim. In the USA, he went in for spectrum analyses but probably did not obtain any special results. He was a knowledgeable and likeable person. After Germany and Russia, living in the USA, in the war time, was not easy for him. When I was in the USA at the end of the 1960s, I tried to find SM (through Miss Ducas—she called Ladenburg's widow, and through Townes) but I did not succeed. Once I told S M Raiskii about this failure of mine and he answered: 'You needn't have gone to America for that, you should have asked me, Levi and me wish each other Happy New Year every year!' Having got SM's address, I wrote to him and sent him off-prints. He answered, it seems, by wishing me Happy New Year. Then our correspondence stopped.

15 October, 1979. I have decided to write more, as I still have not come to the point.

S M Levi offered me a rather strange and, what is the main point, a difficult theme. I think that with the level of technique that we had at that time, I would never have solved the problem. The well-known physicist and fascist, Stark (probably when he still was mainly a physicist), while studying the luminescence of canal rays, came, among other things, to the following conclusion: the intensity of radiation (in separate lines, it seems) along the ray and in the opposite direction is different. Since the velocity of ions and atoms in canal rays is relatively insufficient, this asymmetry cannot be explained by relativistic aberration. My task was to confirm the very existence of this effect experimentally. I sat in a dark (it was also painted black) basement room at the Physics Department (on Mokhovaya street). For some time, V A Kizel shared this room with me. I made a cathode tube, assembled some mirrors, even measured something. But, I repeat, now it is quite clear to me that the experimental level was not at all adequate to the task. I wrote my degree work, defended it and got a diploma with an excellent mark[2] but actually it was a survey containing much information about canal rays and the description of the beginning of the experiment. I also wrote a survey concerning the radiation of canal rays, especially the polarization of the radiation. I sent it to the *Usp. Fiz. Nauk,* they asked me to shorten it and so forth, and eventually I never published it. Not a great loss, at that time little was known about the excitation arising in collisions between atoms. But from all that I derived some benefit: I became acquainted with the theory of collisions and used something somewhere afterwards. At the university, my social work was

---

[2] To be more exact, I am considered to have got it but the diploma itself is something I still have not got. When in 1942, in Kazan, I was defending the doctoral thesis, I had to cook up the following paper: "We, the undersigned professors of the MGU, remember (or know) that VLG graduated, in 1938, from the Physics Department of the MGU..."—and it worked.

organizing the competition of students' scientific works. I remember that some official in the rector's office aroused my righteous indignation and I came down on him writing that 'with his roguish smile' he was not doing something that he ought to do. The man was deeply offended with me. This journalistic debut of mine with harsh words did me a lot of harm and, what is the main thing, I now understand that this 'passionate bolshevist' style does not stand up to criticism.

Considering me a capable student, the chair (G S Landsberg) submitted a demand for my being left at the university for a postgraduate study (on the whole, although not being all that brilliant, I certainly deserved it). At first, however, I was appointed to teach physics at the town of Vireya (or Vereya, as I have never been there). But then the chair managed to save me. However, L M Levin and I were conscripted. I remember the words 'postgraduate student' on our documents but I cannot remember if they had enough time to shave our heads. The MGU pleaded for us, as well as for other postgraduates. Previously, postgraduates had been deferred but that year difficulties arose. Finally we turned out to be the last to be deferred—in 1939, nobody was. This was the second case (after the 'struma') when, thanks to the grace of fate, I had a very narrow escape. In September 1938, when the problem of conscription was still there, I, quite naturally, had no wish to sit in the dark room with the pump working and did not see any point in it. So I started trying to explain the possibility of the effect of asymmetry, which I was exploring. And this is what occurred to me. If the field of an incident ion is resolved into plane waves, these waves could play the same role as light waves, which means that they could produce induced emission. That is why in the direction of the velocity of an ion an excited atom must, owing to the induced emission, radiate more than in the opposite direction. I will not go into details. Summing it up, understanding what induced emission is proved useful. With this idea, I went to I E Tamm—on 13 September, 1938, if I am not mistaken, and so my 'new life' started.

I wrote something about that in my reminiscences about I E Tamm (they were published in *Priroda* [*Nature*] and are to be published in a collection of articles in memory of IE).[2*] By the by, I refer readers to this article and to a short article in the collection in memory of L I Mandelshtam[3*] for some further details which I will not include here (what laziness!). In these memories about IE, I mention, among other things, my preparing, under his guidance, a report at LI's seminar on paradoxes. So as not to focus the attention on myself (on the whole I fought hard against all these 'I', 'me', and the like), I wrote there that 'one of us' solved a paradox at once, giving IE no chance 'to display oneself', thus making him displeased. 'One of us' was me, who cried out on the spot what the matter was (I do not remember the details). I should not have 'stuck out' but I had and have such a manner. It can't be helped. But when I was young, this manner of mine and, in general, my manner of behaving at seminars brought me the reputation of am impudent fellow (that probably was the opinion of many people, among them A A Vlasov, as I remember). Now, so many years later, I behave the same but I do not think anybody considers me impudent. *Quod licet Jovi, non licet bovi.* I am not idealizing myself but I am anything but impudent. Going somewhere out of turn, asking for something and so forth—all that is torment for me, if I ever do it at all. What was (and maybe still is) taken for impudence was a lability of the nervous system, excitability. That is why I usually talk hotly and it seems to be with undue familiarity—for no one knows that my nerves are trembling and my heart is beating. But well then, I knew IE a little, besides, he had given lectures to us. So I waited for the end of his lecture at the Lenin classroom and came up to him in some corner or maybe even at the open space for lecturers not far from the classroom. IE—whether he made an appointment with me or listened to me at once (it seems that it was at once)—burned with enthusiasm.

If I had gone to Landau (I have often thought about this), the result would most probably have been quite different. For my 'idea' was absolutely wrong. Dau either would have noticed it at once and would have thrown cold water on me. Or, in any case, he would not burn. And it was so important for me! I had not yet done anything, not 'tasted' the sweetness of the result, of work. I did not believe in myself, I thought that I could not be a theoretical physicist. And IE talked with me as if I was his colleague, he advised me to have a look at articles on quantum electrodynamics, saying "it is very interesting" or something like that. So I started reading them. And not knowing anything about highbrow matters (quantum field theory and so forth), I miraculously understood something important and interesting. V A Fock and after him A A Smirnov argued in their articles which I somehow came across at that time (maybe on the advice of IE) that in quantum electrodynamics a uniformly moving electron radiates but they did not understand why. I became acquainted with quantum electrodynamics in its most clear form (I still think that it is the most clear), when the field is resolved into waves, the amplitudes of these waves are analogous to the amplitudes of oscillators, and everything reduces to the oscillators (of the field) and their quantization. In such a way, it was set forth in Fermi's article in *Rev. Mod. Phys.*, and also in Heitler's book (especially in its old edition). This is absolutely clear. In my book 1975 *Teoreticheskaya Fizika i Astrofizika* [*Theoretical Physics and Astrophysics*] (additional chapters) (Moscow: Nauka),[4*] I set it forth in this way, which may be mocked by contemporary 'field specialists'. Indeed, this is not enough. The physics of fields has gone a long way. But does it mean that simple and clear things, comprehensible not only to highbrow theoreticians but to every physicist should be discarded? However, this is a different subject. The fact is that V A Fock was very strong on mathematics and his level of rendering quantum mechanics was very high for those times. But the radiation of a uniformly moving electron surprised him. Whereas I, with the help of the equation for the oscillators of the field, understood where the main problem was—it was in setting the task, in the initial conditions. I achieved A A Smirnov's results in the classical way, which let me understand something important. In any case, so I think, though up until now this question has not been thoroughly described anywhere and remains in the shadows. For more details of the gist of the matter, see chapter 1 of my book mentioned earlier.[4*] I also understood, by the by, that my idea of induced emission in collisions was erroneous, for the velocity of light is c, and the charge velocity is $v < c$, and its field resolved into waves is not equivalent to the radiation field. However, here it is impossible to write much about physics. The fact is that I quickly wrote four articles: three for *Doklady Akademii Nauk* [*C. R. Acad. Sci. URSS*] (they were communicated by V A Fock, who was a Member of the Academy) and a small article for *Zhurnal Eksperimental'noi i Teoreticheskoi Fiziki* about the Coulomb gauge. Surprisingly, at that time, neither I E Tamm nor V A Fock knew about this gauge and its benefits, nor was there anything about it in Heitler's book (in the first edition),[5*] and he virtually went out of his way using the usual Lorentz gauge and then excluding the longitudinal field. Afterwards I came across the Coulomb gauge, it seems, in Kramers' book. So it had been known before me but there must have been no clarity and the fact is that my article on this subject was published in 1939 [*Zh. Eksp. Teor. Fiz.* 9 981].

The same oscillator equation continued to serve me faithfully. Shortly before that (in 1937) the theory of Vavilov–Cherenkov radiation had been built (by Frank and Tamm) and, naturally, I was interested in this subject. And I built the quantum theory of effect (I quantized the field in the medium with the index of refraction n using the same method of reduction to the set of oscillators). Then I solved the problem of Vavilov–Cherenkov radiation for crystals. Doing that in the classical way was impossible (in any case, at that time such problems were not solved for an anisotropic medium). Oscillators work in a crystal perfectly well—this resolution into normal waves is adequate for the problem. While working at this problem (and, it seems, even before) I also deduced the Frank–Tamm formula using a different method (the condition for the VC effect as a condition for resonance, calculation of intensity as the change of field energy per unit time: see article 2 in this collection). I also solved (for the first time) the seemingly elementary problem of the radiation of an oscillator residing in an anisotropic medium (in a crystal). Afterwards, all that was chewed over and generalized in plenty of works. It has not always been remembered that it was me who started all that but that is another question and not so important. What is important is that in one year (the 1938/39 academic year) I wrote seven or eight articles and, what is the main thing, I somehow found my true self, I was happy, I understood that I could work. And all thanks to the oscillator. For me it played the same role as the bass string in the popular legend about Paganini. All these results comprised the contents of my candidate degree thesis, which I defended in 1940 (in the spring), though I had done everything in one year, i. e. earlier, and in 1940 I already had started other works. But even so, I was probably the first postgraduate who defended his thesis (PhD degree) ahead of schedule, which brought me the reputation of being a talent. There was an intention to leave me at the MGU but on 1 September 1940 I, most fortunately[3], moved to the P N Lebedev Physical Institute to start working for my doctorate of science under the guidance of I E Tamm. In postgraduate studies, my official tutor had been G S Landsberg, who had put up with my apostasy.

I am again being carried away but I cannot help remembering GS he had a strong character and he did not bow down to anybody. Luckily, he was not put into prison. For some reason, I remembered a 'debate' in the Great Physics Classroom with certain 'mechanicists', Tseitlin, Timiryazev and others, on one side, and I E Tamm, GS, Gessen and others, on other side. The 'mechanicists' insisted that the propagation of electromagnetic waves was impossible without some 'mechanical movement'. Among other things, Tseitlin attacked IE and 'advised' him to learn 'mechanical movement' in the way Daphnis and Chloe had learnt it—by watching mountain goats. IE was indignant. GS was calm but did not retreat a step. I think he rather liked me, though with him I somehow happened to be especially silly and tactless. Once E L Feinberg and I called on GS right after the celebration of his 60th anniversary (it was about 1951) and I blurted out: Andronov refused to be present at his 50th anniversary, as he put it, at the rehearsal of his funeral. GS did not turn a hair, while I was burning with shame. There were some other cases like that. His effect on me must have been a bit like that of a boa on a rabbit ('a bit' seems to have been

---

[3] It is clear that many people at the Physical Department hated me as a 'hanger-on' of Tamm, Landsberg, and others. I would not have got on well at the Department's research institute, to say nothing of the fact that in 1941 all of them joined the irregulars and went to the front. My decision to leave the Physical Department was to a considerable extent related to a certain episode. I went to the banquet arranged by Borya Geilikman on the occasion of his having defended a thesis and Olya (my first wife) was a postgraduate of S G Kalashnikov, who had defended his dissertation the same day (or at least the two banquets were taking place on one day). Naturally, Olya went to Kalashnikov's banquet. And there, probably after too much drinking, some scum from the Physical Department named Turovskii (before the revolution he was Troitskii, then he became Trotskii but he had to change his surname one more time) came and sat next to her. He said to Olya: 'How come that you, a good Russian girl, live with Ginzburg, this hanger-on (or something like that) of these counter-revolutionary professors, Tamm and Landsberg.' Of course, Olya was indignant, it seems that she even made an official complaint to the authorities. But certainly the affair was not set going, whereas I understood that it was necessary to flee from the Physical Department. One more happy twist of fate.

GS's favorite expression). S M Raiskii once well said about GS: 'He cannot be re-educated, he can only be re... "re-conceived".' But, of course, SM was well aware of GS's merits. About his drawbacks I do not want to write—everyone has drawbacks but to me he did only good. There was quite a small episode which, for some reason, is engraved in my memory. In 1953, I was elected a Corresponding Member of the Academy (it happened quite accidentally: IE was in favor of it and I V Kurchatov did not object and even confirmed my merits).[6*] And I do not know what the matter was but GS did not congratulate me and even passed by when seeing me on the staircase. He might have not voted for me and felt too embarrassed to congratulate me. Or there was something else. I did not take offence and I would have forgiven GS (and understood him) if he had been against my being elected in general (for there were people of merit who were older than me). But I just remember my surprise. In general, I see that trying to describe GS here I have failed, as giving a more detailed analyses would mean writing about his weaknesses and I realized that it would be unpleasant for me to write about that.[7*] The same can be said about N D Papaleksi. IE also had weak points, to say nothing of Dau. But I am not one of those who 'foul their home nest' and I have always been tolerant of friends and people of an older generation who belong to our camp (Mandelshtam's school and others). The only person in whom I did not notice any weak points and whom I admired was A A Andronov—what a dominant personality, what breadth—words fail me. But it does not mean that my love for IE was any the less. By the by, my conscience is clear—never, not only in my actions but also in my thoughts, have I been unfaithful to IE. Nor have I played a double game if my attitude to someone was more critical. Anyway, I will not go into criticism here—though there would be enough people to be criticized, not to mention the scoundrels and all sorts of scum. But the purpose of these notes is quite different, though it is already almost forgotten among all sorts of small details.

The main thing that I wanted to say is the following. Without having, as it seemed to me, the necessary qualities and prerequisites, I have become a theoretical physicist and a rather well-known and successful one at that. Speaking about success, I do not mean my becoming a Corresponding Member (1953), then Academician (1966) and award winner (of the Lenin and the State Prizes), and also having been awarded foreign distinctions. All that is fairly conventional, and even complete nonentities formally achieve a lot. However, scientific results are a different matter, they are something objective. And here I believe that I have achieved many important and rather high-class results. It goes without saying that a man cannot judge himself. But everyone has the right to have their opinion. And my opinion is that I have done much. All that is described in the bibliographical 'reference book' *V L Ginzburg* (Moscow: Nauka, 1978) and in the articles in the *Usp. Fiz. Nauk* in connection with my 50th and 60th birth anniversaries

[1966 *Usp. Fiz. Nauk* **90** 195 (1967 *Sov. Phys. Usp.* **9** 782); 1976 *Usp. Fiz. Nauk* **120** 323 (1976 *Sov. Phys. Usp.* **19** 872)]. Of course, there are exaggerations and, perhaps, wrong emphasis, as is practically always the case in such matters, But still the gist of it is that in the field of superconductivity, superfluidity, ferroelectricity, the VC effect and transition radiation, radioastronomy, the origin of cosmic rays, the diffraction of light (and also a number of separate subjects) I have done fairly much.[8*] The question is: why? Of course, first of all people refer (so they think and say) to my abilities. But it is not so or not quite so. In my opinion, my ability for mathematics is simply below average, I have never been good at using the apparatus. I have never been good at solving problems (I mean those from books of mathematical problems). The memory, especially for formulae, is bad. True, I have a rather good memory for ideas and literary references. I did not take an examination in the Landau theoretical minimum and, even if I passed it, it would have been very difficult for me. Indeed, very often I have somehow felt myself to be a deceiver. You are asking questions of a student or a postgraduate while you yourself do not know how to deduce a formula and so forth.

What is the reason for success? Firstly, there is some hunch, an understanding of physics, tenacity, a grip on combinations and associations. Secondly, there was a great wish 'to think up an effect', to do something. Why? I think that it came from an inferiority complex.

It is a sin to complain but, on the whole, life was rather hard on me. Father was a superb man but he was 53 years older than me, besides, we had various difficulties at home. I had few friends, no brothers or sisters, no good school. Later I was not especially brilliant in the Physics Department: I just studied well, no more. So when it started 'to work', I was happy. I wanted to do something again and again. I felt both self-assertion and a great joy, happiness when I managed to think up something. What is the meaning of ambition and vanity? These qualities are considered to be rather despicable and anybody writing about himself tries, unwittingly, to repudiate them. I am also not sure of my being able to write the whole truth. However, I incline to make a distinction between a 'good' ambition, on the one hand, and ambition, in general, and vanity, on the other hand. A 'good' ambition is something that I certainly have—I mean an aspiration to do my work, to do it well, and to have it acknowledged and acclaimed. But I would not like to win acclaim at someone else's expense, a groundless acclaim. When I was repeatedly nominated for the State Prize—and, of course, not awarded it—I remained absolutely indifferent, not to mention not raising a finger in order to get it. As for the Lenin Prize, it was not me who thought about our nomination but Abrikosov and others, who nominated us (from their Institute) and included me, which was well grounded as the work was on superconductivity. Kapitza, who hated me, tried, through Artsimovich, 'to rebuff' me from this prize under the pretext that "Dau will be hurt" (Dau was not included, as he had already been awarded the Lenin Prize and it is not possible to be awarded it a second time—an example of a reasonable regulation, rare for us). However, Dau already was, alas, severely disabled and the prize was for him the last thing to think about, besides he did not know anything about all that. That was the only case when I asked G F Zharkov to compose a letter to the Lenin Prize Committee, it was signed by I E and others. And, by the by, we won the prize in spite of the decision of the Commission of Experts (N G Basov and Co.)! This was accomplished by the activity of A A Abrikosov (I did nothing at all), I E's letter and probably the well-disposed attitude of A P Aleksandrov and M V Keldysh. I can well stand not being awarded a prize and remain absolutely calm. Another matter, if I were 'rebuffed' for my work or under conditions when I certainly deserved the prize, and others got it—here I would suffer. But is it vanity? I doubt it. By the by, I sometimes think about the Nobel Prize. I do not hope to get it, there is too much of a scrum, besides, not being a megalomaniac, I do not think that I must be awarded it, only that I could be awarded it. But I am writing this now because if it is just 'no', it is one thing but if others get it for what I should get (which happens), I will feel hurt; for example, if the prizes were to be given to Abrikosov who solved the Ginzburg–Landau equations or to Shklovsky for radioastronomy and the origin of cosmic rays, though, in any case, I have done not less.

As far as election to the Academy was concerned, I have not done anything either. For the Corresponding Membership I have already said that I did not raise my finger. Nor did I do anything when I was being elected Academician[4] and was calm every time I failed. When, in 1966, I did not pass in three rounds, I sent a telegram to Nina who was making a tour along the Enisei: "Loo repaired, Academicians blackballed". True, when the fourth tour was allowed and I was being 'exchanged' for Basov, while Kapitza was hampering in all possible ways and drawing in I Lifshitz, I lived through a rather unpleasant couple of hours. But I have again 'turned aside'. Whereas the main line is that I wanted very much to achieve results, to think up even small effects, I took great pleasure in it and tried hard. Maybe even with too grueling effort but who can tell what is too much here? As an example (as I see it, not of excessive effort but of one of the possibilities) I will say that I went in for 'brainstorming'. I remember the first time was in 1941 or 1942. Being ill, I was bored and decided that I would think for several minutes and find some small effect. And looking through the possibilities, I thought it that the terrestrial magnetic field must influence the propagation of radio waves in the atmosphere in connection with the presence of a magnetic moment in

---

[4] Moreover, there was a lot about my behavior which created irritation (in Fock, Fesenkov, etc).

the O_2 molecule. Here a decisive role is played by induced emission, quite significant for the radiofrequency range even at room temperature. Having developed this idea, I published an article (*Dokl. Akad. Nauk SSSR* [1942 *C. R. Acad. Sci. URSS*] **35** 302). Another story. In 1964 I was going from Kislovodsk by train, was alone in the compartment, it was an awful bore, and I started 'to brainstorm'—to run over possibilities for superfluidity and superconductivity in space. And I found a number of possibilities. The most interesting of them was the superfluidity of neutrons in neutron stars. I did not know the sign of the forces between neutrons in the s-state, and so the conclusion was conditional. On arrival, I consulted D A Kirzhnits, we looked into this question, and published a small article. I reckon that it was the first clear indicator in this field. We also touched upon the question of vortex lines in the rotation of a star. When I was delivering a report about this in Novosibirsk, A B Migdal said that he had 'already paid attention' to the superfluidity of neutron stars and gave me the reference. In fact, in one of his articles of 1959, there was merely a phrase saying that the superfluidity of nuclear matter can manifest itself inside stars or something like that—not a word about neutron stars. In a way, it is clear that ABM overlooked this possibility, while the superfluidity of the nuclear matter had been talked about before him. But Migdal, being a 'prioritizer', made the most of the whole thing. Many works started to appear which wrote about superfluidity and superconductivity (for protons) in neutron stars, in them I began to notice references to Migdal, while we were completely forgotten. Generally speaking, I could not care less about that, especially as we did not develop this work, because of laziness and being busy. But having once met D Pines, I told him that in the article that I had written together with D A Kirzhnits, there was no reference to ABM merely because we had not known about this work: I said this because I did not want Pines to assume the opposite. And what turned out? Pines, quite obviously, had not even seen Migdal's article but (as I guess) at the request of Migdal started to refer to him and, as so often happens, Migdal's priority was 'adopted by repetition'. In general, many people fight for priority, strive for being quoted and so forth. As for me, if I ever fight, I do it, as a rule, only by way of referring to my works myself but persuading, asking, reproaching I consider shameful (I write about this in the article 'Kto i kak sozdal teoriyu otnositel'nosti' [Who and how created the theory of relativity].[9*] I had several more experiences of successful 'storming', e. g. several effects in the field of the transition radiation and scattering (see the article "Perekhodnoe izluchenie i perekhodnoe rasseyanie" [Transition radiation and transition scattering] in *Priroda* [*Nature*]; this article, like the one already mentioned, is published in my collection 1979 *O Teorii Otnositel'nosti* [*On the Theory of Relativity*] (Moscow: Nauka)). But what is interesting—in recent years I have tried to 'storm' and, in general, without any result! What is the matter? Of course, I am now not as sharp-witted as before. But I think this is not the main reason. The point is that a 'storm' means mostly considering variants and opportunities on the basis of what you already know, in the circle of close, related ideas. But this circle has already almost stopped widening and the old ideas have already been considered. So, it no longer come off—sad but it is a fact.

In my article on the occasion of Dau's 60th anniversary (in fact, it was an obituary),[10*] I wrote that once I had told Dau that he could have done more. And Dau answered distinctly: "I have done what I could". Now, thinking about that, I came to the conclusion that Dau was both right and wrong (the celebrated dialectics). Of course, he 'realized himself' and probably could not have done anything of a still higher class (in comparison with his greatest achievements). But he was phenomenally talented, on the whole, he was not lazy (worked a lot) and could virtually crack any problem. But somehow he did not need to. I think that he had deep complexes (in some ways the inferiority complex as well) but, of course, it was not that he doubted whether he could be a theoretical physicist. He knew his strength. This is why he did not force himself in the field of physics. We spoke about this with E Lifshitz who came to see me in the hospital the day before yesterday (14 October, 1979). Of course, Dau did not resort to 'brainstorming' and so forth (Lifshitz knows that). And if he had done it and, in general, if he had 'forced' it (it seems to be an apt word), he would have written 200 and not 100 works. By the by, I have written more but among them there are quite a few trifles, and I have long been well aware that it does me harm. Dau used to say: "One mustn't write everything one knows". I also felt that my being such a prolific writer was directly condemned. But I had (and still have) the need to write. It is somehow 'hanging' over you until you have written it but when you have written you see 'the fruit'. Here probably is again the same element of self-assertion. In any case, I repeat, I write much not because I try to become famous in such a way. On the contrary, I understood and understand (I repeat) that my 'graphomania' often does me harm. However, I still do not believe that it really is graphomania.

Now it is 4.40 a.m. of 16 October, I cannot sleep, maybe because they promise that today I will be discharged. I am feeling well (touch wood, as here I have become entirely superstitious).

Well, something else to finish this piece of writing. One of my merits is to strive to bring what I am doing to a certain end, often not to the real end but at least to some conclusion or quasi-conclusion. So I feel that I would like to finish writing these notes, which nobody needs, and type them. There is also health. Touch wood again but apart from neurasthenia and bad moods (no small things indeed, though), I have very rarely been ill and then not seriously. And it is essential. For example, take Zhenya (E L Feinberg). I am deeply convinced that his abilities are, in any case, no less than mine; I am sure that they are greater in mathematics and he is certainly an excellent physicist. So why has Zhenya done less than me (I think that it is so and there is no lack of modesty)? To a great extent, because of his illnesses and the need to take care of his family. There probably are some other factors but the ones mentioned are also very important.

By the by, something about myself again. The only talent that I admit in myself is that of an orator. Here I really have something God-given. I become nervous, I am preparing myself, it is important for me that my performance should be a success. And all that gives results. Even when I am speaking in English, I usually manage 'to hold' the audience.

So what can I say summing up? Firstly, irrespective of any summing up, I had nothing to do and felt like writing a bit about the things which I had often thought about but about which it is not quite customary to speak. Besides, even people who are close to me (Zhenya F., for example) for some reason do not believe that I am sincere when I am 'self-flagellating'. To be more exact, not self-flagellating but telling the truth. This truth, important and striking for me and so referred to fairly often, is that I have somehow managed 'to put the pencils together'.[5] Having very little to my name, I have, in general, become prosperous, meaning certainly not the titles and material benefits but some success in science and the satisfaction related to it.

Secondly, the conclusion is that, say what you like, it is difficult to understand what is necessary for success in science (even a limited success, as I do not at all claim to be a great scientist, I am not sure that Dau would have given me as much as the third grade according to his scale). In any case, what matters is not only formal abilities (like fluency in the necessary mathematics, quick counting, memory for formulae and so forth—I do not have any of that). Some major role is played by chance and luck, hunches, tenacity, the aspiration to do something, to achieve a result, to think up 'a small effect'. When, apart from all that, there are also great abilities (formal abilities, as I conventionally call them), we have a really big person. But, of course, to someone really big (a stupid word), to say nothing of giants (Einstein), we should apply some completely different standards.

Thirdly, a tremendously important role—in any case, for many people, me including—is played by a friendly attitude and support at the initial steps. Such was the role which I E Tamm played for me. This I have never been able to forget and will not forget and, what may be less trivial, I have drawn from it some deep (as I hope) conclusions. I have always tried (it was not difficult, it worked automatically) to be nice to beginners,

---

[5] It seems that this was Volodya Berestetskii's expression, who also remembered the words of M Bronshtein 'Spins are not pencils', which probably meant that spins and pencils are not put together in the same way. This reminds me of Bronshtein. He died too early for me to know him. A very talented man indeed, he was eliminated quite young, whether because of his surname or because of his being related to Trotskii.

students, and postgraduates. Here again I am a bad supervisor in some respects. I had great difficulty doing even my own work (calculations) and now calculations just make me sick. That is why I have never 'counted' for my students and postgraduates. 'My' system is merely being nice to them and supporting them. True, I easily give advice, themes and so forth, and this also helped. And the results, as I think, were good: both the theses were successful, as a rule, and the relationships remained good (a couple of exceptions in Gorky, I hope, do not spoil the picture).

Fourthly, everything that has been said may be trivial. Life is complicated. The roads for people and science are not straight but winding and often go in darkness. Yes, it is so. But I do not claim anything. Here I am sitting in my ward no 143, listening in to the BBC in English and hoping that today I will be discharged and will come back home.

I want it very much!

16 October, 1979.

Comments

1*. In 1979 I had to undergo a rather serious operation and wrote these notes at the end of my stay at the hospital. Naturally, they have never been published but their contents were used to some extent in article 16 of this collection and in the paper "Notes on the occasion of a jubilee" (see my book 1995 *O Fizike i Astrofizike* [*On Physics and Astrophysics*] (Moscow: Byuro Kvantum) [2001 *The Physics of a Lifetime* (Berlin: Springer) p 285]). Here these notes are published without any changes, to say nothing of insignificant corrections. Having re-read these notes in connection with the preparation of this issue, I arrived at the conclusion that their publication was a mistake. Such a candid text is more appropriate for posthumous publication. But now it is too late and the absence of these notes in the new edition might be misinterpreted as a renunciation of what is stated therein.

2*. See the previously mentioned book *On Physics and Astrophysics*, p 350 [2001 *The Physics of a Lifetime* (Berlin: Springer) p 351], and *The Memories of I E Tamm* (there were three editions) (Moscow: 1981, 1986, 1995).

3*. See the same book *On Physics and Astrophysics* p 363 [2001 *The Physics of a Lifetime* (Berlin: Springer) p 365].

4*. The last edition 1987 (Moscow: Nauka). For the English translation, see [5] in article 2 of this volume.

5*. Heitler W 1940 *Kvantovaya Teoriya Izlucheniya* [*The Quantum Theory of Radiation*] (Moscow, Leningrad: Gostekhizdat); 2nd edn: 1956 (Moscow: IL).

6*. Now I understand that it was not at all accidental. My 'second idea', used in creating the hydrogen bomb, was highly appreciated (see article 18 in this collection). This reminds me that I read somewhere (unfortunately, I do not remember where) that I E Tamm suggested, in 1953, that A D Sakharov should be elected a Corresponding Member rather than immediately becoming an Academician. I V Kurchatov decided differently—certainly, out of political considerations. In such a situation, electing me a Corresponding Member was quite natural.

7*. See the article "O Grigorii Samuiloviche Landsberge" (About Grigorii Samuilovich Landsberg) in the book *On Physics and Astrophysics* p 405 [2001 *The Physics of a Lifetime* (Berlin: Springer) p 411]. However, in this article, I do not write about GS's weaknesses either (the reasons are given in the text of these notes).

8*. See the article "Opyt nauchnoi avtobiographii" (A scientific autobiography—an attempt) in the book *On Physics and Astrophysics* p 312 [2001 *The Physics of a Lifetime* (Berlin: Springer) p 309].

9*. See the same book, p 178 [2001 *The Physics of a Lifetime* (Berlin: Springer) p 217].

10*. Ginzburg V L 1968 *Usp. Fiz. Nauk* **94** 181 [1968 *Sov. Phys. Usp.* **11** 135]. See also this collection, article 10.

PART III

Article 20

Answers to the *Physics World*'s Millennium survey questionnaire[1*]

Magazines and even newspapers marked the end of the 20th century and the advent of the Third Millennium with publications that marked the occasion, naturally enough, by summarizing the progress in science. Among other things, questionnaires are conducted and their results are published. I think they are not devoid of certain interest.

One such questionnaire called the "*Physics World* Millennium survey" was run by a science-popularizing magazine *Physics World* published by the Institute of Physics in Great Britain. This magazine is similar to the American *Physics Today*: it reminds Russian readers of *Priroda* [*Nature*] but it is devoted exclusively to physics and related branches of science.

Seven questions were sent to physicists (including astrophysicists) in many countries; and the answers to these questions were processed and a brief comment published in *Physics World* no 12, 1999. The editors of the magazine sent questions to 250 addresses and received 130 responses. The answers as such have not been published and only a small fraction of their contents are quoted in the published comment. I will not touch on this aspect (with one exception) but will quote the questions and my answers to them. I hope that this information will be of interest to a number of readers of *Priroda*. So here are the questions and my replies.

Question 1. What have been the three most important discoveries in physics?

The questions only look simple. How do we define a discovery? We may define it as something particular, concrete. For example, the discovery of the electron, of radioactivity, etc. We may limit the definition to the discovery of new laws and principles. I tend to consider the discoveries of unifying concepts, approaches and theories that generalize experimental observations as more important than particular discoveries. With this approach the most important are:

- The creation of classical mechanics (Galileo, Newton);
- The creation of special and general relativity (Einstein);
- The creation of quantum theory (Planck, Einstein, Bohr, Heisenberg, Schrödinger, and Dirac).

As for particular discoveries, I would single out the discoveries of the electron, the photon, and quarks. However, aren't the discoveries of the neutron and the positron (1932) just as momentous?

Question 2. Which five physicists have made the most important contributions to physics?[1]

Newton and Einstein without doubt. But the choice amongst Aristotle, Archimedes, Copernicus, Galileo, Kepler, Faraday, Maxwell, Boltzmann, Planck, Bohr, Rutherford, Heisenberg, Schrödinger, Dirac...? I would settle for Galileo and Bohr. But who will be no 5?

Question 3. What is the biggest unsolved problem in your field?

I still consider physics as a unified field. I wrote on this in detail in the paper published in 1999 *Usp. Phys. Nauk* **169** 419 [1999 *Phys. Usp.* **42** 353].[2] In view of this, strictly speaking, there is no field I should describe as 'my field'. To a degree, superconductivity is closer to my heart than the rest of them. The most important problems in this field are high-temperature and room-temperature superconductivities (HTSC and RTSC). What is the superconductivity mechanism in already known HTSC materials (cuprates)? Is RTSC feasible?

Question 4. What is the biggest unsolved problem in the rest of physics?

The most important among the unsolved problems in physics is undoubtedly the creation of a unified theory of all interactions including the gravitational one. It can also be reformulated as the creation of the quantum theory of gravitation, including quantum cosmology. It appears that physicists are very far from this goal. Calling superstring theory the 'theory of everything' was, from the start, either falling infatuation or a piece of advertising. I describe it in detail in the article for *Soviet Physics–Uspekhi* mentioned earlier.[2]

Question 5. Would you study physics if you were starting university this year?

At the moment I believe I should choose physics again. However, the first place among sciences is currently occupied (with qualifications, of course) by biology. Hence, if I were young, I might have chosen biology (I refer you again to my *Uspekhi* article).[2]

Question 6. If you were starting your research career in physics again, which areas of physics would you go into?

[1] The editors of *Physics World* clarified this for me: this and further questions refer to the entire history of physics, not only, say, the physics of the 20th century.
[2] See article 1 in this book.

As my answer to Question 3 shows, I have always disliked narrowly limited specialization. The number one thing for a physicist is to know and love physics without limits, the wider the better. Of course, one has to choose something narrower, especially if one is an experimentalist. However, even an experimentalist should not, if at all possible, be tied for life to a single narrow field. A theoretician both can and must work in (or at any rate be interested in) various fields. This is the reason why, if I were starting my life in physics, I would choose theoretical physics, even though I have the greatest respect for experimenters and was always weak in calculations and in mastering mathematical tools. Nevertheless, one should be conscious of the difference between theoretical and mathematical physics.

Question 7. Stephen Hawking has said that there is a 50:50 chance that we will find a complete unified theory in the next 20 years. Do you agree that the end of theoretical physics is in sight?[3]

Many times in the history of physics, physicists have believed that the solution of certain profound problems was just around the corner. In fact, the solution was either very long in coming or has not yet been achieved. My estimate of the 'chance that we will find a complete unified theory in the next 20 years is that it can hardly be above 1%.

My answers are given here without any corrections. Here I want to improve upon my answer to Question 2. For physicist number 5, I, after much thought and without hesitation, nominate Maxwell. In addition, among physicists of the 20th century I forgot to mention Fermi and probably Landau, Pauli, de Broglie, and Feynman. Somehow, I happened to concentrate on physicists of the 20th century and, therefore, omitted to mention such people as, for example, Gibbs in the general list.

I also want to add to my answer to Question 7. In one of his articles (I refrained from searching for it now), Max Plank (1858–1947) described how, having graduated from university, he asked one respected physics professor for advice on which field to choose. And the professor said something like "I feel sorry for you, young man, because all you can do nowadays is remove dust from existing physics instruments". In other words, this physicist— like quite a few others—thought that the main discoveries in physics had already been made.[2*] The conversation took place before the discoveries of X-rays, radioactivity, and the electron, before the creation of the theory of relativity and quantum mechanics! This blindness is encountered even today, despite the fact that physics still faces huge unresolved problems.

There is, however, one point in the previously mentioned comment by the editors of *Physics World* which might be of interest to readers; namely, they give the distribution of 'votes' received by physicists in answers to Question 2. Namely, 61 physicists received at least one vote (nomination). The list, in the order of decreasing number of votes received is

---

[3] S Hawking 1982 Is the end in sight for theoretical physics? *Priroda* [*Nature*] no 5 489.

(the names of living physicists are in italics): Einstein (119 votes), Newton (96), Maxwell (67), Bohr (47), Heisenberg (30), Galileo (27), Feynman (23), Dirac (22), Schrödinger (22), and Rutherford (20). Boltzmann, Faraday, and Plank received 16 votes each, Fermi got 13, M Curie received 6 votes, Bardeen and Landau—4 each, Bell, *Bethe*, and Gibbs—3 each; 2 votes each were cast for Archimedes, Copernicus, P Curie, *G 'tHooft*, Hubble, Kepler, Pauli, Shockley, J Thomson, *Ch Townes*, *S Weinberg*, and Yukawa; and finally C Anderson, Aristotle, *Bennett*, *Binning*, Bloch, Carnot, Clausius, Democritus, Doppler, Edison, Euclid, Eddington, Euler, *Hawking*, D Hilbert, Langevin, Lorentz, Michelson, Onsager, Payne-Gaposhkina, Rayleigh, *M Rees*, *Rohrer*, Roentgen, Shawlow, Turing, *J Wheeler*, K Wilson, and *Yang* received one vote each.

I will be honest and confess that I do not know who Bennett, Binning, and Rohrer are. Since all of them received just one vote, I do not feel very ashamed.[3*] I must also remark that a number of other physicists included in the short list of candidates for those five who contributed most to physics 'of all times and all peoples', make me raise my eyebrows. I even suspect that the editorial board included all those who were simply mentioned in responses (e. g. the list contains all those names that I mentioned in my answer to Question 2 even though I have not included all of them in the coveted five). In addition, I suspect that not all respondents understood that physicists of other than the 20th century were eligible. However, this is not really important. The first ten in the list are illustrious and quite a few brilliant names follow.

Among the physicists in the first ten, I was acquainted personally, even if barely so, with Bohr and Dirac and closer than that only with Feynman. And it is the rank given to Feynman that causes me to express doubt. Richard Feynman was an excellent physicist and a brilliant personality: I have already written about this.[4] However, as Landau emphasized in his time, the role and rank of physicists should be classified (if classifying is necessary at all) only 'by the weight of their achievement', not by any other attributes, such as volume of knowledge, oratorical talents, textbooks and so forth. From that point of view, the actual scientific achievements of Feynman, however large they are, seem to me to be inferior to all the other physicists mentioned among the first ten and also to some who got a more modest rank in the list. However, one need not assign too much importance to these 'places' even though such 'lists' are a curious thing.

I think that if conducted at a serious level, a questionnaire of the type carried out by *Physics World* would be a very proper undertaking in Russia too.

---

[4] See V L Ginzburg 1995 "About Richard Feynman—a remarkable physicist and a wonderful man" in *On Physics and Astrophysics* (Moscow: Bureau Quantum) p 430 [2001 *The Physics of a Lifetime: Reflections on the Problems and Personalities of 20th Century Physics* (Berlin: Springer) p 443].

Comments

1*. This article was published in the magazine *Priroda* (no 3, 3, 2000) under the title "Physics: Past, Presence, and Future (Replies to the Questionnaire of the Magazine *Physics World*)". The first part of this title was provided by the editors of *Priroda* and I consider it rather pretentious for such a short text. Therefore, I have left here only the second part of the title.

2*. An article by Ye M Klyaus, added as an appendix to the collected papers of M Plank 1966 *The Unity of the Physical Picture of the World* (Moscow: Nauka), largely confirms what I say here about Max Planck—Planck did get advice not to go into theoretical physics. See also A F Ioffe 1983 *Encounters with Physicists* (Leningrad: Nauka) p 64.

3*. I felt ashamed, nevertheless, and this is what I discovered: G Binning and H Rohrer received the Nobel Prize in physics in 1986 for developing the scanning tunneling microscope. As for Bennett, the respondents probably mentioned the American physicists known for his work in classical and quantum theory of information.

Article 21

Why Soviet scientists did not always get the Nobel Prizes they deserved[1*]

Everything related to Nobel Prizes is widely discussed in the press. I do not know very much about Nobel Prize affairs and notice only what happens to come to my attention. Still, I can name books [1–3] and articles [4–11] on the subject.

The extraordinary prestige Nobel Prizes enjoy is justified but we should remember that nothing done by humans can be raised to an absolute. Some Nobel awards in physics and chemistry are, in my opinion and that of some of my colleagues, either undeserved or dubious. But such cases are few. Anyone who knows how difficult it is to select a winner must think twice before blaming Nobel Prize Committees for unobjective selection or prejudice. The Committee on Physics, for example, sends out more than a thousand invitations to nominate a prize candidate (or candidates). I do not know the number of replies received but I think there must be a few hundred every time, giving a few dozen names. Yet no more than three winners may be selected and the prize may be divided into no more than two different subjects.

My own experience in the matter is confined to membership of commissions of experts awarding the gold medals and prizes of the Russian Academy of Sciences. But it is enough for me to understand how difficult it is sometimes to make the right choice. I recall one of the contests for the L D Landau Prize. We had five wholly deserving nominations and only one prize. Here, sympathy, antipathy, lobbying, and the like came to the fore. I repeat, therefore, that the work of Nobel Prize Committees is extremely difficult. Besides, as far as I know, unprecedented pressure is exerted on them and they do their utmost to be impartial.

Yet it was common in the Soviet Union to revile the work of Nobel Prize Committees (see, *inter alia* [9, 21]). Of course, in post-Soviet Russia, the tone has changed radically; but, regrettably, the activity of Nobel Prize Committees is still occasionally clumsily criticized. In one case [7], I have already given my opinion [8]. This time, I am responding to Yu I Solov'ev's article in the *Herald* [11]. Solov'ev said that the distinguished chemist V N Ipat'ev was not awarded the Nobel Prize because "the Nobel Committee on chemistry did not award the prize to him because he had taken an active part in Russia's economic development in 1918–1927. The decision turned on political partialities" [11, p 630]. And Solov'ev emphasized: "No small part in the decision may have been played by the powerful influence of the IG Farbenindustrie concern".

Suspicions of this sort are not novel. In 1930, the Nobel Prize in physics was awarded to Indian scientist C V Raman for his work on the scattering of light and the discovery of the effect that was given his name. Yet, a no smaller part in the study of light scattering and in the discovery of the effect of combinational scattering (now usually called the Raman effect) was played by Soviet physicists G S Landsberg and L I Mandelshtam. The story of the discovery of the combinational scattering of light is described in detail in the brochure by I L Fabelinskii [12]. I will only say that, in fact, Landsberg and Mandelshtam were the first to obtain really conclusive data—lines of combinational scattering—a week before this was done by Raman and Krishnan. Furthermore, the Soviet physicists obtained a clearer result with quartz and Iceland spar crystals while the Indian scientists worked with liquids. This leaves no doubt that the two groups worked independently. But Raman and Krishnan promptly published a number of reports in *Nature* (31 March, 1928, and later), while the Soviet physicists published theirs a little later: their first report in *Die Naturwissenschaften* was dated 6 May, 1928, and appeared in the journal on 13 July, 1928. The same report was submitted to the *Zhurnal Russkogo Fiziko-Khimicheskogo Obshchestva* and appeared in vol 62, p 335, 1928. It needs to be added that the delay in publication was due to circumstances of a personal nature—a relative of Mandelshtam's was arrested at that time in one of Stalin's early purges—which distracted the attention of the scientists from the matter of research and priorities.

Despite what I think was a short delay, knowledge of Landsberg's and Mandelshtam's achievement spread promptly in the Soviet Union and abroad (for the Iron Curtain had not yet dropped). For this reason, there was a widespread opinion in Soviet Russia that Landsberg and Mandelshtam were not awarded the prize jointly with Raman (as noted earlier, three people could receive it under Nobel Prize rules) because of some gross mistake which was due to anti-Soviet sentiment among Nobel Prize Committee members and those associated with them. I must confess that I too saw no other explanation for a long time. It turned out, however, that this conclusion was incorrect or at least highly doubtful. The papers of Nobel Prize Committees are kept secret, and only some of them are made available for historical research some 50 years later. Thus, who nominated whom for physics and chemistry prizes in 1901–37 have at last become known [13].

And what did these materials reveal?

Raman was nominated for the 1929 Prize by two persons, including Niels Bohr, who enjoyed enormous prestige, while no one nominated Landsberg and Mandelshtam. However, the 1929 Prize was awarded to another physicist (de Broglie). Raman had ten nominators for the 1930 Prize, mostly prominent physicists, while Mandelshtam was nominated by O D Khvolson and N D Papaleksi, and Landsberg by Khvolson alone. The latter was also among the nominators of Raman and, thus, made the most acceptable recommendation on awarding the Nobel Prize to all three— Raman, Mandelshtam, and Landsberg.

Khvolson was a distinguished Russian and Soviet physicist and for his scientific services, especially in teaching and popularizing physics, was made an Honorary Academician (member) of the USSR Academy of Sciences.[1] At that time, Papaleksi was not yet a member of the Academy which shows that invitations to select nominees for the Nobel Prize were sent to rather a large number of Soviet physicists in the 1920s and 1930s. It is, therefore, safe to say that invitations to name prize nominees were also received by the so-called leading Soviet physicists and, especially, Academicians. But none of them nominated Landsberg and Mandelshtam although they knew perfectly well about their discovery and its paramount importance.[2*]

I mention no names, because the Nobel Committee's invitations to nominators are confidential. And that is correct. It is for everyone to decide for oneself whom one thinks worthiest. But I too have the right to conclude that Landsberg and Mandelshtam did not share the Nobel Prize with Raman above all because of the neglect of their colleagues, the Soviet physicists. In the second place, I would name foreign physicists. They too had known of the work of Landsberg and Mandelshtam which had been published in the West [12] but nominated Raman alone. Possibly, some of them acted thus because reports of Raman's work appeared earlier than those about the discovery of Soviet physicists although this argument does not seem to me convincing. Finally, I name the Nobel Prize Committee itself, the third party responsible for the incorrect decision. After all, Landsberg and Mandelshtam were nominated and the number of nominators should have no bearing on the final decision. Nor is the number usually considered. The distinguished German physicist A Sommerfeld, for example, was nominated 72 times in the course of 20 years [5] and did

---

[1] When someone congratulated Khvolson (1852–1934) on having been elected an Honorary Academician (1920), he remarked that the difference between Honorary Academician and simply Academician was the same as between dear Sir (milostivyi gosudar') and simply Sir (gosudar'). A witticism, of course. Many deserving men and distinguished scientists had been Honorary Members of the USSR Academy of Sciences and its predecessor (the Russian Academy). It is another story that the rank of Honorary Academician was discredited when such 'scientists' as Stalin and Molotov were added to their number. Subsequently, this led to the abolition of the rank of Honorary Academician.

not ever receive the prize. Who the nominators were is a different matter: Raman was nominated by ten distinguished physicists including Bohr and Rutherford. In these circumstances, the mistake of the Nobel Committee can be understood but not forgiven.

But what had anti-Soviet sentiment to do with all this? There is all the less reason to refer to it because in 1930, prior to the orgy of Stalin purges and Lysenko's obscurantism, the relations between Soviet and foreign natural scientists had not been strained. And one more point. Neither Mandelshtam nor Landsberg had been members of the Communist Party and there was no political reason why they should not be awarded the Nobel Prize.

It is time, however, to return to Ipat'ev. In 1931, German chemists Friedrich Bergius and Karl Bosch were awarded the Nobel Prize for their contribution to high-pressure chemistry. In Solov'ev's opinion [11], this was unfair because Ipat'ev had done more in the same field and had done it earlier. I know practically nothing about chemistry not to mention high-pressure chemistry. I can only say that Solov'ev's arguments in support of Ipat'ev seem to me well grounded and convincing. But what has the Nobel Prize to do with it? For some reason, Solov'ev does not say that the Nobel Committee could make one of two decisions: to award the prize to all the three (Bergius, Bosch, and Ipat'ev) or to Ipat'ev alone. But neither was possible because no one ever properly nominated Ipat'ev for the Nobel Prize for chemistry ([13]: this refers to the period before 1937).

The Nobel Committees, like all committees I know, choose winners from among nominees.[2] And this is absolutely right. I have no doubt that a large number of Soviet chemists (in Russia and the Soviet Union chemistry was always of a high standard) were invited to nominate winners of the Nobel Prize in chemistry. But as can be seen from the foregoing, none of them nominated Ipat'ev. Nor was he nominated by any foreign chemist. So, what had anti-Soviet sentiments, not to mention Farbenindustrie intrigues, to do with it? In principle, the Nobel Committee could have abstained from awarding any prize for high-pressure chemistry in 1931 and awarded it for something else (there were many other recommendations [13]), securing Ipat'ev's nomination for the next year's prize. But could anyone expect, much less demand, anything of the sort in the aforementioned conditions— in the total absence of Soviet and foreign chemists' support for Ipat'ev's nomination? I cannot help thinking of the Russian proverb, "Blame not the mirror if the face isn't right".

I want to take the opportunity to comment (in regard to physics and physics alone) on the frequent complaints in our press that Soviet physicists are undeservedly ignored because there are 68 Americans among the Nobel

---

[2] I am sorry to say that for a time (prior to 1987), the USSR Academy of Sciences managed to sidestep this obvious requirement when awarding gold medals and prizes [14; 15, p 402].

Prizewinners and only seven Soviet and Russian ones (up to 1997 inclusive; see [10], with seven laureates added for 1995–97) and only four prizes in physics (I E Tamm, I M Frank, and P N Cherenkov in 1958; L D Landau in 1962; N G Basov and A M Prokhorov in 1964; and P L Kapitza in 1978). Could Soviet (Russian) physics really be ten times weaker than American physics? Of course not. At least not before the considerable 'brain drain' after 1991. The number of Nobel Prizes and Nobel Prizewinners reflects the number of discoveries and the level of basic physics rather than the level of physics in the country in general (number of physicists, number of publications, and the like). And here we do lag and have been lagging behind the United States considerably if not by ten times. The reasons are many: greater emphasis on applied science and defense, insufficient funding of basic physics (especially the building of large constructions, such as accelerators, telescopes, and the like), and ideological considerations— the all too often idiotic secrecy, the difficulty of going abroad and even of exchanging information, etc.

I will not dwell further on this subject and will answer just one question: how many more Nobel Prizes in physics could the Soviet Union and Russia have received? In my opinion, we lost two prizes in the Soviet period: the aforementioned prize for the combinational scattering of light which should have gone to Landsberg and Mandelshtam, and also the prize for the discovery by E K Zavoiskii in 1944 of electronic paramagnetic resonance. True, Soviet physicists may be credited for a number of other top-rate works but speaking exclusively of the results that unquestionably deserve a Nobel Prize, I can recall only the cosmological investigations of A A Fridman (see the splendid book [16] on the subject). But Fridman's articles on the general theory of relativity and its application in cosmology were published in 1922–24 and, in 1925, he died at the age of only 37. That is why there was, in effect, no opportunity for nominating him for a Nobel Prize, not to mention the fact that the significance of his work was not properly appreciated until years later (Nobel Prizes are not awarded posthumously). Out of still earlier top-rate studies, I could mention that of P N Lebedev (he died in 1912), who succeeded in measuring the pressure exercised on bodies by light. Khvolson nominated him in 1905 and V Win in 1912. Indeed, those were the sole nominations (aside from Landsberg and Mandelshtam) made by Russian physicists from 1901 to 1937 [13].

In Stalin's time, all ties with the Nobel Committees were cut off and Soviet physicists did not begin nominating candidates again until the mid-1950s. I have already written about my part in nominating Tamm, Frank, and Cherenkov for the discovery and explanation of Vavilov–Cherenkov effect [15, p 406]. In 1958, the prize was awarded. Zavoiskii was nominated first in 1964 and again in 1975 [17]. There may have been other nominations. I began receiving invitations to nominate physicists for the Nobel

Prize after 1966, when I was elected a Full Member of the Academy. I nominated Zavoiskii for the 1974 and 1976 Prizes and intended to continue doing so later as well but Zavoiskii died on 9 October, 1976 at the age of 69.

Paraphrasing the well-known maxim, "In Russia, one must live long", I could say that, generally, to receive a Nobel Prize, one also must live long. P L Kapitza was awarded the Nobel Prize in 1978, at the age of 84, and for work he had done 40 years before. Had Zavoiskii lived longer, he would probably have won a wholly merited Nobel Prize. No one had received a prize for the discovery of electronic paramagnetic resonance, so in this case there is no reason whatsoever to speak of any obvious discrimination of Russian science.

To sum up, the Soviet Union and Russia could or, more precisely, should have had six or seven Nobel Prizes in physics and, accordingly, 10 or 11 winners. A number of the Nobel Prizes received by Americans seem to have been awarded for far from top-rate work. And if anyone wants to know, I think the 'true' number of physics prizes awarded to Russians and, accordingly, the number of prizewinners, is not of the order of ten but rather of five lower than those awarded to Americans.

What is the situation today? As I see it, three or four Russians could contend for the Nobel Prize in physics. Whether any of them will receive it or not is a matter of conjecture.[3] I am sure of one thing, however: if the expected prize is not awarded, the sole reason will be keen competition and some other factors but certainly not any anti-communist sentiment or Russophobia.

And what can be said of the more distant future? The sole dependable way of receiving Nobel Prizes in science is to build up basic research and to provide conditions for fruitful work in a favorable environment, especially for young people. Unfortunately, this often calls for considerable funds to construct gigantic installations among other things. The times when major discoveries in physics (including astrophysics) could be made in a small laboratory on a shoestring are, by and large, long over; but, I think, not entirely. The history of well-known discoveries shows that gifted and tenacious physicists (and not physicists alone) can sometimes reach their goal in the most modest of conditions. It is only right, therefore, to advise our scientists not to give up, in spite of all the difficulties Russian science faces today, and to continue their search.[4] I wish them good luck[3]* (see

[3] Russian physicist Zh I Alferov became one of the three Nobel Laureates in physics 2000. The prize was awarded with one half jointly to Alferov and the German physicist H Kroemer "for developing semiconductor heterostructures used in high-speed- and opto-electronics", and with one half to American physicist J S Kilby.

[4] The recent paper [19] throws light on the procedure concerning the awarding of Nobel Prizes in physics. The number of Nobel Laureates in physics since 1901 to 2000 (inclusive) totalled 162 persons. More than 2000 letters were distributed among possible qualified persons mentioned in the Nobel Statutes when selecting the candidates for the

also comments 4 and 6).

Comments

1*. This article was published in 1998 *Vest. Ross. Akad. Nauk* **68** 52. The translation (though not very adequate) can be found in the English version of the journal (1998 *Herald Russ. Acad. Sci.* **68** 56).

2*. This statement is confirmed by the data reported in [13] and cited in [18] in more detail than in the present article. Suffice it to say that Khvolson and Papaleksi were not the only Soviet scientists who nominated candidates for the 1930 Prize. However, none of them proposed Landsberg and Mandelshtam even though each nominator is entitled to recommend an unlimited number of candidates.

3*. The present article was motivated in the first place by the desire to ridicule the typical manner in which the failure of V N Ipat'ev to become a Nobel Prize winner used to be interpreted in the Soviet press [11]. Indeed, the author of [11] maintains that Ipat'ev did not receive the prize because of anti-Soviet sentiments among members of the Nobel Committee on chemistry. Moreover, a reference is made to the alleged interference from the known German concern AG Farbenindustrie. The truth is, however, that none of Ipat'ev's colleagues including Soviet ones proposed him as a candidate for the award!

Certainly, I came to be interested in the mechanism underlying decisions of the Nobel Prize Committee regardless of Ipat'ev' story. Roughly speaking, I belong to the school of Mandelshtam and Tamm. Therefore, we always (even when students) knew about the injustice done to G S Landsberg and L I Mandelshtam in not awarding them the Nobel Prize for the discovery of combinational light scattering in 1928. As previously mentioned, after Raman independently described this phenomenon in the same year, his name was attached to it (as the Raman effect), and he alone received the Nobel Prize in 1930. We regarded (and still regard) this decision of the Nobel Committee as a crying injustice. Of course, we had no idea about considerations behind that decision and were even inclined to suspect political involvement. The publication of [13] made it absolutely clear, as reflected in my article, that there is no cause to ascribe political motives to the said decision (even though they cannot be logically excluded). There is every reason to believe that Landsberg and Mandelshtam failed to win the Nobel Prize (to be precise, to share it with Raman) owing to the indifference and disregard of their fellow scientists, in the first place in

Nobel Prize in physics 2000. The commonly received answers (nominations) comprise approximately 15% of those delivered, i.e. about 300 proposals regarding prize candidates have reached the Nobel Committee for physics, from which the latter selected 10–15 for future discussion. Paper [19] also reports on some closely related items.

the Soviet Union. See [18, 26, 27] for details. However, no less significant was the part played by Raman's behavior.

It is worthy of note that, according to the information provided by A M Blokh, an authority on the history of Nobel Prize awards (see also [18]), a critical argument against considering Landsberg and Mandelshtam as candidates for the 1930 Prize was the late publication of their reports compared with Raman's papers. This might well be the case. Even then, however, the decision of the Nobel Committee taken on this ground is unfair, the two groups having made the discovery quite independently.

I decided to add this comment for the following reason. This article of mine turned out to be grossly misunderstood. By way of example, I received a letter from colleague X who interprets my motives for publishing it in the following way: first, Ginzburg is longing for a Nobel Prize; second, Ginzburg is disappointed in his hope to be nominated by his colleagues, first of all by Russian physicists. Both these statements are not true. I am by no means one of that rare sort of unwordly scholars who do not care about the recognition of their labours including that in the form of the Nobel Prize. However, I am far from 'longing' for the prize, the more so that I do not think such an award, if given to me, would be beyond question. I wrote about it in an earlier article dated 1979 which is included in this volume (article 19). I am 87 now and have no reason whatever to act against conscience nor do I consider it immodest to express my opinion. I do believe I deserve the Nobel Prize for what I have done in physics and astrophysics. But there are more candidates or, translating this thought into the language of sports reporters, there is a long bench of them. Hence, there is no reason to feel hurt if the prize goes to others. Certain indisputable cases, such as the discovery of combinational light scattering, is quite a different story however. There is no doubt that the authors of such discoveries must be rated as the most likely prizewinners.[5*] Therefore, I take no offence at those who could but did not nominate myself as a candidate for the Nobel Prize. However, I am aware that some physicists here in Russia (formerly in the USSR) and abroad have proposed me as a nominee for more than 25 years. I speak on this matter so insistently for fear that other readers may share the opinion of colleague X. In fact, I would like to refute it.

4*. Since the second edition of my book had been issued, the problems raised in this article have also been dealt with in other publications [20–24]. References [22, 23] are confined to the 1930 Prize alone awarded for the discovery of combinational light scattering. A few remarks concerning Nobel Prizes in physics can also be found in article 1 and comment 3 to it in this volume; see also my paper 2002 *Usp. Fiz. Nauk* **172** 213 [2002 *Phys. Usp.* **45** 205].

5*. The sole more or less serious argument justifying the disallowed nomination of G S Landsberg and L I Mandelshtam arises from the fact that Raman and Krishnan published their results somewhat earlier than the former authors. It is worthy of note, however, that the long paper by Landsberg and Mandelshtam [25] contains the following remark: "We observed the appearance of satellite lines of light scattering in quartz before Raman and Krishnan's publications". This established fact was confirmed in [12, 18]. The Nobel Committee either did not or could not have such evidence at its disposal (see, however, [26]). However, it had no cause to exclude the contribution of Landsberg and Mandelshtam from consideration taking into account the fact that they had worked not with liquids, like Raman and Krishnan but with solids. Moreover, other convincing proofs were available that Landsberg and Mandelshtam had made their discovery independently (see [12, 18, 21, 22]). Last but not least, Landsberg and especially Mandelshtam were scientists of spotless reputation and there was no reason to doubt their statement in [25]. Meanwhile, to the best of my knowledge, the choice of Nobel Prizewinners does not at all depend on the time of publication. Therefore, I see little reason to deny that the Nobel Committee made a mistake attributable to what is disclosed in my paper and also to the behaviour of Raman whose efforts to agitate whoever he could for his nomination were not lost on the scientific community [22]. This issue is discussed at greater length in the paper "Once again on the history of the discovery of combinational light scattering" by V L Ginzburg and I L Fabelinskii ([26]; next article in this book), and in [27].

6*. The Nobel Prize in physics 2003 was awarded to A A Abrikosov, V L Ginzburg, and A J Leggett "for pioneering contributions to the theory of superconductors and superfluids". My Nobel Lecture was published in [28]. This lecture, as well as my autobiography, were placed in [29].

References

[1] Cholakov V 1987 *Nobelevskie Premii. Uchenye i Otkrytiya [Nobel Prizes. Scientists and Discoveries]* (Moscow: Mir)

[2] Sulman R 1993 *Zaveshchanie Alfreda Nobelya: Istoriya Nobelevskikh Premii [Alfred Nobel's Will. History of the Nobel Prizes]* (Moscow: Mir)

[3] Crawford E 1984 *The Beginnings of the Nobel Institution: The Science Prizes, 1901–1915* (Cambridge: Cambridge University Press)

[4] Mott N 1979 Nobel Prizes in science *Contemp. Phys.* **20** 227

[5] MacLachlan J 1991 Defining physics: The Nobel Prize selection process, 1901–1937 *Am. J. Phys.* **39** 166

[6] Crawford E, Fime R L, and Walker M A 1997 Nobel tale of post-war injustice *Phys. Today* no 9 26

[7] Blokh A 1994 Nobel rejects *Poisk* no 49

[8] Ginzburg V L 1995 Once more about the Nobel rejects *Poisk* no 5

[9] Blokh A 1997 Alternative prizes: Nobel in the distorting mirror of the Soviet press *Nezavisimaya Gazeta* 28 June (see also 1995 *Izvestiya* 2 December)

[10] Lishevskii V P 1995 Winning immortality: The life and destiny of Alfred Nobel *Vestn. Ross. Akad. Nauk* **65** 820

[11] Solov'ev Yu I 1997 Why didn't Academician V N Ipat'ev become a Nobel Laureate? *Vestn. Ross. Akad. Nauk* **67** 627 [1997 *Herald Russ. Acad. Sci.* **67** 295]

[12] Fabelinskii I L 1982 *K Istorii Otkrytiya Kombinatsionnogo Rasseyaniya* [*The History of the Discovery of Combinational Scattering*] (Moscow: Znanie); 1978 *Usp. Fiz. Nauk* **126** 123 [1978 *Sov. Phys. Usp.* **21** 780]

[13] Crawford E, Heilbron J L and Ullrich R 1987 *The Nobel Population 1901–1937: A Census of the Nominators and Nominees for the Prizes in Physics and Chemistry* (Berkeley Papers in History of Science 11; Uppsala Studies in History of Science 4) (Berkeley, CA: Office for History of Science and Technology, University of California)

[14] Ginzburg V L 1998 Kak provodit' vybory v Rossiiskuyu akademiyu nauk (How should the elections to the Russian Academy of Sciences be organized?) *Vestn. Ross. Akad. Nauk* **68** 148 [1998 *Herald Russ. Acad. Sci.* **68** 68]

[15] Ginzburg V L 2001 *The Physics of a Lifetime: Reflections on the Problems and Personalities of 20th Century Physics* (Berlin: Springer)

[16] Tropp E A, Frenkel V Ya, and Chernin A D 1988 *Aleksandr Aleksandrovich Fridman: Zhizn' i Deyatel'nost'* [*Aleksandr Fridman: Life and Works*] (Moscow: Nauka)

[17] 1995 *Kurchatovets* (*Organ of the Russian Scientific Center Kurchatov Institute*) no 3 2

[18] Fabelinskii I L 1998 Kombinatsionnomu rasseyaniyu sveta—70 let (Seventy years of combinational (Raman) scattering) *Usp. Fiz. Nauk* **168** 1341 [1998 *Phys. Usp.* **41** 1229]

[19] Rodgers P 2000 *Phys. World* **13**(10) 10

[20] Crawford E 2001 Nobel population 1901–50: Anatomy of a scientific elite *Phys. World* **14**(11) 31

[21] Blokh A M 2001 *Sovetskii Soyuz v Inter'ere Nobelevskikh Premii* [*Soviet Union and Nobel Prizes*] (St-Petersburg: Gumanistika)

[22] Singh R and Riess F 2001 The 1930 Nobel Prize in physics: A close decision? *Notes Rec. R. Soc. London* **55** 267

[23] Feinberg E L 2002 *Rodonachal'nik* (*O Leonide Isaakoviche Mandelshtame*) [*The Forefather* (*About Leonid Isaakovich Mandelstam*)] 2002 *Usp. Fiz. Nauk* **172** 91 [2002 *Phys. Usp.* **45** 81]; 2003 *Epokha i Lichnost'. Fiziki. Ocherki i Vospominaniya* [*Epoch and Personality. Physicists. Essays and Memoirs*] (Moscow: Fizmatlit)

[24] Friedman R M 2001 *The Politics of Excellence: Behind the Nobel Prize in Science* (New York: W N Freeman/Times Books); 2001 *Nature* **414** 690; 2002 *Phys. Today* **55**(3) 61; 2002 *Phys. World* **15**(8) 33

[25] Landsberg G S and Mandelshtam L I 1928 *Zs. Phys.* **50** 769 [Translated into Russian 1978 *Usp. Fiz. Nauk* **126** 155]

[26] Ginzburg V L and Fabelinskii I L 2003 *Vestn. Ross. Akad. Nauk* **73** 215 [2003 *Herald Russ. Acad. Sci.* **73** 152]

[27] Fabelinskii I L 2003 *Usp. Fiz. Nauk* **173** 1137 [2003 *Phys. Usp.* **46** 1105]

[28] Ginzburg V L 2004 *Rev. Mod. Phys.* (to be published)

[29] 2004 *Les Prix Nobel* (*The Nobel Prizes*) *2003* (Stockholm, Sweden)

Article 22

Once more about the history of the discovery of combinational light scattering[1,1*]

It is well known, at least here in Russia, that the combinational scattering of light, a very interesting physical phenomenon of great practical value, was discovered in 1928 practically simultaneously by G S Landsberg and L I Mandelshtam in Moscow and C V Raman and K S Krishnan in Calcutta (India). However, the Nobel Prize for this discovery was awarded in 1930 to C V Raman alone and the phenomenon in question is usually referred to as the Raman effect. To the best of our knowledge, all Russian physicists have always regarded this decision of the Nobel Committee as a crying injustice.

What was behind it? The mechanisms underlying the Nobel Prize award for a concrete work are made public, even if not in full detail, only 50 years after the decision was taken. Therefore, why the Nobel Committee gave preference to a single author for the discovery of combinational light scattering remains a matter of conjecture. The general thinking of many in the former USSR was that the prize was not given to our scientists because of anti-Soviet attitudes among members of the Nobel Committee and other authorized participants of the decision-making process. In 1987, however, the Nobel Committee materials were published concerning the first 37 years of its work [1]. It became clear that, in all probability, there had been no political motives behind the decision to exclude Landsberg and Mandelshtam from consideration as potential prizewinners. We believe [2, 3] that the main reason was the disregard that our Russian physicists had for the work of their compatriots. A lack of objectivity in the assessment by foreign physicists and the Nobel Committee itself was another although less important reason, to say nothing about Raman's specific behaviour.

---

[1] The authors of this article are V L Ginzburg and I L Fabelinskii.

C V Raman was nominated for the 1930 Prize by ten persons including N Bohr, L de Broglie, J Perrin, E Rutherford, and C T R Wilson whereas only O D Khvolson and N D Papaleksi proposed Landsberg and Mandelshtam (indeed, Papaleksi proposed only Mandelshtam but in a broader context, without setting him off against Landsberg). Three other Soviet physicists who sent the Nobel Committee their recommendations proposed scientists other than Landsberg and Mandelshtam even though the number of candidates to be presented by each nominator was unrestricted. The fact that Landsberg and Mandelshtam reported their results later than Raman had might also influence the decision of the Committee. We would like to emphasize from the very beginning that we have always considered this decision to be absolutely incorrect and we did our best to prove this point as convincingly as possible in [2–4]. Nevertheless, we decided to touch on this issue once again after it had been discussed anew in the recent papers by R Singh and F Riess [5] and A M Blokh [6]. It is worth noting that [5] made us somehow change or, more explicitly, re-formulate our views. In order to make what follows easier to understand for the reader who may be unfamiliar with papers [3, 4], to say nothing about the original publications on the subject, we shall try to throw some light on the history of research conducted in Moscow and Calcutta.

In 1926, S L Mandelshtam and G S Landsberg initiated experimental studies on the molecular scattering of light in crystals. In particular, they sought to observe the splitting of the Rayleigh line caused by scattering on thermal acoustic waves as predicted by the former author in 1918. This phenomenon was later called the Mandelshtam–Brillouin (MB) effect. While the study was underway and yielding some interesting results, the researchers quite unexpectedly discovered the combinational scattering of light, i. e. the appearance of additional lines, satellites, in the scattered light spectrum with a frequency modulated so that it was three orders of magnitude greater than the one expected for the MB effect. In other words, the researchers looked for a change in the scattered light caused by the acoustic frequency component but found instead its change by the optical frequency component. Naturally, they undertook to study the new phenomenon and postponed the search for the MB splitting. Both a historical review and the results of the studies on the MB effect are omitted here as having been considered elsewhere [7].

In his answer to a written question from O D Khvolson, L I Mandelshtam said that "we first paid attention to the appearance of new lines on 21 February, 1928. The new lines were already clearly seen on the negative of 23–24 February (15 h exposure)". Both the letters exchanged between Khvolson and Mandelshtam and the spectrum of combinational light scattering in quartz are reproduced in [4]. Also, it is known that the authors reported their discovery at a colloquium on 27 April 1928 and, thereafter, at the Sixth Congress of the Association of Russian Physicists held 5–15

August and attended by nearly 40 scientists including 21 foreigners with such persons of prominence among them as M Born, L Brillouin, C Darwin, P Debye, P Dirac, R Pohl, P Pringsheim, and Ph Franck. In their short notes about the work of the Congress, Born [8] and Darwin [9] informed the scientific community about the discovery of Landsberg and Mandelshtam and emphasized that it had been made independently of Raman and Krishnan (see later). Landsberg and Mandelshtam mailed their brief communications to *Naturwissenschaft* on 6 May 1928 [10] and to *The Journal of the Russian Physico-Chemical Society* on 10 May [11]. A more comprehensive paper dispatched to *Zeitschrift fur Physik* [12] was received there on 12 August 1928. The latter publication contained a detailed description of the experimental device and the results of the study on combinational scattering of light in quartz and Iceland spar together with a clear and accurate explanation of the nature of the phenomenon reported in brief in [10, 11]. It is worthy of note that [12] contains the following passage: "we had observed the appearance of satellites in the spectrum of light scattered in quartz before we came to know the publication of Raman and Krishnan [here, the authors referred to the papers cited later—VLG and ILF] who described a change in the wavelength of light scattered by certain vapours and liquids".

This makes the situation perfectly clear. But, strange as it may seem, neither a comprehensive historical study [5] nor the resolution of the Nobel Committee referred to in [5] cites this detailed article [12]! We shall discuss this question later; meanwhile, let us turn to the works of the Indian authors.

Raman and Krishnan, based essentially on an analogy with the Compton effect, suggested that light scattering should give rise to a low-frequency component.[2] To verify this hypothesis, they observed, using selective filters, the scattering of sunlight in a number of vapors and liquids. They interpreted the results thus obtained as evidence of the predicted low-frequency component and published them in *Nature* magazine (31 March 1928) [13]. It is this publication dated 16 February, that is usually regarded in the foreign literature as the first report of the discovery of combinational light scattering. We can not share this view. To begin with, it is certainly impossible to observe combinational satellites in a continuous spectrum of solar radiation and the authors did not pretend to have seen any. Secondly, it is well known at present that the light of the summarized combinational scattering makes as low a contribution as a few percent of the Rayleigh light scattering to the total flow of scattered light in a liquid. A detailed analysis in [4] has demonstrated that such a faint luminescence could hardly be observed by eye as described in [13]. Raman and Krishnan postulated

---

[2] It is worthwhile to note that Raman and Krishnan's 'idea' about a certain analog with the Compton effect is totally baseless (see, for instance, [26]).

the existence of some low-frequency radiation and they did 'see' it. True, this 'finding' prompted them to make relevant spectroscopic observations which they reported in further publications [14–16]. To avoid splitting hairs (see also [3, 4]), suffice it to cite the words of Raman himself [17]: "Spectral lines of new radiation were first seen on 28 February, 1928. The observation was given publicity the next day." In other words, the lines of combinational light scattering were first recorded (i. e. the discovery was made) by the Indian physicists one week after their Moscow colleagues saw them (see earlier). We are far from attaching great significance to this fact and mention it only by way of protest against the pseudopriority of the Indian scientists. Where they outdid the Russians was in publishing their findings. This is, first and foremost, attributable to the quite different attitudes of Mandelshtam and Landsberg, on the one hand, and Raman, on the other hand, to science, in general, and to priority matters, in particular. Moreover, a tragic event intervened. A relative of L I Mandelshtam was arrested and sentenced to death wherefore Mandelshtam had to spent most of his time in trying to save this man (and succeeded in his efforts, the event luckily taking place in 1928, not in 1937–38). Far be it from us to justify the indolence of our scientists as regards publication of their results, nevertheless, for all we know, the exact date of publication is not as critical as the other considerations that guide the decisions of the Nobel Committee.[3]

As said earlier, paper [12] containing a detailed description of the experiment was mentioned neither in [5] nor in the resolution of the Nobel Committee referred to in the latter article. We for some time believed that the paper had been simply unavailable to the Committee. However, A M Blokh kindly informed us that he received from the Secretary of the Committee a file containing copies of documents concerning the consideration of G S Landsberg and L I Mandelshtam as nominees for the 1930 Nobel Prize. The file does contain paper [12]. Surprisingly, the resolution of the Nobel Committee cited in [5] mentions only a brief communication [10] and even misrepresented its substance! This fact needs to be discussed at greater length. The last phrase in [10] on which [5, 6] are focused reads as follows: "Ob und wieweit die von uns beobachtete Erscheinung mit der von Raman [2] erst kurzlich beschrieben im Zusamemmenhang steht, konnen wir zur Zeit noch nicht beurteilen, weil seine Schilderung zu summarisch ist" (Reference [2] here: Raman and Krishnan, *Nature*, 31 March 1928; 21 April 1928).

Paper [4] cites the Russian version of the same sentence from [10]. In

---

[3] The truth of this assertion is confirmed by the history of the development of masers and lasers described in the book of C Townes [18]. In 1964, the Nobel Prize in physics was divided between C Townes who received half of the prize and N G Basov and A M Prokhorov (the other half) for their pioneer studies in this field but nobody has ever tried to clarify whose maser was the first to work.[2*]

back translation into English it reads as follows: "We are unable to judge at present how closely the phenomenon we observed is related to that described by Raman because his description is rather a general one". We believe this statement calls for no comment. But, to avoid misunderstanding, here is the last phrase from a note by the same authors [11] published in the same year: "At present, we find it difficult to see the relation between the phenomenon we discovered and that observed by Raman and Krishnan in liquids and vapours as briefly described in their letters to *Nature*".[4] It should be noted that Reference [6] offers an inaccurate translation of this sentence. Reference [5] cites it translated into English as it appears in the resolution of the Nobel Committee where it is used to draw the following conclusion: "However Raman's and Krishnan's letters to *Nature* of March 31st as well as that April 21st gave a very clear explanation of the nature of the phenomenon (both cited by Mandelshtam and Landsberg). *Under these conditions, Mandelshtam and Landsberg cannot argue to have obtained their experimental results independently*" [emphasized by us—VLG and ILF]. The papers by Raman and Krishnan published on 31 March and 21 April are our References [13] and [14].

We have already spoken about paper [13] which at best gave a hint of the existence of combinational light scattering because the authors failed to observe spectral lines. Reference [14] reported the observation of such lines but the authors noted that "the position of the principal modified lines is the same for all substances...". At the same time, it is known that the positions of satellites for individual substances should have been different. No wonder, under these conditions, Landsberg and Mandelshtam had to exercise much caution in identifying their findings with the results of the Indian researchers. But the main thing was that the Nobel Committee took an absolutely illogical and unwarranted decision (to say the least) by concluding that the Moscow scientists "cannot argue to have obtained their experimental results independently" (as emphasized earlier). It is appropriate to cite here O D Khvolson who, in his presentation of Landsberg and Mandelshtam's work to the Nobel Committee, emphasized that they "observed and explained the same phenomenon [i. e. combinational light scattering—VLG and ILF] on February 21". In other words, even if the Committee, for one or another reason, chose to ignore paper [12], it had both article [10] and information from Khvolson at its disposal. With those on hand, the statement that Landsberg and Mandelshtam have not "obtained their experimental results independently" (see earlier and [5]) is akin to a charge of plagiarism! We do not even wish to comment on such an insinuation. As regards the exact time of publication, we consider it, in the present case, to be a different matter of secondary importance.

---

[4] This paper mentions the publication in *Nature* dated 5 May besides the aforecited references to [10].

In [2, 3] we pointed out the three most likely reasons why Mandelshtam and Landsberg did not receive the Nobel Prize. These were the indifference of their Soviet colleagues, active support of Raman by foreign scientists, and a mistake by the Nobel Committee. Certainly, the carelessness of Mandelshtam and Landsberg themselves played a role, in striking contrast to the behavior of Raman. The great importance of the latter circumstance became evident after the publication of an article [5] written by foreigners whom we have no reason to consider Russophilic (one of them even appears to be an Indian). It turns out that Raman distributed 2000 reprints of his paper on combinational light scattering dated 16 March [17] to "all eminent physicists including those who worked in the field of light scattering in France, Germany, Russia, Canada, and the USA and also to research institutions all over the world, thereby securing his priority" (see [5] where the source of this information is cited). Moreover, Raman took other measures to popularize his discovery and gain the support necessary to win the prize. Specifically, it appears from [5] that he asked N Bohr, C T R Wilson, E Rutherford and, in all likelihood, some other Nobel Prizewinners to propose him as a candidate. No wonder, the Nobel Committee had to give an attentive ear to the advice of such an assemblage of prominent scientists rather than heed the voice of Khvolson and the inarticulate representation of Papaleksi. Needless to say, Mandelshtam and Landsberg, being persons of culture and education, could not think of self-advertising. Sure enough, they never asked anybody for support nor could they agitate for themselves.

We fully agree with Singh and Riess [5] that "The example of Raman demonstrates that contacts with known scientists play a decisive role in the nomination for a Nobel Prize. The nomination of Raman by known physicists and such Nobel Prizewinners as Rutherford, Bohr, and Stark strengthened his case whereas prospects for Landsberg and Mandelshtam (nominated by their countrymen alone) were poor". It is appropriate to remember that most of their countrymen did not actually nominate Landsberg and Mandelshtam. Interestingly, Indian physicists behaved in a similar manner, i.e. they did not propose Raman [5]. It appears that, at least in this case, an old saying "a prophet is not without honour, save in his own country" proved true in Russia as well as in India. In addition, Raman won "the description of a person lacking tact in dealing with the public" [19]. It should be noted that the working climax in the discovery of combinational light scattering in India was carried out by Raman in cooperation with Krishnan, an able and skilled physicist. His contribution is confirmed by the fact of joint publications [13, 15, 16]. However, Raman chose to ignore the contribution of Krishnan, in disagreement with established practice.

Recently, an interesting book [20] was published in this country by A M Blokh, a foremost Russian authority on matters related to the statutory rules governing Nobel Prize awards and their history. The author collected many interesting material, some of which illustrate the attitude of Soviet officials to Nobel Prizes. However, we strongly object to his manner of attributing "Nobeliana" (as he calls it) to G S Landsberg and L I Mandelshtam in [6]. The reason is perfectly clear. A M Blokh justifies the decision of the Nobel Committee whereas we consistently oppose it. The arguments have been given in preceding paragraphs. Let the reader censure them in his/her wisdom.

In conclusion, we would like to use this opportunity and make a few remarks on the developments related to Nobel Prize awards for physics which are a subject of considerable public interest. W Roentgen was the first to receive such a prize in 1901 for the discovery of the X-ray. To the best of our knowledge, he had no co-authors. Thereafter, prizes continued to be awarded only to those who were directly involved in the research. In the course of time, however, the number of both nominees and nominators increased [1]. As many as 38 nominators proposed 21 candidates for the 1930 Prize that was eventually awarded to C Raman. The Nobel statutes give little opportunity for knowing much about the work of Nobel Committees over the past 50 years. However, P Rodgers [21] reports that the Committee on physics sent out over 2000 invitations to nominate candidates for the 2000 Prize. About 300 proposals were received from the respondents of which 10–15 were selected for further consideration. According to the interesting information presented in [22], some candidates were nominated many times during 1901–50. The record-holders are O Stern who was nominated 81 times and eventually awarded the Prize in 1943 and A Sommerfeld who was also nominated 81 times but who never received the prize and died at the age of 82 in 1951. It is known that no more then three persons are entitled to receive the prize in one year. However, only four years are on record throughout the first 24 years (up to 1924) when the prize was awarded to more than one person (H A Lorentz and P Zeeman in 1902, A Becquerel and the Curies in 1903, G Marconi and F Braun in 1909, and the Braggs, father and son, in 1915). Conversely, only four persons have been the sole awardees during the last 24 years (1979–2002) whereas in all other cases the annual prize was shared jointly by two or three winners. Moreover, in many cases, the Nobel Prize was awarded to the heads of large research teams. The authors of original papers sometimes represented several groups of physicists and engineers numbering tens of specialists of whom only three should have been selected. This gives an idea of the great challenges that such a situation presented to the members of the Nobel Committee. Its work remains exceedingly difficult to-day when, at least in certain cases, it appears to be closely akin to refereeing sport competitions in running or swimming [23]. This observation should not be interpreted as pique. What is said only reflects objective changes and trends in the character of physical and astronomical studies. Many, if not most, of the topical problems are beyond the power of individual researchers to address and they can be solved only by the collective efforts of large research groups. In this situation, it is only natural that preference should be given to the leaders. There seems to be no other way to award Nobel Prizes. An example is the 2002 Nobel Prize. R Giacconi who received one-half of it is characterized by R A Syunyaev, his colleague and a known astrophysicist working in Russia and Germany, in the following way: "He is a great scientist and truly big American manager" (*Izvestiya* [*News*] 9 October, 2002). Indeed, the joint work of Giacconi, Gurskii, Paolini, and Rossi [24] led to the discovery of the first bright 'X-ray star', Scorpius X-1, which in turn lent a powerful impetus to the advancement of X-ray astronomy. To our knowledge, however, the first leading figure in this field was Bruno Rossi, a distinguished physicist who deceased in 1993. This, of course, does not belittle the achievements of Giacconi, all the more so that he afterwards directed the work of the Hubble Space Telescope. Such outstanding organizers and scientists as he play an exceedingly important role in modern physics and certainly in other disciplines (in our country, S I Vavilov, M V Keldysh, and I V Kurchatov acted in this capacity). We are aware that Mandelshtam, one of the highly reputable theorists in this country, was of the same opinion.

To conclude, what has been said here has by no means been aimed at casting a shadow on the Nobel Prizes in science in general. Our sole purpose was to show that their meaning and value should be understood in the context of a particular situation and to warn against making a fetish of them and their winners.[5] New times call for new songs.[3*]

Comments

1*. This paper was published in 2003 *Vestn. Ross. Akad. Nauk* **73** 215 [2003 *Herald Russ. Acad. Sci.* **73** 152] in co-authorship with I L Fabelinskii whom I thank for the permission to include it in this volume. It is worth noting that after its publication I L Fabelinskii once again analyzed pioneer works on combinational light scattering [26]. Some debates concerning this question are also reported in the newspaper *Poisk* (suffice it to mention reference [27]).

2*. There is little doubt that the maser constructed by C Townes and co-workers 'started to work' much earlier than a similar device by N G Basov and A M Prokhorov [18].

3*. What is said in the end of this article about changes in the selection of candidates for the Nobel Prize in physics during the last decades does not apply to the one received by Raman in 1930. Setting aside the role of K S Krishnan which remains unclear to me (neither Raman nor anyone else nominated him), the limitation of the number of Laureates to three

---

[5] In this regard, a recent book [25] may be of interest.

was insignificant in this case. In other words, there was no reason why the Nobel Prize for the discovery of combinational light scattering could not be awarded to Landsberg, Mandelshtam, and Raman as was officially proposed by O D Khvolson.

References

[1] Crawford E, Heilbron J L, and Ullrich R 1987 *The Nobel Population 1901–1937: A Census of the Nominators and Nominees of the Prizes in Physics and Chemistry* (Uppsala Studies in History of Science, Vol. 4) (Berkeley, CA: Office for History of Science and Technology, University of California)

[2] Ginzburg V L 1998 Pochemu sovetskie uchenye ne vsegda poluchali zasluzhennye imi nobelevskie premii? [Why did Soviet scientists not always receive deserved Nobel Prizes?] *Vestn. Ross. Akad. Nauk* **68** 51 [1998 *Herald Russ. Acad. Sci.* **68** 56] (see article 21 above)

[3] Fabelinskii I L 1998 Kombinatsionnomu rasseyaniyu sveta—70 let [Seventy years of combinational (Raman) scattering] *Usp. Fiz. Nauk* **168** 1341 [1998 *Phys. Usp.* **41** 1229]

[4] Fabelinskii I L 1978 Otkrytiye kombinatsionnogo rasseyaniya sveta [The discovery of combinational scattering of light (the Raman effect)] *Usp. Fiz. Nauk* **126** 124 [1978 *Sov. Phys. Usp.* **21** 780]

[5] Singh R and Riess F 2001 The 1930 Nobel Prize for physics: A close decision? *Notes Rec. R. Soc. London* **55** 267

[6] Blokh A M 2002 'Nobeliana' Grigoriya Landsberga i Leonida Mandelshtama ['Nobeliana' of Grigorii Landsberg and Leonid Mandelshtam] *Priroda* 6 73

[7] Fabelinskii I L 2000 Predskazanie i obnaruzhenie tonkoi structury linii Releya [The prediction and discovery of Rayleigh line fine structure] *Usp. Fiz. Nauk* **170** 93 [2000 *Phys. Usp.* **43** 89]

[8] Born M 1928 VI Kongres der Assoziation der russichen Physiker' *Naturwissensch.* **16** 741

[9] Darwin C G 1928 The sixth congress of Russian physicists *Nature* **122** 630

[10] Landsberg G and Mandelstam L 1928 *Naturwissensch.* **16** 557. This brief communication translated into Russian is reproduced in [4]

[11] Landsberg G S and Mandelstam L I 1928 *Zh. Russk. Fiz.-Khim. Obschestva Ch. Fiz.* **60** 335

[12] Landsberg G S and Mandelstam L I 1928 *Zs. Phys.* **50** 769; this paper translated into Russian is reproduced in the Appendix to [4]

[13] Raman C V and Krishnan K S 1928 A new type of secondary radiation *Nature* **121** 501

[14] Raman C V 1928 A change of wave-length in light scattering *Nature* **121** 619

[15] Raman C V and Krishnan K S 1928 The optical analogue of the Compton effect *Nature* **121** 711

[16] Raman C V and Krishnan K S 1928 The negative absorption of radiation *Nature* **122** 12

[17] Raman C V 1928 A new radiation *Indian J. Phys.* **2** 287

[18] Townes C H 1999 *How the Laser Happened. Adventures of a Scientist* (New York: Oxford University Press)

[19] Bhagavantam S 1978 The discovery of the Raman effect, reminiscences of Sir C V Raman *Proc. Sixth Int. Conf. on Raman Spectroscopy, Bangalore, India, 4-9 September, 1978* vol 1 eds E D Schmid *et al* (London: Heyden) p 3

[20] Blokh A M 2001 *Sovetskii Soyuz v Inter'ere Nobelevskikh Premii [Soviet Union in the Interior of Nobel Prizes]* (St-Petersburg: Gumanistika)

[21] Rodgers P 2000 Countdown to the Nobel Prize *Phys. World* **13**(10) 10

[22] Crawford E 2001 Nobel population 1901–50: anatomy of a scientific elite *Phys. World* **14**(11) 31

[23] Ginzburg V L 2002 O nekotorykh uspekhakh fiziki i astronomii za poslednie tri goda [On some advances in physics and astronomy over the past three years] *Usp. Fiz. Nauk* **172** 213 [2002 *Phys. Usp.* **45** 205]

[24] Giacconi R *et al* 1962 Evidence for X-rays from sources outside the solar system *Phys. Rev. Lett.* **9** 439

[25] Friedman R M 2001 *The Politics of Excellence: Behind the Nobel Prize in Science* (New York: W N Freeman/Times Books)
Elzinga A 2001 *Nature* **414** 690
Nielsen H 2002 *Phys. World* **15**(4) 46

[26] Fabelinskii I L 2003 Otkrytie kombinatsionnogo rasseyaniya sveta v Rossii i Indii [The discovery of combination scattering in Russia and India] *Usp. Fiz. Nauk* **173** 1137 [2003 *Phys. Usp.* **46** 1105]

[27] Ginzburg V L 2003 There might be no mistake *Poisk* 7, 21 February (the title was given by the Editorial Board; my letter to the newspaper was entitled 'There were a mistake and lack of objectivity (about the history of Nobel Prize unawarded to G Landsberg and L Mandelshtam')

Article 23

Back to the Middle Ages—where soothsayers, wizards, and witnesses of miracles, all through television and newspapers, lead us[1][*]

You switch on the TV and you see something right from the Middle Ages: astrologers, alchemists, chiromancers, and witnesses of all sorts of miracles. And the newspapers are not very far behind. 'Charged' portraits, allegedly capable of healing everyone, stare from the pages of the newspaper *Vechernyaya Moskva* [*Evening Moscow*]. Horoscopes are published, somewhat shamefacedly, under the title "Believe it or check it out". The newspaper *Pravda*, on 2 January, 1991, suggested building communism with the help of flying saucers (more about this later).

In short, a wave of pseudoscience or, more often, simply mysticism and antiscientific hogwash has struck our mass media. It is usually said that such phenomena are typical of hard times. To this, in Russia, should be added the misinterpretation of the meaning of freedom of the press and 'glasnost' which has opened the door to a fecund deluge of antiscientific speculations and disinformation.

It would seem that a high dam should have been 'built' by scientists and all other educated people to stop this deluge. Unfortunately, this has not happened for many reasons. First, today science in Russia is out of honor and is defamed by the highest echelons of power. It is claimed that science is to blame for shortages of food, for the absence of a reasonable economic policy, etc. For this reason, many scientists feel depressed, so to say, and choose to keep a low profile. And they have other troubles to worry about. Second, fighting pseudoscience is a dirty and non-prestigious job. Third, culture (especially scientific culture) in our society is, on the whole, at a very low level. As a result, pseudoscience has found a rich soil and scientific arguments are accepted with difficulty or simply ignored.

This being the situation, it was impossible to remain silent. I have decided to call on scientists and all other educated people not to ignore the orgy of pseudoscience but to fight it actively.

The reader will recall that, roughly a century ago, humans knew nothing about radio or aviation, not to mention television, genetic engineering, aeronautics, nuclear power engineering and much more without which today the life of a civilized society is impossible. All these achievements are based on science and its advances. Hence, the antiscientific climate in our society seems like a lapse into savagery and obscurantism. The pitiful state of our agriculture is not the result of the bad doings of biologists and soil scientists. The blame must be placed on the repression, collectivization, an unsound agricultural policy, and support given to the pseudoscience of Lysenko by Stalin (and Khrushchev, too).

Physics in the USSR was harmed less due to its role in creating the atomic bomb. Not everyone knows, although this has been reported in the press, that there was a plan in March 1949 to remake physics in the Lysenko manner: a meeting similar to the notorious 1948 session of VASKhNIL (All-Union Academy of Agricultural Sciences) was scheduled. As far as I know, Igor V Kurchatov, the leader of the Soviet nuclear program, told the high and mighty that an atomic bomb (such a bomb was soon to be tested) could not be built on the basis of pseudophysics.[2*] And so faith has it that this most awful weapon saved whole areas of science and technology in the USSR from destruction and, maybe, many of us, too.

Today Soviet science and the entire country are in crisis. There is a shortage of money and equipment, the brain drain is significant, and there are many organizational difficulties. However, we have many capable and well-educated people and I am sure that science is not a bottleneck. If it proves possible to preserve a powerful federal (united) state instead of letting it become an ill-assorted conglomerate of separate republics, science is sure to flourish and achieve world standards.[3*] However, even if the general political and economic situation is favorable, the effective development of science will not proceed automatically. Many well-conceived measures have to be taken for this to happen. One such measure is the fight against pseudoscience, so that never again will there be any ridiculing of genetics, cybernetics, and works in economics, agriculture, philosophy, and history and their authors. Another closely related measure would be to show intolerance to antiscientific research, which is often concealed under the mask of secrecy. This aspect has been described in an article by E P Aleksandrov, titled "Black Science" and published in the journal 1991 *Nauka i Zhizn'* [*Science and Life*] no 1 56. I highly recommend this fine article, whose author, a well-known physicist, is one of the few who is actively fighting pseudoscience.

Another important condition for the progress of science is society's commitment to education. This, of course, is a special and comprehensive issue and I will confine myself to just one remark. I believe that a cultured person is considered to be someone who has mastered the curriculum of our inadequate secondary school system. With universal obligatory school education, there would seem to be many people who meet this criterion. Unfortunately, this is not so, especially in the field of scientific knowledge. Just ask a person about the reasons for the rotation of seasons. I often used this test and have come to the conclusion that many people, even those with higher education, either do not know the answer at all or explain the phenomenon by the seasonal changes in the distance between the Earth and the Sun. The correct answer has been known for several centuries: the Earth's axis of rotation is inclined with respect to the Earth–Sun line. How can one consider himself or herself a cultured person and not know the answers to this or similar questions concerning the structure of the atoms or the Universe, the mechanisms of heredity, and the like? It is hard to imagine a person not knowing the authors of *Yevgeny Onegin* and *War and Peace* and yet claiming to be cultured. Yet absolute illiteracy in the field of science seems to be acceptable. One often hears the phrase "I am incapable of learning mathematics". But nobody is talking about mathematics. We are talking about the basics of scientific knowledge. Of course, all this stems from a delayed development in a social consciousness, which has yet to grasp the role that science and the current scientific revolution plays in modern life. This situation cannot be tolerated and the mass media must concern itself with the dissemination of scientific knowledge instead of promoting the gibberish of charlatans.

The range of knowledge given in school is quite sufficient for understanding the antiscientific nature of astrology. Ideas about the effect of celestial bodies on a person's fate emerged thousands of years ago and were still dominant hundreds of years ago, when even the structure of the solar system was a mystery and humans felt defenceless against the forces of nature. The high priests, and not only they, cast horoscopes: some believed in this, others made money on it. But today (more precisely, two or three centuries ago), it has became quite obvious that astrology has nothing in common with science. Stars, to say nothing of galaxies, are so far away that light from them has to travel many, many years, even millions and billions of years, before it gets to us. Planets are, of course, closer but their motion has been thoroughly investigated and the law of universal gravitation, which is studied in school, makes it easy to calculate how negligible the force is with which they act on objects on the Earth. Physics has left no place for astrology in science. Naturally, numerous comparisons of the forecasts provided by horoscopes with events in real life have only corroborated this fact. The reader can find detailed information about this in V G Surdin's useful and comprehensive article titled "The foolish daughter of wise astronomy", which appeared in the journal 1990 *Vestn. Akad. Nauk*

SSSR **60**(11) [1990 *Herald Russ. Acad. Sci.* **60**(6)].[1] But the circulation of this journal is limited and cannot be compared with the great numbers of people watching television and reading newspapers, where astrology is so popular. The playing field is not even.

The arguments of N Morozova, the author of the previously mentioned article "On a flying saucer into communism" (*Pravda* of 2 January, 1991), are also on the astrological level. In her article, Morozova advocates the idea of something called rotational gravitation. Essentially it amounts to the possibility of 'generating gravitational energy by using a rotating solid body'. At first I thought that this was a humorous item timed for the New Year holidays. But then I realized that this was no joke, that what I was reading was 'serious' antiscientific hogwash with a good-sized dollop of demagogy. Words fail to convey my feelings when I finished it—this is a masterpiece that deserves to be placed in a museum of curiosities. I highly recommend it to everybody. Here I will only remark that the 'scientific' part of the article promotes the ideas of B P Groshaven, who seeks to create what he calls nature–machine energetics, which utilizes the 'dense energy fluxes of the Earth's gravitation'. Groshaven only requests that he be allowed "to develop his project, carry out experiments, and build experimental devices. And all that he needs is money, laboratories, and a facility for experiments. But he has been given nothing".

And why are there such malevolent people who do not allow Groshaven and Morozova to "integrate themselves into the cosmic fluxes, bathe in the fluxes, bask in them, and the like"? Now "a bureaucrat from scientific circles enters the debate: 'What? You say rotational–gravitational energetics? You can't be serious! Why, it contradicts the second law of thermodynamics, the third law of somebody else, etc'. Obviously, Groshaven has met an educated bureaucrat who knows the basic facts of physics, instead of the criminal sponsors of black science that feature in Aleksandrov's article mentioned earlier. Then N Morozova calls the ill-fated bureaucrat, who knows the basics of mechanics, a 'hundred-percent anticommunist'. "What is the future of planet Earth to him if he draws a big paycheck every month?", etc. One has such an urge to go on quoting, since, as I said before, we have a masterpiece. But I will confine myself to just one more quote: "I am now engineer Groshaven's friend. We have a common cause. We are working for communism."

For about 50 years I have been receiving letters and manuscripts from various subverters of the basic laws of science and simply amateurs who are trying to explain the riddles of nature but I must say, with regret and amazement, that I do not recall a single case (among the hundreds) of finding some valuable ideas. (I am talking about what is known as fun-

---

[1] See also Surdin V G 2000 Why is astrology a pseudoscience *Nauka i Zhizn'* no 11 79; no 12 130.

damental science, since the situation with inventions and innovations is quite different.) Generally, we are, of course, speaking here of physics and astronomy, the oldest sciences. Many generations of astronomers and physicists, among them the greatest minds of all times, the cream of mankind, spent all their time and energy in studying the laws of nature and gathered an enormous amount of information. To master it, students in physics departments spend roughly five years, then do graduate work and accumulate more knowledge. It comes as no surprise, then, that a retired engineer or even a person who is actively working and, hence, has little time to study physics cannot come even close to the frontiers of science.

Theoretically, a frontier of this kind can be reached, since the mind of a genius operates in a special way but I know of no such cases. This is true, of course, only of recent times. In ancient times, and even a few centuries ago, the sum total of all knowledge was much smaller than it is today and only a few dozen people were involved in scientific research, so that amateurs had an easier life. In fact, the dividing line between professionals and amateurs was hazy, if it existed at all. But today, I repeat, this line is clear-cut, at least in physics. And so I answer the letters and reject another 'wonderful idea'. Almost all the replies to my letters accuse me of being insensitive and unconvincing and demand proof. Often in such letters and the literature, one comes across the phrase: "This cannot happen because it just can't happen". My answers, the correspondents claim, are reduced to this phrase. Of course, nobody writes directly in such a way but, if one ignores stylistic nuances, the essence of the responses is just that. For many centuries, attempts were made to build perpetual motion machines and always mistakes in the design were found. Scientific and governmental sanctioning bodies in various countries decided more than two centuries ago to refuse to correspond with anyone claiming to have invented a perpetual motion machine. This was a correct move, since confidence in the validity of the law of conservation of energy only grew with time.

What we have just said is widely known; hence, today nobody directly suggests projects for building perpetual motion machines. But the idea of generating energy by utilizing the "dense energy fluxes of the Earth's gravitation" is no better than a perpetual motion machine. The laws governing the mechanical motion of bodies, terrestrial or celestial, have been known ever since the time of Newton. They are used to design all machines, to launch artificial satellites, etc. The simple verdict that a project violates the laws of mechanics is sufficient, since, indeed, such a project just can not 'happen'. In her article, Morozova exclaims emotionally: "To stubbornly ignore the ideas of alternative energetics is, in my opinion, a crime. So what if it has not been proved or verified! Let us check and recheck again and again but let us act. Soon it will be too late." What naivety! What total ignorance of the methods and place of science! The laws of mechanics and gravitation have been checked many times over the centuries and no contradictions have been found.

Such an assertion often meets the following counter-assertion. Since we cannot attain absolute truth and absolution accuracy, the established laws are approximate, so there is always space for possible refinements and changes. And indeed, the laws of physics have certain limits but the area within which they remain valid is well defined. For instance, the laws of classical (Newtonian) mechanics are valid in our solar system to within one part in a million. Seventy-five years ago, however, more exact laws governing the motion of planets were discovered, the corollaries of Einstein's general theory of relativity. Today the calculations done in space navigation take into account the results of this more general and accurate theory.

One more remark of a general nature is in order. Sometimes pseudo-scientific statements are made by people with various degrees and titles. Of course, the words of an Academician or Doctor carry more weight than those of an unknown engineer. But we also need hardly mention that degrees or titles are alone not a guarantee that these words contain any scientific truth. First of all, degrees and titles, as everything created by humans, may be awarded mistakenly. Second, past achievements, even real, do not guarantee a sensible view and scientific competence, especially late in life. So that the only way to resolve a questionable case is to rely on collective opinion. In our days, there are so many qualified people, at least in the main scientific areas, that it is easy to get several qualified opinions. In these conditions, the possibility of a mistake is, generally speaking, close to zero (in cases where the opinion of the referees is unanimous). Of course, reviewing and refereeing is a cumbersome and time-consuming job but this is, generally, the only way by which science develops, as a result of direct or, more often, indirect collective discussions. In the latter category (indirect discussions), I put discussions at seminars and conferences of works published and works being prepared for publication. In this way, all information that is valuable and useful is preserved, while all information that is erroneous or barren is simply discarded and forgotten.

Of course, here I have in mind largely physics but the same is generally true of all natural sciences and mathematics. For instance, if we are speaking of chemistry, medieval alchemy has had no place in it for a long time. The situation becomes much more complicated if we turn to biology and, to be more specific, to physiology and psychology. Not being a specialist in this field, I will refrain from further comment but a few words must be said in this connection. When a would-be healer claims that he transfers something he calls energy over great distances, 'charges' objects, etc, this is either a figment of a sick imagination or simple charlatanry (I think there is more truth in the latter since, according to reports in the mass media, such healers never mistake scrap paper for paper money). Hypnotic suggestion is quite different, of course, since it is real. As difficult as it is for me to believe in this, possibly there are people who can experience autosuggestion when they look at a photograph. To my knowledge, the hypnotist A Kashpirovsky limits himself to performing hypnotic suggestion, without referring to any sort of mystic energy, etc. He articulates (or repeats) the following hypothesis: under the influence of suggestion, the human body produces certain substances that have a therapeutic effect. Is this actually the case? I do not know but this hypothesis does not contradict the laws of logic and physics. Of course, this is not enough to justify holding TV sessions.

On the whole, the potential of the human body and psychology should, unquestionably, be studied scientifically. As far as I know, interesting research in this area is currently conducted at the Institute of Radio Engineering and Electronics of the USSR Academy of Sciences and, possibly, in a number of other reputable research institutions.[4*] Unfortunately, the mass media are little interested in the results of such research. Instead they fling the doors wide open to various blatant charlatans and crooks or, in any case, people that talk unimaginable nonsense. The blame for this must also rest on scientific organizations and, for one thing, on the USSR Academy of Sciences and other academies and on the All-Union Society for the Dissemination of Knowledge, *Znanie*. And what have these done to fight pseudoscience and all types of mysticism? If something is done (I know nothing of this), it is clearly not on a wide enough scale.

At the beginning of this article, I mentioned the reasons that led to such a situation. These reasons are real but that in no way justifies passivity. Those who cherish science and culture, who wish to see the Soviet Union a modern, prospering, and democratic country must do everything to block the aggression of pseudoscience and mysticism.[5*]

Comments

1*. This article was published in the newspaper *Izvestiya* no 45 of 21 February, 1991. Unfortunately, it has not lost its relevance. Therefore, I have decided to include it in the present collection. Note that the same article in a somewhat modified form has been published in the journal 1996/97 *Zdravyi Smysl* [*Common Sense*] no 2 43.

One of the readers of the present article has asked why I failed to mention religion. The question is justified. Indeed, religions (Christianity, Islam, Judaism) definitely involve miracles (the Immaculate Conception, resurrection of the dead, afterlife, etc). In this respect, religion is no different from astrology or some other pseudoscience. I am a staunch atheist, i. e. I deny the existence of any god and assume that the only thing that exists is nature (matter) as cognised by man. Such cognition is the goal and essence of science. At the same time, the problem of how science, art, and religion interact is extremely interesting and rather complicated (e. g. see the article "Science, Art, and Religion" by E L Feinberg in the journal *Voprosy Filosofii* [*Problems of Philosophy*] no 7, 1997, and his book 2004 *Dve Kultury. Intuitsiya i Logica v Iskusstve i Nauke* [*Two Cultures*] [Expanded 3rd Russian edn (Fryazino: Vek 2)). This is not the place, of course, to discuss the various aspects of this problem. Suffice it to note that, as far as I know, the most important religious organizations in Russia, such as the Russian Orthodox Church, do not support astrologers or various charlatans or pseudoscientists. Hence, there is no reason to link the fight against pseudoscience and various manifestations of charlatanry and obscurantism closely with antireligious activity (or, if you wish, antireligious educating). It is a different matter that the established church (we mean, of course, all denominations) should not try to influence freedom of choice with respect to religion or atheistic views. The church must (and in this respect, in particular) be separate from the state. This condition, as we know, is stipulated in the current Constitution of Russia although, unfortunately, it is not fully observed. In this connection, see articles 31–34 in the present collection.

2*. As mentioned in article 26 of the present collection, the reasons for calling off the meeting are not known for certain (i. e. no documents have been discovered). However, this fact is unimportant in the sense that it in no way alters my assertion that only the need to use modern physics in building the atomic bomb saved physics from destruction (i. e. as we have said before, physics was not redone in a Lysenko manner).

3*. This was written before the collapse of the Soviet Union. However, the same argument is true for Russia, since most of Soviet science was concentrated in Russia. Of course, if science in Russia is to flourish, the country must cope with its economic crisis.

4*. As is known, the USSR Academy of Sciences became the Russian Academy of Sciences (RAS). The Institute of Electronics and Radio Engineering of the RAS is continuing this research.

5*. The relevance of this article to the situation in Russia, noted in Comment 1 in the first Russian edition (1997), far from decreasing, has even become more evident. Suffice it to say that as of its September 1999 issues, *Izvestiya* began to publish astrological forecasts. And that was the same paper that, in 1991, published the present article! This is as if such forecasts had appeared in *The New York Times*, since *Izvestiya* is one of the most prestigious newspapers in Russia. To my letter of 20 September 1999 addressed to the Editor-in-Chief M Kozhokhin protesting against such shameful actions, I did not even get a reply.[2] The newspaper has also published other articles of a pseudoscientific nature. The fight against the

[2] I must note than in 2001 *Izvestiya* stopped publishing astrological forecasts.

various forms of pseudoscience is so important for Russia that, in 1998, the Russian Academy of Sciences (RAS) set up a special committee to combat pseudoscience and falsifications of scientific research. In this connection, the Presidium of RAS issued a special statement beginning with the words:

> Today this country is witnessing grand-scale and totally un-opposed dissemination and propaganda of parascientific and para-normal beliefs, such as astrology, shamanism, occultism etc. Attempts are continuously being made to implement senseless projects at the expense of the government, for example, building torsion generators. The population of Russia is brainwashed by television and radio programs, articles in newspapers, and books with explicitly antiscientific content.

(For more details see the newspaper *Poisk* no 23, 1993, and the journal 1999 *Vestn. Ross. Akad. Nauk* **69**(10) [1999 *Herald Russ. Acad. Sci.* **69**(5)].) In view of this situation, I have included in the present collection articles 24 and 25 (see also the following articles: Aleksandrov E B and Ginzburg V L 1999 Concerning pseudoscience and its propagandists *Vestn. Ross. Akad. Nauk* **69** 199 [1999 *Herald Russ. Acad. Sci.* **69** 118], Ginzburg V L 2000 Concerning pseudoscience and the need to fight it *Nauka i Zhizn'* no 11, 74; see also 2001 *Vestn. Ross. Akad. Nauk* **71** 702 [2001 *Herald Russ. Acad. Sci.* **71** 355], the journal *Komp'yuterra* [*Computer Land*] no 41, 2001); see also my articles "Demagogues and ignoramuses in the fight against scientific appraisal" in the newspaper *Literaturnaya Gazeta* [*Literary Newspaper*] for 16–22 October, 2002, "A discussion with only one goal" in the newspaper *Rossiiskaya Gazeta* [*Russian Newspaper*] for 29 January, 2003, and "On misunderstanding in the problem of pseudoscience and on the interrelation between science and religions" in 2003 *Vestn. Ross. Akad. Nauk* **73** 816 [2003 *Herald Russ. Acad. Sci.* **73** 560]. After that I published several other articles concerned with the same subject. Since all these papers were published only in Russian, I will not list them here. Most of them like the previous ones, can be found on the Web at www.ufn.ru, part "UFN Tribune".

Article 24

Should we fight pseudoscience?[1*]

The Russian Academy of Sciences (RAS) has recently set up a special committee to combat pseudoscience and the Presidium of RAS has issued a special statement in which it urged the print and electronic media "to refrain from preparing and disseminating pseudoscientific programs and publications and to bear in mind their responsibility for the spiritual and moral education of the nation". Earlier the newspaper *Izvestiya* (17 July 1998) published to the same end a letter titled "Science condemns pseudoscience" and signed by more than 30 scientists.

Unfortunately, not everyone understands in what respects science differs from pseudoscience. Some believe that there is no sense in fighting pseudoscience and that this could even be counterproductive, since it would lead to the suppression of new ideas and to the triumph of dogmatism. The main arguments of those who oppose the idea that fighting pseudoscience is necessary and important are the following. The first argument can be formulated as a question: "And who will be the judges?" Why should we believe those 'elderly people' from the RAS rather than the astrologer Globa, the would-be healer Chumak, and the like, who radiate health? The second argument goes like this: The history of science tells us that in the past scientists sometimes made errors, e. g. refused to acknowledge that meteorites fall to the Earth or that there is such a phenomenon as hypnosis. So why should we not assume that the pundits are mistaken when they say, for instance, that astrological predictions are hogwash? The third argument is actually close to the second. As is known, for a long time in the Soviet Union and its satellites, the unscientific activities of Lysenko (he elaborated on Michurin's theories of hybridization or 'Michurinism') were highly favored by the authorities, which resulted in genetics being rejected and declared a pseudoscience together with cybernetics and several other areas of science. So how can we guarantee that the special committee to combat pseudoscience will not do something similar?

Such ideas are characteristic of people who do not know or do not

want to know (or have forgotten, to put it mildly) the system of the totalitarian Soviet regime and also know nothing about the state of modern
science and scientific information. Nor do they know the modern methods with which specialists appraise and referee scientific ideas, suggestions,
and projects, methods adopted today in all civilized, democratic countries.

Lysenko's doctrines and claims and similar phenomena in the Soviet
Union stemmed entirely from the bolshevik dictatorship. Stalin considered himself competent in biology (after all, he was declared 'the chief
expert on all the sciences') and simply gave Lysenko and his vassals free
rein to impose their unscientific assertions and take full control of the biological sciences in the country. Those who disagreed (the majority of
scientists) were fired and often arrested (among the latter the most prominent one was Nikolai I Vavilov, who perished in prison). Incidentally,
Khrushchev also supported Lysenko: it was only after Khrushchev was
ousted from power at the end of 1964 that scientific biology was again able
to develop freely. The losses that the Soviet Union incurred in connection
with Lysenko's doings were enormous. However, so much has been written and said about this that there is no need to dwell on this topic any
further. The point I would like to stress here is that something similar
to Lysenko's doings would be possible in Russia only under a dictatorship.

Another logical possibility would be domination by the Church, something similar to the Inquisition in the Middle Ages. But this is clearly
impossible, since today both the Catholic and Orthodox Churches have
rejected opposition to science, for more details see my article 'Reason and
faith' in the journal 1999 *Vestn. Ross. Akad. Nauk* **69** 546 [1999 *Herald
Russ. Acad. Sci.* **69** 271]; it has also been published in the present collection as article 31.

Now about the judges. In the cases in which we are interested these
are professional scientists, whose selection (one could even say 'natural
selection') is a prolonged, rigorous, and multifaceted process. Take the case
of a physicist that today is an Academician of the RAS. First he or she
must graduate from, say, a university, then earn a PhD, then a doctorate
of science, then be elected first a Corresponding Member of the RAS and
then, finally, a Full Member. At all stages the academic degrees and ranks
are awarded by bodies of scientists (academic councils or the appropriate
Department of the RAS) by secret ballot. Naturally, the candidate has
dozens if not hundreds of published papers, which have also passed various
reviewing stages. Therefore, the filter is just about perfect.

Nevertheless, as in everything humanity does, mistakes and mishaps
are possible. In a dictatorship, they are even more possible (after all, Lysenko was elected to the Academy of Sciences as a Full Member). However,
even with a perfect selection system, there is the possibility that a person

becomes, say, an Academician owing to his/her achievements but then, due to sickness or the effects of old age, loses his/her qualifications. A qualified person can also make mistakes, though. So what conclusions should we draw from all this?

First, we should not pay too much attention to academic degrees and ranks. For one thing, all articles submitted to scientific journals undergo rigorous refereeing, irrespective of the rank of the author. Second, in most more or less non-trivial cases, the refereeing is done by several specialists rather than by one. Commissions of expects serve just this purpose.

But this refers to the more complicated cases. Usually, when pseudoscience finds its way into the media, there is no need for a commission. For instance, astrology has been criticized for several centuries and its inability to predict anything definite has been proved without doubt. Even a high-school student knows that the forces with which other planets, let alone stars, act on a person are negligible compared, say, to a whiff of air. Statistics, too, argue against astrology. Analysis of an enormous number of horoscopes has shown their falseness and the rare cases when astrological predictions are confirmed are clearly random events. Still other arguments indicate that astrology is a hoax. The only way to believe astrological predictions would be to follow the well-known dictum of the Latin ecclesiastical writer Tertullian: *"Certum est quia impossibile est"* [It is certain because it is impossible].

And a word about mechanics, whose laws were established by Isaac Newton in the 17th century. These laws are used, for instance, to compute the paths of satellites and, as we know, to ensure with extreme accuracy the docking of spacecrafts in orbit (here small corrections introduced by Einstein's theory of relativity developed in the 20th century are also taken into account). No commission is needed to dismiss the attempts to bypass Newton's laws, just as there is no need for a commission to establish the totally unscientific nature of the doings of the would-be healer Chumak, who 'charges' various objects with his powerful aura and then sells them. And if astrology, in view of its long history and the terminology it uses, can at least be called a pseudoscience, the various fortune-tellers, seers, and healers of the Chumak type do not even refer to science. This is not pseudoscience but simple charlatanry.

I would also like to mention the huge increase in the speed at which information is exchanged today, particularly through e-mail and the Internet. Hence, reports about discoveries are checked very rapidly and in many places, and the truth is established promptly and for certain. The possibility of a prolonged erroneous estimate of fundamental scientific assertions is reduced to a minimum. The latest striking example was a report released in the United States in 1988 concerning 'cold' nuclear fusion, fusion in the electrolysis of light atomic nuclei with release of energy. This report was immediately checked in Russia and many other countries and it was found that such fusion had never been achieved.

To conclude this item, I would like to make a remark concerning the behavior of the media. Should a newspaper that considers itself a serious and honorable publication disseminate astrological predictions and the announcements of Chumak and the like? I believe it must not. For this reason, I recently rejected the offer to be a correspondent of a newspaper, which is considered a serious publication, to answer questions when I learned that it publishes astrological predictions. In trying to persuade me, the correspondent said that he and, even more so, the editorial board have no faith in such predictions but that readers want them. I believe many readers would also be very pleased to see pornography and crude advertising, which serious newspapers do not publish (in contrast to the yellow press).

I am not for censorship but I am for self-censorship. The publication of materials of a pseudoscientific nature is as unworthy as the propaganda of violence and the publication of pornographic stories and drawings.

Comments

1*. The item was published in the newspaper *Moskovskaya Pravda* on 11 September 1999. Here I would like to pay my respects to this newspaper, which later published two articles (on 18 September and 12 November 1999) by V Surdin, who showed how great the gap between astrology and science is. Unfortunately, such examples of this attitude in the Russian press toward science are very scarce. In fact, the newspaper *Izvestiya* on 5 January 2000 published an article by L Leskov supporting pseudoscience that was titled "Mouse on a mountain", which describes the activity of a special committee to combat pseudoscience set up at the Russian Academy of Sciences and headed by Academician E P Kruglyakov. Neither *Izvestiya* nor the newspaper *Rossiiskaya Gazeta*, which carried a similar article by Leskov, published Kruglyakov's response. Such are the morals of newspapers in Russia today. In a note titled "Counterattack of Pseudoscience" and published in the newspaper *Poisk* no 7 of 18 February 2000, I cast some light on this problem and, to a certain extent, responded to Leskov's articles. See also the journal 2000 *Nauka i Zhizn'* [*Science and Life*] no 11 74, 2001 *Vestn. Ross. Akad. Nauk* **71** 702 [2001 *Herald Russ. Acad. Sci.* **71** 355], and article 25 in the present collection.

Closely linked to the problem of pseudoscience (or, to be exact, of countering the propaganda of pseudoscience) is the question of objectivity in illuminating the history of science. I would like, in this connection, to direct the reader to some of the articles published in my book 1995 *O Fizike i Astrofizike* [*On Physics and Astrophysics*] (Moscow: Byuro Kvantum) [English translation 2001 *The Physics of a Lifetime* (Berlin: Springer)] and to my article titled "Regarding some pitiable historians of physics" and published in 2000 *Voprosy Istorii Estestvoznaniya i Tekhniki* (Problems of the History of Natural Sciences and Technology) no 4 5.

Article 25

What is happening to us?[1*]

The reader of *Uspekhi Fizicheskikh Nauk* needs no explanation of what
pseudoscience is: for example, it includes astrology, designs for perpetuum
motion machines (even if masked in some way or another), information
transfer at a speed faster then light, mythical 'torsion generators', and so
on. In Soviet times, some pseudoscientific 'theories' were proclaimed and
propagated openly (the most striking example is Lysenko and his supporters) while others flourished under the aegis of dialectic materialism or a
veil of complete secrecy and the possibilities for scientific criticism and
appraisal were, for some reason, obstructed.

In spite of the exceedingly important achievement of the post-Soviet
epoch of freedom of speech but, to some extent, because of the misuse of
this freedom, the propaganda of pseudoscience is unfortunately still blos-
soming today. Suffice it to mention the appearance of astrological forecasts
in newspapers and on television.

Opposing this disaster is a necessary but very difficult task. What
is there to say when even the government newspaper *Rossiiskaya Gazeta*,
with a circulation of half a million copies, published a number of articles
in 1997 and 1998 advocating the fraudulent 'research' on torsion technologies. Before the disintegration of the USSR, the authors of these
'research works' received impressive sums of money under the cover of
secrecy and now they try to revive their activity by employing ignorant
and unscrupulous journalists. Being exasperated by such pseudoscientific activity, E P Kruglyakov, an actively working physicist from Novosibirsk, succeeded in publishing an article "On the other side of science" in
1998 into *Rossiiskaya Gazeta* on 19 May. In the same edition, mud was
flung at him by A Valentinov, head of the newspaper's Science Department.

In our attempts to protect E P Kruglyakov and, which is even more
important, to clarify the distinction between science and pseudoscience,
E B Aleksandrov and I failed to have our corresponding paper published in

Rossiiskaya Gazeta or in any other newspaper. We had to place our note "About pseudoscience and its propagandists" in the *Vestn. Ross. Akad. Nauk* (the journal of the Russian Academy of Sciences issued in a limited number of copies) 1999 **69**(3) 199 [1999 *Herald Russ. Acad. Sci.* **69**(2) 118]. Such is the present-day situation in Russia and we cannot tolerate it.

Here, in the small book in question, the author sets an example to all of us of how to counteract the pseudoscience and ignorance that have also been thriving in some of government's highest echelons. The book is a compilation of 12 papers, one of which ("Does *Rossiiskaya Gazeta* understand freedom of the press correctly?") is devoted to the previously mentioned attempts by E P Kruglyakov, E B Aleksandrov, and me to interact with *Rossiiskaya Gazeta*.

In the other papers and the interview "The sleep of thought gives birth to pseudoscientists", which he gave to a correspondent of a Novosibirsk newspaper, the author cites a number of striking precedents of disgraceful goings-on caused by the ignorance that now inhabits various organizations up to the President's Security Service and one of the Committees of the State Duma. The author reminds us quite justly of the necessity of examining all kinds of projects and suggestions rendered to government bodies, which is often neglected.

In this connection, he pertinently quotes (p 37) the words of Peter the Great: "All projects should be well ordered lest the treasury be badly ravaged and the motherland be caused damage. Those who will devise projects carelessly will be stripped of rank and whipped—to teach their descendants a lesson." As we know, their descendants have not learnt the lesson.

The wide circles of physicists do of course distinguish between science and pseudoscience and charlatanism but they are not active enough in working against these phenomena because of the natural feeling—like that disgust experienced by sewage disposal men, as well as tiredness under the exhausting conditions of life and when it is very difficult to find the truth (suffice it to say that newspapers merely ignore protests as a rule). Such a standpoint can be understood but not justified. One should not keep silence but oppose the enemies of science and progress.

The efforts made by the Russian Academy of Sciences in this direction are insufficient. Some of the so-called Public Academies have become breeders or, at any rate, a sanctuary for pseudoscience.

Now, a special committee to combat pseudoscience has at last been set up in the Russian Academy of Sciences and has already done something but it will not make much progress without the support of wider scientific circles. E P Kruglyakov demonstrates that one should not remain indifferent to a pseudoscientific attack and how one should (and can) fight against it. It is, therefore, not out of place to draw the reader's attention to the book by E P Kruglyakov and to remind of his/her duty to protect science.

Do not pass by!

Comment

1*. These comments on the book by E P Kruglyakov 1998 *What is Happening to Us?* (Novosibirsk: Izd. Sib. Otdeleniya RAN) were published in 1999 *Usp. Fiz. Nauk* **169** 358 [1999 *Phys. Usp.* **42** 295]. Several papers from this book were included in a new, more extensive book by E P Kruglyakov 2001 *'Uchenye' s Bol'shoi Dorogi* [*Highway 'Scientists'*] (Moscow: Nauka). Articles 23 and 24 of the present book also touch upon pseudoscience.

Article 26

Communists again stand as the paramount menace for Russia and the entire world[1,1*]

1. It is nearly five years now since the self-acclaimed 'world socialist system' crashed and the USSR broke into pieces. One might think that the grandiose experiment over the lives of hundreds of millions of people that the bolsheviks (communists) launched in 1917 had run its full course and the total impotence of the communist ideology and practices had become obvious. Unfortunately, there is a lot of wisdom in the saying that the only lesson we learn from history is that history teaches us nothing at all. The short span of human memory, the built-in shortcomings of the democratic form of government, and certain crude blunders by the ruling élite in Russia resulted in a realistic communist and fascist threat for the future of our country and of the whole world. The historian, A Yanov, compared today's situation in Russia with that in Germany in the 1930s directly before the fascists gained power (1995 *Moskovskie Novosti* [*Moscow News*] no 85). Alas, I do see a basis for this analogy (with communists substituted for fascists). The danger is all the more frightening since Hitler had no nuclear weapons while Russia may again turn into an aggressive nuclear power.

In this situation, the momentous significance of the presidential elections on 16 June 1996 is obvious to everyone. If communists win, there will be foundations for the blackest predictions; alas, many even some very decent people fail to understand this. This factor provided the stimulus for my writing this text. I firmly believe that a just and decent person, in my interpretation of these two qualities, cannot sympathize with fascists. The fascist ideology, with its racism, antidemocratic stance, chauvinism etc is self-proclaiming and generates nausea. Communists, however, proclaim in their mottoes the shining ideals of justice and for equality all people and

---

[1] This article was written in 1996 when the presidential elections were approaching. This is obviously expanded in comment 1 but I better mention it immediately.

peoples, denounce arbitrary rule and lawlessness. As a result, those who do not know better or turn a blind eye to history's lessons are quite capable of swallowing the bait of the communist propaganda with its shameless, brash, and covert lies. This is why anyone who is not blind to the past and believes that communism must be resisted should make a contribution to this fight.

I do not doubt that very many people understand this and share the view that only democratic evolution can bring a happy future to Russia. Hence, I need not bother. Nevertheless, my opinion may not be devoid of interest. The reason is that political writing is typically done by journalists and sociologists or people with 'lyrical' inclinations (I refer to the image, popular in the USSR, of the 'lyricist *versus* physicist' confrontation) who were always viewed with suspicion in our country. In contrast, I am a physicist and partly astrophysicist, i. e. I belong to a 'favored' group. Moreover, even if this is accidental, previous political powers have not overlooked me: I received the Stalin Prize of first degree (later renamed the State Prize), the Lenin Prize, some orders, was elected a Corresponding Member of the Academy of Sciences of the USSR (now the RAS) in 1953 and a Full Member (known as an Academician) in 1966. I could boast of a number of awards from this Academy and from a number of Academies abroad. In other words, I am not among those to whom the Soviet (communist) power either made active life difficult or whom it even imprisoned. Thus, this is not a case of vindication for me. Furthermore, I was a naive believer in communist ideals for many years, contrary to all facts and experiences. I will return to this aspect later. At the moment I only wish to explain that having understood at the sunset of my life the awful fallacy of the communist ideology and practices, I strive to help others not to commit old mistakes, not to be blind and deaf to the facts of the past—not that it is such a distant one either.

2. Several years ago I watched a videotape about Robert Oppenheimer and the development of the atomic bomb in the USA. One detail caught my eye especially and stayed in my memory. All the 'nice guys' were liberals (some of them, if I am not mistaken, were communists) but they understood absolutely nothing about the practical side of communism in Russia or about Stalin's dictatorship. On the contrary, the unpleasant guys, some of them in the secret services, proved to be considerably more sagacious. This is very typical. Quite a few well-known and respectable writers and scientists in the West supported the USSR and condoned, up to the second half of the 1940s—some even longer—everything that its rulers, with Stalin at the helm, were doing to the population. Suffice it to mention Romain Rolland and Frederic Joliôt-Curie in France. In 1937, at the very peak of the terror, Lion Feuchtwanger visited the USSR but left having understood nothing.[12*] How about the number of personalities in the West who became KGB informers on ideological grounds? We can mention K Fuchs who

leaked atomic bomb data, and S Efron who ultimately brought himself and his wife Marina Tsvetaeva (the Russian poetess) to a sad end. Infinite is the list of all misled souls.

A fairly popular opinion is that the communists' successes and the support accorded them by liberals in the West stemmed from a lack of information. This is definitely not true. The butchery carried out by Stalin's pack was sufficiently documented as early as the end of the 1930s and the 1940s. Public trials and the absurd confessions of former leading bolsheviks, their executions, the assassination of Trotskii and other acts of 'heroism' by the KGB outside the borders of the USSR, the murder of nearly the entire top echelons of the Red Army—this is a lot more than an 'awl in a sack' that, as the Russian saying goes, will out. But those who refused to see chose to close their eyes to it, blinded by a nearly religious faith in communism and, at least to the same degree, by the opposition of the USSR to fascist Germany.

Archives have been opened in Russia in recent years and we were able to find out much of what remained hidden. Among these materials are, for example, the Ribbentrop–Molotov German–Soviet Pact, Lenin's previously top-secret bloodthirsty 'instructions' (such as those on murdering Orthodox priests), and the scale of the terror. Much new evidence of the lawlessness and repression have been published, to complement that already contained in Solzhenitsyn's invaluable *Gulag Archipelago*. The latest estimates indicate that more than 21 million people were shot or died in prisons and concentration camps. This is roughly the total population of the Scandinavian countries (Denmark, Norway, Iceland, Sweden, and Finland)![2] We have also learnt that the spiritual and organizational leader of all this repression was Stalin in person, this truly bloodthirsty bandit who showed mercy to no one. For example, it is his signature that was found under the 'decision' to shoot thousands of captive Polish officers in Katyn in 1940. It appears that by now only the blind and deaf may refuse to conclude that Hitler and Stalin are virtually brothers in blood, that fascism and communism are totalitarianisms of the same mint (see also the *post scriptum* at the end of this article).[13*]

3. Why has it taken so long to digest these truths?

Three causes emerge. First, as I have already emphasized, fascist mottoes produce a negative response as a matter of course. Contrary to this, communism's appeals to social justice, internationalism etc are very seductive. Second, Hitler and Co. openly broadcast their troglodyte plans. True, even fascists tried to hide their ugliest deeds such as the gas ovens in Oswiecim (Auschwitz). On the whole, though, their true face and ideology were exposed for all to see. As for the communists, they played their hypocritical game skillfully and consistently from beginning to end. The

---

[2] Other sources give differing estimates (see article 27 of this book).

list of examples is endless. Suffice it to mention the brutal murder of the Tsar's entire family, including young teenage girls and a terminally ill boy. Lenin kept this crime secret. I have read somewhere recently that responding to incessant queries about the fate of the Tsar's family that kept coming from A A Ioffe, the Russian representative in Germany, Lenin gave roughly the following instruction: 'Tell him that the family have not been shot; this will make it easier for him to lie'.[3] In fact, this is a relative 'detail'; hypocrisy and lies infused everything that the bolsheviks ever did. Consider the previously mentioned Katyn. I remember very well how a specially formed commission, which included a known KGB agent, the Orthodox metropolitan Nikolai, published the conclusion that the crime was committed by the Germans. The truth was acknowledged only very recently—50 years later!

The third reason for the popularity of communism in the West was that the alternative form of governance—a democracy—is not an ideal system, not by a long shot. Furthermore, capitalism is disgusting in many of its facets, and familiar lines from the Soviet poet-laureate Vladimir Mayakovsky are quite true to life:

> A doughnut to one
> > and the hole from its middle to another
> This is what the democratic republic
> > is all about.

This is quite correct[2*] but the tragedy was that in the socialist society (a 'developed socialist society' of the 1970s on top of that) that communists in the USSR and in its satellite states pretended to have constructed, it was only on paper that social equality existed. In actual fact, in the absence of human rights, lies, despotism, arbitrariness, and often terror ruled the land. I heard that Mayakovsky, the author of the previous quote, ultimately understood it, which was one of the reasons for his suicide in 1930. In addition to Mayakovsky, we immediately recall other outstanding poets whose physical ordeals at the bolsheviks' hands have no question mark on them. Nikolai Gumilev was shot, Osip Mandelshtam died in a concentration camp, Marina Tsvetaeva was humiliated, lost all roots, and ultimately hanged herself, Anna Akhmatova and Boris Pasternak were persecuted, Iosif Brodsky was arrested and then exiled abroad. Recalling all this is frightening and painful, especially if one keeps hearing eulogies from dirty liars or incorrigible fools about 'communism's humanism'.

4. We used to read tons about 'the rot setting in' in capitalism, about its forthcoming doomsday, about the economic advantages of the 'socialist system', and its inevitable victory. Let us not forget too that in the past

---

[3] On the atrocities initiated by Lenin, on his cruelty and hypocrisy, see Latyshev A G 1996 *Lenin Declassified* (Moscow: MART).

communists did accept the 'criterion of practical test', which is an inherently clear statement that any theoretical construct must be judged—and its fate evaluated only by comparison with reality. By the way, experimental testing (i. e. the same 'criterion of practical test') and its decisive role forms the basis of all natural sciences. Theories are almost always in abundance while there is only one reality. Communist practices crashed in flames, getting no help from the vaunted cliches of the 'proletariat's dictatorship' and 'democratic centralism' which proved to be nothing less than dictatorship by a pack of leaders (führers) and then by a single führer cum 'genius of all sciences'. The economic system of 'socialism' ended in complete failure: a country with the most abundant resources became a poor beggar. The victorious country that paid for its victory with tens of millions of lives of its citizens, fell to the humiliating low of receiving humanitarian aid from the previously vanquished and destroyed Germany. The latter rose from the ashes owing to the democratic society that replaced fascism and thanks to the help provided by other democracies. Democracy proved capable of modifying and adjusting its rule, curing the drawbacks of capitalism where possible, fighting racism (one example is the successful fight to exterminate racial discrimination in the USA) etc.

The conclusion that one has to draw from applying the practical criterion—what we call history's verdict—is obvious. One of its versions was excellently formulated by Winston Churchill: democracy is a very bad sort of rule but nobody has suggested anything better.

I also heard an opinion that Russia was particularly unlucky in that Stalin became its leader. True, such monsters are not produced very frequently but not so rarely either (remember Hitler and Pol Pot). However, there is no ground to believe that the fate of the country in a broader perspective would be significantly different if Trotskii or someone else had happened to be in power. The basis for this conclusion is the story of 'socialism' in East Germany, Czechoslovakia, Albania, Poland, Bulgaria, Romania, and, finally, Cuba. In the last case, there was not even a threat of direct occupation by the USSR but the self-proclaimed 'Freedom Island' proved to be a 'Prison Island' from which its citizens flee in despair to the USA. Another example is Yugoslavia. Bros Tito rejected Stalin's pressure and created his sort of 'socialism' in his own manner. The tragic fate of his country is here for all of us to see. Unfortunately, as I had occasion to write before, so many people refuse to see the obvious.

5. So much has been written about communism, its essence, and its fate that I could not hope, nor attempt, to add anything new. I will only remind the reader of a few points. Also, in this context, I cannot help mentioning Leo Tolstoy's revelation. I was brought up in the Soviet period, my interests did not concentrate on great literature, so that I regarded Tolstoy as a great writer and knew nothing of him as a great thinker. Moreover, I always was and still remain an incorrigible atheist, hence Tolstoy's re-

ligious pursuits leave me indifferent. However, in an excellent article by the recently deceased I D Konstantinovsky (in a Moscow magazine 1990 *Ogonek* no 45 7) and in his book 1990 *As a Candle from a Candle... An Experimental Biography of an Idea* (Moscow: Moskovskii Rabochii) I learnt of Tolstoy's unfamiliar facet. It may be of interest to very many people.

Long before the revolution, Tolstoy was acutely aware that violence could not create a just and flourishing society. For him, attempts to grab power and then build a 'socialist society' by any sort of 'equidistribution', forced production quotas, drawing 'lists' for food rationing etc were a social utopia. Tolstoy asked:

> Why do you think that the people who will form a new government, the people who become directors of factories and land ... will not find a way to get hold, just as now, of a lion's share and leave to the simple and the meek only the bare essentials ... People who are governed in life only by the worries of their own well-being will always find a thousand ways of corrupting the social arrangements.

Elsewhere Tolstoy wrote:

> If what Marx predicts does materialize, all that would happen would be for the despotism to shift: where capitalists were lords, administrators of workers will be lords.

Lenin's predictions were different: "The entire society will be one office and one factory, with equal labor and equal pay". Isn't it obvious that Tolstoy was right while comrade Lenin and his pack were ignorant of human nature and had no success in trying to force it to change—by shedding blood.

6. For reasons that I hope will become clear soon, I wish to reveal some details of my own life.[3*]

My father was an engineer of the old generation, never a party member. We had no one among our relatives or close friends was active politically or at least would be capable of comprehending the real situation in the country. I was not personally acquainted with a single person who suffered from Stalin's repression. I was surrounded by communist banners, by prayers sung to the 'Great Stalin', and by information on the truly appalling feats and gambles of the fascist side. We should not forget that the Soviet authorities have unquestionable achievements as well. Suffice it to mention here the elimination of illiteracy and unemployment, the absence—in the pre-war decades—of racial discrimination (and specifically of state-supported anti-Semitism), the possibility to get education. Consequently, I am not going to repent, even today, that in 1937, at 21 years of age, I enrolled in the Young Communist League (comsomol). There was not a shadow of career-making in it: non-party members were allowed into postgraduate courses of the Physics Department and that was as far as my plans stretched. Neither am I ashamed of joining (or rather becoming a candidate for) the Communist Party in 1942. It happened in Kazan on the Volga to which the predominant part of the Academy of Sciences of the USSR was evacuated: that was the period when the German armies were closing in on the Volga. I never tried to avoid mobilization, I tried to volunteer twice, and I had no 'shield' from being conscripted. However, there must have existed some sort of obscure instruction: do not accept into the army, without special reason, young scientists without previous military training. Frankly speaking, this was a wise decision which proved very beneficial after the war, when the destroyed industry was being reconstructed and new technologies were emerging. An aside: I have never occupied any party positions of any distinction whatsoever (like membership in a local party bureau etc).

In 1942, working in the P N Lebedev Physical Institute of the Academy of Sciences of the USSR (FIAN), I defended my DSc degree in Physics and Mathematics but, although I had plenty of energy to teach, I could not find a position in a Moscow institute. Consequently, when I was invited in 1945 to become a part-time professor in the Radiophysics Department of the University of Gorky (now Nizhnii Novgorod) State University (GGU), I accepted.[4] There I met Nina Ivanovna Yermakova and we married in 1946. I mention this fact here only because Nina was not a mere ordinary Soviet citizen: she was in exile. Her father was a communist, an outstanding engineer; in 1938 he was arrested and sent to serve his 15-year sentence in the far North. However, it proved possible to collect a large number of supporting signatures from his colleagues (in those times this was an exceptional, very non-trivial feat) and I P Yermakov was returned to a Moscow prison for an 'additional inquest'. At this moment the Second World War erupted on the borders of the USSR, the Butyrka prison was relocated to Saratov and Nina Yermakova's father died there of hunger, allegedly in the same prison cell as the well-known biologist Nikolai I Vavilov. Nina was at that moment a student at Moscow University's Mechanics and Mathematics Department, and was acquainted with other students whose parents were prosecuted ('repressed'). At that moment our glorious 'services' decided to frame a group of youngsters who were chosen for the roles of 'insurgents' who, allegedly for vindictiveness, planned to kill the Great Leader and Teacher himself. Alas, Nina lived in a flat on Arbat street through which this leader sometimes 'drove in five cars' (I am using poet B Slutsky's imagery). Well, the young terrorists were said to be plotting to shoot Stalin from the window of that Arbat room. However, the script-writers from the NKVD-KGB had

---

[4] I worked at the Radiophysics Department of the GGU and at the Radiophysics Institute (NIRFI) and later kept in contact with them for many years. Quite a few physicists of a high caliber emerged from these places.

not bothered to check their facts. The rooms in the Arbat flat in which the Yermakovs lived had been passed over to 'reliable inhabitants' while Nina and her mother were left in a single room whose window faced the inner yard. The KGB thus had to change the accusation of terror to an 'innocuous' one of 'counter-revolutionary discussions' and of participation in a 'bad' group of terrorists (Clause 58-10, 11 of the Criminal Code of the time). The absence of windows open to the Arbat roadway and the refusal to 'confess' guilt, despite ten days in a severe-regime single cell with a ban on sleep, led to a 'sentence', by the Special Trial Unit, that in those times was regarded as unbelievably mild: three years of hard labor minus the nine months already spent in jail (from July 1944). The war ended at this very moment and an amnesty was announced: for inmates with no more than three-year terms, it covered Clause 58 prisoners as well (there were so few 'counter-revolutionaries' with only three-year sentences that their cases, I believe, were simply overlooked when the amnesty was being defined). The amnesty was half-hearted: people were let out of labor camps but banned from settling in large cities. This is how Nina found herself in the village of Bor, on Volga's banks across from Nizhnii Novgorod: a lucky combination of circumstances allowed her to enroll in the Polytechnic Institute. A curious stroke to the picture: after marriage, I would apply to the KGB once each year (I was not allowed to do it more often) requesting that my wife be permitted to move to Moscow. These applications were seconded by the directors of FIAN, S I Vavilov and D V Skobeltsyn, well-known physicists and public figures. I had no success—but learnt that these two directors also had exiled relatives and equally failed to wangle official permissions for them to stay in the capital.[5] To exhaust this topic, I will add that my wife was able to return to Moscow only in 1953, as a result of the new amnesty that followed Stalin's death. Another nice feature: when my wife received the official 'rehabilitation' (notification that she had committed no crime), this operation needed a visit by a KGB officer accompanied by witnesses who were to compose a protocol to the effect that the windows of the room where Nina used to live did indeed give into the yard. Truly, there is but one step from the tragic to ridiculous.

But let us return to Stalin's times. Clouds were gathering over my head. I was a party member who had lost his class sensitivities to such an extent that I had married an exiled person, being himself a Jew to boot (this became an important factor by that time) and, finally, a 'cosmopolitan' and a 'sycophant'. I doubt if the younger generations remember this last rabid label but in the second half of the 1940s it was all over the pages of newspapers and agendas of meetings. And lo and behold: there were indeed bad apples—literary critics more than anybody else but quite a few

[5] See "A story of two directors (S I Vavilov and D V Skobeltsyn)" that was published in the magazine 1992 *Nauka i Zhizn'* [*Science and Life*] no 1 and in my book 2001 *The Physics of a Lifetime* (Berlin: Springer).

others as well—who 'began to kowtow to everything foreign', to 'lick the boots of foreign science' and, at the same time, 'belittle national values and national interests'. A great many, I do not know how many, were subjected to the so-called 'trials of honor' and some were repressed. On 4 October 1947 (my birthday, by the way) an article was printed in *Literaturnaya Gazeta*, with a title "Against sycophantism": it was signed by the director of the Agricultural Academy, B Nemchinov. The article was an attack mostly against certain 'enemies of truly scientific' Michurin–Lysenko biology (A R Zhebrak was the main culprit) but I was also branded a 'bootlicker'. Some time later I managed to find out how my name and my 'sins' got into this article but this is not the right place for this story.[4*] I cannot help mentioning that 11 of the most prominent Soviet physicists sent an official protest to the paper, characterizing the accusations against me as an 'insulting slander'. The communist party cell of the FIAN treated the article in *Litgazeta* similarly and passed a special resolution to clear my name. As was normal for the times, no retractions were printed in the paper. The ideological 'crusade against sycophants' was getting stronger by the day. Before I could bat an eyelid, the Superior Attestation Commission rejected my application for a professorship degree for me, submitted by Gorky University. My name started its odyssey as a negative specimen through various executive orders and articles and I was dropped from the Learned Council of the FIAN 'for the purpose of strengthening it'. I describe all this not for revenge (that would be ludicrous) but to conjure up the colors of the time and, most of all, to formulate my firm conviction: in view of the combination of circumstances, I was a sure candidate for arrest. If anything saved me, it was ... the hydrogen bomb.

7. In 1947, the gigantic effort of producing the Soviet atomic bomb, headed since 11 February 1943 by Igor V Kurchatov, was in full swing. There was still a long way to go to the test blast (it took place on 29 August 1949). Nevertheless, the possibility of developing the hydrogen bomb (i.e. thermonuclear weapons) was already a matter for thought. Did it happened by a local initiative or was it stimulated by intelligence data— I do not know (A D Sakharov in his *Memoirs*[6] writes in chapter 6 that intelligence information was the likeliest stimulus). Preliminary calculations carried out by Yakov B Zeldovich and his group at the Institute of Chemical Physics of the Academy of Sciences of the USSR led to fairly pessimistic conclusions. As was customary in such situations, Kurchatov

[6] Sakharov A 1990 *Memoirs* (New York: Chekhov). A very recent article by Khariton Yu B, Adamskii V B and Smirnov Yu N (1996 *Usp. Fiz. Nauk* **166** 201 [1996 *Phys. Usp.* **39** 185]) describes for the first time and in considerable detail the history of development of the Soviet hydrogen bomb. (An even more precise and detailed description was given by Goncharov G A (1996 *Usp. Fiz. Nauk* **166** 1095 [1996 *Phys. Usp.* **39** 1033]). The history of the development of atomic and hydrogen bombs in the USSR can be found in Holloway D 1994 *Stalin and the Bomb* (New Haven, CT: Yale University Press). See also 1996 *Phys. Today* **49**(11).

decided to throw the same problem to a parallel group of physicists at the FIAN, headed by Igor E Tamm. It was most probably quite difficult to get permission for this since Tamm's reputation with the authorities was not too good (he was a Menshevik before 1917 and, in the 1930s, his brother was shot; I wrote about this in an article devoted to Tamm's 100th anniversary in 1995 *Vestn. Ross. Acad. Nauk* **65**(6) 520 [1995 *Herald Russ. Acad. Sci.* **65**(3)] (see also article 8 of this book). I believe that Kurchatov was able to involve Tamm only because the problem did not seem to be too pressing; rather, it appeared fairly hopeless. For the same reasons, Tamm was able to incorporate me into his group, despite my shortcomings already outlined. Sakharov was included in the group, as he writes himself, because he was supported by FIAN director S I Vavilov who used the occasion to somehow provide him with accommodation in Moscow. Truly there is but a single step from the great to the ludicrous.

I will not describe the details of our work since Sakharov has done so in his excellently written *Memoirs*. It is only important to mention that, in 1948, Sakharov came up with what he called 'the first idea' and then I suggested 'the second idea' and these made the creation of the H-bomb possible. Even in 1989, when Sakharov had a chance to amend his book, these ideas were still strictly classified: 40 years after their emergence! This absurdity, so typical for our recent past, was overcome (if this expression is permissible) only after Sakharov's death (see 1990 *Priroda* no 8, pp 10 and 20). Expanding the gist of the two ideas is unnecessary here.[5*]

In 1950, Tamm and Sakharov were sent to the 'Ob'ekt' (which at that time was deeply classified but nowadays is known as Arzamas-16) in order to direct the actual development of the hydrogen bomb. However, I was not allowed to participate in this work and had stayed behind, in Moscow, heading a small 'support group' guarded by the security service: I still had only restricted access to classified research, even though my classification was 'Top Secret. Special File'. I was obviously tremendously lucky: I had some sort of ensured protection and, at the same time, was able to do research 'for the soul's sake' (I had enough time left for it) and to visit my exiled wife. In fact, menacing complications resumed in 1951: I lost even access to some of my own research reports (on controlled nuclear fusion). Sakharov wrote in his *Memoirs* (chapter 11) that, at the beginning of 1953, 'cargo trains were ready for the deportation of the Jews, and the propagandist texts were already printed to justify the measure'. Unfortunately, I have no other reliable information on the matter. I do not know what fate awaited me: to stay behind in a 'sharazhka' (prison research unit) as a still 'useful Jew' or depart with a doomed train. Fortunately, Stalin failed to implement his final insane schemes.

After the first hydrogen bomb was successfully tested on 12 August 1953, my awards were a class lower than those of Tamm and Sakharov but were quite high nevertheless. However, the most important thing was that after the arrest and execution of Lavrentii Beria and his henchmen, the KGB's hands became shorter and the climate in the country mellowed in many respects. This article is not the place to describe it all. I should only remark that until the Khrushchev's famous revelation in 1956 I, like so many others, remained ignorant of the true role played by Stalin in unleashing the now exposed outrageous atrocities. I am very ashamed of this blindness of mine. These revelations were so painful that I became very careless and soon attracted the KGB's attention. Some of our acquaintances began to avoid me and my wife's company: we found out later that they had been invited to 'where one does not dare to refuse invitation' and demanded information about me. The menacing hand with the sword had weakened, however, and people, at least of my station in life, were not thrown into jails or lunatic asylums for mere loose talk among fellow Soviet citizens. The only field in which the damage was done was with respect to permission to travel to scientific conferences abroad. Using secrecy rules as a pretext, I was not allowed abroad and lost a great deal from missing conferences to which I was regularly invited. That the secrecy was a pretext was beyond doubt since people who knew incomparably more about classified matters had much greater freedom of travel (fortunately, Tamm had). Still, I did visit some places in the 1960s and a couple of times even together with my wife (that was a rare privilege in Soviet times).[6*] But even this phase came to a close, first as a result of clandestine reports of 'well-wishers' and then because I was, let us say, too close to Sakharov. I do not mean any personal closeness but the fact that since 1971 I had headed the Tamm Department of Theoretical Physics of FIAN and Sakharov was on the staff of this department from 1969 till his death. I never signed any 'letter' directed against him, I tried to help him, and visited him twice in Gorky during his exile.[7] I am not trying to boast: I simply explain why I did not enjoy any favors from the authorities.

In 1989, I was elected People's Deputy from the Academy of Sciences of the USSR to the Supreme Soviet and tried to be conscientious about my duties: I was a fairly active member of the Parliamentary Commission Against Privileges. Now that I write this, it is a sad joke to remember that B N Yeltsin was just about the originator of this struggle. In politics I was, not without qualifications, on the side of Gorbachev and Yakovlev, and I do not regret this. I will complete the autobiographical part of this article by saying that I resigned from the Communist Party at the beginning of 1991 and have completely stopped any political activity since the dismissal of the Parliament at the end of 1991.

---

[7] For details, see my article "On the Sakharov phenomenon" published in the magazine 1990 *Svobodnaya Mysl'* [*Free Thought*] nos 14, 15 and in the already mentioned book 1995 *On Physics and Astrophysics* (Moscow: Byuro Kvantum) [2001 *The Physics of a Lifetime* (Berlin: Springer)]. The article was also published in 1996 *He Lived among Us... Reminiscences about Sakharov* (Moscow: Praktika).

8. Soviet science and technology made outstanding achievements: suffice it to mention the launching of the satellites and the space research, the highest level achieved in physics, mathematics and in some other sciences. Evidence of this is the fact that, having ultimately gained the freedom to travel and, at the same time, suffering from low wages, a number of our specialists (some members of the Russian Academy of Sciences among them) were able to find excellent employment in the West, some as full professors. Note that this happened against the background of obvious insufficiency, for instance in the USA, of vacant positions for young American scientists. An opinion was formed in this connection that science under 'socialism' was in full bloom. This judgement is at least very one-sided. Physics, mathematics and some other fields were in a good state. But biology, which was doing quite well after the 1917 revolution, was virtually quartered, especially at the sadly immortalized session of the Agricultural Academy in 1948. The point was that Stalin, and Khrushchev after him, chose to imagine themselves to be experts in biology as well and had supported the ignorant and aggressive brute Lysenko and suppressed, sometimes physically, all his opponents. Cybernetics (computer sciences) and cosmology suffered a similar fate, even if in less dramatic circumstances. As for the social sciences (economics, history, literature and language studies, etc), there was not even a shade of freedom of opinion there, everything was dictated by the Marx–Engels dogmas and by the latest instructions from the Central Committee of the Communist Party.

In fact, physics had matured for a 'Lysenko treatment' as well: the all-USSR Conference was in preparation, to take place on the 21 March 1949 but was called off at the very last moment. No documents about this have survived or have not yet been uncovered. The most plausible version is this.[8] The atomic bomb had not yet been tested and Kurchatov replied to Beria's question that no bomb could be produced without relativity theory and quantum mechanics. Beria responded, very disturbed, that the bomb was all important and the rest was rubbish ('the rest' meant sniping at modern physics for its allegedly idealistic orientation). Presumably, Beria immediately reported to Stalin about the situation and Stalin ordered the conference being organized by the Central Party Committee to be cancelled: no one else would have dared to do it.[7*]

This event points to the obvious: physics, mathematics, and some other sciences were supported only inasmuch as they were required to develop bombs, rockets, and other weapon systems, for the industrial effort, etc. To achieve such aims, nothing was spared. I recall how our salaries were raised several-fold after the Americans exploded atomic bombs in Japan: imagine the golden rainfall enjoyed by the leaders of important

[8] About this and the preparation of the previously mentioned conference, see Sonin A S 1994 *Physical Idealism: A History of One Ideological Campaign* (Moscow: Fizmatlit).

projects after their successful completion. However, man lives not on bread alone and, on the whole, our life was quite bitter. The ideological repression,[9] censorship, the ban on using copiers unless in a special room behind iron bars, the impossibility of free travel abroad and of free exchange of information with our colleagues in the West. And of course, brutal repression. A brilliantly talented physicist Matvei Bronshtein (husband of L Chukovskaya) was shot, the outstanding physics theorist S Shubin died in a labor camp, no one knows where. An incomparable physics experimentalist L Shubnikov, and several other physicists from Kharkov with him, were shot. The great Lev Landau spent a year in jail and miraculously avoided death. In fact, the recently published 1995 *Tragic Fates: Scientists of the Academy of Sciences of the USSR in Gulag* (Moscow: Nauka) lists 105 Full and Corresponding Members of the Academy of Sciences of the USSR, incarcerated in different periods (the list is not definitive).

I should remark that even in this first stage of democracy in Russia, censorship has been removed, we talk freely, we can travel abroad, and our science could flourish were it not for financial difficulties. World science today is a typical democracy with all its advantages and disadvantages. Sometimes scientists get a doughnut and sometimes the hole in it, there is no equality and no justice, and there cannot be. Some people get Nobel or other prizes (doughnut) while some (just as deserving) do not (hole). Some are elected to various academies, others (equally deserving) are not. Some occupy excellent positions, others fail to find good employment. Some publish their results easily, others struggle, etc. This all stems from human nature and is typical of democracy. A lot can be improved and indeed needs improving and this ability to improve is also a typical trait of democracy. However, it is impossible to reach the ideal state (complete justice) and, quite often, it is not clear what the ideal is. Still, there is a great difference between sometimes getting the hole of a doughnut and for an innocent person to be thrown into jail or be exiled.

9. It is high time, however, for me to return to the main topic of the article. I have already stressed that we now possess important fruits of democracy: freedom of speech, free elections, etc. Plus, of course, there are shelves full of food and merchandise in shops. I remember too well not only empty shelves but also rationing cards, so the abundance of food and stuff never ceases to amaze me. Alas, here I have to stop the pros and start the cons. The goods are there, indeed but a considerable part of the population cannot afford to buy them. Doctors, teachers, and all

---

[9] See, among others, the book by A S Sonin cited in note 8 (see also the article by V P Vizgin in 1999 *Usp. Fiz. Nauk* **169** 1363 [1999 *Phys. Usp.* **42** 1259]). I would also like to recommend E Kolmen's 1982 *Our Lives Should Have Been Different* (New York: Chalidze). This is an autobiography of one of the active 'fighters for purity of the Marxists–Leninist ideology', who only understood what was what by the end of his long life.

other 'state budget' people draw wages that are patently inadequate for a decent life. Pensioners and invalids fare much worse. In January 1996, the total number of citizens with an income below the official minimum was 37.3 million people, which is a quarter of the entire population of the country. Add to this the crazy and unbelievably unscrupulous delays of many months in paying wages and pensions. This takes place against the background of growing social inequality. 'Nouveau Russians' and a certain, not quite decipherable to me, layer of the very rich attend night clubs and restaurants, drive expensive foreign cars, have super-luxurious (by our standards, anyway) country villas, spend holidays at spas abroad, or even buy real estate in the West. How could they get hold of the money? What part of the rich (I fear that a very high proportion) are crooks? The situation is aggravated by the inane waste of resources by the authorities: enormous expenditure on repairing the 'White House' and buying equipment for it and for the Federal Assembly, the cost of the upkeep of former and current lords of the Duma (Russia's parliament) (see 1966 *Argumenty i Fakty* no 7 2), of countless, mostly meaningless, trips of various highly placed officials with their retinues, let alone Duma members, and powerful administrators. I was never able to see published figures of all these 'expenditure items'. In contrast, as far as I could find out, in the UK there are only about 20 state-owned cars for all ministers and civil servants. Papers reveal again and again how ministers in the USA and other countries resign after being caught using taxpayers' money for private purposes (such as going to a sanitarium with a family). In fact, numerous examples could be found that show how the 'people's servants' are monitored by their electors and dare not misuse public money or resort to parliamentary immunity to escape criminal investigation.

I cannot help mentioning, when describing the current situation in Russia, the rampant crime wave and the corruption that is eating away at the administration and law-enforcement branches. Finally, we all know about the blunders in Chechnya and the aberrations in the army.

Is it any surprise then that the 'ruling party' has received a mere 10% at the elections of 17 December 1995, while the LDPR (Zhirinovsky's ultra-right Liberal-Democratic Party of Russia) and other organizations of this type (referred to as the 'left' in Russia, for some reason, although friends and analogues of Mr Zhirinovsky are labelled right everywhere else) together collected about 15% of the votes and communists of all colors received 25%. Clearly for most of those who do not have enough to eat, and not only for them, these negative sides of our lives now cannot be outweighed by the freedom of speech and the availability of foreign travel (provided money is available).

10. I am not an expert in economics, finances, or methods of governmental control and, thus, have no right nor desire to offer advice in these fields. In fact, I am not of those who claim 'to know how everything should

be done'. It appears admissible, though, and in the context of this article to some extent necessary, to make several remarks and express an opinion on select issues.

What is the purpose of the reforms carried out in Russia and of the entire historical process of passing from a 'well-developed (or advanced) socialism' to something different? The aim is to build in Russia a Western-type, stable, affluent society, i. e. a society characterized by a market economy and strong democracy. Sometimes people speak of 'advanced capitalism' but for myself, as indeed for quite a few people who lived for a long time under the Soviet system, the word 'capitalism' is fraught with very negative connotations. It is associated with grim factory blocks and workers who are jaded with poverty and hard labor. Marx may have imagined a picture like this but nowadays it has very little in common with reality.

The most important aspect that permits one to regard this aim as realistic is the fact that it has already been achieved in a number of countries and, thus, it is not a communism-like utopia. Of course, each country has its specifics and Russia has its specifics raised to power two, if I may say so. Nobody in his or her senses suggests blind copying.

Alas, our reforms were poorly prepared and were carried out in haste and confusion. This may explain how very grave errors were committed. 'A voucher to everyone' operation seems ludicrous to me (I have sold mine for 20 US dollars), while many other privatization acts appear to be simply criminal. A typical example can be found in a recent interview with the former mayor of St Petersburg Anatolii Sobchak (*Izvestiya* 20 February 1996). He said:

The Baltic Shipping Company' was in fact sold out as a result of careless privatization. It was bought for a song as a rented enterprize and then sold off piecemeal for debts. About 15 people got very rich, bought real estate abroad and emigrated while a huge shipping company, one of the largest in the world, dominating the Baltic Sea, disappeared.

Such follies, uncountable in number, bled the country and generated this phenomenon of 'new Russians'. Tens of billions of dollars fled from the country and was anybody punished? At the same time, Russia struggles for credits from the West (these will have to be paid back) of considerably smaller size.

The President and the government looked indifferently on the creation of crooked banks of MMM type and of various 'enterprizes' erected to rob gullible investors. Who of the organizers of this thievery has been prosecuted? The previous Duma discredited itself by saving its member Mavrodi (the MMM–Mavrodi) from trial by a revolting shield of parliamentary impunity. Now the authorities, seeking populist support, discuss possible compensation for the duped investors. At whose expense? I am of the opinion, by the way, that if compensation is undertaken, it is only the pensioners and the poor who must be given anything. As for those who simply wished to get rich 'while the getting was good' (I think that their number is not less than that of the former group), there is no reason to restore their investments at the expense of the rest of the population who had clear heads on their shoulders.

When we read about the life of the military in the army and the navy, the heart virtually misses a beat (the latest evidence known to me appeared in *Izvestiya* on 14 February 1996 about the situation on the *Admiral Gorshkov* aircraft carrier). Russia, armed with its nuclear arsenal, is not threatened by any power: it is absurd to keep, feed, and arm an enormous army that the country cannot afford. The only argument I have heard for maintaining this army is the unbelievable, unheard of anywhere else in the world—the number of generals who insist on commanding a respectable number of subordinates. As a result, instead of urgent reductions in the army, the government extends the term of conscription and sends students to the army! Using the words coined by Napoleon's minister Talleyrand, "This is worse than a crime—this is a mistake".

I have already mentioned the delays (sometimes of many months) in the payments of wages. Entering the election campaign on 15 February 1996, B Yeltsin promised to do away with this disgusting despotism. However, what precluded him from declaring this earlier? Why was it only the threat of failing at the election crucible that awoke the President from his lethargic sleep? (or was it his many courtiers who woke him up once they realized that the threat of losing power was quite real?)

This list of shortcomings would be very incomplete if I forgot to mention the absolutely insufficient fight against the cancerous corruption and crime. Papers often revealed that the police and security services know nearly all the criminal 'authorities' and 'thieves-in-law' virtually by name. Their meetings are broken into but ... the culprits were immediately released! Numerous assassinations and murders remain unsolved. Courts let obvious criminals and fascists off the hook.

I will allow myself two more remarks. I still remember the destruction of the Christ the Savior cathedral in Moscow and definitely regard it as a barbaric act, quite in the spirit of other feats of Stalin and his ilk. However, why is it necessary now, amid a raging financial crisis, to reconstruct the cathedral at the state's expense? We hear, though, that some church money has also been attracted but the proportion is hushed up and that is reason enough for knowing the true source of financing. This flirting with the church and direct financial support contradict the separation of the church from the state. If the government has money to spare, it must build hospitals and homes for the aged, not reconstruct the churches. By the way, our modern communists, whose fathers, led by Lenin, persecuted priests in the most brutal fashion, nowadays flatter the Orthodox church and even denounce atheism (I have heard this remark from Zyuganov), even though atheism is not at all against freedom of religious beliefs.

My second remark is aimed at the KGB. In the Czech republic and in Germany, the archives of the KGB's counterparts are open to the public. Each citizen can find out who had been writing slanderous reports to the KGB on him or anyone else. I do not speak of vengeance: simply, exposed informers must be prohibited from occupying positions in law-enforcement structures. By the way, our media has informed the public several times that some of the highest-rank priests of the Russia's Orthodox church co-operated closely with the KGB. Do not the Christians have the right to know the truth about this? In short, I believe that the archives of the former KGB must be completely opened for all Russian citizens. This act would be the best guarantee that the FSB (Federal Security Service) will never turn into something equivalent to the KGB.

11. Why did I need to enumerate the open sores of our society and to recall the errors committed? Obviously, to ask a question: how can we cure these sores, how to live on, how to reach the goal of building here a rich democratic society? Could it be that this requires yielding power to the communists whose critical arrows partly aim at the same target and who have plenty of experience in destroying a society they dislike? My answer to this question is, categorically, NO!

In principle, our current constitution and the presidential type of government make it possible to eliminate the shortcomings and speedily move in the predetermined direction.

What party is capable of this? A tough question; I can only confess that, in the elections to the Duma in 1993, I voted for Grigorii Yavlinsky's 'Yabloko' [*Apple*] combination of parties although almost all those close to me preferred Yegor Gaidar's 'Russia's Democratic Choice' (DVR). As these friends, I do regard Gaidar and his team as honest and truly democratically minded people.[10] Alas, Gaidar and his team made many errors while Yavlinsky, to say least, has had no chance of making them. I consider him to hold honestly to democratic principles and to be a highly professional economist. These are the reasons why I would like to see him at the helm. In the elections to the Duma in 1995, I also voted for Yabloko, although my vote in the single-mandate electoral region went to the representative of the DVR. I plan to vote for Yavlinsky at the presidential elections as well, unless it becomes unmistakably clear that he will fail to pass on to the second ballot. In this case I may have to vote for Boris Yeltsin. If even Yeltsin does not make it to the second round, I will have a single option: to

---

[10] In addition to this, Yegor Gaidar is undoubtedly a very knowledgeable and well-educated person (see his book 1995 *State and Evolution* (Moscow: Eurasia). However, now that I had a chance to reread these lines when proof-reading this edition, I cannot help remarking that over the years, my opinion of our democrats have taken quite a beating (see comments to article 28 in this book).

vote against both opponents (say, Zyuganov and Zhirinovsky). Indeed, if the number of second-round voters that vote 'against everybody' is found to be greater than that supporting any of the candidates, the results of elections will be declared null and void.[8*]

Should the communists win in the presidential elections, the country will be retreating quickly into the past. This conclusion follows, on one hand, from knowledge of communist (bolshevik) rule between 1971 and 1991. On the other hand, the available documents of the Communist Party of the Russian Federation (CPRF) point in the same direction.[11] A detailed analysis would be very instructive but I cannot fit it in; luckily, there is no need. O Latsis has already done it in his brilliant articles in *Izvestiya*, 14 February 1995 and 10 February 1996.

The program of the CPRF (which is the most liberal among all currently active communist groups acting now in Russia) does not have a single word of reproach regarding the crimes of the VKP(b)–KPSS, Stalin's terror, the persecution of dissidents, including the use of psychiatric hospitals, in the post-Stalinist era. There is not a sign of a seemingly inescapable and anticipated call for repentance. I was truly shaken when I found in the text of the election platform of the CPRF (p 5) a quote from Lenin: 'political parties must be judged not by their promises but by the results of the implementation of these promises'. Don't we remember what the communists promised in 1917, and many times later, and what they actually did? If the communists win now, the story would repeat itself. Indeed, communists 'forgot nothing and learnt nothing'. Formally, my last statement could be challenged. Indeed, I have not found in the publications of the CPRF a demand for a proletariat dictatorship, nor loyalty to Marxism–Leninism, and these were a trademark of the documents of the CPSU. I think, nevertheless, that this is no more than a smoke screen needed to avoid frightening away the voters by reminding them of the past; moreover, these mottoes are not that important in practical activities. Their practical steps will involve nationalization, restriction of the freedom of speech, and freedom of the press (I can advise looking at the statements of CPRF's member of Duma Yu Ivanov, published in *Izvestiya* on 15 February 1996). Once communists inevitably-fail-to achieve economic success on a 'socialist' path, they will immediately start a witchhunt for the culprits (this will again generate 'enemies of the people'), and law-enforcement branches will start on the familiar path of the VChK, OGPU, NKVD, and KGB. The overtures we see even now between the CPRF and various nationalist-chauvinistic and brownshirt elements will grow stronger, the army will not be reduced, foreign policy will get tougher by using demagoguery about NATO's aggressiveness, and so forth. On the whole, the communists' electoral victory

[11] 1995 *Third Congress of the Communist Party of the Russian Federation (relevant documents)* (Moscow: Informpechat); 1995 *For our Soviet Motherland. Election Platform of the Communist Party of the Russian Federation* (Moscow).

will be their revenge, a leap into the past (I quite agree with the prediction of O Moroz in *Literaturnaya Gazeta* of 14 February 1996) and a terrible threat for Russia and the entire world.

The responsibility of the Russian voter is overwhelming. Everyone to the polling stations! I suggest you should vote exclusively for a democratic candidate or a candidate supported by the democratic parties.[9*]

PS. A N Yakovlev, formerly member of Politburo of the CPSU and one of the initiators and leaders of the 'perestroika', is currently the Chairman of the Commission, set up by the President of the Russian Federation, on the Rehabilitation of the Victims of Political Repression. In this capacity, he had access to an enormous amount of reliable data, numbers, and facts that characterize communist (bolshevik) rule in Russia and the USSR. He has recently published a book 1995 *Like the Relic, Like the Holy Oil* (Moscow: Eurasia), based on this archive. On the whole, we know it all but one keeps discovering many important details, many new examples. The general statements and conclusions in this article and in Yakovlev's book are in excellent agreement (see comments 10, 11, 14, and 15).

Comments

1*. This article was published in the first edition of the book (1997). This is how the paper came to be written.

The political situation in Russia became especially menacing at the beginning of 1996. If anything, the results of elections to the Duma (17 December 1995) were a defeat to democrats. The communists, in contrast, strengthened their position and for a number of reasons (such as imperfections in the election mechanism) began to dominate the Duma. At the same time, Boris Yeltsin's rating fell very low. The country faced the very real threat of a communist's victory in the presidential elections in June–July 1996. Like many others, I regarded this threat as deadly. However, not everybody accepted this conclusion—far from it. Thus, people pointed out that the victory of former communists in Poland, Lithuania, and some other countries failed to entail catastrophic results. In fact, they ignored the fact that 'our' (Russian) communists, in contrast to their Western counterparts, never renounced their bolshevik heritage. I was especially perturbed by the possibility that the so-called Western liberals would, as in Stalin's times, prove to be blind and deaf to the communist menace. Therefore, I chose to behave in an extraordinary manner while at the Astrophysics Conference in Copenhagen celebrating I D Novikov's 60th anniversary. Namely, when invited to deliver a speech at the Jubilee banquet (11 January 1996), instead of cracking jokes and extolling the virtues and achievements of the protagonist (this is the custom), I delivered my anticommunist diatribe. I simply could not do otherwise, especially after I learned that Novikov's father and that of N S Kardashev, also present at the table, were imprisoned and shot at the end of the 1930s. Many a participant regarded my speech as misplaced but there were also those who completely approved of my behavior. Among them were John A Wheeler and Martin Rees, both of whom I greatly respect. However, what is the worth of a short speech in front of a small audience, and in poor English too? Therefore, I decided to write an article on this communist menace—and I offer it here to the readers. The paper was too long for a newspaper and was thus never published in full before the elections. However, on O R Latsis' initiative, a shorter version abridged by himself and his editors did appear in *Izvestiya* on 23 May 1996 (the heading and the subheading were the paper's). Another abridged version of the article (rewritten by G E Gorelik) was published in *Novoe Russkoe Slovo*, a newspaper in Russian published in the USA, on 17 May 1996.

The circulation of *Izvestiya* is more that 600 000 and, therefore, I entertain a hope that I made some kind of contribution to the resistance to the communist menace. However, this contribution was probably a very little mite because most of the readers of *Izvestiya* needed no persuading. However, could I do more?

Note that the full text of the paper was also posted to A Voronel— editor-in-chief of an Israeli magazine *Twenty Two* published in Russian in Tel-Aviv. The article was indeed published by this journal (no 101, 136, 1996) only slightly cut, even if under the a rather cliché title (provided by the magazine) "I can't be silent". But the important point was that the article was accompanied by a comment written by a member of the editorial board of the magazine, M Kheifets, in which he completely demolished my arguments. In M Kheifets's opinion, my article was counterproductive and could even be used by the Russian Communist Party to prove that they were doing the right thing! To put it bluntly, M Kheifets immensely disliked my article; and he thought that since I was not a professional politician (which is, of course, true) it would be much better if I kept my mouth shut. It is rather difficult to argue against such accusations—in fact, I was given no chance to respond since I saw Kheifets's comment only in the issue already printed. To be honest, I have no desire to start any polemics with Kheifets even now. What I wanted to tell the reader was already in my text and they (the readers) were to make their own conclusions.[12]

A copy of the Russian version of the article was also sent in March 1996 to Martin Rees (Cambridge, UK) who forwarded it to the Institute of Physics magazine *Physics World*. I completely forgot about it (the presidential elections were already history) when I received at the end of 1996 the information that this magazine intended to publish my paper in a very

---

[12] Perhaps M Kheifets was guided by the same sort of argument that I mentioned in my article about D A Kirzhnits (article 12 in the present book).

abridged version. After certain corrections, this was done (1997 *Physics World* **10**(4) 17). Furthermore, the full version of the article (differing only insignificantly from the text here) was placed on the Internet (*Physics World* Web site, http://www.iop.org/Mags/PW).

Communists failed to win that time—in mid-1996—but even then comrade Zyuganov did gather about 30 million votes. Therefore, this struggle between democracy and communism in Russia may continue for a very long time. This is why I believe that publishing my article here is justified. I decided not to make any significant changes to the text of this article, as in other articles of this book (several bibliographical references do not count; I also wrote some comments).

I also want to make some general remarks. We all know that the communist ideas are anything but new. They were borne a very long time ago as a response to material and legal inequalities. People were dreaming of a society with social justice and devoid of the inequalities that we find in the slave, feudal, and capitalist forms of society. This ideal society of total justice became known as communism. Millions of people believed and probably continue to believe in the possibility of building a communist society—this is a kind of religion of which communism occupies the place that God holds in traditional ones. Faith in communism, as faith in God, is the right of each individual. However, the right of believers (whether in communism or God) to exterminate or at least humiliate or in any way discriminate against heretics (i. e. non-believers) is a very different matter.

I would refer to people who categorically reject such rights of believers (in communism or God) as democrats. The time of inquisition was left far behind and, ignoring for the moment Islamic fundamentalists and some relatively few sects, believers in God are not a great danger for a democratic society. However, bolsheviks believing in communism started from the very beginning in 1917 on the way of dictatorship, terrorism, and hypocrisy. They essentially followed the old Jesuit principle: the aim justifies the means. This is how the path to a declared target (communism) proved to be covered in blood and tears, led to the death of many millions of people, and degraded society. Ultimately this pseudosocialist society in the USSR and its satellites crumbled under the weight of its crimes and ideological insolvency.

Pope John Paul II justly remarked: "Communism fell due to its internal weakness as it was found to be a medicine more dangerous than the illness". To be specific, Russia was suffering from many ills under the Tsar but bolshevik medicines did prove more dangerous, more appalling, and more harmful than these illnesses.

Bolsheviks, who called themselves communists, were in power very recently; hence, when we refer to communists we mean not those idealists who believe in the future communist kingdom of justice but the heirs of the CPSU and related parties. These heirs, who form the CPRF and some smaller parties, have not renounced their bolshevik past and flaunt the word 'communist' even in the name of their parties. When we speak about the communist menace, we mean precisely the possibility of such bolshevik communists returning to power. As for the communists of the idealist breed, they, in most cases, realized a long time ago how discredited the terms 'communist', 'communist party', etc are. Therefore, they spare no effort to distance themselves from communists of the bolshevik ilk and join various socialist and social-democratic parties and groups.

Something else I want to mention. Very many people joined the Communist Party for material gains or to build a career, or because they were pressured or frightened. However, the number of people who, especially after the revolution and during the Second World War, became party members for ideological or patriotic reasons while failing to understand the true nature of the bolshevik society, is probably at least as large. There is no doubt that all party members, even the most passive of them, bear responsibility for the actions of the Communist Party. Nevertheless, when we speak of party rank and file, of many millions of people, I see no reason to set them apart from non-party masses. The moral rot and fear (Daniel Granin gave an excellent analysis of the anatomy of this fear recently in 1997 *Literaturnaya Gazeta*, 12 March, p 10)[13] which reigned in the totalitarian society were great equalizers for party members and the rest of the people. This becomes amply clear if one looks, for example, at the signatures under the 'letters' that denounced Akhmatova, Zoshchenko, Pasternak, Solzhenitsyn, Sakharov, 'the cosmopolitans', the 'poisoning doctors', etc. In fact, non-party members probably had it slightly easier than party members when not signing such 'letters' or when declining to make a speech at the specially organized meetings—but they did both.

On the whole, I am convinced that people can and must be denounced for specific crimes, dastardly acts, and repulsive behavior regardless of whether or not they were carrying a party membership card. I see no foundation for declaring people heroes simply because they had not enrolled in the Communist Party in the USSR or the National Socialist Party in Germany. Heroes who went beyond simple silence but actively fought for their human dignity were, unfortunately, the tiniest minority. I did not belong to this minority. However, I have no intention to confess, as the previously mentioned M Kheifets and some others would like me to, to sins that I have never committed.

The topics that I have touched upon here are related to the aspect of the work of scientists (physicists in particular) on the creation of weapons, especially atomic weapons, in the USA, Germany, the USSR, and any other countries. In this field, everything (or nearly everything) depends on specific conditions and circumstances: location, time, and situation. It is one

[13] For details, see Granin D 1997 *Fear* (St Peterburg: Inform. Tsentr Blitz).

thing to develop the atomic bomb because it is anticipated that a merciless enemy (Hitler, to be specific) may be the first to have it. It is another thing if the bomb is developed as a weapon for aggression. We already know a great deal about this matter, with respect to the USA (and their allies), Germany, and the USSR. The volume of published literature is enormous.

My participation in creating such weapons (the hydrogen bomb) is completely described in this article and in articles 17 and 18. I feel happy that I never had to work on specific (bombing) aspects or to take part in tests and discussions. The question about responsibility arises nonetheless. I will first of all refer to E L Feinberg's book 1999 *The Epoch and the Personality. Physicists. Essays and Memoirs* (Moscow: Nauka) [2nd edn: 2003 (Moscow: Fizmatlit)]. On the whole, I share the judgement of E L Feinberg, my closest friend, on moral issues concerning the behavior of physicists.[14] At that moment all physicists, not only those in the USSR with whom I was in contact and who would discuss these topics (before the role of Stalin and his regime was exposed), were of the opinion that the development of atomic weapons in the USSR was justified by the need to create some sort of equilibrium with the USA and to prevent a nuclear war. The possibility that Stalin could abuse the possession of the frightful weapon at the first convenient chance and use it as a means of blackmail never even came into my head; and I have not heard a hint at it from anyone else. Neither I nor anybody else who was frank with me understood that, as Churchill wittily remarked, Stalin and Hitler 'differed in the shape of moustache only'. Now (in fact, for a long time) I clearly understand that everything is possible under a totalitarian, dictatorial regime whether it be Hitler, Stalin, Mao Tse-Dung, or lower-rank dictators and even the most frightful weapons (atomic, bacteriological or any other) are very likely to be used. If we do not accept the equilibrium of fear, only a democratic form of governance and international cooperation are capable of preventing this eventuality and, to some extent, guaranteeing people's security.

2*. This is true only for 'classical capitalism' as it occurred in developed countries (say, in England) in the 19th century. Since then, the social inequality in developed 'capitalist' countries has been greatly smoothed out. Even the unemployed, the disabled and pensioners in the USA, UK,

[14] In connection with the article about W Heisenberg in E L Feinberg' book, I need to point to a document published in *Physics Today* in July 2000. I would like to add that I was unpleasantly shocked (I avoid a stronger expression) by the information given in G Holton's article in this issue of *Physics Today* (p 38) that Heisenberg publicly reproached Einstein for assisting in the creation of the American atomic bomb. This is truly an attempt to shift the blame, as we say, from a sick head to a healthy one. Einstein was a true democrat and antifascist. I know nothing that would give any base for reproaching him in the matter of atomic bomb creation. However, the behavior of Heisenberg, one of the supervisors of the German 'Uranium project' is still unclear today and is a matter of discussion (see E L Feinberg's book, the issue of *Physics Today* mentioned and 2001 *Physics World* no 11 9).

Germany, Sweden, etc enjoyed a much higher standard of life than the workers and peasants had in our country during bolshevik rule. I do not even touch on the 'non-materialistic' aspects of life—freedom of movement, freedom of the press and other human rights.

3*. To a large extent, these paragraphs repeat the contents of articles 17 and 18 of this book. However, for reasons already mentioned (e. g. in comment 9 to article 18) I consider it inadvisable to abridge the text.

4*. See comment 10 to the article 18 of this book.

5*. See article 18 of this book and the articles from *Uspekhi Fizich-eskikh Nauk* (*Physics–Uspekhi*) cited in footnote 6.

6*. The Academy of Sciences of the USSR wrangled permission for Full Members of the Academy (the Academicians) to travel abroad with their wives, probably referring to their respectable age. Therefore, after 1966, when I was elected to Full Membership, I was several times allowed to travel with my wife. In reality, in 1984 and 1985, I had already been stripped of this privilege (see my article "On the Sakharov phenomenon" already mentioned in footnote 7).

7*. Actually, other versions of the causes that resulted in the cancellation of this meeting can be found in the literature.

8*. I can reveal that on the first ballot of the presidential elections I voted for Yavlinsky even though I realized that he could not win. I was criticized sharply for this by many people close to me because they were afraid of Zyuganov's victory. However, I was certain that Zyuganov could not carry the first ballot and, therefore, voting for Yavlinsky would not help Zyuganov in any way. I voted for Yeltsin on the second ballot for obvious reasons: I had to.

9*. As we know, the communists failed to squeeze through but the victory over them was by no means easy. I am of the opinion that the important point was that the pre-election struggle proved very useful: it clearly emphasized the weak points of the current social structure and displayed a stern warning to the regime. At least the most dire shortcomings have to be removed in the following years (at most four, until the next presidential election) otherwise the way for a communist (bolshevik) or fascist dictatorship would be open. I may be a captive of my idealistic illusions but I believe that the victory of democracy and the rebirth of our country on this basis are still possible.

10*. I can cite a number of new publications of 2001 that expose the communist regime. They are: the books by A N Yakovlev 2000 *Krestosev and Memory's Whirlpool* (Moscow: Vagrius) and collected papers: *Russia in the XXth Century. Documents* edited by A N Yakovlev (Moscow: Democracy International Foundation). The way science was bossed about in Soviet times is clear from the collection of documents *Academy of Sciences in the Decisions of the Political Bureau of the Central Committee of the RCP/b–USSRCP/b–CPSU. 1922–1952* (This is volume 1, the sec-

ond is being prepared for printing) (Moscow: Rossmen, 2000). See also Zh Medvedev and R Medvedev 2001 *Unknown Stalin* (Moscow: Prava Cheloveka).

11*. The previous comments (except for comment 10 and part of comment 1) were written in 1997. What would I change or add at the beginning of 2001? The most important factor in the context of this article is the weakening of the position of communists in Russia. In the elections of the new president in the spring of 2000, the option of communists coming to power was already unrealistic even though their candidate was still in the second place. Despite this, V V Putin won on the first ballot. Nevertheless, communists are still a strong force in Russia and the communist (bolshevik) ideology has not been buried. This is why I think that my article is still relevant in this new edition of the book, quite apart from the feeling that it retains a certain historical interest. I want to use this opportunity once again to stress that an unusual asymmetry between 'the left' and 'the right' noted in the article (see section 26.2) has not disappeared at all. The so-called 'left' (by international criteria, these are pacifists, 'Green Party', liberals, etc; however, communists are also referred to in Russia as a party of 'the left') are very much preoccupied with arranging the trial of the 84-years old Pinochet (whom I do not even think deserves defending) but refuse to unmask the crimes of communists and their protectors from the CPSU Central Committee in the same Chile. The reminder recently appeared in an article "How to end the cold war" by V Bukovsky, this true fighter for human rights, in 2000 *Izvestiya* of 23 August.

12*. *Poisk* no 34–35, 30 August 2002, published an article by R Medvedev "Charmed pilgrims. Why did even the greatest and acutest writers of Europe regard Stalin as a sincere and decent human being?". B Shaw, E Ludwig, H Wells, H Barbus, R Rolland, L Feuchtwanger, R Alberty, and M T Leon were received by Stalin and had long conversations with him between 1931 and 1937. All of them left fascinated by his personality. For instance, Herbert Wells wrote that he never met a man more sincere, or decent, or honest; that there was nothing dark or menacing about Stalin, and that it was this quality that should explain his huge power over Russia ... that Stalin was a Georgian completely devoid of slightness and craftiness... that his sincere orthodoxy was the guarantee for the security of his companions-in-arms.... Isn't this an astonishing characterization of a bloody murderer who several years later had most of his brothers-in-arms and friends tortured and shot. The best and shrewdest writers in Europe proved to be utterly lacking in perception. Isn't it high time for progressive, excellently educated people to learn at last to understand something about demagogues and to protect true democracy not only against scoundrels and criminals but also against the kindest but brain-dead idiots who profess unbridled 'political correctness' which is essentially demands non-resistance to evil.

13*. I also want to cite S Courtois 1999 *Le Livre Noir de Communism: Crimes, Terreur et Repression*, published in Russian with a foreword by A N Yakovlev (Moscow: Three Centuries of History Publishing House). Hitler's and Stalin's bloodthirsty nature is vividly painted by A Neumaier 1997 *Dictators in the Mirror of Medicine* (Rostov-on-Don: Feniks). Stalin's vileness and treachery are vividly described by A Larina-Bukharina 2002 *Never to Forget* (Moscow: Vagrius). Despite the express wish of the author of the book—the widow of N I Bukharin, shot in 1938—her memoirs give further evidence of the complete bankruptcy of the bolshevik ideology and practice.

14*. Has anything important changed by the beginning of 2002? The main factor was, of course, the terrorist act of unprecedented scale committed by Islamic fundamentalists in the USA on 11 September 2001. I fully understand, with my heart and my mind, the reasons behind the war declared against these contemptible murderers. Only the merciless punishment of bandits and all their assistants will be a just measure and may hold the promise of saving this civilization. I completely supported the position of President V Putin in this matter when he chose close cooperation with the USA and with the rest of the civilized world. I think that the false 'lefties' and pacifists may be the main hurdle to victory over terrorists. London's Mayor Ken Livingston gave an excellent example of such people when he proposed not to prosecute British citizens who went from the United Kingdom, which gave them refuge, to Afghanistan to help the Taliban and Bin Laden's terrorists (2001 *Izvestia* 26 November). If you believe Mr Livingston, these Muslim fundamentalists were stimulated to murder innocent people by 'deep feeling of injustice". The heinous crimes of Hitler, Stalin, and other tyrants were justified by some in their time in a very similar manner. I suspect that Mr Livingston is merely a scheming politician trying to gain dividends by his false protection of the 'humiliated and insulted" bandits. Unfortunately, there did exist and still exist quite honest and noble people who failed to understand that wonderfully commendable feelings of sympathy and protection of human rights are not always justified. Indeed, having watched on TV screens, mobs of people overcome with delight, dancing after learning about the deaths of thousands of innocent human beings in New York skyscrapers on 11 September 2001, I feel no compassion for these delighted dancers. It goes without saying that people should not be prosecuted for just expressing their ecstasy and compassion to bandits; however, neither can I see any reason in specially protecting them. Nothing good ever came out of pacifying an unhappy mob and its depraved leaders by never-ending concessions, bribes and sops.

As for Russia's internal politics, this is a very topical issue and hardly a matter for this book which may see light at best only several months after I write this comment. I will only refer to the last comment to article 28.

15*. I am writing this comment on 15 September 2002 while proof-reading the Russian book. Only four days have passed since the first Anniversary of the tragic events of 11 September 2001. The date was marked in a very dignified manner in the USA. The terrorist acts proved to be so barbarous that the majority of those who preach Islam and the previously mentioned political correctness in the USA cannot or dare not speak of forgiveness or of tolerance to the bandits. The situation is quite different in Europe. According to the data quoted by O Osetinsky (2002 *Izvestiya*, 13 September) the terrible acts of Bin Laden were supported in Germany, UK, and France by 68, 75, and 80% of Muslims (these are millions of people) who reside in these countries! This is a staggering reply to the hospitality that Europe has shown to refugees from Muslim countries. Papers tell us that the first anniversary of the terrorist act in New York was even marked in the UK by open meetings of supporters of the terrorists. I happened to read a declaration by a large group of European scientists calling for relations with Israel's scientists to be broken because, in their opinion, Israel responds too harshly to Palestinian terrorism. This is not the appropriate place, of course, to go into the details of challenges thrown by terrorists and anti-globalist hooligans to modern civilization. I will allow myself only a few remarks.

As a first approximation, we can classify various 'leftist' liberals, supporters of unlimited political correctness etc into three categories: the ultra-kind, the brain-dead, and, ultimately, rascals and villains. I am really worried only about people from the first of these categories and I would like to help them. This paper has been addressed to them, as I emphasized at the beginning. It was people like this who allowed the bolshevik propaganda to carry them along, who believed that merciless violence can lead to a just society. They paid a terrible price for their illusions or, rather, many of them paid it but so did many more people whom they helped to enslave. History must teach us something! Hitler could be stopped—but the Munich agreements were signed, Hitler's hands were untied and millions of human beings paid for it with their lives. Even if it were very difficult, Stalin could also have been stopped and unlimited power need not to have been gifted to him. This has not been done and the price was again many millions of human lives.

Even today communism and fascism rule in Cuba, North Korea, Zimbabwe and are an umbrella for various 'liberation' movements all over the world. Iran's ayatollahs and others like them are no better. Chinese communists are potentially dangerous too.

However, it became clear after 11 September 2001 that the main danger now is Islamic fundamentalism or some of its militant fractions. The place of Hitler and Stalin has been taken by Bin Laden and his followers. They are the people that need to be repelled first of all; it would be better if this happened before they get hold of chemical, bacteriological, or even atomic weapons. In reality, though, colossal damage can be achieved without dumping something from a airplane or spreading anything but just by a terrorist act on an atomic power station or another plant in this industry.

I do strongly support democracy and the protection of human rights but I realize with regret that a very high price has to be paid in today's world for maintaining civilization and the level reached in the leading countries. Of course, O Osetinsky goes too far when he claims that "the most foul religion today in this world is the so-called political correctness. As a narcotic, political correctness robs a person's view of the world of clarity" and later: "Civilization does have the right of preventive self-defense! To fight the Devil without delay, rapidly and efficiently—is the new fundamental doctrine in relations with the West and the terrorism in the XXI century!". This is true to some extent but the question is now: where should we draw the line? Comrade Stalin would simply deport all Muslims from the United States and Europe: this is out of the question, of course. However, I would deport all who profess terrorism, even if they did not succeed in implementing their plans. Of course, if this approach is declared, all potential terrorists would stop advertising their views and intentions at each street crossing. I believe that it is permissible to deport and punish, in general, for preparation and approval of criminal actions: you do not wait until the dog bites you. However, this is when the 'cursed questions' emerge: what if the dog only barks but may not bite? I do not know the answers, the response mostly depends on specific circumstances; however, I would definitely not extend material help to those who approve of the actions of the 11th of September, and I am against attempts to pacify them. This will not help, this will only multiply the number of innocent victims.

Article 27

Juridical sentence upon bolshevik communism[1*]

The Nuremberg Trial (1945–46) pronounced sentence on fascism and, more specifically, on national socialism. Regrettably, communism has never been the object of similar legal proceedings. However, a picture has emerged largely due to the publication of various books and materials, in particular those from formerly inaccessible archives. Even the carefully concealed crimes of bolshevik communism, like the mass execution of Poles in Katyn', have come to be known, though only quite recently. In essence, the sentence on world communism has already been pronounced and especially so on its Russian version, which is best referred to as bolshevik communism. This sentence is borne out by A I Solzhenitsyn's *The Gulag Archipelago, 1918–56* and by numerous other books, among which I would like to emphasize the *Rossiya XX Vek* [*Twentieth Century Russia*] a book series edited by A N Yakovlev as well as the books written by A N Yakovlev himself (*Omut Pamyati* [*The Whirl of Memory*]—the most recent of them—was published in 2000). However, as far as I know, evidence of bolshevik communism crimes in the juridical sense of the word is still insufficient to make the picture complete for legal proceedings. We are concerned with an analysis of the bolshevist judiciary and of the judicial and extrajudicial practice of repression in the USSR performed by jurists. Such a book has finally emerged. The case in point is *Politicheskaya Yustitsiya v SSSR* [*Political Justice in the USSR*] by V Kudryavtsev and A Trusov (Moscow: Nauka, 2000).

As repeatedly noted, the practices under all totalitarian regimes turn out to be astoundingly similar even at first glance. To illustrate this, the following is an extract from the speech of the Chief Prosecutor of the United States of America during the Nuremberg Trial established to indict and try former Nazis as war criminals:

With all administrative offices in Nazi control and with the Re-

ichstag reduced to impotence, the judiciary remained the last obstacle to this reign of terror. But its independence was soon overcome and it was reorganized to dispense a venal justice. Judges
were ousted for political or racial reasons and were spied upon
and put under pressure ... jurisdiction over treason cases was
transferred to a newly established 'People's Court' consisting of
two judges and five Party officials .... Special courts were created to try political crimes, only Party members were appointed
judges, and 'Judges' letters' instructed the puppet judges as to the
'general lines' they must follow. (The book under review, p 17)

A deep analogy with the practices of bolshevist judiciary is compelling
to anyone who has ever heard of the methods of work of the VChK–GPU–
NKVD-KGB and courts in the USSR. But of course, a comprehensive
well-documented investigation of the political judiciary in the USSR (or,
more exactly, in Russia since 1917) has been absent. However, the book
by V Kudryavtsev and A Trusov, which numbers 365 pages, is concerned
with precisely this subject.

The book numbers ten chapters, an introduction, and conclusion,
which harbor a great body of data, including historical data (for example, on the experience of Judicial Reform of 1864 in Russia). Very much,
however, is intended for jurists and, in general, for people related to the
activities of court, prosecutor's office, etc. But the pivotal intention in this
book is an analysis of the origin, flourishing, and break-up of the political
judiciary. The conclusion drawn is as follows:

The main crime of Stalinism—the extermination of absolutely
innocent people—cannot be justified by any historical or pseu-
dohistorical reasons. The totalitarianism gave rise to fear and
disbelief in the authorities. The orgy of political terror signified
the disruption of the foundations of law and order, the triumph of
arbitrary rule and the liquidation of independent judicial power.
The methods of shadowing and informing against, the persecution
of differently minded people undermined the respect of people for
each other and fostered the decay of public morals. The country lost the respect of progressive world community and has not
regained it up to the present. Finally, the terror fostered the
degradation of the party and state leaders, the very one to have
given birth to this way of 'guidance' of the society. Not only did
the world revolution never take place, but it was also 'removed
from the agenda'; socialism has never been built in a separate
country, and the great country disintegrated in a few days. It had
failed to provide not only the well-being of the whole people, but
merely the normal human life as well. The totalitarian system

lost the slightest support of the society and became obsolete, and with it the political judiciary. (p 350)

The book also provides a wealth of specific examples of the felonious activities of different repressive agencies. As an illustration I will cite two fortuitously selected 'cases'. Here is a passage from V E Meierkhol'd's letter of 13 January 1940 addressed to the Procurator-General Vyshinskii:

> I was laid down on the floor, my face down, and they whipped my legs with a cable. The following days, when these areas in the feet were amply flooded with internal hemorrhage, they would whip these red-blue-yellow bruises with this cable again and again, and the pain was so strong that it seemed as if water on the boil was poured over the painful areas (I was screaming and crying of pain). They would strike me in the face .... The interrogator would threaten me, telling me over and over again: "If you wouldn't write, we would beat you again and again, leaving your head and right arm unbroken, and turn the rest of your body into a shapeless bloody mass. And so I signed everything before November 16, 1939. (p 238)

In the February of 1940, Meierkhol'd was shot. The life of 'ordinary people' was not sweetness and light either. For instance, after the trial of Tukhachevskii in the August of 1937, a worker Zh told his friends that "it was wrong to try Tukhachevskii and others behind closed doors, and workers are unaware why he and his accomplices were shot". This single phrase was interpreted as "resentment against repressing enemies of the people" and qualified as counter-revolutionary propaganda, for which Zh was shot by the decision of a 'triplet' attached to the NKVD Department for the Moscow Region (p 169). Such examples are innumerable. However, some numbers do exist. In particular, 839 722 humans were shot over the period from 1918 to 1958! This is one of the estimates given by the authors on p 314 of the book. It is also noted in this case that the figures are underestimated and number of 1 165 000 executions by shooting is more probable. Over the same period 6 165 000 people were convicted and imprisoned. By the way, the annual distribution of the number of executions is instructive (p 314). In particular, 1118 humans were shot in 1936, as many as 353 074 in 1937, 328 618 in 1938 and, once again, 'few'—2601 humans—in 1939. (Somewhat different data are given in chapter 12 of *Omut Pamyati* by A N Yakovlev.) For comparison, the entire population, say, of Finland amounts to 5.2 million people and of Denmark to 5.1 million.

The following phrase was adopted as the epigraph to the book: "By forgetting the past we doom ourselves to experiencing it over again". This is in truth a profound statement and the book by V Kudryavtsev and A Trusov is a significant and valuable work, which will be an aid in retaining the past, particularly to those destined to build the new Russia.

They, like all Russian citizens, should remember that the "political judiciary broke up with the cessation of activity of its parent and guardian—the Communist Party of the Soviet Union (CPSU)" (p 356). But its successor is the Communist Party of the Russian Federation (CPRF) and only those who are stone-blind and stone-deaf to the laws of history can be unaware of what awaits the country should communist (or, generally, totalitarian) rule be restored in Russia. To my great regret, I regard the adoption of Aleksandrov–Mikhalkov's hymn as the National Anthem as a step towards such a restoration. I consider this as an outrage upon the memory of the many millions of victims of bolshevik communism. At the same time I am very glad to be able (may still be able!) to write freely about it. While I fear falling once again into idealism in my old age, I believe in a bright, democratic, and civilized future for Russia. The authors of the book under review conclude it by discussing precisely "the conditions necessary to make impossible or at least unlikely the recurrence of past tragedies" (p 358). The most important conditions are, in the authors' view, first, improvement of the cultural and educational level of the people and, second, internationalization of social and state life. I agree with them but would like to add a requirement for the strict observance of the freedom of conscience (i. e. the right to freely decide between the faith in God with adherence to any religion or confession, atheism, and agnosticism), the irreconcilability with pseudoscience and all kinds of quackery. Needless to say, this list should be complemented with vigorous efforts to ensure the observance of human rights, the triumph of democracy, and legality (specifically, we mean guaranteeing the independence of judges and checking the activity of prosecutor's office).

Comment

1*. This article (review) entitled "The Shooting of Meierkhol'd and others" was published in 2001 *Literaturnaya Gazeta* no 14 04–10 April. On the lawlessness that reigned during the bolshevist rule, much was written also in the preceding Article 26 and, chiefly, in the literature mentioned therein.

Article 28

Inner weakness (what has been forgotten by Academician Zhores Alferov)[1*]

I cannot help responding to the article by my colleague, Academician Zhores Alferov 'My dolzhny vinit' sami sebya' [*We should blame ourselves*] published in the *Literaturnaya Gazeta* [*Literary Newspaper*] no 39, 2001 in the column *Ten years that shook*.... In this column, in no 22 of the newspaper, my interview was published, titled 'Ya syt po gorlo kommunisticheskim gumanizmom' [I am fed up with communist 'humanism']. For me, as well as for Alferov, Russia is not 'this country' but 'our country'.[1] That is why I cannot but feel deeply upset and indignant in connection with the poverty of the greater part of the country's population and the plunder of public property by the 'new Russians', especially all sorts of oligarchs, and also with the decay of our science. But Alferov and I give different answers to the perennial question: "Who is to blame?" He sees the root of all evil in the activities of 'the population of young reformers'—although not CIA agents these are either idealists or mercenary people with poor qualifications. These people committed glaring errors, both in economics and in politics, which led to the collapse of the USSR and the degradation of Russia. As for me, I believe that the fundamental and principal cause of all these disasters is the bolshevist–communist totalitarian regime which was set up in Russia as a result of the coup d'etat in October 1917.

In Alferov's opinion, as a result of the rule of this regime, by 1985 (regarding this year as a starting point of the reforms) things in the USSR were not all that bad, so he is "convinced that the only thing the country needed by 1985 was a new economic policy". True, this quotation is not from Alferov's article in the *Literaturnaya Gazeta*, which is being discussed

---

[1] In the current Russian language, some persons understand and use the expression 'this country' to emphasize that they would like to dissociate themselves from what is going on in their own country.

but from his interview published in the newspaper *Sovetskaya Rossiya* [*Soviet Russia*] on 30 December, 2000. But his article in the *Literaturnaya Gazeta* says basically the same, merely in a more verbose way. As I see it, such is also the stand of the Communist Party of the Russian Federation (CPRF), of whose faction in the State Duma Alferov is now a member. Of course, the CPRF admits that before 1985 the Communist Party of the USSR made some 'mistakes' not only in economics but that, in their opinion, is a matter of secondary importance. However, in fact by 1985 the USSR, apart from the economic crisis, was in the throes of the cruel Cold War with the whole Western world, of the unjust and bloody war in Afghanistan and of serious ideological decay. There was only one party (the CPSU), which governed everything by dictatorial methods, there was only the puppet Supreme Soviet of the USSR, which rubber-stamped everything suggested by the CPSU. The critics of the regime were to go to prison, exile, or madhouses. In particular, it is beyond doubt that A D Sakharov was in exile in Gorky, and not elsewhere, only because by that time he had already been awarded the Nobel Peace Prize. I think that if it had not been for this internationally acclaimed prize, Sakharov, even in spite of his possessing three titles of Hero of the Socialist Labor (these titles were taken away from him, though), would have found himself not in the city of Gorky but much farther or in a madhouse. Freedom of speech was out of the question, and even using photocopiers was permitted only under surveillance and only on a small scale (see, for example, my article in the book 1988 *Inogo Ne Dano* [*There Is No Other Way*] (Moscow: Progress). The same situation was found the restrictions on the 'freedom of movement' or, putting it simply, on the freedom in travel abroad. Even going abroad to scientific conferences was allowed only to the select few, though in 1985 it was already easier than at the times of Stalin. By the by, it is in connection with the opinion of "the insufferable non-freedom aggravating the guilt of the communist regime" that Alferov mentions me in his article. From the context one can see that what he means is mostly "the cruel dosing of international exchange". And indeed, the scientists of my generation (I graduated from the Physics Department of Moscow State University in 1938) lost a lot because most of them did not have the possibility of going abroad and—for about ten years, from 1947—for publishing their articles in English even in the USSR. But of course, it is a secondary question in comparison to that non-freedom in general. Here I also remembered the well-known thesis 'existence determines consciousness'. Alferov's parents belonged to the minority of old bolsheviks who did not become victims of Stalin's terror. And Alferov himself started working in the post-Stalin years and must have been able to go abroad freely enough. Anyhow, he worked for several months in the USA. I do not know whether he would have won, together with two American colleagues, the Nobel Prize—which he quite deserved, as far as I can judge—if he had been born earlier and, in general, had had better acquaintance of Stalin's freedom of movement. Since what has been said might sound as if being prompted by some personal offense, let me note that there is no such feeling in my heart—my destiny in science turned out to be quite happy. And if I have already started to speak about myself, I shall say that my convictions have especially much in common with those of social democrats, with their concern for social justice and dislike for capitalists. Or, more specifically, now I belong to the adherents of the International Humanist Movement. But writing here about that would be irrelevant (see my article "Mezhdunarodnoe gumanisticheskoe dvizhenie i 'Manifest 2000'" [*International Humanist Movement and 'Manifesto 2000'*] written together with V A Kuvakin and published in the 2000 *Vestn. Ross. Akad. Nauk* 70 483 [2000 *Herald Russ. Acad. Sci.* 70(3) 223] and in the 2000 *Neva* journal no 8, 141; it is article 35 in this book).

Let us come back, however, to the question of the causes of the current situation in Russia, which is a crucial question for us now. Having nuclear weapons and a powerful mechanism of repression, the bolshevik regime in the USSR, in spite of its overwhelming difficulties mentioned earlier, would have been able to last much longer. This can be well illustrated by the lasting of Castro's regime in Cuba and the regime of Kim Chong Il in starving North Korea. But, fortunately, M S Gorbachev, having become the bolshevik tsar—the General Secretary of the Central Committee of the CPSU—took the road of reforms and not the preservation of the tottering regime.

But who could he lean on for support?

Exterminated in the years of Soviet power was a very broad stratum of the intelligentsia: the people who could prepare and carry out radical reforms had been eliminated. In recent years, we have already seen quite a number of publications on this subject. I will confine myself to mention of the fact that Lenin took the trouble to ensure, as early as in 1922, that 'about three hundred most prominent Russian scientists were put on ... a barge? ... no, on a steamer, and sent to the European dump'. (A I Solzhenitsyn *The Gulag Archipelago* vol 1, chapter 10; see also Selezneva I N 2001 *Vestn. Ross. Akad. Nauk* 71(8) 738 [2001 *Herald Russ. Acad. Sci.* 71(4) 421]). And this could have been considered very humane, as absolutely guiltless people were, without argument, only exiled. In other cases, Comrade Lenin did not show such soft-heartedness. For instance, in his secret letter of 19 March 1922, published only in 1990 (*Izvestiya TsK KPSS* [*Bulletin of the CPSU Central Committee*] no 4, p 192), he says, 'The more representatives of the reactionary clergy we manage to shoot for this reason, the better'. The instruction of the leader was carried out—among those shot were 32 metropolitans and archbishops. Here, I think, it is also worth recalling that from 1918 to 1958 in our country 839 722 people were shot, with 353 074 of them in 1937 and 328 618 in 1938 (see Kudryavtsev V and Trusov A 2000 *Politicheskaya Yustitsiya v SSSR [Political Justice in the USSR]* (Moscow: Nauka); see also article 27 in this book). Some authors give even higher figures. Altogether more than 20 million people died in concentration camps and prisons. But according to Alferov, "in spite of all its negative elements, it was the first more or less successful (as it held for 70 years) attempt to create a state of social justice". Having nothing to comment on here, I shall go on to note that Stalin steadfastly continued Lenin's policy of getting rid of the intelligentsia, of economists in particular. For example, in the 1930s, the renowned economists N D Kondrat'ev and A V Chayanov were shot. In the book 1995 *Tragicheskie Sud'by: Repressirovannye Uchenye Akademii Nauk SSSR [Tragic Fates: the Scientists of the Academy of Sciences of the USSR Who were Subjected to Repression]* (Moscow: Nauka), one can see the names of 105 members of the Academy subjected to repression in different years, this list being incomplete. Besides, among the scientists who died in such a way there were, of course, many non-members of the Academy—as, for instance, the outstanding physicists L Shubnikov, S Shubin, M Bronshtein, and A Vitt, to name only a few.

I also cannot help recalling here that once, in the USSR, tank and aircraft units for the German Wehrmacht were trained. There were also special schools or camps where fighters of foreign so-called liberation movements were educated or, putting it bluntly, many future terrorists were prepared. I think that it will not be an exaggeration to regard the notorious Bin Laden as follower of a Comrade Stalin's cause.

As noted by Pope John Paul II, "Communism died on its own because of its inner weakness, as it turned out to be a medicine more dangerous than the disease itself". Indeed, the world and tsarist Russia, in particular, were ill but the bolshevist Leninist medicine—an attempt to achieve communist justice by flooding everything with floods of blood turned out to be ineffectual. And 70 years later, the collapse of the communist system followed. Not only in the USSR but in the whole world. Perhaps only China went down a particular road. But a transition from a totalitarian regime to a democratic one is, on the whole, a shock and there were few places where it occurred peacefully and quietly, as, for instance, in Czechoslovakia (I mean the so-called 'velvet revolution'). In the USSR, with our country's one-party system and the complete lack (for the reason mentioned earlier, the extermination of the differently minded) of some organizing force, such as a group of qualified economists or some other progressive specialists, the collapse of the communist regime was mostly chaotically, without any elaborate plan. As a result, we have got what we have now. Putting the blame on 'the population of young reformers' is the simplest thing to do and it is unfair. Who could have taken their place, who else could have carried out the reforms, and who gave them power? Why did the communist rule guided by the CPSU Central

Committee fall to pieces like a house of cards? In short, there is only one answer to the question: "Who is to blame?"—the blame should be put on the inefficiency of the totalitarian, tottering communist regime. Changing from this regime to a democratic one is a most complicated and difficult process, it does not happen automatically as soon as some form of parliament is created and power is snatched from the hands of decrepit old party bosses and the young arms of discredited comsomol leaders. In such a way, as I have already said, we have come to what we now have.

Of course, now there is not much to be happy about, the difficulties are great and the shortcomings are numerous. But we have found freedom of speech and freedom of movement, we have found, though imperfect, an elected parliament—the State Duma. Among the deputies of this Duma there are quite a few people, like Shandybin, Zhirinovsky, Mitrofanov, and a number of others whom many people intensely dislike, and I really loathe. But the Duma also has some qualified specialists, both economists and representatives of other professions. The Duma is working, and even though its work is far from being perfect, can it be compared with the silent, as a matter of fact, puppet Supreme Soviet of the USSR. In the person of President Putin, the place of 'Tsar Boris' with his 'family' (as President Yeltsin and the people closest to him were called) has been taken by a man both hardworking and, as far as I can judge, sincerely aspiring to the prosperity of the country. True, I do not altogether consider myself competent in questions of economics and government. So I will neither defend nor criticize the numerous measures being proposed by the president and his team. Some options of reforms which they have chosen might not be the best. But I can only say that of all Putin's actions that I know of I strongly resent only the adoption of the national anthem created in Stalin's epoch,[2*] with the new text written by Mikhalkov. I regard this fact as a slap in the face of the millions of the still living and an outrage against the memory of the millions of the dead. Besides, the mention of God in the anthem of a secular country is glaringly out of place. But still, these are questions of secondary importance and, if I were younger, I would hope to witness the prosperity of our country on her way to democracy, civilization, and progress.

Comments

1*. This article was published in the 2001 *Literaturnaya Gazeta* [*Literary Newspaper*] no 42 on 17 October. To the newspaper I sent the article under the heading "O chem zabyl ili zakhotel **zabyt'** akademik Zhores Alferov" (What Academician Alferov forgot about or wanted to forget) and its present title was given by the editorial staff. Maybe it is better and I decided not to change it. More important is the fact that at the editorial office a number of references were left out and they have been restored.

The readers have brought to my notice some critical observations and I want to answer them in brief.

The first observation is as follows. The main thing about Alferov's article is that he bitterly states that our country is now in a difficult situation and looks for ways out of the crisis. Whereas I in my answer focused on the question: "Who is to blame?" and on criticism of Alferov's stand in this respect. Yes, it is true but in a newspaper article, especially if it is merely a response to the previous article by Alferov, one cannot speak about everything at once. I stressed at the very beginning that I feel 'deeply upset and indignant in connection with the poverty of the greater part of the country's population and the plunder of public property'. That is why I focused on the no less important question, "Who is to blame?", in the answer to which Alferov and I fundamentally disagree.

The second observation is, roughly speaking, that not all 'new Russians' are cheats, among them there was and there are quite a few honest people, who managed in the new conditions to show their abilities and initiative, useful for the development of the country. Of course, I agree with that.

The third observation is that I ought to have specially pointed out the pernicious role of the bolshevist regime with regard to the development of a number of scientific trends—especially of biology, where the illiterate adventurer Lysenko held complete sway supported by Stalin and Khrushchev. It stands to reason that I agree with this observation.

Finally, the fourth observation, which is the most essential. I am quoting the phrase by Alferov where he gives the following characteristic of the Soviet power: "In spite of all its negative elements, it was the first more or less successful (as it held for 70 years) attempt to create a state of social justice". Before citing this passage in my article, I recalled that in the time of the existence of this "more or less successful state of social justice" about a million people were shot and more than 20 million died in concentration camps and prisons! That is why I said that here I have 'nothing to comment on'. But actually there is something on which to comment, as was noted by one of my readers. Indeed, if at the horrible cost of the million shot and many millions who perished in prisons and camps the population of the USSR remaining alive (about 287 million at the beginning of 1989) had started to live happily, one could say that an outrageous price was paid but a socialist, or maybe even communist, state of 'social justice' was created. Such was, I believe, the reckoning of Lenin and his accomplices. It goes without saying that such a cannibalistic stand, based on disdain for the lives of masses of people, is ineffectual in itself, if only for moral reasons. It is certainly no less important that the possibility of building socialism on blood and terror is utopian and undoubtedly has not been realized. By the by, the fact that communist ideas are utopian was absolutely clear to many people, including Leo Tolstoy, about whom I write in section 26.5. Indeed, in spite of the bloody repression, the population of the USSR, on the whole, was poor or even very poor in comparison to the workers in the developed countries of Europe and the USA. And, what is the most important thing about the question now being discussed, social justice in the USSR, i. e. any similarity between the living standards of ordinary citizens, workers, and peasants, and of different party, Soviet and comsomol authorities, was absolutely out of the question. The inequality in this respect was glaring, and everyone who lived in the USSR knows it. True, there was a sharp increase in the living standards of research workers granted after the atomic bomb appeared in the USA, in 1945, this measure being considered expedient for the success of the arms race. But can this fact be evidence of the triumph of social justice in the USSR? The appallingly enormous scale of the social inequality in Russia today and, on the whole, of the gap between the rich and the poor (see, for instance, the interview with N Rimashevskaya in the *Vek* [*Age*] newspaper no 30, 2002) arouses indignation. I would welcome any reasonable, even if strict, measures for the elimination of the shortcomings we now have, especially those of crime, corruption, and bureaucracy. But it is certainly not by returning to the criminal bolshevist past that any progress on this road can be achieved.

Now three more observations on my own behalf.

Alferov's reasoning that the "attempt to create a state of social justice" can be considered successful because this state "held for 70 years" arouses surprise, to put it mildly. The possibility of the lasting existence of dictatorial, repressive regimes can in no way be evidence of their being just, which I have already stressed in the text of the article. The cited examples of Cuba and North Korea speak for themselves. In this regard, I can add that in our country up till now, in spite of a whole number of publications showing the criminal character of Castro's regime, his stronghold, this 'prison island', is sometimes still habitually called 'the island of Freedom'. Reading such things one just feels ashamed.

The second observation relates to the delights of the notorious, allegedly progressive Stalin's national policy. I wish the partisans of the Soviet regime would remember the deportation of whole nations, Chechen among them. Our country is still paying the bloody tribute for this manifestation of social justice. And just think about the state anti-Semitism with its culmination, 'the case of the medical doctors'. This, together with the deportation of many nations, ranked the USSR with the criminal Hitlerite Nazi regime.

Finally, the third observation relates to misunderstanding, or often simply juggling, in the usage of words 'communism' and 'communist'. Actually, I have already written about this in comment 1 to article 26 in this book. The point is that when communism and communists are criticized, what is meant is, without any reservations, not primitive communism nor Utopian communism, nor Campanella with his *The City of the Sun* but the bolshevist communism of Lenin and Stalin with its bloody dictatorship and outrage against human rights. That is why when the critics of bolshevik communism are accused of allegedly being opponents of social justice, partisans of capitalism and so forth, it is not true or, to be more exact, usually not at all true. Let me say about myself. I am certainly a partisan of social justice and, generally speaking, I agree with the communist formula "from everyone according to their abilities, to everyone according to their needs"—or, in its milder form, "from everyone according to their abilities, to everyone according to their work". I had long believed in the possibility of something like that, in general, and in the USSR, in particular, but my eyes were opened, rather late and with difficulty. It was only by chance that such blindness did not turn out to be a tragedy for me. I have written about this in other articles of this book and here I do not want to repeat myself (in particular, see article 26).

I am convinced that mankind can expect a radiant or, at least, relatively radiant future only in case of its development on the road of democracy and humanism (for details, see article 35). Where the current situation in Russia is concerned, I believe that Putin's policy and specifically, his orientation to the civilized world is absolutely right. In the internal policy, I would like to pin my hopes, among other things, on the creation of the United Social Democratic Party of Russia (SDPR(u)) headed by M S Gorbachev, A N Yakovlev, and K Titov. By the by, reading about the creation of this party in *Izvestiya* [*News*] newspaper, on 26 November 2001, and watching a corresponding TV program, I paid attention to the fact that at the party's constitutive congress V Zhirinovsky was present and made a speech. If he had been invited to the congress, for me it would have meant that it was impossible to deal with the new party. In this regard, I specially found out that nobody had invited Zhirinovsky, he came of his own accord. At the nearest election, I am going to vote for the SDPR(u), I would like it to pass to the Duma. Of course, it is desirable but certainly not at the cost of compromise and all sorts of concessions to disruptive forces. Russia does need, after all, a party of social democratic type, honest, and true to its principles.

2*. Now that I am writing this commentary in September 2002, I, regrettably, have to add two more critical observations. Firstly, I would like to point to the fact that the state support of religion and, in particular, of the Russian Orthodox Church, is becoming stronger and stronger. I speak about this question in more detail in articles 32–34 of this book. I am convinced that clericalization of the country is deeply reactionary, it is throwing us back. I wish the advocates of this course would remember that in the tsarist Russia orthodoxy was the official ideology. It is well known that this fact did not at all stop the collapse of the tsarism, the atrocities of the civil war, and later the excesses of militant non-believers and the horrors of Stalin's terror. Secondly, it has become obvious that corruption and simply plundering the country's natural resources have achieved a truly monstrous scale, this being known to the authorities and covered up by them (a glaring example is cited in the articles "Mafiya i more" (The mafia and the sea) by B Reznik in the 2002 *Izvestiya* on 18, 19 and 20 July and also on 9 October). However, still more important is enormous gap now existing in Russia between the incomes of the poorest and the richest strata of the population. According to the article in the newspaper *Vek* no 30, 2002, already mentioned, the income of 5% of the richest is one hundred times more than the income of 5% of the poorest (these incomes are $3000 and $30 a month respectively). Such a situation jeopardizes the very existence of the Russian Federation, at least with the form of government corresponding to the 1993 Constitution. Unfortunately, this suspicion has become stronger after the mayor of Moscow, Yu M Luzhkov, who trims his sails to the wind, suggested, in the middle of September 2002, that the monument to Dzerzhinskii should be brought back to Lubyanskaya square. Such an act would be an outrage against the memory of the millions who died from the bolshevist terror and might be a signal to the restoration of the totalitarian regime in Russia.

Article 29

I do not understand these so-called patriots (a self-interview)[1*]

I sometimes give an interview but usually I am not satisfied with the result. The text looked as if it was not my own and reworking it was difficult. So when a certain periodical asked me for an interview, I wrote it myself, even before I met with the journalist, both the questions and answers—thus producing some sort of a self-interview. It seems to me that such an interview form also has the right to exist. Let the readers judge whether it really does.

Question 1. You are a physicist but in different newspapers and journals we have seen a whole number of your articles on subjects having little to do with science. What are your social or political interests at the present time?

Answer. I have never been and I am not a politician but apart from science (which, of course, is the main thing for me) I am interested in social problems. I am not indifferent to them. So I have been trying to do something in this respect, especially since the beginning of *perestroika*. For example, I took part in the book 1988 *Inogo Ne Dano* [*There Is No Other Way*], once much talked about (Moscow: Progress). It was a collection of articles, which included my article "Protiv byurokratizma, perestrakhovki i nekompetentnosti" (Against bureaucratism, playing safe and incompetence). Just to think that as far back as 12 years ago, one had to argue that using photocopiers should not be impeded! Then I was a deputy in the Supreme Soviet of the USSR (where I represented the USSR Academy of Sciences). There I was, for some time, a member of the well-known inter-regional group and, what is still more important, a member of the Anti-Privileges Commission. But instead of digressing, let me answer your question. At present I would specify five problems in which I am particularly interested and rather worried about. They are the position of science in Russia; a certain aspect of the issue of the freedom of the press; the necessity to combat pseudoscience; atheism, and religion—and the most important thing, the situation in Russia in this respect; and finally, the eternal question of anti-Semitism.

Question 2. We are especially interested in the issue of the freedom of speech. Many journalists are afraid lest restrictions in this field appear again.

Answer. You may be surprised but I am actually for some censorship of the freedom of the press if it is understood to mean permissiveness. Before explaining my point of view, I will take the liberty of noting that I did render some services, however small, to the cause of the freedom of the press in Soviet times. I will give one example, because it is this example that looks especially funny. For a number of years I was the Editor-in-Chief of the specialized journal *Radiofizika* [*Radiophysics and Quantum Electronics*]. Our work was permanently hampered. There was paper to increase the volume of the journal and thus accelerate the publishing of articles but we were not allowed to do that, because the volume had been set somewhere and it cannot be changed. Nor can the price be changed and so forth. And so, in 1964, I wrote an article criticizing all this together with a humble, as was proper for a Soviet man, proposal to change some things. I sent the article to *Izvestiya* [*News*]. It was accepted for printing and already set up but suddenly I was told that the Glavlit[1] forbade its publication. After that I sent the article to the *Literaturnaya Gazeta* [*Literary Newspaper*] and the Glavlit forbade it again. Basically this was quite understandable, for the article criticized the organization of the publishing business, with which the Glavlit was connected. Well, then I, in our usual Russian hope for a good tsar, sent a corresponding letter addressed to Comrade Khrushchev. Of course, no one would think of answering it but in October 1964, Khrushchev was dismissed. And virtually two weeks later I had at once two telephone calls from the officials of the CPSU Central Committee. One of them told me, literally, the following (I cannot forget it though it was almost 40 years ago): "It is not true that your article has been forbidden by the Glavlit. We summoned the head of the Glavlit, Comrade Romanov, and he said that the Glavlit had not forbidden anything, he only had strongly recommended not to publish the article." I remember this exactly—"had not forbidden" but "had strongly recommended". I was also told something about the possibility of publishing this article in the future. But it became possible only five years later—the article appeared in the *Literaturnaya Gazeta* of 18 June 1969. However, even then it could not do without a ridiculous ending. In the original, the article ended with the phrase: "If a scientific journal is not criticized for its work and has more and more articles waiting to be published, the editorial staff must have the right to decide annually on the volume of the journal, depending on the current situation. It should be done quickly and efficiently, without making

[1] The abbreviation for the governmental censorship committee.

an extreme effort comparable to that which is necessary for the flight to the Moon." But while the article had been lying in the editorial office, the date of the American flight to the Moon had become closer, someone's 'vigilant eye' decided not to worry Soviet readers and mention of the Moon was crossed out. You may grudge giving so much space to this old story but it might be a reminder to those who have already forgotten in what conditions of 'the freedom of the press'—which was officially proclaimed in the then Soviet constitution—we lived.

Of course, now the situation is quite different. But I must say that endless cries about suppressing the press, which one hears on the television and sees in newspapers, have, to a certain extent, already started to arouse my indignation. Of course, the freedom of the press, one of the most important achievements after the liquidation of the bolshevist dictatorship, should be defended. Words fail to express our sympathy and indignation, for instance, at the murder of the journalist L Yudina, who criticized the khan-like Kalmyk president. But we should also look at the other side of the coin and remember some saying like 'You can have too much of a good thing'. What do I mean?

First, the freedom of the press every so often turns into permissiveness, unscrupulousness, and contempt for the evident duties of an honest mass media. Secondly, some mass media will not observe the decencies—at any rate, as I understand them. For example, the mass media are now extensively dealing with, and, one can say, propagating all sorts of anti-scientific rubbish, like astrology. It is probably for the sake of a trifling profit (some increase of circulation) that they resort to fooling people. At one time I wrote letters about this to the editors of the newspapers *Argumenty i Fakty* [*Arguments and Facts*] and *Izvestiya*—but that did not work, they did not even answer. The Russian Academy of Sciences (RAS) has set up a special commission to combat false science but its work, far from being helped, is being impeded in every possible way. I am not speaking about astrology, its falseness can be exposed by any competent person (for details see the journal *Nauka i Zhizn'* [*Science and Life*] no 11, 2000). But there are more delicate issues, like creating some generators of torsion fields. Meanwhile, criticism of the corresponding antiscientific activities, which are carried out at public cost, has been adamantly opposed in the governmental newspaper *Rossiiskaya Gazeta* [*Russian Newspaper*] and even in *Izvestiya*. Scientists are being criticized but their articles with refutations (namely the articles by Academician E P Kruglyakov) nobody is going to publish. The RAS once addressed the then prime-minister E M Primakov asking him to call to order at least the governmental *Rossiiskaya Gazeta* but with no result. Academician E B Aleksandrov and I also tried to publish a critical article but, you see, in it, we hurt the feelings of Mr A Valentinov, head of the *Rossiiskaya Gazeta's* department of science. Accordingly, "Can a colleague be criticized?"—and newspapers have never published our article. It appeared only in the small circulation journal 1999 *Vestn. Ross. Akad. Nauk* **69**(3) 199 [1999 *Herald Russ. Acad. Sci.* **69**(2) 118].[2*]

I am not for restoration of the Glavlit and so forth—this is out of the question. But unfortunately, we already know only too well what democracy turns into if it is not defended and the laws are not observed. In the same way, freedom of the press should be defended from unscrupulous people, who do not care a straw for the decencies and interests of the country and its population. How to do it? I like very much the well-known phrase by Aleksandr Galich (the author and performer of songs, who was not officially recognized but very popular in the Soviet epoch): "Most of all beware of the one who will tell you that he knows how it should be done." The question of judicious control over the mass media is very delicate and I do not know "how it should be done". I do not know it but I have an opinion, which are different things. It seems to me that we need some organ which would act openly and enlist the services of the general public. But this organ must be able to demand that the mass media observe certain norms and rules. Specifically, such things as the propagation of undoubtedly false sciences, like astrology, or refusal to publish refutations, at least those coming from people or organs authoritative enough, like the RAS are not to be tolerated.

Now about the decencies. Apart from what has already been said, I consider it indecent to scoff at the lawfully elected president of the country. I am speaking, among other things, of the 'Kukly' [*Puppets*] TV program at the NTV channel. In 'Kukly', Putin is so openly scoffed at and humiliated—that it is really the limit. Let me say at once, in order to avoid being regarded as a sycophant, that it was Yavlinsky, and not Putin, for whom I voted. Besides, 'Kukly' and 'Itogo' [*In All*] by Viktor Shenderovich are my favorite TV programs, their author is very talented. But such people probably have their own specific tasks, like trying to be as sharp and biting as possible. But I wonder what the leaders of the NTV think about. If I am not mistaken, it was Napoleon who said, "A politician has his heart in his head". I am not a politician, my heart is not in my head, and I must say that if I were scoffed at in such a way, being portrayed as a malicious idiot and so forth, I would hate these 'critics' and my wish would be to respond in the same manner or, maybe, even worse. Putin's behavior shows that he is a politician or that he so far does not have the opportunity to get even. But I repeat that I do not understand what the masters of the NTV want, even though on the whole I highly value this channel. Defense of one's interests and dignity is a good and legitimate business but I think mockery is counterproductive. However, as is known, "politics is a dirty business" and, maybe, I just do not understand the rules of the game.[3*]

Speaking my mind frankly here, I am happy that it can be done without being afraid of "administrative consequences". This is what the freedom of the press is. But to be honest, I am afraid that someone may get me wrong. So, to be on the safe side, I will stress that I am not for censorship and bans, such as ban on 'Kukly'. I am for self-censorship, like the one which prevents decent publications from publishing pornography and using the so-called four-letter words.

Question 3. In your previous answer, you have already touched on the necessity to combat false science and now I would like to ask you what you meant when you mentioned religion and atheism.

Answer. I was, in Soviet times, and I certainly am now, a convinced atheist. Unfortunately, now in our country some former atheists and the so called 'militant atheists' have changed colors and even hung crosses on their necks. Well, such people are not worth speaking about. But what should attract public attention in this field are the following points.

Firstly, some ill-informed people identify atheists and 'militant atheists', i. e. those who persecute believers and support repression of clergymen. By the by, it was Lenin who was the main figure among such militant atheists in our country. In Lenin's secret letter of 19 March 1922, published only in 1990 (*Izvestiya TsK KPSS* [*Bulletin of the CPSU Central Committee*] no 4, p 192, 1990), he says, among other things, "The more representatives of the reactionary clergy we manage to shoot for this reason, the better". So they were shot in great numbers. As for atheism, i. e. non-belief in God, it is completely compatible with the conviction that complete liberty of conscience is necessary, i. e. that every person has the right to believe or not to believe in God or, maybe, to be an agnostic. Identifying an atheist with a 'militant atheist', also called 'militant non-believers', is as absurd as identifying a Christian believer with advocates of the Inquisition. Incidentally, such an anniversary as the 2000th jubilee of Christianity is now often being mentioned in our country. But we should also remember another jubilee, that of Giordano Bruno being burnt at the stake in 1600. However, it is perfectly obvious that the atrocities of the Inquisition do not refute the existence of God. At the same time, in my opinion, though this is not a refutation, if God existed, he would not have allowed such atrocities, nor the wars or genocide.

Secondly, though the right to liberty of conscience and the principle of secularism are included in the Constitution, in fact the Orthodox Church in Russia is becoming the state religion, as in the times of the tzars. I find it scandalous that different buildings and functions are publicly sanctified and huge sums of money are spent on the restoration of Christ the Savior Temple. Its demolition indubitably was an unpardonable manifestation of the bolshevik barbarism. But when we have so many poor people, when pensioners are not able to buy medicines, it is they who should be cared for, and not the building of churches. Anyway, this is not the business of the state.

Thirdly, on television, for example, there are a lot of religious programs but the voices of atheists are not heard at all. If I am not mistaken, such a 'subject' as theology is being introduced into the curricula not only of theological colleges but even in state higher educational establishments. Equally, why not introduce 'scientific atheism', though I consider both these things absolutely inappropriate for a secular country. More details of my views on the questions of science and religion, the mind and faith, can be found in my recent article published in the journal *Nauka i Zhizn'* [*Science and Life*] (no 7, 2000). See also articles 31–34 in this book.

Question 4. What would you like to say about the question of anti-Semitism, although it is hardly possible to say anything new about.

Answer. Yes, basically you are right but we are witnessing more and more examples in this respect. For instance, the invention of the Deacon A Kuraev, who has announced, without any grounds whatsoever, that International Women's Day, the 8th of March, is a Jewish venture, since Clara Tsetkin allegedly was a Jew and the 8th of March coincides with the day of some Jewish holiday. In reality, neither the former nor the latter assertion corresponds to the facts (this has been sought out and explained in the press by V G Kadzhaya). As I am a physicist, accustomed to the necessity to observe the laws of at least elementary logic, what arouses my particular indignation and irritation are anti-Semitic sorties of Kuraev's sort, or, to be more precise, of Kondratenko's sort. As you know, N Kondratenko, the governor of Kuban, puts the blame for all difficulties of life and work in Kuban and in Russia in general not just on somebody, and not even on the Jews but on the Zionists. This statesman does not even know (or, rather, does not want to know) what a Zionist is. As is known, the Zionists once aspired to create a Jewish national hearth, and this has been created—it is Israel. I do not have a clear enough idea of what the Zionists want today.[6*] Apparently, they want more Jews from all over the world to move to Israel and they want this country not to be obliterated by Islamic fundamentalists and the like. In any case, I cannot even imagine any connection between Zionism and the economy of Kuban. However, Minister Sergei Shoigu has recently said that it was a pity that such a wonderful person as Kondratenko had allegedly decided not to run for the next term of office as the governor of Kuban. For me, such a statement was quite enough to see Shoigu's true face. Too bad, he seemed to be a worthy person. But I have digressed.

Before answering the question about my attitude towards anti-Semitism and nationalism, I would like to say the following. I have always been and certainly am a convinced internationalist, i.e. I consider representatives of all races and nations to be equal and to have equal rights. It is out of the question that some nation should be regarded as a chosen one, with special rights and so forth. The nationalists, in contrast, believe that 'their' nation has such rights.[4*] That is why a nationalist will justify representatives of his nationality, help them, and, to a certain extent, even defend swindlers and scoundrels of his nationality. Such an approach is totally alien to me. Actually, it is just the opposite of my views, as I am now going to explain. The thing is that nationalism should not be mixed up with national feeling, with the awareness of belonging to a certain nation and of this nation being in some measure near and dear to you. That does not in the least contradict internationalism. I will illustrate this by my own example. I was born in Moscow as far back as under the tsarist regime. My mother tongue is Russian. I do not know any Jewish language (Hebrew or Yiddish). I regret it but I am not good at languages and there was no need for me to learn Hebrew. I think that if I had lived in a country where there was no anti-Semitism, I would probably have been assimilated. But I knew from childhood that I was a Jew and, sometimes, a Yid. I knew about the persecution of the Jews and about pogroms in particular. Besides, there were some elements of the Jewish culture in our family. Thus, a the national feeling came into being and still remains. In my case, the most vivid manifestation of this feeling is surprising for myself. That is to say, when I come across (not necessarily in person) a Jew who is a scoundrel or some other reprehensible person, I am ashamed. I feel responsibility. Another sentiment is less strong but it also exists—I am pleased when it turns out that some very worthy person is a Jew. For instance, I am glad that Einstein was a Jew. I am unable to account for this feeling but I do not see anything shameful in it. It is important that the love or, putting it more correctly, some sense of belonging to one's nation should not be, so to speak, at someone else's expense—in particular, at the expense of other nations. So I am glad that the State of Israel exists, where Jews from all over the world have found a place to live and can find defense and shelter. But I myself have never wanted to move to Israel, though I have had such offers and on good conditions, in spite of my age. I would also like to illustrate what has been said by the example of A I Solzhenitsyn, as it is more interesting than my example. The thing is that I repeatedly heard that he was accused of anti-Semitism. And indeed, in some of his writings he perhaps overemphasizes the fact that this or that scoundrel is a Jew. But here it is very difficult to say what is the right proportion. Sometimes representatives of persecuted nations (certainly not only Jews) are so sensitive that they cannot stand the mere mention of them. Anyway, since I have been acquainted with Solzhenitsyn for about 40 years (though it is far from being a close acquaintanceship), it is interesting for me to know his frame of mind, although I have never believed in his anti-Semitism. It was from this point of view that I read his latest book, *Rossiya v Obvale* [*Russia is Crumbling*] (1998). It was not an easy task for me—the language is archaic, which I do not like. But it is important that I have come to a firm conviction that he is not an anti-Semite. What he manifests exactly is national feeling, in this case, regarding the Russian nation. It is his nation, he loves it and suffers from seeing that his nation lives badly. However, it could live well—it would seem that there are all the prerequisites for that, both in respect of the natural resources of the country and of the level of its culture and education (true, here certain reservations should be made but let us leave them aside for now). But Solzhenitsyn, as far as I understood, does not in the least blame others, Kondratenko's mythical Zionists in particular, for the sufferings of the Russian people. He criticizes and suggests ways out but not at the expense of other peoples.[5*]

So, what is the future of the Jews in Russia? First, Stalin's and the subsequent state anti-Semitism in Russia has done great damage to the country. As has already been said, I certainly do not think that Jews are better than the others. But it is a fact that, for certain historical reasons, the Jewish population has a high percentage of educated people, who are useful for the country. Such also was the situation in Germany before the Nazis came to power—it is common knowledge what a huge detriment to Germany Hitler's anti-Semitism was. For the same reason, Russia has already lost many citizens who were able to work, to emigration to Israel and other countries. And now, if bans on going abroad do not return (which, I hope, is hardly possible), there are two possible prospects. Either our authorities will start to observe our laws forbidding racial discrimination and will curb kondratenkos, mokashovs and the like, as, for instance, the newly made Kursk governor, Mikhailov—or the country will lose even more capable people. I do not understand these so-called patriots. The country is in a difficult situation, so why not try to really help it by your work. But instead they say that Zionists are stopping them—and all the rest, as in the popular song about a beautiful marchioness, is quite well.[6*]

Question 5. You have not yet said anything about the position of science in Russia, which was among the problems enumerated at the beginning.

Answer. This is a major topic of conversation, deserving undivided attention. Here I will confine myself to the remark that I do not share panic assertions about the death of Russian science. Indeed, there are many difficulties and, first of all, an acute shortage of finance. But we still have plenty of highly qualified scientists and our first-rate higher educational establishments are not at all dead. That is why, if we get over the economic crisis soon enough, and I do not understand why this is impossible, Russian science will very quickly be able to flourish again.

Comments

1*. The article (slightly abridged) was published in the newspaper *Literaturnaya Gazeta* (no 47, 22–28 November 2000).

2*. In *Izvestiya*, astrological forecasts are no longer published. Some articles and interviews criticizing false science have been printed in different journals. The book *'Uchenye' s Bol'shoi Dorogi* [*Highway 'Scientists'*] by

E P Kruglyakov has been published in Moscow, at the Nauka Publishing House, in 2001. I can also mention my speech at the session of the RAS Presidium (2001 *Vestn. Ross. Akad. Nauk* **71**(8) 702 [2001 *Herald Russ. Acad. Sci.* **71**(4) 355]) where I reported, among other things, on the letter from E B Aleksandrov, V L Ginzburg, and E P Kruglyakov to President Putin with a proposal for a package of measures in order to combat pseudoscience and corresponding pernicious phenomena. At our request, the RAS President, Academician Yu S Osipov, has in person handed this letter over to President Putin's administration as early as March 2001. However, up to the present time (I am writing this in December 2001), there has been no response. In September 2002, we still do not have any reply (the same in May 2004).

3*. I would like to recall that this 'Interview with myself' was written before the NTV channel changed hands. I am glad that I was able to feel in good time the insincerity of some executives of the 'old' NTV who were defending not so much freedom as their personal interests. Up till now I have not noticed that the 'new' NTV have limited their criticism of the authorities.

4*. I have come across another definition of nationalism, namely that a nationalist is just a person safeguarding the right of 'his' nationality to self-determination (under certain conditions), to some cultural autonomy, and so forth. Such nationalism, in fact, does not differ from national self-identification and the national feeling described in my text. Unfortunately, in practice, it is difficult to draw a line between such 'innocent' nationalism and an 'aggressive' nationalism professing the superiority of one's nation over the others.

5*. In the middle of 2001, Solzhenitsyn published his book 2001 *Dvesti Let Vmeste* [*Two Hundred Years Together*, Part I] (Moscow: Russkii Put'), exploring the history of the Jewish question in Russia. This book immediately gave rise to a discussion of the question of whether Solzhenitsyn is an anti-Semite. By the by, in one article he was called an 'enlightened anti-Semite' but I do not understand what this means. Basically, as in most other arguments, first of all it is necessary to agree about the definitions. It seems to me that anti-Semitism—the following definition is certainly not original—is a particular case of racism denying the equality of all or some races and nations. For such a racist/anti-Semite, a Jew is bad irrespective of his/her individual qualities, social status, religion and so on, merely by virtue of his/her being Jewish. Such were Hitler and some of his myrmidons. I have difficulty understanding and defining all this more exactly, as even contemporary genetics, as far as I know, is not yet able to tell a Jew from an Arab, for instance. But 'zoological' anti-Semites think that they can somehow manage to do this.

Having written this paragraph, I felt that I am not able to go deeply into the topic of anti-Semitism, its roots, essence, forms, and so on. I will confine myself to the remark that if someone is not very keen on Jews, it does not necessarily mean that s/he is an anti-Semite. Secondly, as has already been said in this article, in Solzhenitsyn's book *Rossiya v Obvale* I have not seen any signs of anti-Semitism. Meanwhile, if he were an anti-Semite, in this book he would have found quite a few occasions to show himself as such. As for the book *Dvesti Let Vmeste*, I have merely looked through it, and not read the whole of it. It was simply because I read slowly, my eyesight is already rather bad, and I have a lot of work to do and interesting material to read. But a detailed examination undertaken by V G Kadzhaya (published in 2001 *Mezhdunarodnaya Evreiskaya Gazeta* [*International Jewish Newspaper*] nos 38 and 39 of October 2001 and in the journal *Novoe Vremya* [*New Times*] of 6 January and 24 February 2002 conclusively shows a number of mistakes and obvious tendentiousness. However, an incorrect treatment of the role of Jews in Russian history and in any other issue in general is not in itself convincing evidence of anti-Semitism. A I Solzhenitsyn is not a 'zoological' anti-Semite. And as for more delicate questions arising, in particular, in the light of the book by V N Voinovich 2002 *Portret na Fone Mifa* [*Portrait against the Background of a Myth*] (Moscow: Eksmo), I do not have clear answers.

Here it is perhaps worth explaining why I am interested in Solzhenitsyn and write about him. It may be clear from my article "Zametka ob A I Solzhenitsyne, A D Sakharove i 'Bokovom Vetre'" (Notes on A I Solzhenitsyn, A D Sakharov and the 'Crosswind') (V L Ginzburg 1995 *O Fizike i Astrofizike* [*On Physics and Astrophysics*] (Moscow: Byuro Kvantum) p 499) [2001 *The Physics of a Lifetime: Reflections on the Problems and Personalities of 20th Century Physics* (Berlin: Springer) p 507], the content of which I will not set forth here. When Solzhenitsyn came back to Russia, in 1994, I sent him a letter asking his permission to publish the previously mentioned article, or, rather, a note and he gave his consent by telephone. Further on, we occasionally exchanged some literature by mail and spoke over the telephone and, as I was one of those on whose initiative he was elected, in 1997, to the Russian Academy of Sciences, we have had several chance and very brief meetings at the Academy. I am giving all these small details certainly not in order to show that I do not have anything to do with Solzhenitsyn nor, conversely, to recall my old acquaintance with him. I only want to explain that I do not have any interesting information about him and his views which would not be known to readers of his books.

6*. As one of my readers has told me, at present the essence of Zionism is reflected in a legal regulation now in force in Israel, which gives to all Jews and persons of Jewish origin, as well as to some of their relatives, the right to move to Israel and receive shelter there (the exact conditions of the repatriation are unknown to me).

By the by, I can say that some topics raised in this "interview" are also tackled in my notice "Neskol'ko zamechanii ob ateizme, religii i evreiskom natsional'nom chuvstve" [*Several observations on atheism, religion, and the jewish national feeling*] published in the newspaper *Evreiskie Novosti* [*Jewish News*] no 4, August 2002.

7*. After the Russian edition of this book was published, a lot of material appeared concerning Part I of Solzhenitsyn's book 2002 *Dvesti Let Vmeste* [*Two Hundred Years Together*] and, what is still more important, its Part II was published (Moscow: Russkii Put'). Entitled *V Sovetskoe Vremya* [*In Soviet Times*], it recalls this period. Speaking of publications connected with Part I, I would first of all like to note the new edition of the half-forgotten work (notes) 'Evrei v Rossii (Neskol'ko zamechanii po evreiskomu voprosu)' [*A Jew in Russia (Several Observations on the Jewish Question*] by the 19th century renowned Russian writer N S Leskov.[2] As V G Kadzhaya justly stresses,[2] Leskov's notes convincingly show that some points of the history of the Jews in Russia have been misinterpreted by Solzhenitsyn. As far as Part II of his book is concerned, it has also been subjected to sharp criticism in a number of articles, of which I shall mention here only the articles by V G Kadzhaya published in the newspaper 2003 *Evreiskie Novosti* nos 2, 3, and 4 (see also the articles by the same author in the 2003 *Mezhdunarodnaya Evreiskaya Gazeta* nos 11–12 and 13–14). Some evident misrepresentations of historical truth by Solzhenitsyn, which were mentioned in these articles, are simply beyond my understanding. A more detailed discussion of Solzhenitsyn's attitude to anti-Semitism and the Jewish question in Russia would probably be out of place here and especially merely in the commentary to article 29. In general, I regret having touched upon these issues in the article published in 2000 before the publication *Dvesti Let Vmeste*. However, following my principle which I have declared many times, I have not discarded this article nor changed it under the influence of the new material which has appeared. I can only note that now I see that, unfortunately, I misunderstood Solzhenitsyn's position—now I would not write that "Solzhenitsyn, as far as I understand, does not in the least blame others, Kondratenko's mythical Zionists in particular, for the sufferings of the Russian people".

[2] This work written by N Leskov in 1883, together with V G Kadzhaya's articles "Pochemu evreev ne lyubyat?" [*Why are Jews not liked?*] and some other ones, has made a book published in Moscow, in 2003, by the publishing houses Most Kul'tury and Gesharim (Jerusalem).

Article 30

Answers to several questions[1*]

Question 1. Does a scientist bear moral responsibility for taking part in projects creating weapons of mass extermination? Many prominent scientists who participated in such projects have pondered over what a 'horrible, inhuman business' they were engaged in (Sakharov, Tamm). In fact, scientists worked for the dictator. By the way, Sakharov in his reminiscences questions the thesis of 'good physics'.

Answer. In my opinion, a scientist, generally speaking, bears responsibility for working on weapons of mass extermination. However, there can be no universal judgement, that is why I have inserted the words 'generally speaking'. For instance, Einstein initiated, to a considerable degree, the beginning of the work of creating the atomic bomb in the USA. Were his activities justified? Undoubtedly, they were, as he and his colleagues were aware of the terrible threat if Nazi Germany made the bomb earlier than the Allies. In this connection, I was recently indignant when I read in *Physics Today* (July 2000, p 38) that Heisenberg ventured to reproach Einstein in public for this initiation of the work. Virtually, 'the pot calling the kettle black', as Heisenberg was one of the leaders of the German 'uranium project'. And it is not at all clear yet why the German physicists were not successful (they committed several glaring scientific errors and attempts to explain these by their conscious or subconscious unwillingness to make the bomb are a moot point).[2*,6*] All Soviet physicists who I know (including Tamm and Sakharov) justified their taking part in the 'atomic project' by the necessity to catch up with the USA, i.e. to stop the USA from having a monopoly on the bomb. And indeed, when terrible weaponry is possessed not by one but by two or several countries, an 'equilibrium of fear' emerges, which stabilizes the situation to a certain extent. I myself took part to a degree in creating the Soviet hydrogen bomb (1948–53) but at that time it never occurred to me that the Soviet Union might use it in an aggressive way. I am sure that Tamm thought the same, as well as everyone with whom I chanced to speak absolutely frankly. Here it is also essential that we did not see Stalin's real worth. True, I have recently learned that a certain well-known Soviet physicist worked on the bomb although he knew what Stalin was worth. The motive for doing this was fear, which, indeed, could not be considered as unfounded. But at that time he kept silent and I do not blame him (still, I would not like to give his surname). Your colleague M Kheifets, as well as some others, does not believe that we could have been so blind[3*] but this is undoubtedly true. It would take too long to describe all that. You will understand this if you read, for instance, the book by A N Yakovlev 2000 *Omut Pamyati* [*The Whirl of Memory*] (Moscow: Vagrius) (see, for example, chapter 4). Now I understand and, of course, I have understood for a long time that Stalin was an archbandit and would have used, without any hesitation, any terrible weapon if he considered it necessary for achieving his cannibalistic goals. Mankind was extremely lucky that Stalin and Hitler did not take possession of atomic weapons first.

Coming back closer to your question, I can repeat: the measure of responsibility depends on many factors and first of all, on understanding the purpose of creating a weapon. I certainly justify the creation of a weapon for defense of a country from aggressors and bandits.

Question 2. What is the true reason for Soviet physicists' rare enthusiasm in the 1940s and 1950s—was it a thirst for knowledge, 'technical sweet', or a trivial fear for their existence?

Answer. I specially touch upon this subject in the second edition of my book *O Nauke, o Sebe i o Drugikh* [*About Science, Myself, and Others*], which is now in print.[1] A renowned Anglo-American physicist, who visited the USSR at the end of the 1950s, said to explain the enthusiasm of Soviet physicists, 'But they do not have anything else!' He meant that from the point of view of economic conditions our life was difficult, contacts with abroad were almost non-existent, and we lived in a totalitarian state. That is why such things as work and science were everything for us—serving as a safety-valve and acting as a drug. I must confess that, for a long time, I agreed with this diagnosis. But now I understand that it was the truth but by far not the whole truth. Indeed, in spite of all the drawbacks of our life, in today's Russia we can go to conferences and in general we can communicate freely with our colleagues all over the world, there is no our censorship and, on the whole, we are free. However, the majority of people no longer have our former enthusiasm. This can be explained by the sharp decrease in the financing of science and the change of attitude towards it in the country. In the USSR, physicists were held in high esteem. Going in for physics (unlike going in for the truly scientific biology) was prestigious. And now a scientist in Russia is, on average, far worse provided for than a

[1] See articles 12 and 13 in this book.

secretary in some firm and scientific work is not prestigious. That is why many scientists move abroad or go into business or so forth. This situation is true even for people who ardently love science.

Question 3. Do you condemn politicians (Truman, who gave the order, and Churchill, who sanctioned it) for using the atomic bomb?

Answer. As far as I know, without use of atomic weapon the Allies, and the Japanese as well, would have suffered tremendous losses until the end of the war. If this is really so, I do not see any reasons to reproach Truman and Churchill. However, apparently it could have been possible to use only one bomb and a warning of the possibility of dropping a second one.[4*]

Question 4. A scientist and a politician—what do these two professions have in common and why do they differ from one another? A scientist and a politician in one person—is this a nonsense?

Answer. The spectrum of scientists and politicians is very wide and I do not see any gap between them. I do not regard it as a law of nature that among politicians there are more cheats and dishonest people in general. Among scientists, too, there are many unworthy people. And as for common traits, there are a lot of them in all intellectual activities.

Question 5. Why are fundamental scientific studies necessary in our time? In the 400 years of the rapid development of science and technology, man has not become more moral and noble. Moreover, technical progress leads to the spiritual degradation of personality. Young people are reading less and articles of mass culture are supplanting the classics.

Answer. The achievements of science in 400 years have resulted in a tremendous progress in the life of society. Man has not become more moral and noble but he has eliminated many diseases and started to live longer, a lot of new possibilities have opened before him. However, the development of science is based on fundamental studies. Such studies are necessary for the development of mankind. I am convinced that certain negative aspects of the development of science are secondary and, so to speak, local. It is a big subject, which requires a detailed analysis. But in short, I consider antiscientific tendencies, by and large, to be groundless and reactionary.

Question 6. Does society need religion? Is there anything which could substitute for religion? The Moral Code of the builder of communism substituted for the Ten Commandments. In a democratic society, there are no generally accepted ethical norms, except those contained in the Bible. Soldiers in Israel swear their oath by putting their hand on the Torah. In one of your interviews, you said that Pavlov and Einstein, in fact, were not believers. However, there was another scientist, no less great, who sincerely believed in God and thought that science and religion are not in conflict. I mean Pascal. If religion is important for society, it appears that the attitude of a scientist to it should be tolerant.

Answer. I am a convinced atheist. However, I can understand agnostics and those who believe in some cosmic (universal) mind. The latter is something like deism, while atheism seems to me, in fact, indistinguishable from pantheism. But what I consider utterly absurd for a contemporary, really educated person is attachment to some religion in the literal sense, i. e. to believe in miracles like the Immaculate Conception, the Resurrection, and so on, and so forth. In ancient times, when there was no science, it was impossible to tell miracles from reality—because a miracle is something that is not confirmed by scientific analysis. It was then that religion appeared, having imbibed in its structure certain moral and ethical norms. But the latter can well be accepted even if we throw off the decrepit religious envelope. That is exactly what modern atheists do.

Numbered among them are secular humanists (including myself). This is an international movement headed by the International Academy of Humanism. For more details, see the article by Ginzburg V L and Kuvakin V A in 2000 *Vestn. Ross. Akad. Nauk* **70**(6) 483 [2000 *Herald Russ. Acad. Sci.* **70**(3) 223].[5]* Incidentally, about Pascal—he died in 1662, when the level of scientific knowledge was quite different from what it is today. Finally, about tolerance. I have never been nor am I now one of the so-called militant atheists. On the contrary, I am a convinced partisan of the freedom of conscience, which means that every person has a complete right to freely believe in God, to be an atheist, or to be an agnostic. But convicted of the fact that religion is archaic and the aspiration to propagate scientific knowledge and atheism is something quite different.

Question 7. Can society consist of creative people only? A creative person always arouses interest. However, a true creator is always a parasite living at the expense of other society members. As a rule, a creative person is terribly selfish—cherishing his work and slighting his near and dear ones.

Answer. Your question reflects some inadmissible, from my point of view, division of people into the élite and the plebeians. All people have equal rights. Those endowed with great abilities (by nature or by the 'play of genes', if you like) have no right to be parasitic on other society members but, in fact, in most cases the capable, creative people achieve success by their work rather than by slighting and exploiting others. I am convinced that no matter what abilities and talents people might have it does not give them the right to behave as if everything was allowed them. I despise such people, I can see no excuse for their selfishness, haughtiness, and so forth. Another matter, among those who are very talented there are those for whom living in the society in which they have to live in is difficult. A sympathetic attitude and attention to such people is entirely justified but it does not mean that their boorish behavior should be tolerated.

Comments

1*. The Israeli journalist V Shapiro e-mailed me several questions. The answers were published in V Shapiro's article "Unlimited responsibility" printed in the *Okna* [*Windows*] supplement to the newspaper *Vesti* [*News*] (Israel) on 12 April 2001. Given here are the questions and answers.

2*. In this connection, see Feinberg E L 2003 *Epokha i Lichnost'. Fiziki. Vospominaniya* [*The Epoch and the Personality. Physicists. Reminiscences*] (Moscow: Fizmatlit). In the article about W Heisenberg, the author expresses an opinion which differs from many other writers' opinions.

3*. See comment 1 to article 26. M Kheifets thinks, for some reason, that we understood far more than we actually did.

4*. I would like to stress once again that the situation is not quite clear to me and the capitulation of Japan might have probably been obtained merely by some demonstration of the power of an atomic explosion instead of the example of Hiroshima.

5*. See article 35 in this collection.

6*. In this connection, see the article Durrani M 2001 *Phys. World* **14**(11) 9, and also 2001 *Phys. World* **14**(12) 35; and 2002 *Phys. World* **15**(3) 7.

Article 31

Reason and faith[1*]: some remarks concerning the encyclical letter *Fides et Ratio* of the Supreme Pontiff John Paul II

The encyclical letter *Fides et Ratio* [*Faith and Reason*] by Pope John Paul II was published on 15 October 1998. Incidentally, the following day, 16 October 1998, was 20 years since Cardinal Karol Wojtyla had been elected Supreme Pontiff of the Roman Catholic Church and the previously mentioned latest encyclical (his 13th) appeared to mark that anniversary.

1. 'Faith and Reason are like two wings on which the human spirit rises to the contemplation of truth; and God has placed in the human heart a desire to know the truth—in a word, to know himself—so that, by knowing and loving God, men and women may also come to the fullness of truth about themselves' [1]. These are the opening words of the encyclical, devoted to the relationship between faith in God, religion, and theology, on the one hand, and reason, science, and philosophy, on the other. The Pontiff is 78 years old and the lengthy encyclical *Fides et Ratio* (which has seven chapters, besides an Introduction, broken into 108 paragraphs) may have been designed as his theological testament. It offers a vision of the status and content of Catholicism on the threshold of the third millennium AD. Certainly, it is not for me to judge how theologians will respond to the encyclical. As for philosophy and divinity scholars, they must undoubtedly give much attention to the encyclical letter of John Paul II, an outstanding and highly educated person.

2. I will take the liberty of making a few remarks concerning the encyclical—or, rather, in connection with it—although I am but a dilettante in philosophy and the more so in theology. However, attitudes to dilettantes depend in no small measure on the area in which they voice their views. For example, in the 60 years of my professional work as a

physicist, I have never found anything of value in the countless proposals of dabblers in physics who offer their hypotheses about the structure of matter or space and time. Of course, as the saying goes, not God but man makes pot and pan, and professional physicists are not celestial beings. Simply, in our day, the cutting edge of physics has moved far ahead and is separated from secondary-school graduates or engineers by a wide margin mined with a wealth of factual information and mathematical formulae. Negotiating this strip requires years of work even for very capable people. The study and interpretation of historical or theological texts likewise necessitates a great deal of preliminary hard work. At the same, time I am convinced that every educated person can (and even must) form his/her own views on reason and faith, atheism and religion. Perhaps what I have said is all too obvious but I am apprehensive lest I should be reproached for straying outside my field. I have done so, however, only under the pressure of circumstances.

3. The collapse of the bolshevik (Leninist–Stalinist) regime is known to have produced something of an ideological vacuum in our country. As a result of this, the 'militant atheists' ('militant non-believers') of Soviet times have given way to the Russian Orthodox Church (ROC) and other religious organizations and cults. At the same time, very many people have distanced themselves completely from all ideology, except, possibly, from thieves' cant. The ROC has been particularly successful: its spokesmen often appear on television screens and their articles and religious information in newspaper columns. As for the voices of atheists, they are hardly heard at all. Some fear to act contrary to the spirit of the times and vogue; others are probably denied the chance to voice their opinions. The only exception I know is the journal *Zdravyi Smysl* [*Common Sense*] published by the Russian Humanitarian Society, based at the Philosophy Faculty of the Moscow State University. But while circulation of the journal is only 1000 copies, an Orthodox metropolitan broadcasts on the ORT (Public Russian Television) channel every week and addresses an audience of many millions.

In such a situation, it was evidently decided in certain clerical quarters that atheism had been defeated in Russia. Be that as it may, the newspaper *Literaturnaya Gazeta* [*Literary Newspaper*] of 8 April 1998, printed an interview with an ROC clergyman who declared that "an atheist is today an extremely rare creature listed in the Red Data Book". The absurdity of such a statement is obvious but, I think, not to everyone. Therefore, E L Feinberg and myself felt that such an assertion should not be left unchallenged [2]. For certain reasons I believed it to be my duty to contribute two more articles to the newspaper *Poisk* [*Quest*] defending the atheistic outlook [3, 4].

4. The latter of these articles [4] is actually a review of the book *Science and Christian Belief: Theological Reflections of a Bottom-Up Thinker*

by the physicist and Anglican clergyman J Polkinghorne [5]. Readers of this book learn of the existence of the theological trend called *natural theology*. Its aim is to try to achieve a certain cognition of God via reason or on the basis of general experience, for science and theology are share the conviction that there exists a certain truth of the nature of things that can be discovered and comprehended .... Of course, science and religion deal with different aspects of truth that relates to a single world—the world of human experience. Scientific research is concerned with objective phenomena, which can be verified experimentally whereas religion addresses the suprapersonal reality of God, i.e. a rendezvous where research must give way to trust and where the human response consists of obedience, besides understanding [5, p 6]. Natural theologians recognize science as enabling us to comprehend the grand picture of the physical universe and its history. Nonetheless, Polkinghorne says, he is even more impressed by what seems to him even more significant religious revelations making it possible to behold the Divine Reason and Will that lie beyond anything that science can reveal [5, p 6]. A new book by Polkinghorne on the same subject has appeared recently under the title *Belief in God in an Age of Science* [6]—this speaks for itself.

5. An interesting review of this book [6] has been written by the cosmologist G Ellis [7]. The reviewer believes that atheists—scientists and philosophers—demonstrate a *great rigidity* when they express an absolute conviction that they are right; because the issues are metaphysical, definite scientific assertions about them are out of place. However, representatives of natural theology (both physicists and philosophers) are, in the opinion of Ellis, flexible and have a better understanding of the relationship between science and religion. Such an opinion seems to me to reflect a misunderstanding of the essence of atheism.

An atheist denies the very existence of God, of anything outside the realm of nature. Needless to say, such a conviction is an 'intuitive opinion' [8, 9], which cannot be proved. Faith in God is likewise an intuitive opinion. However, there is a substantial difference between these two opinions: whereas the atheist's opinion is based on science, on the study and analysis of natural phenomena or experiments, religion accepts the possibility of miracles and is based on certain miracles (i.e. propositions that cannot be verified and are in conflict with scientific data). So, what kind of flexibility should the atheist display? What s/he can do is to criticize theological exercises and endeavor to somehow reconcile religion and science. The theologians are in a different position since religious dogmas provide them with a certain rigid framework. Accordingly, theologians have somehow to adapt themselves to ever-advancing science. Of course, many centuries ago religion could afford to ignore (or almost ignore) the arguments of science which was then in infancy. If there is no science, the very notion of a miracle loses all meaning: everything is possible and one can believe in anything. The door is thereby opened to dogmaticism, fabrications, and blind faith.

6. Hence, Tertullian's well-known formula of the religious doctrine: "It is certain because it is impossible". But this dates back to the second or third century AD. Equally typical was Cardinal Baronius' remark: "The Holy Spirit's intention is to teach us how to ascend the heavens—not how the heavens move". But already in their battle with Copernican principles and especially with Galileo (1564–1642), the church dignitaries had to invoke sophisticated arguments in their defense of religious dogmas (highly interesting remarks were made in this context by Cardinal Bellarmine, the 'admonisher' of Galileo; see, for example, [10–12] and the references therein). How theologians 'interacted' with scientists in those days is clearly evident from Galileo's message to Christine, Grand Duchess of Lotharingia:

> Professors of theology must not arrogate to themselves the right to administer by decree in professions that are not within their domain, for opinions about phenomena of nature cannot be imposed upon the natural scientist.... We advocate new teachings not to sow discord in people's minds but to enlighten them; not to destroy science but to place it on a sound foundation. Our opponents, on the other hand, dismiss as false and heretical everything they cannot refute. These hypocrites employ pious religious zeal as a shield and humble the Holy Scriptures by using them as a tool for achieving their personal aims..... To direct professors of astronomy to rely on their own resources in seeking defense against their own observations and conclusions as if all these were mere deceptions and sophistry would mean confronting them with a more than impossible demand—it would be tantamount to ordering them not to see what they do see, not to understand what they understand, and to infer from their studies the very opposite of what is obvious to them.

Incidentally, these words had a very topical ring throughout almost the entire recent Soviet period—naturally, if we substitute professors of Marxism for professors of theology and Marxism–Leninism for the gospel.

7. In the four centuries that separate us from the days of Galileo and Kepler, science has made strides that are difficult even to appreciate. Here is just one example: as recently as 1618 even Kepler, in his book *Epitome Astronomiae Copernicanae*, expressed the view that there exists a sphere of stationary stars, which "consists of ice or crystal". Or take the advances made since then in physics, biology, medicine, and engineering! We live in an entirely different world compared with the Middle Ages, let alone the beginning of the Christian Era.

Meanwhile, religion (specifically, Christianity) has actually remained just as it was two millennia ago. To be sure, a great deal of effort (as we would now say, man-hours or rather man-years) has gone into debating various theological issues, into disputes (at times involving the shedding of blood) over, say, how to cross oneself: with two fingers or three and in what direction. Christianity has divided into several confessions, the most important of which are Catholicism (850 million people), Protestantism (400 million), and Orthodoxy (150 million).[1] All these confessions—and this is particularly true of Protestantism—are far from being monolithic, and there is struggle within them, at times quite fierce. For example, the differences in theological views and styles among some Orthodox churchmen are simply astonishing.[2]   Incidentally, believers in God or, to be exact, religious people (i. e. people professing this or that faith, e. g. the Christian faith as opposed to pantheism, deism, etc) should, as I see it, be surprised by the diversity of creeds and trends even within one and the same religion. If God existed, he would seemingly instill the same faith and the same ideas in the hearts of believers and would not allow gory religious wars, the Inquisition, or the breakup of Christianity into a variety of confessions and sects. Of course, this is not proof that God does not exist—such an intuitive opinion cannot be proved at all [8]. However, one cannot but be struck by the contrast this presents to the situation in science which in our days is marked by profound unity: there is a single world science which recognizes no national frontiers, let alone ethnic boundaries. To be sure, in science too there sometimes arise differences and debates but they are resolved in the process of development leaving far behind various pseudoscientific constructions, such as astrology (for a more detailed discussion of the status of science and pseudoscience see, for example, [16]).

8. But it is time to return to the main subject of this article: evolution of the views of the church on the relationship and links between religion and science, faith and reason. According to a legend in the New Testament, Jesus Christ gave no answer to Pontius Pilate's question, "What is truth?". The Christian Church decided that it knew the answer and for centuries tried to dictate to science her own understanding of what is true and what is false. In the foregoing, this was illustrated by the conflict of the church with Galileo who was convicted by the Inquisition in 1633. Earlier, in 1600, Giordano Bruno was burned at the stake—likewise for disagreement with 'truths' proclaimed by the church. (The year 2000 will thus see not one but two major anniversaries in the history of Christianity.) It is only John Paul

---

[1] The figures in parentheses indicate the number of followers of a given confession (specifically, the number of people baptized; of course, the figures are very approximate) [13].

[2] To illustrate this point, let me cite a book by Antonii, Metropolitan of Surozh [14], an educated and humane person, and a letter full of enmity and hatred by another Orthodox metropolitan, likewise Antonii [15].

II who has turned a new page in the history of Catholicism, by the way, not without the influence of the Italian physicist A Zichichi and other European physicists [17]. The Pope reviewed Galileo's 'case' and exonerated him in 1992.

But let us not dwell endlessly on the past. It is the stand taken by the church today—and, probably, the stand it will take in the foreseeable future-as reflected in the encyclical *Fides et Ratio* which, as far as I can see, continues the course charted by the Second Vatican Council (1962–65). Here I can mention only the principal propositions of the encyclical.

9. The word used in the encyclical most frequently is probably *truth*. In the words of the Pontiff, "One may define the human being, therefore, as *one who seeks the truth*" (paragraph 28—here and hereinafter the figures refer to paragraphs in [1]). The human being is capable of learning the truth and

> The more human beings know reality and the world, the more they know themselves in their uniqueness, with the question of the meaning of things and of their very existence becoming ever more pressing. This is why all that is the object of our knowledge becomes a part of our life. The admonition *Know yourself* was carved on the temple portal at Delphi, as testimony to a basic truth to be adopted as a minimal norm by those who seek to set themselves apart from the rest of creation as 'human beings', i. e. as those who 'know themselves'. (Paragraph 1)

The church acknowledges the role of reason (specifically, of philosophy) in learning the truth but considers that a full understanding of the truth about human beings and the reality (visible and invisible) that surrounds them cannot be achieved by reason alone. This is its main, central postulate.

The Pope discusses and criticizes various philosophical trends (eclecticism, historicism, scientism, pragmatism, nihilism, etc), which cannot, in his opinion, be of help to human beings in finding correct answers to questions concerning life or in explaining faith. But there is also (or should be) a 'good' philosophy, which will be "an indispensable help for a deeper understanding of faith and for communicating the truth of the Gospel to those who do not, yet know it" (paragraph 5). However, an often repeated thought in the encyclical is the already mentioned proposition that philosophy (reason) alone is not enough—there must also be revelation for "the truth attained by philosophy and the truth of revelation are neither identical nor mutually exclusive" (paragraph 9). As for God's intentions, they are defined as follows: "As the source of love, God desires to make himself known; and the knowledge which the human being has of God perfects all that the human mind can know of the meaning of life" (paragraph 7). At the same time,

Based upon God's testimony and enjoying the supernatural assistance of grace, faith is of an order other than philosophical knowledge which depends upon sense perception and experience and which advances by the light of the intellect alone. Philosophy and the sciences function within the order of natural reason; while faith, enlightened and guided by the Spirit, recognizes in the message of salvation the 'fullness of grace and truth', which God has willed to reveal in history and definitively through his Son, Jesus Christ. (Paragraph 9)

To avoid misunderstandings, it should be pointed out that the scientific approach does not rule out either the recourse, in certain cases, to extralogical, intuitive notions [8]. However, they have nothing in common with the "supernatural assistance of grace" or revelation which, according to religious thinking, is contained in the Bible or the Koran.

10. Hopefully, the meaning of the encyclical is already clear from the foregoing. The church today advocates the necessity and unity of two modes or directions of cognition: scientific based on reason, experience, and philosophy and religious based on faith. The two modes are not opposed to one another:

This truth, which God reveals to us in Jesus Christ, is not opposed to the truths which philosophy perceives. On the contrary, the two modes of knowledge lead to truth in all its fullness. The unity of truth is a fundamental premise of human reasoning, as the principle of non-contradiction makes clear. Revelation renders this unity certain, showing that the God of creation is also the God of salvation history. It is the one and the same God who establishes and guarantees the intelligibility and reasonableness of the natural order of things upon which scientists confidently depend, and who reveals himself as the Father of our Lord Jesus Christ). (paragraph 34)

11. The Pontiff's encyclical is a document of Catholicism. We have seen, however, that the Anglican Church believer Polkinghorne, adheres to much the same views [5, 6]. Since in Russia the most widespread religion is Orthodoxy in which the ROC plays the leading role, we cannot but be interested in its attitude to science. On this score, I am unfortunately familiar only with a brief announcement concerning the church hearings on the subject "Faith and Knowledge: Science and Technology at the Turn of the Century" [18]. Patriarch Aleksii II, in his speech there said *inter alia*: "The criterion for separating the wheat from the chaff in this complicated area [i.e. science and technology—VLG] can be the spiritual experience and spiritual guidance of the church". And further:

Amidst his labors, the scientist must be appropriately humble and reverential before God bending his efforts to facilitate, to the best of his ability, the realization of God's design for the world and man . . . People producing the most advanced scientific learning and newest technologies are in need of firm support, spiritual tradition of Orthodoxy.

It follows from this statement by the Patriarch and from certain other information at my disposal that the ROC has not taken any aggressive stand toward science. Nor has it, as far as I know, any cause for repentance toward scientists—it did not condemn Galileo or burn Bruno at the stake. However, when the ROC was recently confronted with demands of modern science, it turned away from science. I refer to the identification of the remains of the tsar's family. As is known, experts in genetics reliably established the authenticity of those remains—in full agreement with other expert studies. But the ROC declined to recognize the validity of scientific assessments. Undoubtedly, it was motivated by internal considerations (possibly, rivalry with the Orthodox Church abroad, etc) but this disregard for scientific data is nevertheless highly symptomatic.

As for the Patriarch's claim that the Orthodox Church is in a position to help scientific research and scientists by its 'spiritual experience' etc, I have no information on this score. I am convinced that Orthodoxy, like any other religion, is alien to science and any benefit here is out of the question. There are, however, other opinions on the subject. Specifically, with respect to Orthodoxy, such opinions are held by Academician Yu S Osipov, President of the Russian Academy of Sciences [18, 19], and Academician V E Fortov, former Minister for Science and Technologies [18]. It would be inappropriate to engage in polemics with them in this article. Let me just say that the major Christian churches—the Catholic, the Protestant (to which the Anglican may likewise be referred), and the Orthodox—have today adopted a common position with respect to science: science (reason) is acknowledged but, at best, as something enjoying an equal status with faith in God and the dogmas of Christianity.[3] In plain words, the church proposes to science to 'be friends' and even to cooperate. Such cooperation is possible within certain limits for the church is defending 'its' miracles, protecting its territory, and combating all charlatanry, superstition, astrology, diabolism, and the like. And religious miracles are not as harmful to society and people as is recourse to quacks, astrologers, etc.

12. Such a situation should suit believers but for an atheist there is no common ground here for discussion since s/he denies the very existence of God and deems it impossible to profess a religious faith. At this point, however, there is a need to clarify some concepts and terminology. Faith

---

[3] Here, as probably in several other instances, I simplify things (for a more detailed discussion of these matters, see [20]).

in God is taken to mean the 'intuitive opinion' whereby, apart from nature and the world around us, there exists something else. This may be some absolute intelligence which does not fit into anything mundane but which created the Universe (somewhat conventionally we can speak here of deism). Sometimes the term *believers* also embraces the pantheists who identify God with nature but I fail to see in what ways pantheism differs from atheism.[4]

On occasion, I have encountered the term *cosmic religion*. This was the term employed by the great Einstein who noted that this "cosmic religious feeling knows neither dogmas nor a God created in man's own image"; and that "it does not lead to any consummate conception of God at all or to theology".

Einstein was an atheist and used religious terminology only in a conventional sense. For example, he wrote: 'I simply cannot find a better word than "religious" to describe faith in the rational nature of reality... What the deuce is it to me that priests make money by playing on this feeling?' (All the quotations from Einstein's writings are taken from [8], where references to the originals are also given.)[2*]

Finally, there is the term *theism*—faith (again an intuitive opinion) in the existence of God who created the Universe and humans, who intervenes in worldly life, and who is capable of performing and who does perform miracles. Theism specifically covers Christianity, which believes in the holiness of the Scriptures, the resurrection of the dead, an afterlife, etc. Christianity, like Judaism and Islam, is a religion. Needless to say, a religious person is a believer but not *vice versa*: pantheism, deism, etc are not theism or religion (at any rate, according to the previous definition).

13. Going to church or, say, to a synagogue is regarded as a sign that a person is religious and believes in God. But, generally speaking, such a conclusion is mistaken. A striking example of this was provided by the great Russian physiologist I P Pavlov. He attended church services systematically and, moreover, addressed a letter to the Council of People's Commissars, pleading that the Cathedral of the Holy Trinity in St Petersburg not be demolished; he gave up his chair at the Military Medical Academy in protest against the expulsion of clergymen's children from the student body, etc. He would thus appear to have been an Orthodox believer but this reputation was based on false information. In fact, Pavlov "was, of course, a complete atheist and could not have been anything else". This is a quotation from the reminiscences of M K Petrova, Pavlov's closest associate and friend.[5] Petrova also quotes Pavlov's following words:

---

[4] To avoid complicating the picture, I shall not touch upon agnosticism (positivism). This is, in its way, a consistent outlook but an agnostic cannot be called a believer in God.

[5] Unfortunately, the reminiscences of this remarkable woman, which have been kept in the Party Archives since 1949, were not published because of a ban imposed 'personally'

"The human mind seeks the cause of all things, and when it grasps the final cause, it is God. In its pursuit of the cause of everything, it goes so far as to invoke God. But I do not believe in God, I am a non-believer." Pavlov went to church "not because of a religious motivation but because of pleasant contrasting emotions. As the son of a churchman, he had been very fond of this holiday [a reference to Easter—VLG] in childhood. He attributed this fondness to particularly joyous feelings during the festive days that followed Lent." As for Pavlov's defense of believers and the church, this was prompted by quite understandable considerations of justice and freedom of conscience, by a protest against bolshevik barbarism.

It is thus quite clear that not only religious people attend places of worship. Nor is it only deeply religious people, devoted to the canons of the church, who perform various religious rites and pray. My own father sometimes prayed and even went to the synagogue but when I, in the arrogance of youth, demanded an explanation and proof of the existence of God, his reply was very simple: "When I pray, I recall my parents and my childhood; I want some distraction from the hardships of this life." And, mind you, he was an educated person (an engineer of pre-revolutionary, custom-made 'production'). What then about people of scant education who bear a heavy load and set all their hopes upon God alone.

14. This article is not a scholarly treatise—in fact, I even find it hard to define its genre. Be that as it may, I shall take the liberty of writing about my own views, too. I am not a 'militant non-believer' and never was one. What is more, I realize that faith in God and religion can be of help to people, especially at difficult moments. Religion can also help to strengthen morals and the observance of ethical standards. This, above all, dictates the need for freedom of conscience, i. e. freedom to believe or not to believe in God, freedom to choose one's creed (of course, this does not apply to some kind of fanatical cult), and, naturally, freedom to adopt other views (atheism or agnosticism). But this also implies the requirement that the church be separated completely from the state.

In Soviet times, the Constitution contained such a provision but, in reality, the state oppressed believers and hampered them in discharging church rites. In post-Soviet Russia, the separation of the church from the state is likewise proclaimed in the Constitution but, once again, the state is violating the Constitution, albeit 'with the opposite sign'—by propagating religion and patronizing it, especially the ROC. This takes the form of customs privileges, the transfer of property (specifically, of former churches long ago turned into museums), and the restoration of demolished churches (e. g. Christ the Savior Cathedral). When the cathedral was blown up, this was, undoubtedly, an unforgivable act of barbarism. But its restoration

by M A Suslov. Only in 1995, were selections from these reminiscences published [21] and it is from these that I quote. I hope that Petrova's reminiscences will be published in full—this will be more than justified.

funded by no means by believers alone but also, indirectly, by an enormous amount of money collected from the entire population, is in my view likewise impermissible at a time when people are impoverished, when teachers and doctors are going hungry. Or what can one say about episodes of the 'consecration' of new buildings and barracks shown so often on television or of religious programs on state radio and television? Meanwhile, atheistic enlightenment has been completely forgotten.

All this is, however, a different topic. Here, I do not hesitate to say that I often envy believers—they find it easier to 'write off', as the will of God, instances of flagrant injustice and all the bitterness of life that we experience daily and observe in our society. They find it easier to think of the hardships of old age and inevitable death. Thrice blessed is he who believes (although I am not sure I am using these words in their usual meaning). Intelligence, however, has been bestowed upon humans not to succumb to emotions or be guided by superstitions or the timeworn beliefs of the hoary past. My acquaintance with theology has served only to fortify my atheistic convictions, i. e. the intuitive opinion, that there exists only Nature together with the laws governing it which the intellect—and science under its direction—comes to know or, at any rate, seeks to know.

15. The enormous advances made by science in understanding the world around us and, at the same time, the realization that there is so much that we still fail to understand can suggest (and actually suggest) the idea of the existence of some absolute intelligence or, if you like, God. While repeating and endorsing the famous words of Laplace that he 'needs no such hypothesis', I nevertheless understand, to a certain extent, those who adhere to such a faith in an abstract God. Evidently, this is deism, although the name hardly matters. But a theistic faith in a God who intervenes in earthly affairs, who performed and performs miracles, belief in an afterlife and in the holiness of the Bible, etc appear to me to be simply a survival of ancient and medieval times. How all this can be believed on the threshold of the 21st century is beyond my comprehension, if we are speaking of educated men and women. I also absolutely fail to understand how theists can explain the multiplicity of religions (see the foregoing) and, most important of all, retain faith in an Almighty God who tolerates genocide, gazes with indifference at the bestiary bared teeth of fascism and bolshevism, and at the suffering of many millions of people. Theologians have 'explanations' for all this but these are scholastic exercises. At any rate, that is how I see even the views formulated on this score by John Paul II [22]. Theism and religion are incompatible with the scientific outlook or with scientific thinking (as discussed in somewhat greater detail, for example, in [3, 4]). Only science can bring one closer to the truth; atheists are convinced that cognizance of the truth through revelations or the reading of holy books (the Bible, etc) or the writings of church dignitaries is utterly impossible.

16. That there are still many believers is due primarily to the fact that the vast majority of the six billion people inhabiting the Earth are uneducated and far removed from science. To see planes flying overhead, to listen to the radio, and to watch television is not enough to become part of modern civilization. Furthermore, even those whom we call intellectuals have, on the whole, a very one-sided education and their understanding of scientific knowledge is often still at a medieval level. This is aptly illustrated by the fact that even the question of why the seasons change very often elicits a wrong answer [2]. The stratum of broadly educated people is very thin. Significantly, a survey conducted in the United States in 1996 among people listed in a certain register as scientists showed that believers made up about 40% of the respondents [23]. At the same time (in 1998, to be exact), only 7% of the members of the US National Academy of Sciences surveyed said that they were believers [24].

The more educated a person is, the less likely he or she believes in God. The Pope hopes [1] that men and women, on the basis of faith in God, will come to an ever fuller understanding of what is truth. I am convinced of the opposite: that, with the passage of time, the crisis of religion will only become more acute and that, with the triumph of education and science, fewer and fewer people will believe in miracles, the sacraments, etc. In the past 400 years, humanity has made great progress in this direction. But one cannot fail to see that lately this process of discarding prejudices, pseudoscience, and religion has slowed. Unfortunately, I see no grounds for particular optimism in this respect, especially in Russia. But that does not mean that atheists should sit with arms folded: it is their duty to promote atheistic enlightenment and to oppose the growth of clericalism.

17. John Paul II begins his encyclical, as stated earlier, by likening faith and reason to two wings "on which the human spirit rises to the contemplation of truth". To an atheist, such a metaphor appears unfounded, since reliance on one of those wings, faith in God, cannot bring an understanding of truth. If the nevertheless attractive image of a bird is to be used as a symbol of progress, the following formula may be proposed: reason and good will are the two wings on which human civilization and culture will soar aloft. The role of reason is clear but it is insufficient to prevent human society from straying from the correct road and, say, lapsing into totalitarianism. A will, a strong good will, is also needed to protect the fruits of reason (science), democracy, freedom, and progress.

Comments

1*. This article was published in 1999 *Vestn. Ross. Akad. Nauk* **69**(6) 546 [1999 *Herald Russ. Acad. Sci.* **69**(4) 271] and also in 1999 *Zdravyi Smysl* no 13 51. The editor of the latter publication substituted the word 'will' by 'good will'; I consented to the change. It should be recalled that the article was written in 1998 when the Pope was 78 years old as mentioned in the text.

2*. I wish to emphasize that, by calling Einstein an atheist, I expressed my understanding of pantheism as essentially equivalent to atheism. Einstein did not call himself an atheist (see [25]). When asked if he believes in God, Einstein once answered (in 1929): "I believe in Spinoza's God who reveals himself in the harmony of what exists but not in a God concerned about the people's fate and affairs". See also article 33 for Einstein's views of religion.

It must be noted that I have recently published an article in defence of atheism in a popular science magazine [26].

3*. The article in [26] contained an appendix offering the readers a short questionnaire that they were asked to fill in. A total of almost 200 completed forms were returned; and the results of their analysis were made public in 2001 *Nauka i Zhizn'* [*Science and Life*] no 9 16. Almost 60% of the respondents labelled themselves as atheists; 20% stated that they believe in God, profess a religion, attend church services from time to time, etc; 10% believe in God without being affiliated with any particular creed; 10% are agnostics. Finally, only six persons claimed to be 'militant non-believers' and insisted on the necessity to combat religion (without suggesting repressions and violation of the liberty of conscience). Atheism and religion make up subjects of the following three articles (32–34). See also my paper 2003 *Vestn. Ross. Akad. Nauk* **73**(9) 816 [2003 *Herald Russ. Acad. Sci.* **73**(5) 560].

References

[1] 1998 Encyclical Letter *Fides et Ratio* of the Supreme Pontiff John Paul II to the Bishops of the Catholic Church on the Relationship between Faith and Reason (Summarized briefly in the Russian Catholic newspaper *Svet Evangeliya* no 37 11 October)

[2] Ginzburg V L and Feinberg E L 1998 We, atheists, are not so few *Literaturnaya Gazeta* no 22 3 June (Without the cuts made by the paper, the article appeared under the title On atheism, materialism, and religion in the journal *Zdravyi Smysl* no 9 54 (1998))

[3] Ginzburg V L 1998 Faith in God is incompatible with scientific thinking *Poisk* nos 29, 30

[4] Ginzburg V L 1998 Once again about religion and science *Poisk* no 38; 1999 Not faith but reason and will *Poisk* no 35

[5] Polkinghorne J C 1994 *Science and Christian Belief: Theological Reflections of a Bottom-Up Thinker* (Gifford Lectures for 1993–1994) (London: SPCK) [Translated into Russian under the title 1998 *Vera Glazami Fizika* (*Bogoslovskie Zametki Myslitelya "Snizu-Vverkh"* (Moscow: Bibleisko-Bogoslovsk. Inst. Sv. Apostola Andreya)]

[6] Polkinghorne J C 1998 *Belief in God in an Age of Science* (New Haven, CT: Yale University Press)

[7] Ellis G 1998 Are science and religion compatible? *Phys. World* **11**(9) 49

[8] Feinberg E L 2004 *Dve Kultury. Intuitsiya i Logica v Iskusstve i Nauke* [*Two Cultures*] expanded 3rd Russian edn (Fryazino: Vek 2) [German translation 1998 *Zwei Kulturen* (Berlin: Springer)]

[9] Feinberg E L 1997 Science, art, and religion *Vopr. Filos.* no 7 54

[10] Ginzburg V L 1979 The heliocentric system and the general theory of relativity (From Copernicus to Einstein) *O Teorii Otnositel'nosti* [*About the Theory of Relativity*] (Moscow: Nauka) pp 7–61

[11] Idel'son N I 1943 Galileo in the history of astronomy in *Galileo Galilei* ed N I Idel'son (Moscow: Izd. Akad. Nauk SSSR)

[12] Kuznetsov B G 1964 *Galilei* (Galileo) (Moscow: Nauka)

[13] Malherbe M 1990 *Les religions de l'humanité* (Paris: Critérion) [Translated under the title 1997 *Religiya Chelovechestva* (Moscow: Rudomino)]

[14] Antonii, Metropolitan of Surozh 1995 *Chelovek pered Bogom* [*Man before God*] (Moscow: Tsentr po Izucheniyu Religii)

[15] Metropolitan Antonii (Mel'nikov) 1998 An open letter to the priest Aleksandr Men' in *Pravoslavnoe Knizhnoe Obozrenie* (Moscow: Kovcheg)

[16] Aleksandrov E B and Ginzburg V L 1999 Concerning pseudoscience and its propagandists *Vestn. Ross. Akad. Nauk* **69** 199 [1999 *Herald Russ. Acad. Sci.* **69** 118]; see also articles 22–24 in this volume and Ginzburg V L 2000 *Nauka i Zhizn'* [*Science and Life*] no 11 74 and 2002 *Literaturnaya Gazeta* (Literary Newspaper) 16–22 October

[17] 1998 *Phys. World* **11**(11) 11

[18] 1998 *Poisk* no 13 14

[19] Osipov Yu S 1998 Truth is not provable but demonstrable *Voskresnaya Shkola* (a weekly supplement to the newspaper *Pervoe Sentyabrya*) no 22 8

[20] Mitrokhin L N 2000 Scientific knowledge and religion at the eve of 21st century *Vestn. Ross. Akad. Nauk* **70**(1) 3 [2000 *Herald Russ. Acad. Sci.* **70**(1) 1]

[21] Petrova M K 1995 From reminiscences about Academician I P Pavlov *Vestn. Ross. Akad. Nauk* no 11 1016; see also 1999 *Priroda* [*Nature*] no 6 61

[22] Giovanni Paolo II con Vittorio Messori 1994 *Varcare la soglia della Speranza* (Milano: A Mandadori) [Translated under the title *Perestupit' Porog Nadezhdy* (Moscow: Istina i Zhizn', 1995)]

[23] 1997 *Nature* **386** 435

[24] 1998 *Nature* **394** 313

[25] 2000 *Phys. World* **13**(6) 55

[26] Ginzburg V L 2000 Religiya i nauka. Razum i vera *Nauka i Zhizn'* [*Science and Life*] no 7 22 (see comment 3 to this article)

Article 32

Science and religion in the modern world[1*]

The articles "Science and religion must avoid each other" by P Gaidenko and "Science must recognize religion" by A Silin published in *Izvestiya* [*News*] (Science section) on 25 January 2002 impels and, in a way, makes me feel bound to once again turn to the relationship between science and religion in the modern world. Unfortunately, this subject has recently become a burning topic here in Russia in view of the extensive clerical activities initiated in the first place by the Russian Orthodox Church (ROC). A most striking manifestation of this clerical attack is a large number of press publications and broadcast programs, such as a long interview of G Pavlovskii, a political analyst closely connected with the ruling élite (2001 *Nezavisimaya Gazeta–Religiya* [*Independent Newspaper–Religion*] of 26 December). However, the present paper is not designed as a discourse on political problems even though numerous cases of clerical interference in Russian public life in violation of the country's constitution arouse my bitter indignation.

A fruitful discussion of the role of religion and belief in God is impossible without first defining some notions and terminology albeit any such notion is, in a sense, arbitrary. Let us agree to understand religiousness not only as a simple belief in God but also as an affiliation with a particular creed (confession) of some universal religion, e. g. Christianity, Islam, Judaism, etc. In the sense of such an understanding, a religious person believes in a supernatural reality identified with God and confesses faith in a creed and its central tenets, such as (in Christianity) the Immaculate Conception, resurrection from the dead, the holiness of the Bible, etc. At the same time, many believers do not profess any religion. In short, they answer in the affirmative when asked whether they believe in God but express a mistrust of miracles, 'holy' books and 'sacred' scriptures; not infrequently, they even show a negative attitude toward religious organizations. In other words, they make up a class of non-religious believers, deists as opposed to theists.[2*] Another category is agnostics who say: "We don't know", whether a God exists or not. Finally, atheists reject belief in God and argue that only nature exists which, with its structure and driving forces, it is the ultimate goal and purpose of science to explore.

An interesting question is how many people belong to each of the four categories. Unfortunately, there is no detailed information at my disposal. Nevertheless, I have managed to collect some data. My article entitled "Religion and science, reason and faith" and published in the magazine 2000 *Nauka i Zhizn'* [*Science and Life*] (no 7 p 22) contained an appendix offering the readers a questionnaire. We received some 200 answers to the questionnaire from which it appears that 60% of the form-fillers are atheists, 20% religious people, 10% believe in God (without being religious), and 10% label themselves as agnostics (these data were made public in 2001 *Nauka i Zhizn'* (no 9 p 46). It is worthy of note that only 7% of the members of the National Academy of Sciences, USA, enrolled in a similar survey claimed to be believers (1998 *Nature* **394** 313). It would be interesting and instructive to send out such a questionnaire to the members of the Russian Academy of Sciences and all its officers. Regrettably, this was never done and one can hardly expect such a survey to be carried out in the future. I believe there are a mere few percent of religious persons among Russian researchers at the most (there may be more among those whose interests are in the humanities).[3*]

Besides this popular scientific paper, I published the article "Reason and faith (notes on the encyclical letter *Fides et Ratio* of the supreme pontiff John Paul II)" in 1999 *Vestn. Ross. Akad. Nauk* **69**(6) 546 [1999 *Herald Russ. Acad. Sci.* **69**(4) 271], 1999 *Zdravyi Smysl* [*Common Sense*] magazine (no 13 p 51), my book 2001 *About Science, Myself, and Others* (Moscow: Fizmatlit), and this volume (see article 31). The aforementioned encyclical letter of Pope John Paul II appears to be his last theological statement on the means and purposes of Roman Catholicism on the threshold of the third millennium of the Christian era. This kind of testament from a prominent clerical figure has been translated into the Russian language (Moscow: Izd. Frantsiskantsev, 1999). I am unaware of the opinions of many other Christian confessions as regards this document. But what I do know about attitudes toward the relationship between science and religion in the Orthodox and Anglican Churches suggests a consensus in sentiment and thought.

This consensus mirrors radical changes in the views of the church about interactions between science and religion, reason and faith since past times. It is known from a legend in the New Testament that Jesus Christ did not answer the question of Pontius Pilate: "What is truth?". But the Christian Church has for centuries pretended to know the answer and tried to dictate to science her understanding of truth and falsehood. The truth was supposed to come from nowhere but the Bible. Since her early history, the church has tried to impose this truth by force: the most striking example of these attempts was the activities of the Inquisition. In 1600, inquisitors condemned Giordano Bruno and had him burned alive (needless to say, the 400th anniversary of his death was not commemorated in 2000; another anniversary was observed instead). In 1633, the great Galileo had to stand trial before the Inquisition. Only some 350 years later, in 1992, Pope John Paul II opened a new page in the history of the Roman Catholicism by 'rehabilitating' Galileo.

Modern natural science built upon the foundations laid by Copernicus, Galileo, Kepler, and Newton and so brilliantly developed in the 18th and 19th century that it razed 'biblical science' to the ground and greatly promoted the spread of atheism at least in the scientific community and educated laymen. In order to maintain a broader influence on public affairs other than a hold on ignorant folk believing in miracles, the church had to reconsider some of her previous tactics and policies and admit science and reason on an equal footing with faith. The aforementioned Pope's encyclical letter begins with the words:

> Faith and reason are like two wings on which the human spirit rises to the contemplation of truth; and God has placed in the human heart a desire to know the truth—in a word, to know himself—so that, by knowing and loving God, men and women may also come to the fullness of truth about themselves.

Thus, it may be argued with some plausibility that, in central points of the contemporary ecclesiastical doctrine, reason (science) is recognized as a tool whereby to approach the truth about man and the reality of human existence (both visible and invisible) although insufficient for its full understanding. Hence, reason must be assisted by a sacred revelation because "the truth attained by philosophy and the truth of Revelation are neither identical nor mutually exclusive".

I would be glad to understand the proper content of revelation and how it, and religious faith in general, can foster the investigation of the truth of being which is the highest aim of science, at least materialistic science that recognizes the existence of truth. I understand atheism as a denial of God or of the gods and as the rejection of any evidence that faith helps to know the truth. It seems that theologians should seek to produce such evidence of which, for my part, I am totally unaware. P Gaidenko, who recommends that religion should abstain from meddling in scientific matters, appears to be equally ignorant. It is not my purpose here to dwell at any length on the paper by P Gaidenko. It has been already discussed by E L Feinberg with whom I fully agree.[4*] However, I cannot help being indignant against what P Gaidenko thinks to be 'deep and philosophically grounded' observations to the effect that outstanding achievements of modern theoretical physics border on mysticism and occultism. These achievements that are met with great enthusiasm by professional physicists, to an ordinary person appear to be at odds with common sense. Hence, they are not meritorious. There is a nice philosophical idea indeed! By the way, the advice of P Gaidenko that "science and religion must avoid each other" is at variance with what the church proclaims. Ecclesiastical authorities, I noted earlier, now insist that religion can and must help science to know the truth. The position of P Gaidenko on this question is more to my liking.

Now, to the paper by A Silin, DSc (Technology). It seems that a failure to find evidence of God's presence and activity in the technical sciences drove him to search for it in cosmology and biology. It looks like he had a second-rate popular scientific publication as a source of information. Here is how he describes the anthropic principle: "the appearance of man was brought about by Supreme forces". In fact, the so-called anthropic principle is reduced to an observation that life (the existence of all living organisms) in the forms known to humanity is possible not at any physical parameters that characterize matter. In our Universe, these parameters are such that make it a liveable environment. If other Universes (more precisely, Metagalaxies) hypothesized by certain scholars actually exist, some of them may be unsuitable for life as they have different parameters of particle interactions. But what have a God and religion to do with all this? In his appeal to biology, A Silin takes advantage of the well-known difficulties encountered in explaining the rate of evolution of living organisms, etc. This issue as well as the great problem of the origin of life and human thinking is highlighted in numerous studies. Many questions including those posed by A Silin are clarified but the problem at large remains to be resolved in an increasingly multidisciplinary manner. The reader interested in the current state of research is referred to the paper of D S Chernavskii "The origin of life and thinking from the viewpoint of modern physics" (2000 *Usp. Fiz. Nauk* **170** 157 [2000 *Phys. Usp.* **43** 151]). A Silin concludes his article with the words

> Contemporary natural sciences need no God (as they did not need it during Laplace's time) making naturalists daringly suggest hypotheses and verify them by experience. Nonetheless, science must openly admit as true what it in fact recognized long ago, that is the presence in nature of an ideal reality irreducible to material reality.

This "must openly admit" is great indeed! No, not only science 'must admit' nothing of the kind but also the overwhelming majority of scientists do not think of acknowledging it. Editorial notes introducing the articles of P Gaidenko and A Silin as well as the papers themselves refer to certain "prominent scientists" who allegedly "faced with facts and seeing no other way, had to appeal for help to religion". Who are these scientists, our contemporaries? There is no one that I know. As a foreign member of several

Academies of Sciences including such renowned and widely respected ones as the National Academy of Sciences (NAS), USA, and the Royal Society, London, I regularly receive their documents. For many years, there has been nothing in these materials pertaining to religious matters save one booklet issued by a special commission of the NAS for the sole purpose to confirm the groundlessness of scientific creationalism.

Here is one more remark on a popular argument against atheism provoked by the activities of the so-called 'militant non-believers'. Under the criminal bolshevik dictatorship, these people constantly pursued believers, destroyed church buildings, and performed numerous acts of vandalism and even murderous actions. But it would be absurdly obtuse to compare, even more so to identify, atheists with 'militant non-believers'. There is no more reason for that than for identifying faithful Christians with inquisitors or all those professing Islam with Islamic fundamentalists (known also as terrorists). The recognition of freedom of worship and liberty of conscience in association with full respect for the fundamental rights of those with faith is a dominating characteristic of the conduct of a civilized human being and atheists make no exception.

To conclude. What underlies the current revival of religiousness and clericalism in Russia? First and foremost, it is a reaction to the situation created by the former Soviet regime. Secondly, it is a consequence of the low educational level of the population. An educated person must not only know the names of the authors of *Yevgeny Onegin* and *War and Peace* but also have an extensive knowledge of the structure of matter (physics) and living organisms (biology). Unfortunately, true education is a rare thing in this as well as in many other countries. Finally, the last but not least cause of the re-establishment of clericalism in Russia is the official state policy encouraging the spread of religious ideology, in the first place the Orthodox faith, and supporting the ROC to the extent that the word God appeared in the newly adopted national anthem of this secular country! It is my deep conviction that, after the communist (bolshevik) ideology proved to be a shattering fiasco, a reactionary attempt to replace it by the Christian faith as G Pavlovskii and his like-minded associates dream of is doomed to fail. A bright future for mankind cannot be achieved unless the principles of enlightened secular humanism are consistently observed. Readers unaware of the contemporary humanist movement are referred to my article on this subject (in co-authorship with V A Kuvakin) published in 2000 *Vestn. Ross. Akad. Nauk* **70**(5) 483 [2000 *Herald Russ. Acad. Sci.* **70**(3) 223], *Neva* magazine (2000 no 6 141), and my aforementioned book *About Science, Myself, and Others* (see article 35 in this volume).

Regrettably, I am not destined to see the triumph of humanism but my great-grandchildren who are now two years old may hopefully have better luck.

Comments

1*. This article was published in the weekly supplement *Nauka [Science]* to *Izvestiya* (briefly *Izvestiya–Nauka*) on 1 February 2002.

2*. For the convenience of readers, certain terms as defined by an encyclopedic dictionary are given in the comments to the next article (see article 33).

3*. A paper in 2002 *Vestn. Ross. Akad. Nauk* **72**(3) 230 gives estimates of the number of believers in Russia. According to this, ardent believers make up 2% and those who formally observe sacraments and ceremonial rites 20%, supposedly of the sample that actually responded (the conditions of the poll are not specified).

4*. See Feinberg E L 2002 "V zashchitu nauki" [In defense of science] (*Novye Izvestiya* of 22 February). It is noted in the introduction to this article that the primary object of critique is not so much the paper by P Gaidenko as the text of a correspondent. However, in another paper dated 22 February 2002 (referenced to in article 33 of this volume), P Gaidenko does not mention this circumstance. The fact that I am not fully informed about this "affair" makes little difference as far as the essence and significance of the questions under discussion are concerned.

Article 33

Once again about science and religion in the modern world[1*]

Izvestiya–Nauka has recently published a few articles on the interaction between science and religion and related issues. I think that the newspaper does right in taking into account the considerable public interest in matters pertaining to this subject in present-day Russia which is rapidly being overwhelmed by the re-establishment of clericalism. Suffice it to say that a good proportion of television and radio broadcasting time is devoted to religious themes in the complete absence of atheist education. It is not my purpose, however, to tackle here the political aspect of this situation. The present paper contains a few notes designed to amend and clarify my views expounded in an earlier article (see 2002 *Izvestiya–Nauka* of 1 February and article 32). I believe that such clarification and relevant comments (similar comments were made in the article by E L Feinberg in *Novye Izvestiya* dated 22 February 2002) may be useful in the light of later publications.

In a paper in *Izvestiya–Nauka* of 22 February 2002, P Gaidenko rebukes me, in disregard of her previous publication dated 25 January, for being naive and ignorant of the history of Christianity, in connection with my assertion that the church has radically changed its attitude to science since medieval times. My statement was illustrated by references to the encyclical letter *Fides et Ratio* of Pope John Paul II (1998). Unlike P Gaidenko, I am not an expert in religious matters and do not actually know what role was assigned to reason by numerous saints, prophets, and all sorts of blessed simples in different periods of the history of Christendom. But I do know what is common knowledge, i.e. the attitude of the official church to science as far as 400 years ago. To spare readers a description of the situation in my own words, here is a famous passage from a letter by the great Galileo to Christine, Duchess of Lotharingia:

> Professors of theology must not arrogate to themselves the right to administer by decree in professions that are not within their

domain, for opinions about phenomena of nature can not be imposed upon the natural scientist.... We advocate new teachings not to sow discord in people's mind but to enlighten them; not to destroy science but to place it on a sound foundation. Our opponents, on the other hand, dismiss as false and heretical everything they cannot refute. These hypocrites employ pious religious zeal as a shield and humble the Holy Scriptures by using them as a tool for achieving their personal aims.... To direct professors of astronomy to rely on their own resources in seeking defense against their own observations and conclusions as if all these were mere deceptions and sophistry would mean confronting them with a more than impossible demand—it would be tantamount to ordering them no to see what they do see, not to understand what they understand, and to infer from their studies the very opposite of what is obvious to them.[1]

Meanwhile, the situation described in this quotation very closely resembles that in which scientists had to work in the former Soviet Union where Marxists acted the role of professors of theology and Marxism–Leninism was preached as the Gospel. Those who forget history will recall that the church by no means confined herself to discoursing on philosophical matters of faith and reason; indeed, the opponents were violently pursued and not a few suffered death at the stake. No wonder, many guided by the authority of reason alone rather than by revelations chose to keep a still tongue in their heads. To-day, the church admits reason on an equal footing with faith. This change in the church's official standing is explicitly declared in the encyclical letter *Fides et Ratio* as I noted in my earlier paper (see article 31).

Another important issue that I would like to stress emphatically pertains to the necessity of distinguishing between those who simply believe in God and truly religious people (for instance, theists) who are not only believers but also profess faith in a creed (e.g. Christianity). Of course, the dividing line between the two groups is often rather blurred and nuances are possible because the definition of notions is always somewhat arbitrary. One thing is to believe in the existence of God or an Absolute as a supernatural entity, the alleged Creator of the world including life and human consciousness, that does not however intervene in the ways of man (such is deism as I understand it).[2] Religion (theism) with its faith in miracles

---

[1] Galilei G 1943 *Lettera a Cristina di Lorena, sui rapporti tra l'autoritá della scrittura e la libertà della scienza* (Firenze: G.C. Sansoni); 1897 *Galileo a madama Cristina di Lorena* (1615) (Padova).

[2] To recall, an *Encyclopedic Dictionary*[2*] defines deism as 'A religious philosophical doctrine which recognizes God as the world's wisdom that has created a purposeful "machine" of nature, to which it gave laws and momentum, but denies the interference of the Creator with the laws of the universe (i. e. rejects "God's providence", miracles,

and 'holy' books like the Bible or Koran is a different thing. Theists believe in such miracles as the Immaculate Conception, resurrection of the dead, paradise and hell, even somewhat like a material soul, as well as the Second Coming of Christ, the end of the world, etc. Certainly, any theist believes in God whereas a believer is not necessarily a religious person in the previous definition.

I for one think religion (theism) and the faith in God that deists proclaim to be two different things. To my mind, religion (with its belief in miracles) and pseudoscience (such as astrology) are alike, with this reservation that, before Galileo, Kepler, and Newton created classical mechanics, i. e. some 300 years ago (Newton published his *Mathematical Principles of Natural Philosophy* in 1687), astrology could not have been considered pseudoscience for the lack of evidence that the positions of the stars and the planets did not influence the lives and behaviour of individuals on Earth. Yet, nowadays and even long before our time, astrological "predictions" have been disproved as lacking in all rational foundation (see, for instance, an article by V Surdin in the magazine *Nauka i Zhizn'* [*Science and Life*] (2000 nos 11, 12). Religious miracles like those mentioned are akin to esoteric superstition found in astrological speculations. Therefore, they are vigorously denounced by science. An essential difference between astrology and religion (theism) is that the former is pernicious for it leads one into error and can bring much trouble to those who believe in its absurd omens. In contrast, religious miracles are not that bad—what they call for is some naive belief which generally does no harm. What is more important, religion, with the exception of certain sects and such movements as Islamic fundamentalism, calls on people to live the good life required by God and formulates the well-known commandments that prescribe universal norms of conduct and the observance of basic ethical precepts. That is why, being a determined adversary of pseudoscience, I am far from proclaiming hostility toward religion and those who profess it excepting, of course, religious fanatics, such as terrorists and the like. Furthermore, albeit a convinced atheist, I have never been a militant non-believer; quite to the contrary, I am a staunch proponent of the liberty of conscience, i. e. the right of any human to worship God or not and freely express his or her personal convictions. In the last years, I have had to publicly defend atheism just because the state, in defiance of the Constitution of the Russian Federation, supports the spread of religion, in the first place the Orthodox faith, both materially and ideologically. A most shocking extension of this policy is the appearance of the word God in the newly adopted national anthem of this secular country.

Thus, religion is a vestige of superstitions, a rotten fruit of backward thinking and lack of education. With this in mind, atheism must seek to

etc) and takes no other path leading to the knowledge of God but human reason.'

combat ignorance rather than religion, by the rationalist critique of the baselessness of creationism.

Advocates of religion like to allude to the sayings of great personalities. Many of them who lived centuries ago were certainly religious people. It should be remembered, however, that science at those times was distinct from what it is nowadays. Hence, the maxims of such eminent scholar as Pascal who died in 1662 can scarcely serve as arguments when discoursing on the relationship between science and religion in the beginning of the 21st century. If allusions to authorities are actually needed, these must have lived in the 20th century and not before. Regrettably, it must be noted, the situation in this respect to-day is far from what it should be. For example, the authors of numerous papers repeat again and again that Einstein, the greatest physicist of the 20th century, was a religious person. It is only natural that Einstein changed his mind in the course of time but his views are widely known and summarized, for example, in the book entitled *Einstein and Religion* by Max Jammer (Princeton, 1999) and in a volume of collected utterances of Einstein (*Quotable Einstein*, Princeton, 1996). Asked about his religious beliefs in 1929, Einstein said as much: "I believe in the God of Spinoza that expresses himself in the harmony of being but not a God whose presence is experienced in human actions and history". Also, Einstein is known to have used the term "cosmic religion" but when his friends once reproved him for using religious terminology, he blurted out: 'I simply can not find a better word than "religious" to describe faith in the rational nature of reality... What the deuce is it to me that priests make money playing on this feeling?' In short, Einstein was anything but a theist; rather, he should be regarded as a pantheist like Spinoza. Also, I cannot see any great difference between pantheism and atheism.

I P Pavlov, the well-known physiologist, was another prominent figure (in the former Soviet Union) whom religious propaganda strives to portray as a dedicated believer. But here it seems to be wide of the mark again. This is how M K Petrova, Pavlov's closest associate, cites him when she writes in her book of memoirs (prohibited for publication by M A Suslov): "Human mind looks to the source of all things, and when it grasps the final cause, it is God.... But I do not believe in God, I am a non-believer" (Petrova M K 1995 *Vestn. Ross. Akad. Nauk* **65**(11) [1995 *Herald Russ. Acad. Sci.* **65**(6)]). To my knowledge, there is more evidence that I P Pavlov did not believe in God. His protest against the persecution of the church was a natural reaction to bolshevik tyranny. In a paper published in *Izvestiya-Nauka* of 22 March, 2002 Yu Vladimirov quotes P Dirac, one of the most renowned physicists of the 20th century, thus: "Religion is like opium given to people to lull them to sleep with sweet fantasies". Characteristically, this observation almost literally reproduces K Marx's formula: "Religion is the opium of the people" or its variant "Religion is the people's opium". I am no adherent of Marxism but I share this view. True, opium may be sometimes useful and its intake justified. I, for one, envy those who believe in God. Indeed, I am 87 now and realize that life is drawing to a close. Death may be as agonizing as the thoughts of the fate of those who are dear to me. Religious people find some consolation in the belief in the heavenly afterlife. But Man is endowed with reason to enable him to achieve control over the promptings of the emotions, and to avoid self-delusion and faith in miracles. In the mentioned paper, Yu Vladimirov asserts that some known physicists were faithful Christians. I can neither confirm nor disprove this assertion for the lack of relevant information. However, I cannot reckon seriously with the argument by Yu Vladimirov that N Bogolyubov was a religious person because he "did so much to promote the restoration of the cathedral in Dubna". It is appropriate to note that believers are not alone in their protestations regarding the demolition of church buildings.

Another assertion of Yu Vladimirov that "atheism, is also a religion but an alternative one" is worthy of more serious consideration. This issue is really important. The thing is, both atheism (i. e. the denial of God) and theism (a system of belief that affirms the reality of God) are "intuitive concepts". The most suitable terminology for the further discussion can be found in the book by E L Feinberg 1992 *Dve Kul'tury* (Moscow: Nauka) [1998 *Zwei Kulturen* (Berlin: Springer)] and his article published in the journal *Voprosy Filosofii* [*Problems of Philosophy*] no 7, 1997. Intuitive ideas can be neither proved to be true nor refuted. However, there is a great difference between the intuitive experience of an atheist who rejects the existence of God (i. e. a transcendent entity) and the intuitive warrant for God's reality. Atheists draw their knowledge from science, i. e. from exploration of natural phenomena and experiments. Scientific progress and the new information it brings about result in permanent changes in scientific ideas. In contrast, religion is relatively static as shown by a comparison of trends in science and Christianity in recent centuries. The difference is really striking. Religion (theism) accepts miracles; in fact, it is based on faith in miracles, i. e. alleged facts and unverifiable "revelations" that are in conflict with scientific data. Similarly, deism has undergone no notable changes during the same period. And what could have changed? Science certainly does not give an immediate satisfying answer to every topical question as yet unanswered but it develops unceasingly. To engage God in the search for an answer to all impelling problems is to reduce one unknown to another called God. I failed to find any positive content in the articles by P Gaidenko, Yu Vladimirov, and other authors that I happened to read. I sense no need to dwell at any length on these publications here because I have already made my mind known in an earlier paper entitled "Reason and faith" (1999 *Vestn. Ross. Akad. Nauk* **69**(6) [1999 *Herald Russ. Acad. Sci.* **69**(4)], see also article 31). Of course, believers may state that atheists, and I for one, fail to grasp their knowledge. It seems that their prime concern should be to set themselves quite explicitly to articulate the truth conferred by revelation,[3] to show what knowledge about nature and man they acquired through revelation, by way of meditation upon what is true and by other modes of religious perception. It is opportune to note here that I strongly object to a widely popular rule of conduct expressed in the phrase "Given there is no God, one is free to do whatever he (or she) likes". Atheism by no means justifies murder, theft, and all other crimes. Religions too (albeit not all of them and not always) call for the observance of basic ethical norms. But they are not alone in their struggle nor do their action to this end can compromise the basic principles of atheism.[3*]

The absence of or a deficiencies in scientific knowledge in ancient times naturally gave rise to a body of religious and ethical custom and practice which is reflected in the Bible. Bible stories remain and will ever remain a most important historical document and work of narrative fiction. With the development of science, however, the Bible, the Koran, and biblical literature have lost the status of "sacred" books. Those who still believe in holy writings may be identified with believers in astrological divination. The fact that there are still so many religious people reveals the generally low level of educational attainment by a significant proportion of the community. Suffice it to say that about one-sixth of the present world's population, i. e. nearly one billion people, are unable to read and write. Moreover, the overwhelming majority of literate men and women have no basic knowledge of modern physics and biology. Now that civilization, faced with the growing threat of terrorism, probably with the use of non-conventional weapons, is about to confront a severe crisis, theological controversies may seem very obscure stretches of the past. I am no prophet but have an intuitive feeling that the crisis will be overcome and mankind will not be reduced to degradation: it will well survive as a society with no place for religion.

Comments

1*. On 24 and 31 May 2002, *Izvestiya–Nauka* published an abridged version of this paper. It is reproduced in full in this volume. See also article 34 and my earlier paper "Misunderstanding of problems concerning religion and relationship of science and religion" in 2003 *Vestn. Ross. Akad. Nauk* **73**(9) 816 [2003 *Herald Russ. Acad. Sci.* **73**(5) 560]. Also see a volume of collected articles 2003 *In defense of reason (against aggression, quackery,*

---

[3] The *Encyclopedic Dictionary* defines revelation in the following terms: 'In monotheistic religions, revelation is an infallible word of God or knowledge deduced from him as an absolute criterion of human conduct and cognition. The idea of revelation is present in both "Sacred Scripture" (the Bible in Judaism and Christianity, the Koran in Islam) and written "Tradition" (Talmud in Judaism, the writings of Apostolic Fathers in Christianity, Sunna in Islam).'

and paranormal beliefs) in the Russian culture of the century (Moscow) and 2003 *Svetskii Soyuz [Secular Union]* no 2.

2*. See 2001 *The New Encyclopedic Dictionary* (Moscow: Bol'shaya Rossiiskaya Entsiklopediya, Ripol Klassik). This is how it defines three more terms, besides "deism" and "revelation" used in the text:

- *Atheism.* Historically distinct forms of the rejection of religious beliefs and cults which affirm that man and the surrounding world are fully real in their own right. Atheism is displayed as religious free-thinking, etc. Contemporary atheism considers religion to be a product of man's subliminal consciousness.
- *Theism* (from the Greek theus, "god"). A religious outlook based on the belief in the existence of God viewed as the freely creative absolute Being who transcends yet is immanent in the world. By accentuating the transcendence of God, theism sharply contrasts with pantheism whereas deism denies the interference of the Creator with the laws of the universe. The most distinctively theistic religions are the genetically linked Judaism, Christianity, and Islam.
- *Pantheism.* Religious and philosophical doctrines that equate God with the forces and laws of the universe. A pantheistic attitude is especially apparent in the heretical mysticism of medieval religions. It is also characteristic of the natural philosophy of the Renaissance and B Spinoza who held that nature is to be identified with God.

The terms 'theism' and 'religion' used indiscriminately in this paper, may be unluckily at odds with the commonly accepted usage. This, does not, I trust, obscure the essence of the matter. If so, I wish for no better terminology.

3*. A customary argument advanced to illustrate the importance of religion, especially Christianity, and the merits of the church in particular draws on the fact that many outstanding scholars not only believed in God but also lived in monasteries or former monasteries, such as the old colleges in Cambridge and Oxford. This is really true, suffice it to recall Copernicus, Newton, and Mendel. But what does it prove? Nobody denies that Christianity remained the dominant ideology throughout centuries and monasteries were centers of scientific thought. However, the situation changed very long ago. What remains of those times are, for instance, the Pontifical Academy of Sciences and the very small Vatican Astronomical Observatory. Neither of these institutions make any substantial contribution to the development of contemporary science.

Article 34

Russia must not slide into a slough of clericalism[1*]

It is more than ten years now since our country rid itself of the bolshevik totalitarian regime and seemingly embarked on the path of civilized democratic development. Considerable progress has been achieved in certain areas and I am far from belittling its importance. We enjoy freedom of speech (censorship is abolished), conscience (freedom to perform religious rites), and movement (people can go abroad without many formalities and the KGB's permission) as well as elected power (still imperfect but much better than the dummy Supreme Council). Disappointingly, almost all these changes lag far behind expectations. To begin with, censorship gave place to the unrestrained publication of all sort of nonsense and pseudoscientific materials [1] coupled with a reluctance by the mass media to refute many false views. There is a shocking social inequality and stratification. Suffice it to say that the income of 5% of the richest people is 100 times that of the poorest ones (30 and 3000 US dollars a month, respectively) [2]. Corruption is unbelievably widespread and there are no serious attempts to stop it (see, for instance, [3]).

I shall dwell at length on the single issue of the liberty of conscience. In principle, it appears clear: under the Constitution of the Russian Federation that came into force on 25 December 1993, Russia is a secular democratic state in which "no religion can be established as the state or compulsory religion" and "religious unions are separated from the state". Furthermore, "each citizen is guaranteed freedom of conscience, freedom of faith including the right to profess any religion... or no religion at all..." (articles 14 and 28). But, as a matter of fact, Russia is being fast overwhelmed by the re-establishment of clericalism. Despite an acute shortage of children and old people homes and the very low living standards of a wide section of the population, large, sometimes huge, sums of money are being spent to restore churches and monasteries (suffice it to mention the

restoration of Christ the Savior Cathedral[1]) and to give material support
to the Russian Orthodox Church (ROC) at large. A good proportion of
television broadcasting time is devoted to religious themes and various aspects of church life, in the complete absence of atheist education. The most
shocking extension of this policy is the appearance of the word God in the
newly adopted national anthem of this secular country, not to mention the
fact that the anthem itself, with Mikhalkov's text reminiscent of Stalinism
and the music of Soviet times, arouses the indignation of many people (I
happened to write about this in an earlier article [4]).

What is behind the support given to religion by the Russian governing authorities? Marxism–Leninism was the official ideology in the former
USSR. In practice, if not on paper, democrats decisively reject elements of
this 'teaching' including the so-called proletarian dictatorship which was, in
fact, the dictatorship of the ruling clique of the communist party and even
of its leader alone. It grossly violated of human rights and liberties coupled
with an inefficient economic policy as its consequence. This attitude to religion had two aspects. On the one hand, atheism was declared to be the
official government philosophy. On the other hand, religion was suppressed
and the so-called militant non-believers were given full reign. But atheism
is quite compatible with liberty of conscience. Moreover, militant non-
believers acting as hooligans and even bandits discredited atheistic views
in the eyes of poorly educated people. As a matter of fact, the actions of
militant non-believers resemble those of the inquisitors in Roman Catholicism or the persecutors of Old Believers in the ROC. Evidently, neither the
Inquisition nor persecution of Old Orthodox Believers and other 'heretics'
is an argument against religion, in general, and Christianity, in particular.

Similarly, the actions of militant non-believers and all forms of persecution of religion and its followers must be turned down as an argument
against atheism. In today's Russia, however, a natural reaction to the 70
years of the bolsheviks' anti-religious policy serves to foster the popularity
of the church, thus strengthening it. This process is further promoted by the
loss of the 'national idea' that communists (bolsheviks) proclaimed, at least
by word of mouth, in building a socialist and then communist society. Having nothing to offer in substitution of this tempting but utopian idea, our
leaders (I am not aware at which level and how unanimously) chose to set
their hopes on religion and, in the first place, Orthodoxy. Does present-day
Russia really need any particular national idea? I believe we must simply
strive to raise every citizen (in fact, the entire population of Russia) to a
comfortable level of living and working, at least by the prevailing standards
in developed countries. It is certainly a very difficult task. It is much easier
to build and restore churches, sprinkle holy (in fact, tap) water right and

---

[1] Certain sources estimate the construction cost at 500 million US dollars (*Izvestiya*
dated 12 January 2003).

left as if to 'sanctify', senselessly and to no purpose, houses, army barracks, submarines, and spacecraft that everybody knows are as likely to explode after a public prayer service as without it. I ask the reader not to regard this sarcastic remark as mockery of truly religious people. I do understand many of their feelings. I understand that the unfortunate and those who have to stand in the fierce struggle of daily existence may find consolation in setting their hopes on God; I understand that, for them, participation in public prayers and attendance at the church, mosque, or synagogue may serve as a safety-valve for the overflow of redundant tensions and frustrations. The well-known adages "religion is the people's opium" or "religion is opium for the people" (there is some difference between these)—reflect an objective situation which should not be neglected. Therefore, any form of persecution of believers or violation of their rights to express their religious feelings freely cannot be tolerated (of course, I do not mean certain fanatic sects and those who commit terrorist crimes under the guise of religion). Also, most religions urge people, at least so to say officially, to be good and observe certain ethical norms (commandments). Hence, there is no need to combat religion as the militant non-believers call for. The task of an atheist is to promote atheist education and expose the baselessness of creationism. Unfortunately, as mentioned before, the Russian mass media publish very few materials to this effect.

In my previous articles (see, for instance [5]), I emphasized the necessity to distinguish religious people affiliated with a particular confession (i. e. theists: Christians, Muslims, Jews) from persons whose belief in God is somewhat abstract and impersonal. They answer in the affirmative when asked whether they believe in God but I have never managed to learn what their specific belief consists of. Generally speaking, they believe in a supernatural power that created the visible universe. From the philosophical point of view, deists belong to the latter category.[2] In a sense, this position is understandable taking into consideration that modern science is still unable to give a clear answer to the question concerning the origin of life and consciousness (see [6] for current views on this problem). Atheists reject belief in God and argue that only nature exists. This assertion is based on 'intuitive judgement' (see [5] and especially references therein for more details) that can be neither conclusively proved nor refuted. Belief in God is likewise an intuitive judgement but there is a great difference between the two (see [5] for details).

Discussion of deistic and agnostic views is the realm of philosophy and they are open to debate. In contrast, religion (theism) is inseparable from

---

[2] Deism is a religious philosophy (or belief, as you choose) that finds expression in the acknowledgement of the existence of God or an Absolute Being that created and 'constructed' nature and the laws governing it. Deists recognize that God virtually withdrew and refrained from interfering in the processes of nature and the ways of man. Deists do not see any other means to know God but common reason.

belief in miracles (i. e. alleged facts and statements the truth of which is impossible to verify and which are in a striking contrast with scientific knowledge, such as the Immaculate Conception, resurrection of the dead, etc). Indeed, religion and pseudoscience, such as astrology, are essentially alike, with this reservation that religious miracles are usually more or less harmless (indeed, it seems to be of little importance how Jesus Christ was conceived or whether he actually was resurrected some 2000 years ago). More important, as mentioned before, religion, within certain conditions and constraints, calls on people to live a good life and prescribes some positive ethical precepts. It is opportune to note that a widely popularized rule of conduct expressed in the phrase "given there is no God, one is free to do whatever he (or she) likes" is totally unacceptable. Does this mean that atheism justifies murder, theft, and all other crimes? Religions too (albeit not all of them and not always) condemn these lawless acts but this position is not the church's monopoly and, by no means, compromises the basic principles of atheism.

Attempts to revitalize religion, especially Orthodoxy, in Russia are, on the whole, reactionary, a step backwards into the past especially noticeable in view of the brilliant achievements of science both in our time and in the past. Meanwhile, the ROC's struggle to strengthen its position is clearly supported not only by the ruling authorities but also by fashionable political analysts [7] and even functionaries of the Communist Party of Russia (CPRF). I completely agree with the characteristic of the latter given by A N Yakovlev in his excellent book (see [8, p 267]): "Especially dirty and cynical are the present communist party leaders' oaths of adherence to Christian ideals. bolsheviks who demolished thousands of churches and killed hundreds of thousands of priests today represent themselves as the bearers of morality." Then, the author justly asks [8, p 268]: "Why do not highly honored and respected hierarchs of our church pronounce an anathema against the anti-patriotic and anti-Christian party that tried to destroy the church and treated religion as an evil that needs to be eradicated?"

But the ROC does not care. It tries to strike roots in the army [9], institutions of higher learning, and general schools (see, for instance, [10]). In the meantime, it is clear that, in a secular state, school must be completely separated from religion. Parents wishing that their children acquire some religious knowledge must send them to meeting houses (churches, mosques, synagogues, etc), not to schools providing general education and designed to teach physics, biology, and other disciplines. Science has nothing in common with theism and its biblical legends [5]. Religion is a product of past, remote times. To propagate religion in schools and teach children obsolete and false creation science instead of modern true science means to literally push Russia and its people into a mire of clericalism. By the way, adherents of reconstruction of Orthodoxy as the leading ideology in

Russia need to be recalled that they should know better from past experience. Indeed, Orthodoxy was the official and dominant ideology in tsarist
Russia. This, however, did not prevent the atrocities of the Civil War or
the subsequent outrages of the bolshevik dictatorship in the USSR. Terror
in its most brutal forms was organized by a former student of a theological
seminary while both the champions and victims of this policy were, in the
first place, baptized persons.

The future of mankind and of Russia is bound up with the international
humanist movement [11]. The ideals of this movement may seem utopian
in the face of current challenges, such as the growing wave of banditry,
terrorism, and drug abuse. However, I do not think that modern Islamic
terrorism is more horrible and mightier than Hitler's fascism or Stalin's
communism which we managed to defeated in the last century. As long as
terrible nuclear and bacteriological weapons exist, the Earth's population
has no guarantee that difficulties will be overcome and the human mind
escape degradation. Nevertheless, there are good reasons to believe that
progressive forces will gain the upper hand and civilization will be saved.
Of course, colossal efforts are needed to fight poverty, hunger, disease, and
terror. These are daring goals for planetary humanism and it may be
hoped that considerable success in this field will be achieved during the
21st century. It will inevitably result in the rejection of obsolete religious
faiths and the triumph of the scientific world outlook.

Comments

1*. This article was published in the magazine 2002/2003 *Zdravyi Smysl*
[*Common Sense*] no 1 4.

2*. Articles 31–34 expound my views on atheism, science, and religion.
They were written at different times and, to a certain degree, overlap.
Therefore, I found it opportune to publish here a kind of summary, more
precisely abstracts of my report on "Atheism, science, and religion in the
modern world" sent in April 2003 to the Organizing Committee of the
Second International Conference on Theology and Science (Moscow, 14–
18 May 2003) in which I was invited to participate. By the way, I have
never received an answer from the organizers of that conference. In Russia,
representatives of the Church usually prefer to avoid discussion of anything
and seek to penetrate even the schools. The abstracts are as follows:

(1) Both atheism (the denial of God) and the belief in God are intuitive
judgements. Neither can be conclusively proved or refuted like, say, a
mathematical theorem.

(2) Atheism by no means rejects freedom of conscience, i. e. the freedom
to believe or not believe in God. Generally speaking, atheists have nothing
in common with militant non-believers. Identification of atheists with mil-

itant non-believers would be akin to identification of believing Christians with supporters of the Inquisition.

(3) One should distinguish between people whose belief in God as a certain absolute, etc is somewhat abstract (e. g. deists) and religious persons affiliated with a particular confession (e. g. theists). Theists believe not only in God but also in the holiness of the Bible (or Koran), in miracles, etc.

(4) Discussion of atheism, agnosticism, materialism, and deism is the realm of philosophy. At the same time, theism is a typical pseudoscience and, like astrology, completely rejected by atheists. Theism is a decrepit survival of past remote times incompatible with a scientific world outlook and science at large.

(5) Because, generally speaking, theism calls on people to live the good life and observe some positive ethical norms (commandments), there is no need to combat it (unlike a pseudoscience, such as astrology). The task of an atheist is to promote atheist education, in particular to expose the baselessness of creationism and all other antiscientific 'theories'. I would like to emphasize the unacceptability of a widely popular rule of conduct expressed in the phrase "Given there is no God, one is free to do whatever he (or she) likes". Indeed, theism sometimes albeit not always (cf some trends in Islamic fundamentalism) contributes to the strengthening of positive ethical and moral norms. But atheism 'professes' exactly the same views and principles.

References

[1] Ginzburg V L 2002 Demagogi i nevezhdy protiv nauchnoi ekspertizy (Demagogues and ignorant persons against scientific expertise) *Literaturnaya Gazeta* [*Literary Newspaper*] 16–22 October

[2] 2002 Gazeta *Vek* [Newspaper *Century*] no 30

[3] Reznik B 2002 Mafiya i more [Mafia and the sea] *Izvestiya* 18, 19, 20 July; 9 October

[4] Ginzburg V L 2001 Vnutrennyaya slabost' [Inner weakness] *Literaturnaya Gazeta* 17–23 October [This article with comments was included in the 3d edition of the book Ginzburg V L 2003 *O Nauke, o Sebe i o Drugikh* [*About Science, Myself, and Others*] (Moscow: Fizmatlit)]

[5] Ginzburg V L 2002 Nauka i religiya v sovremennom mire [Science and religion in the modern world], Eshche raz o nauke i religii v sovremennom mire [Once again about science and religion in the modern world] *Zdravyi Smysl* no 2 33 and no 3 23, respectively [These articles in a somewhat abridged form also appeared in *Izvestiya–Nauka* 1 February, 24 and 31 May 2002 but special efforts were needed to have them published]

[6] Chernavskii D S 2000 Problems of origin of life and thinking from the standpoint of modern physics *Usp. Fiz. Nauk* **179** 1557 [2000 *Phys. Usp.* **43** 151]

[7] Interview with G Pavlovskii 2001 *Nezavisimaya gazeta–Religiya* 26 December

[8] Yakovlev A N 2000 *Omut Pamyati [The Slough of Memory]* (Moscow: Vagrius)

[9] Mozgovoy S 2002 Klerikalizatsiya rossiiskoy armii: opasnye tendentsii [Clericalization of the Russian Army: dangerous trends] *Svetskiy Soyuz: Al'manakh* (Moscow: Rossiiskoe gumanisticheskoe obschestvo) p 91

[10] 2002 Pastyri na potoke [Production line manufacture of pastors] *Izvestiya* 11 October

[11] 1996–2002 *Zdravyi Smysl* nos 1–25

Borzenko I M, Kuvakin V A, and Kudishina A A 2002 *Osnovy Sovremennogo Gumanizma [Fundamentals of Modern Humanism]* (Moscow: Rossiiskoe gumanisticheskoe obschestvo)

Article 35

International Humanist Movement and *Manifesto 2000*[1,1*]

The development of human society has always involved a search for the ways and means of meeting humankind's material and spiritual requirements in the best possible way. There is no need at this point to delve into human history and we may (by association with the title of the present article) begin by mentioning *The Communist Manifesto*, which as early as 1848 mapped the road to a glorious future. Different varieties of this communist (frequently referred to as Marxist) road of development have been discussed and specified. The variety known as "the dictatorship of the proletariat" was translated into reality in Bolshevism–Leninism that triumphed in Russia after the October 1917 *coup d'etat*. The evolution of bolshevism is well known: under the guidance of Lenin and Stalin, it led to brutal totalitarianism and the extermination of millions of people. A decade ago, international communism collapsed. In its bolshevist Leninist–Stalinist incarnation, it became equivalent to fascism (National-Socialism). The Cheka–GPU–KGB–Gulag, Katyn, deportation of entire population, etc were perfectly consonant with the Gestapo and Auschwitz. Characteristically, it is not a national or ethnic phenomenon. Totalitarianism is essentially the same and has actually produced the same results in Germany, the USSR, China, in the so-called People's Democratic Republics, Cuba, and Cambodia.

They used to say: 'He who is not a communist at 20 has no heart but he who is a communist at 50 has no brain'. In connection with this adage, it should be emphasized that in its genesis and terminology, bolshevik communism differs from fascism. Many honest people believed in the slogans proclaimed by the bolsheviks. But the ideals were trampled upon and turned out to be fraudulent from the very outset. Long years of propaganda glossed over and muted in human minds the horrible truth

---

[1] The authors of this article are V L Ginzburg and V A Kuvakin.

about the first years of bolshevist rule. Today, this truth is well known (see, for example, [1, 2])—Stalin merely raised it to monstrous proportions (see, for example, [1, 3]). It is bittersweet to recall idealistic bolsheviks who lost their lives in Stalin's prisons or in the Gulag or, at best, lived out the rest of their days with broken hearts and sealed lips.

We are firmly convinced that only the blind can fail to see that the road of totalitarianism (including bolshevik communism) cannot bring well-being and happiness to people. Figuratively speaking, humankind today is 50, not 20, years old. What plan and what program should be adopted for the development of human society? Innumerable attempts have been made to answer this eternal question. Among those who propose answers are both liberals and socialists of every stripe, as well as utopians who believe in communism with a human face and sundry religious organizations.

Meanwhile, there is a growing international humanist movement headed by the International Academy of Humanism, the International Humanist and Ethical Union, and other prestigious organizations that maintain close contacts with the UN and UNESCO. Among the leaders of the International Humanist Movement were such outstanding people of their time as Ernst Haeckel, John Dewey, Charles Pierce, Bertrand Russell, Julian Huxley, Albert Einstein, and many others.

The authors support this movement and the aim of this article is to inform readers about it and about its latest declaration, recently published in Russian under the title *Humanist Manifesto 2000: A Call for a New Planetary Humanism*, referred to here as *Manifesto 2000* for short.

The Russian edition of *Manifesto 2000* [4] lists the names of the 90 Russian intellectuals who signed it (including 10 Russian Academy of Sciences members). Particularly important is the conclusion of the *Manifesto*:

> Those who endorsed *Humanist Manifesto 2000* do not necessarily agree with every provision of it. We do however accept its main principles and offer it in order to contribute to constructive dialogue. We invite other men and women representing different traditions to join with us in working for a better world in the planetary society that is now emerging. [4, p 36]

The authors of this article are naturally among the signatories of *Manifesto 2000* but, as pointed out already, they do not necessarily agree with all its propositions. We do not deem it appropriate, however, to discuss the document in detail and, as previously mentioned, seek to elucidate only its main ideas. Meanwhile, wherever there is no direct quotation from the *Manifesto*, we bear the responsibility for the following text. Furthermore, some of its propositions are not contained in the document.

35.1   Background

As a system of moral and civil values and a world view closely associated with science, reason, and ideas of democracy, civil society, and social justice, humanism goes back to Ancient Greece and China, and the Charvaka (Lokayata) movement in classical India. A humanist world view is characterized by exceptional adaptiveness to the changing social and cultural conditions of human life. Historians distinguish between ancient humanism and the humanism of Arabic renaissance (the 11th–12th centuries), between Byzantine humanism (the 14th–15th centuries) and the humanism of the European renaissance (the 14th–16th centuries). Its values were widely recognized in the epoch of the Enlightenment and have developed since then—alongside the progress of science and technology and the idea of human rights—into a deep-rooted permanent factor of European civilization.

For a variety of reasons, the fate of humanism in Russia has turned out to be controversial and dramatic. In our country, for example, the Russian Orthodox Church (ROC) was never reformed or liberalized, serfdom was abolished with obvious delay, and in the 20th century, revolutionary upheavals, wars, and the 70-year communist experiment suppressed and perverted traditional values of humanism in the human mind. This does not mean at all that Russia is alien to them as some religious nationalists or decadent homebred postmodernists believe. Russia can and should be grasped by the mind. Its 'specific character' is, by no means, a lack of intelligence and humanism. It is sufficient to mention M V Lomonosov, L N Tolstoy, and a number of other names whom we will not list for fear of presenting a canonical constellation. All outstanding achievements in science, culture, and art in this country are feats of intelligence and humaneness. It should be emphasized that the phenomenon of the Russian spiritual renaissance in the first quarter of the 20th century, so actively propagandized today, emerged as a liberal reaction mainly by the former so-called legal Marxists to archaic values of absolutism and conventional Orthodoxy and as an attempt to uphold the dignity and freedom of an individual. It was only the brutal and uncompromising pressure of 'proletarian' revolutionaries, the Red Hundred as Berdyaev had aptly called them long before 1917, that caused a deep rift in Russia and forced a large part of the liberal and democratic intelligentsia to fall into the embrace of Orthodoxy and conservatism.

Nor had humanism better luck under Soviet rule. General human values, rejected as allegedly abstract and, therefore, illusory or disguising the pseudovalues of a 'decadent' 'bourgeois' culture, were replaced by class values. That was how the theory of 'proletarian (socialist–communist) humanism' came into being. During *perestroika*, the efforts of M S Gorbachev to restore general human values to humanism did not met with much success. They were suppressed or sidelined by increasingly spontaneous and less controllable developments in the early 1990s. Ironically, the word 'spirituality' that gained popularity at that time was soon monopolized by the ROC and the word 'humanism' gave way to 'humanitarian', as though it referred not only to humanities or humanitarian (i. e. disinterested) aid to people but to everything associated with humanism.

Meanwhile, attempts to establish specialized social institutions promoting the values and ideals of free thinking, a general human morality, and secular culture were made even during the dominance of communist ideology. They usually originated from soberminded scientists and atheists, fully aware that free thinking in the service of the Central Committee's Ideological Department and the Committee for Religion and Church (in fact, a branch of the KGB) was no more than a travesty of free thought and natural moral norms. All these attempts, however, aroused great suspicion and proved futile.

It was not until the mid-1990s when, in response to the rapid expansion of religion into culture, education, politics, and government structures, as well as to the unprecedented dissemination of prejudice, superstition, paranormal faiths, and quasi-scientific ideas, that scientists at a number of Moscow universities and the Russian Academy of Sciences founded a volunteer inter-regional association called the Russian Humanist Society. They began to publish the journal *Zdravyi Smysl* [*Common Sense*] to draft educational and research programs based on the values of reason and critical thinking, a scientific picture of the world, free thought, humane ethics, humanist philosophy, and psychology [5, 6]. This initiative enabled Russian humanists, atheists, and free-thinkers to be drawn into the circle of ideas and organizations of the International Humanist Movement.

Unfortunately, not only the ordinary citizens of our country but also Russian intellectuals are practically ignorant of this important aspect of modern world culture. Without delving into a detailed explanation of the causes of our ignorance, let us point out some of them: the non-commercial character of humanist values, their orientation toward common sense, the lack of any eccentricity, a high level of self-discipline, independence, moral, legal, and civil responsibility, imposed by a humanist world view on its adherents.

It is the mentality and way of life of a truly mature, serious, naturally democratic and, by and large, balanced person, confident in the progress of science and intelligence and the ability of people to meet any particular or global challenges that beset their path. Correlating this with the realities of contemporary Russia, an obvious conclusion suggests itself: in many respects, our present cultural, moral, and psychological atmosphere is, to put it mildly, at odds with the values of humanism. It is true that diametrically opposite conclusions are drawn from it. Some people say that humanism is a luxury that we cannot afford; others claim that Russia has the special destiny of a "Eurasian" power whose principal values are orthodoxy, unity, and spirituality, among others. Humanists, in contrast, are convinced that in the absence of a world view and moral crisis, the most important thing is to do one's utmost to oppose the dehumanization and demoralization of society, protect the values of a scientific world view, show elementary decency and personal and social responsibility.

Determining the vector of post-Soviet Russian dynamics is probably one of the most difficult tasks. But in the context of our subject, the obvious is revealed: in a historic choice arising from the collapse of communism, Russian society has turned to primarily obsolescent and essentially archaic values of religion, nationalism, and big-power chauvinism (i. e. a blend of authoritarianism and somewhat moderated jingoism). Far from seeing the foundation of values that ensures the normal functioning of democratic societies, it has wasted the moral potential accumulated in the Soviet period in the course of everyday non-ideological communication. In basic civil, social, and particularly moral values, Russian society has largely regressed rather than progressed.

The question arises whether Russia can cope with the crisis of values that has so deeply affected both the upper and lower strata. Humanists answer in the affirmative this fundamental question of the newest—turbulent and transient—Russian civilization. Before spelling out this answer, however, it is necessary to describe, from the perspective of secular humanism, a picture of life in the contemporary international community of which Russia is an important member.

35.2   Documents of 20th century humanism

Humanist Manifesto 2000: A Call for a New Planetary Humanism was not the first appeal of international humanist organizations to the international community. In 1933, a large group of European and US scientists and prominent public figures signed the *Humanist Manifesto I* [7] stressing the need for a serious reorientation of values, based on common sense, democratic principles, and socially oriented economic policy. Its 15 paragraphs advocated the ideas of a non-created Universe, the natural origin of a human being as a result of an evolutionary process, rejected the principle of the dualism of mind and body; recognized religion and culture as a whole as being products of humankind's social development in its interaction with nature: "the nature of the universe depicted by modern science", the document pointed out, "makes unacceptable any supernatural or cosmic guarantees of human values" [7, p 24]. *Humanist Manifesto I* proclaimed a society that aimed solely to make a profit to be inadequate and advocated the establishment of a socialized and cooperative economic order.

The signatories of the *Manifesto* were convinced that, in modern society, there is no room for theism, traditional clerical institutions, cults, rites, prayers, and uniquely religious emotions. Meanwhile, it was pointed out that "religion must formulate its hopes and plans in the light of the scientific spirit and scientific method" [7, p 24]. It was essentially an attempt to propose humanism to society as a secular non-theological religion named religious humanism where, according to its advocates, the distinction between the sacred and the secular can no longer be maintained.

Forty years later, in 1973, *Humanist Manifesto II* [8] saw the light of day. It reflected the new realities of world history: the spread of fascism and its defeat in the Second World War, the establishment of the 'socialist camp', the division of the world into two antagonistic poles, the Cold War and the arms race, the establishment of the United Nations, the ever-accelerating progress of science and technology, the development of democracies, and the rise of human rights movements in the West against the backdrop of higher living standards and a better quality of life. The *Manifesto* was signed by scientists and public figures with both social and sociodemocratic orientations. It defended human rights and democratic values, condemned all forms of totalitarianism, racism, and religious and class antagonism. The *Manifesto* left room for both atheistic humanism associated with scientific materialism and liberal-religious humanism that rejects traditional religions, the existence of a supernatural and afterlife and considers itself as an expression of 'sincere aspiration' and 'spiritual experience' inspiring the pursuit of 'supreme moral ideals'. In fact, religion was replaced by a general human morality and ethics free from any theological, political or ideological sanctions. The dignity of an individual whose freedom is in line with a responsibility to society was recognized as the central humanist value. In addition, the *Manifesto* advocated the values of democracy, peace, international security, and cooperation.

In the 1980s, the International Humanist Movement clarified and supplemented its world view platform by publishing the *Declaration of Secular Humanism* (1980) and the *Declaration of Interdependence* (1988). The former stressed the basic difference between religion and secular humanism which reflected the overall trend of the vast majority of humanist organizations (whose number runs into hundreds, cf [9]) emphasizing the independent moral and civil status of humanism. The proponents of this trend advocated the view that secular humanism is a set of moral and scientific values that cannot be equated with religious faith [10].

The *Declaration of Interdependence*, approved at the World Humanist Congress in Buffalo, USA, in 1988, aimed to supplement *The Universal Declaration of Human Rights* (adopted by the UN in 1948) by a code of mutual moral, legal, and civil obligations of an individual and society in the light of the globalization of human relations.

The last few decades of the past century were as dynamic as the previous ones: communism collapsed in the USSR and Eastern Europe, the confrontation of military blocs came to an end, democratization received a new stimulus, globalization of the world economy accelerated, the high tempo of scientific and technological progress was maintained, and important changes occurred in the mass media giving rise to worldwide information, cybernetic, financial, commercial, and cultural realities. Those and other changes necessitated a new integrative assessment of contemporary life and the prospects of the international community in terms of a humanist world view. Such an assessment was made in *Manifesto 2000*.

35.3　Salient features of the contemporary epoch

The appeal of the international community for a new—planetary—humanism was antedated in *Manifesto 2000* by an elaborate and realistic outline of the leading trends in science and technology, world views, world economy, and international relations. The document is imbued with a spirit of optimism and with the confidence that humankind is capable of successfully meeting the unprecedented challenges it meets at the threshold of the new millennium. Such confidence is in direct contrast to both the innumerable apocalyptic scenarios of religious ideologists and charlatans and to the pessimistic predictions of sundry alarmists and postmodernists.

In fact, the situation in the world testifies to progress in many spheres:

- Medical science has considerably improved human health; the average life span has increased; the development of new medical preparations, genetic engineering, and modern surgery are contributing greatly to the development of health care.
- Preventive medicine, modern water supply and sewage treatment, sanitation and hygiene have reduced the risk of epidemics and contagious diseases. Infant mortality has experienced a sharp decline.
- Modern agricultural technologies and food industry have removed the problem of hunger and improved the quality of nutrition in many countries.
- The production of consumer goods has made them widely available; and more and more people can use services and amenities of everyday life.
- Technological innovations have restricted the sphere of hard and exhausting manual work.
- The revolution in transportation has opened up for many people an opportunity to cross continents and boundaries, put an end to geographic isolation, and reduced distances between peoples and cultures.
- Technological achievements have raised modern communication facilities to the highest level by making communication worldwide; and computer technology has radically changed all aspects of social and

economic life.

- Science is steadily expanding our knowledge of the Universe and humankind; discoveries in astronomy and physics, the theory of relativity, and quantum mechanics are offering ever deeper insights into nature, from microparticles to the Metagalaxy; biology and genetics are shaping our views of the biosphere; the synthetic theory of evolution, the discovery of the role of DNA, and studies in molecular biology are revealing the mechanisms of the functioning and dynamics of life; the humanities and social sciences are raising the level of knowledge about an individual and society, the laws of economy and culture.

The stagnation and misery of the remaining non-scientific and anti-scientific views stand out in sharp relief against the backdrop of progress in science. Religious, occult, and other paranormal beliefs and superstitions display unenlightened conservatism; their proponents appeal to human weaknesses, ignorance, and fear of social upheavals and the unknown; they exploit the latest achievements of humankind in order to modernize the appearance of their pseudoscientific and archaic doctrines and practices.[2]

Changes for the better have occurred on the international scene:

- Practically all colonial empires have collapsed.
- The threat of totalitarianism has considerably decreased.
- Many countries in the world have adopted (at least on paper) *The Universal Declaration of Human Rights.*
- The ideas of democracy and freedom are gaining ground in Eastern Europe, post-Soviet countries, Latin America, and Africa.
- Human rights, particularly for women, ethnic and other minorities, are increasing in many countries.
- The world market is drawing new regions into its orbit, thereby opening the opportunities of economic upswing and social development for them.
- Countries with high living standards are successfully solving the problem of population growth; in many countries, it is due not to a higher birth rate but to lower mortality and a longer life span.
- An increasing number of people are gaining access to education and wealths of the world culture.

All these breakthroughs notwithstanding, humankind faces many economic, social, and political problems. A large part of the population is still far from economic well-being. These people are subsisting in poverty and diseases. This is especially true of a number of countries in Asia, Africa,

---

[2] Struggle against pseudoscience, such as astrology, and numerous antiscientific speculations (for details see [11–13] and articles 23–25 of this book) is a topical problem in today's Russia.

and Latin America; and, even in rich countries, there are strata suffering from poverty, malnutrition, and poor health care.

In many countries, the rate of population growth is still so high that it dooms them to degradation and extinction. By 2000, the world's population had topped six billion and, consequently, should the present rate remain unchanged, it will grow by another three billion in the next half-century. The threat of demographic crisis is fraught in some countries with a critical reduction in arable land per capita, depletion of water resources necessary for irrigation, drying of rivers, and shrinking of forest areas.

There is still a risk of adverse effects from global warming caused not only by natural processes but also by deforestation and the growing release of carbon dioxide into the atmosphere in countries that continue to squander their natural resources. Judging by the rate of decline of animal populations and of many types of flora and fauna, we seem to be living through the greatest biological disaster since the extinction of dinosaurs.

There is a broad range of problems associated with overpopulation of large cities, unemployment, and crime. The incidence of diseases previously considered eradicated, such as tuberculosis and malaria, increases again; and the spread of AIDS is virtually uncontrolled.

Important decisions adopted by numerous summits and top-level international conferences have not yet been implemented by national governments and international organizations. Democracy as a global social value is still weak; in a number of countries, it is totally undeveloped; and free elections and freedom of speech are not yet available to many peoples.

Despite the end of the Cold War, the world has not warded off the menace of nuclear destruction: fanatical terrorists, rogue nations or an uncontrolled military can plunge humankind into a global catastrophe. Nor have some basic problems of economic development been resolved; the idea that a free market can as such heal all social ills is scientifically unfounded; and the mechanisms of interaction between market laws and programs of support of social non-commercial sphere and unprotected population strata remain unclear.

Humankind faces problems of utmost gravity. But that does not mean that the prophets of doom and inevitable social and cosmic upheavals should be trusted. Humanists believe that such problems can be resolved by intelligence, critical thinking, and a willingness to compromise, agree, and cooperate.

At the same time, there is a set of problems associated with unscientific, antiscientific, and totally outdated concepts, perpetuating backwardness and ignorance, slowing down social and moral progress, the development of freedom, democracy, and intelligence. The negative effect of retrograde forces is still insufficiently recognized and their danger is underestimated. It is these forces that breed discord, intolerance, and fanaticism thereby preventing improvement in the conditions of human existence. For instance, the stubborn devotion of certain social organizations and population groups to traditional religious beliefs ever more clearly shows that fanaticism fosters an unrealistic and mystical attitude to social life, sows a mistrust of science and, in most cases, tends to support obsolete ways.[3] Religious traditionalists, fundamentalists, and fanatics are hostile to the achievements of science, medicine, and health care, oppose family planning and sex education programs, campaign against the use of contraceptives, oppose the equality of men and women, violate the rights of children, etc.

More and more frequently, the world witnesses violent ethnic conflicts and feuds, based on religious motives; at the same time, political reactionaries and extremists easily play the religious card using primitive psychology, low cultural level, and lack of critical thinking among most of the believers. Increasingly violent are the manifestations of terrorism and genocide behind which, as a rule, are the same ideologies of religious traditionalism, political extremism, and chauvinism. Largely destructive are the forces of nationalism, racism, and fascism: they fuel separatist and isolationist trends; are fraught with intolerance and violence; poison the social atmosphere; and breed mutual fear and mistrust. Additionally, we should not forget about attempts to rehabilitate Stalinism and other forms of totalitarianism.

In many countries, the ideology of so-called postmodernism has gained ground. It is the latest form of irrationalism and nihilism laying claim to a special brand of intellectual elitism, denying the achievements of science, condemning technological progress, undermining confidence in universal human values, and rejecting the value of democracy and human rights. Proclaimed by postmodernism, the slogans of deconstructivism and repressive mania make this ideology counterproductive and pose an obstacle to scientific progress and social justice.

What set of values can best oppose these negative factors and propose an optimistic and positive outlook to humankind? The system that can do it is modern humanism, based on a scientific picture of the world, universal human morality, and secular (non-religious) cultural values.

The unique features of humanism are its open devotion to a scientific world view, i. e. to a set of methodological guidelines associated with objectivity, experimental or observational testing of hypotheses or theories, the self-testing and self-development potential of science, a critical approach, and openness to everything new. Scientific methods are not dogmas. They do not guarantee absolute truths but, in the end, they provide the most reliable tool for acquiring sound knowledge.

The possibilities of science are unprecedented, both in the depth of penetrating into the nature of things and processes and in the spectrum

---

[3] The inter-relation between religion and science has changed as time has passed. The present status of this problem is reflected in the last encyclical letter of Pope John Paul II [14] and in an article by one of the co-authors [15].

of realities subject to study. There is nothing basically inaccessible to scientific thought today. Humanists believe that the methods of science should be extended to all spheres of human endeavor. There should be no restrictions in this respect except when such an application violates human rights. Attempts to prevent rational comprehension and free research on some moral, religious, ideological, or political grounds have always failed. Today this is clearer than ever before. This fact does not obviate humanists from the need to comprehend and evaluate moral, aesthetic, and other cultural forms of human experience.

Devoted to a scientific world view, humanists pay special attention to the lack of sufficiently objective, rational, and experimentally verified evidence of the validity of religious interpretations of reality or assumptions of some supernatural causes; attempts of some scientists to impose on public opinion an interpretation of facts appealing to something otherworldly should be rejected. The time is long overdue for humankind to grow aware of its own maturity and discard the survivals of primeval thinking and myth-making, perpetuating ignorance, superstitions, and human frailties.

The need for a more drastic and goal-directed change of the world view paradigm is motivated by many circumstances. The principal factor is the growing role of intelligence, science, and technology in humankind's life support and well-being.

Humanists are aware that scientific progress may lead to a number of negative consequences in the absence of total control over weapons of mass destruction and the development of industries stimulated by inhuman goals or purely economic benefits: some dangers are involved in discoveries in genetics, medicine, psychology, and mass media. While opposing the uncontrolled use of technologies, humanists, at the same time, vigorously object to attempts to restrict research or censor its results. All problems that may arise should be resolved in open discussions, which implies a certain scientific literacy in their participants. We should recognize that the life of modern humankind is inconceivable without the progress of science and the development of technology. Therefore, while greatly appreciating the potential of science and technology, we must learn to use them wisely and humanely. Particularly important is the development of ecologically safe production and industries raising the living standards of the poor.

35.4   Ethics and reason

The role of the human factor in the life of the international community is becoming decisive. Therefore, it is ethics and reason that constitute the values determining the quality of individual and social life. Humanists believe that the growth of knowledge makes people wiser and more humane, while moral norms contribute to the progress of reason. Reason is the best tool to ensure the proper moral choice and behavior. Humaneness is the best way of ensuring the utilization of humankind's cognitive resources.

Theological moral doctrines based on the notion of a supernatural and superhuman, reflect prescientific notions of the physical world and human nature inherited from ancient times. Theistic commandments are contradictory: various religions hold different views on moral issues. Thus, theists sometimes support and sometimes oppose capital punishment, monogamy, and the equality of men and women. Outbursts of religious intolerance lead to massacres and terror. Many wars of the past and present have been inspired by religious dogmatism and bigotry. No doubt, believers do a lot of good. It is wrong, however, to believe that piety is a serious and—still less— the sole guarantee of virtue.

Humanists advocate the total separation of the church from the state, which should be *secular* and should neither support nor oppose religion. In their opinion, the basic principles of moral conduct arising from real human life are *common to all*, believers and non-believers alike. All moral norms are associated with ordinary and natural interests, desires, needs, and values. We determine how ethical they are by the way in which they promote human welfare and social justice.

What are the key principles of humanist ethics?

- Personal dignity and independence are considered the main value: the humanist ethics aims to maximize the freedom of choice provided it does no harm to others.
- Man should be aware of his responsibility and obligations to others and to the environment.
- The ethics of perfection shared by humanists implies such virtues as creative ability, independent choice, thinking and behavior: it is the duty of man to realize the talents and abilities he is endowed with by nature.
- Sympathy with others and concern about them are essential to moral behavior. Humanist ethics advocates the moral education of children and youth.
- Moral judgements and actions should be based not only on a direct moral motive but also on intelligence playing an important role in analysis and decision-making. In case of moral collisions, a reasonable dialogue and the search for an agreement are of great importance.
- The ethics of humanism deems it necessary to prepare human beings for the correction of their moral principles and in the light of future challenges and possibility of unique new situations.
- Humanists believe that a worthy ethics should be based on principles rather than on goals alone. This means that the end does not justify the means but the type of end is determined by the means: there are limits to the permissible. One cannot remain indifferent to the tragedy

of millions perpetrated by those who dare to justify a great evil by the promise of great future benefits.

35.5 Planetary humanism

The most urgent goal in the realm of human values is the development of a world view in line with the contemporary requirements and possibilities of the international community. This function can be performed by *planetary humanism*, which would safeguard human rights, advocate human freedom and dignity, and point out its duties to a united humankind.

Rapid globalization of life makes us recognize that *the need to respect the dignity of all members of the international community* is the basic principle of planetary humanism. Human duties to traditional social entities (family, relatives, and friends, community, nation, and society) should include one more: responsibility for the future of humankind, for people living abroad. We are physically and morally welded together as never before. Today, when the bell tolls, it tolls for all of us.

The planetary character of this humanism requires that the following principle should be extended *to humankind as a whole*: act so that the sum of human sufferings will decrease and that of well-being increase.

One should abstain from overemphasizing national cultural features because it may cause mutual alienation and be a destructive factor in social life.

Respect for dignity should apply equally *to all* people.

The principles of planetary ethics imply *perspectivism* (posteriorism). In other words, they appeal not only to the international community as it exists today but also to its future. We are responsible for the near and distant future, for people who will live after us.

Each generation must leave a more favorable environment to the next one. This actually means that a contemporary person may represent the future that expects something better from us, just as we need the cultural achievements handed down from previous generations.

The growing might and consequences of human deeds make it incumbent on us to avoid doing anything that may endanger the very life of future humanity.

In other words, *Manifesto 2000* proclaims a new viable humanism stressing the idea of promoting safety and well-being for the present and future world. Its implementation involves the performance of a number of duties formulated in the *Planetary Bill of Rights and Obligations*. This document (see [4]) is based on the principles of "The Universal Declaration of Human Rights" and includes the following propositions:

- We should strive to end poverty and malnutrition and to provide adequate health care and shelter for people everywhere on the planet.

- We should strive to provide economic security and adequate income for everyone.
- Every person should be protected from unwarranted and unnecessary injury, danger, and death.
- Individuals should have the right to live in a family unit or household of their choice, consonant with their income, and should have the right to bear or not to bear children.
- The opportunity for education and cultural enrichment should be universal.
- Individuals should not be discriminated against because of race, ethic origin, nationality, culture, caste, class, creed, gender or sexual orientation.
- The principles of equality should be respected by civilized communities, and in four major senses: equality before law, equality of consideration, basic needs (food, shelter, security, health maintenance, culture, and education), and equal opportunities.
- It is the right of every person to be able to live a good life, pursue happiness, achieve creative satisfaction and leisure in his or her own terms, so long as he or she does not harm others.
- Individuals should have the opportunity to appreciate and participate in the arts.
- Individuals should not be unduly restrained, restricted, or prohibited from exercising a wide range of personal choices, including the right of privacy.

35.6   A new global agenda

The new global situation that has arisen by the new millennium calls for *a new global agenda*. Its priorities are:

- *to maintain lasting peace and security* at national and international levels;
- *human development*, i. e. the acceleration of human progress on the global scale leading, above all, to the levelling of social, economic, and cultural development in all regions of the world;
- *control over activities of global commercial and industrial conglomerates* that tend to disregard the wishes of individual governments and avoid taxes by exporting profits;
- *consolidation of social justice*, i. e. the need to re-emphasize that international documents pertaining to human rights, particularly those of women, children, minorities, and indigenous people, be ratified and applied by all countries;
- *development of a system of international law* that transcends the laws of the separate nations and guaranteeing protection to any citizen,

i. e. it is necessary to transform a lawless world into one that has laws everyone can understand and abide by; and

- *protection of the environment,* i. e. measures to protect the environment need priority for the planetary community.

The most urgent problem of the 21st century is the establishment of *global institutions* to resolve global problems. Unfortunately, there is a huge gap between the opportunities of a number of international institutions, such as the UN, WHO, etc, and the needs of the international community. Today, more than ever before, we need a world organization that would represent the interests of people rather than those of individual states. It is necessary to raise the efficiency of the UN by transforming it from an assembly of sovereign states into an assembly of sovereign peoples.

If we intend to resolve our global problems, individual states would have to delegate part of their national sovereignty to a system of transnational power. Such a system will undoubtedly give rise to political opposition everywhere, particularly on the part of nationalists and chauvinists. And yet it should and eventually will be established if we seek planetary accord and the well-being of people, wherever they may live.

With a view to establishing a system of transnational power, the following reforms are proposed:

- the election by the world's population of a world parliament representing the interests of people, not their governments;
- the establishment of a workable security system to resolve international conflicts—the veto by the Big Five needs to be repealed;
- the establishment of a world court and an international judiciary with sufficient power to enforce its rulings;
- the establishment of a planetary environmental monitoring agency on the transnational level;
- the introduction of an international system of taxation in order to assist the underdeveloped sectors of the human family and fulfill social needs not fulfilled by market forces;
- the development of global institutions to elaborate procedures for the regulation of multinational corporations and state monopolies;
- keeping alive a free market of ideas, respect diversity of opinion, and cherish the right of dissent.

Humanists insist on abstaining from any form of censorship by authorities, uncontrolled by society, advertisers or mass media proprietors. It is necessary to support *bona fide* competition between the mass media and the establishment of volunteer societies to inspect the quality and content of broadcasts and publications. There is a special need to maintain free access to the mass media. Neither powerful oligarchic associations of global mass media organizations nor state power can dominate the information media.

A worldwide democratic movement is required to ensure the diversity and mutual enrichment of cultures and a free exchange of ideas.

Planetary humanism sets great goals for humankind. Modern humanists would like to cultivate and develop optimism, a sense of surprise, and admiration in view of our great opportunities to provide a better and more valuable life for ourselves and the coming generations.

Ideals produce the future. Optimism can be productive, provided it is based on a realistic assessment of the existing opportunities and on a determination to overcome the difficulties confronting us. Humanists reject the nihilistic philosophy of doom and despair, sermons denying reason and freedom: they do not accept fears and somber forebodings of apocalyptic scenes of doomsday. Today, humanists as vigorously as ever urge people not to search for salvation from above. We alone are responsible for our own destiny. The best thing we can do is to mobilize our intelligence and determination to translate our noble aspirations and hopes into reality.

35.7 Closing remarks

Much of what is contained in *Manifesto 2000* sounds utopian. In a number of paragraphs, it is a declaration of intent combined with good wishes. We are certainly quite aware of this and believe that the authors of the *Manifesto* understand it, too. The humanist movement, however, has deep historical roots and, as we saw earlier, even administratively it is international. Published on the eve of the new century, *Humanist Manifesto 2000: A Call for a New Planetary Humanism* caused broad public repercussions and was published in most countries of the world. It was signed and supported by many modern scientists and public figures, as well as different international and national democratic and civil rights organizations. Yes, *Manifesto 2000* is in a sense a declaration of intent. Yes, the implementation of many things that we want and dream of is still far away. But you cannot move ahead without knowing and seeing the goal, however far it may be. Skepticism, irony, and mockery, let alone censure, are not constructive and usually serve to justify a lack of principles or inaction.

With respect to our reality, we hope that *Manifesto 2000* will inspire serious thoughts about humanism and democracy in Russia and its place in the world family of nations.

Russia is, by no means, entering the 21st century in the best shape. In its history, the 20th century was probably both its most dynamic and most dramatic one. Many people today yield to apathy and pessimism and, instead of facing the future to help solve their problems, turn to the past. But, as is known, it teaches us, as a rule, only what we wish to learn and mainly tells us only what can confirm our sentiments and expectations when we delve into history. The present is something else. It is insufficient to search in it for what confirms our beliefs. In the present, people live and act. To make life worthwhile in our own eyes, we must search for positive facts and processes that we face, right here and now. In light of an endless series of outrages and crimes, utterances about some encouraging prospects seem utopian. Nevertheless, such prospects do exist. It is necessary not only to describe them but to act and do our bit on their basis. Such is a humanist and life-asserting approach.

Perhaps the decisive positive aspect of Russian reality is the acquisition of unprecedented freedom by our society. Obviously, the first decade of liberty in Russia can be described as wild. But insofar as the citizens of Russia[3*] have retained their intelligence, freedom is beginning to bring its first, barely discernible fruits. And if we look 10 to 15 years back, we will see how greatly we have matured. A degree of realism in perceiving the reality, the printed word, and political statements has changed beyond recognition compared to the 'Soviet' attitude to life. Equally dramatic have been changes in the character of self-appraisal. The school of freedom is very costly. Many people are still trying to avoid its lessons, relying on 'the strong arm', big-power chauvinism, or prayers. But everything seems to indicate that this stern approach to democracy cannot be avoided.

A great country, such as Russia, with its millennial culture, vast natural resources, and the unique viability of its peoples, is simply doomed to coping with an economic, social, moral, and psychological crisis. There are sufficient subjective grounds for discontent with the existing state of affairs, the more so since the lifespan of each of us seems to be less than the tempo of expected changes, given a great desire for instant changes for the better. But there are no miracles in nature or in society. That does not mean that you can stand aside and wisely shake your head or—what is even easier—become a cynic, a whiner, or pessimist. Difficult as life in present-day Russia may be, there is still room for self-affirmation, creativity, and good deeds.

As optimists, we believe that, in the new century, Russia will increasingly develop on the basis of democracy, reason, goodwill, and just laws, and its citizens longing for a normal and worthy life will finally be able to build it on the basis of independence and responsibility for their destiny.[2*,4*]

Comments

1*. This article was published in 1999 *Vestn. Ross. Akad. Nauk* **69**(6) 546 [2000 *Herald Russ. Acad. Sci.* **70**(3) 223] in co-authorship with the philosopher Valerii Aleksandrovich Kuvakin, Professor, Moscow State University. I am grateful to V A Kuvakin for permission to include this article here. It was also published in the monthly 2000 *Neva* no 6 141.

2*. Having read the proof of the article in September 2000, I certainly understood that it did sound completely utopian. But it is so nice to dream! I therefore did not withdraw the article.

3*. Foreigners usually apply the English word *Russian* to any citizen of Russia just as the word *Americans* is used to designate all citizens of the United States regardless of their descent. However, in its original Russian usage, the same word refers to a member of the dominant ethnic group of the country accounting for 80% of its total population. After the collapse of the USSR, citizens of Russia irrespective of their ethnic affiliation tend to be called *Rossiyane*. As a result, two words are currently in use, *Russian* and *Rossiyanin*. Hence, there are frequent misunderstanding when pieces of literature or various documents are translated from the Russian language into the English one.

4*. Naturally, this paper written in 2000 makes no mention of the infamous terrorist act of 11 September 2001 and subsequent events. My attitude towards such evil deeds is expressed in comment 15 to article 26.

References

[1] Yakovlev A N (ed) *Rossiya XX vek. Dokumenty [20th-century Russia. Documents.]* See the following volumes: 1997 *Kronstadt 1921* (Moscow: Mezhdunarodnyi Fond "Demokratiya"); 1997 *Filipp Mironov* (Moscow: Mezhdunarodnyi Fond "Demokratiya"); 1997 *Katyn* (Moscow: Mezhdunarodnyi Fond "Demokratiya"); 1999 *Vlast' i Khudozhestvennaya Intelligentsiya [Power and Creative Intellectuals]* (Moscow: Mezhdunarodnyi Fond "Demokratiya"); 1999 *Lavrentii Beria* (Moscow: Mezhdunarodnyi Fond "Demokratiya"). See also Yakovlev A 2000 *Krestosev* (Moscow: Vagrius); Jakovlev A N 2000 *Omut Pamyati [The Whirl of Memory]* (Moscow: Vagrius)

[2] Latyshev A G 1996 *Rassekrechennyi Lenin [Lenin Declassified]* (Moscow: MART)

[3] Courtois S 1997 *Le Livre Noir du Communisme: Crimes, Terreurs et Répression* (Paris: R. Laffont) [Translated into English 1999 *The Black Book of Communism: Crimes, Terror, Repression* (Cambridge, MA: Harvard University Press)] [Translated into Russian 1999 *Chernaya kniga kommunizma. Prestupleniya. Terror. Repressii* (Translated from the French) (Moscow: Three Ages of History)]

[4] 1999 Manifesto 2000. A call for a new planetary humanism *Zdravyi Smysl* no 13

[5] 1997 Research center of the Russian humanist society *Zdravyi Smysl* no 5; 1999 *Zdravyi Smysl* no 12

[6] Kuvakin V A 1998 *Tvoi Rai i Ad. Chelovechnost' i Beschelovechnost' Cheloveka (Filosofiya, Psikhologiya i Stil' Myshleniya Gumanizma) [Your Paradise and Hell, Humanity and Inhumanity of a Human Being. The Philos-*

ophy, Psychology, and Mentality of Humanism] (St Petersburg–Moscow: Aleteiya)

[7] 1996 Humanist Manifesto I *Zdravyi Smysl* no 1

[8] 1997 Humanist Manifesto II *Zdravyi Smysl* no 3

[9] 1966 Organizations and journals of modern humanism *Zdravyi Smysl* no 1

[10] 1998 A declaration of modern humanism *Zdravyi Smysl* no 7

[11] Kruglyakov E P 1998 *Chto zhe s Nami Proiskhodit?* [*What Is Happening to Us?*] (Novosibirsk: Izd. SO RAN); 2001 *'Uchenye' s Bol'shoi Dorogi* [*Highway 'Scientists'*] (Moscow: Nauka)

[12] Aleksandrov E B and Ginzburg V L 1999 O lzhenauke i ee propagandistakh *Vestn. Ross. Akad. Nauk* **69**(3) 199 [1999 Concerning pseudoscience and its propagandists *Herald Russ. Acad. Sci.* **69**(2) 118]

[13] 1999 Problemy bor'by s lzhenaukoi. Obsuzhdenie na Presidiume RAN [Problems of fighting pseudoscience. A discussion at the RAS Presidium] *Vestn. Ross. Akad. Nauk* **69**(10) [1999 *Herald Russ. Acad. Sci.* **69**(5)]

Ginzburg V L 2001 *Vestn. Ross. Akad. Nauk* **71**(8) 702 [2001 *Herald Russ. Acad. Sci.* **71**(8) 355]

[14] John Paul II 1998 Encyclical letter '*Fides et Ratio*' (Vatican City: Libreria Vaticana) [Translated into Russian 1999 (Moscow: Izd. Frantsiskantsev)]

[15] Ginzburg V L 1999 Reason and faith *Vestn. Ross. Akad. Nauk* **69**(6) 546; Reprinted in 1999 *Zdravyi Smysl* no 13 51 [1999 *Herald Russ. Acad. Sci.* **69**(4) 271]. See also article 31 in this book

Curriculum vitae of Vitaly L Ginzburg

Born 4 October 1916 in Moscow, Russia. Graduated from Moscow State University in 1938. Defended candidate dissertation (PhD) in Moscow State University in 1940. From 1940 until now works in P N Lebedev Physical Institute of the USSR Academy of Sciences (now the Russian Academy of Sciences). In 1942 defended DSc dissertation. From 1971 till 1988—head (director) of I E Tamm Department of Theoretical Physics, from 1988 till now adviser in Russian Academy of Sciences and head of the theoretical group in P N Lebedev Physical Institute (known as FIAN). From 1945 for several years was visiting professor at Gorky State University and from 1968 till now professor at the Moscow Physico-Technical Institute. From 1998 Editor-in-Chief of *Uspekhi Fizicheskikh Nauk* (*Physics–Uspekhi*) journal.

In 1953 was elected a Corresponding Member of the USSR Academy of Sciences (now RAS) and in 1966— Full Member.

Decorated with USSR and Russian Orders. Was also elected to the following institutions: 1969—member International Academy of Astronautics, 1970—associate Roy. Astron. Soc., 1971—foreign honorary member American Acad. Art and Science, 1977—foreign member Royal Danish Acad. Sci. and Letters, 1977—honorary Fellow Indian Acad. Sci., 1981— foreign fellow Indian Nat. Sci. Acad., 1981—foreign associate National Acad. Sci. USA, 1987—foreign member Roy. Soc. (London), 1990— member Academia Europaea.

Was awarded: Mandelshtam prize 1947, Lomonosov prize 1962, State prize 1953, Lenin prize 1966, M Smoluchovskii Medal Polish Phys. Soc. 1987, Gold Medal Roy. Astron. Soc. 1991, Bardeen prize 1991, Wolf prize 1994/1995, Vavilov Gold Medal 1995, Big Lomonosov Gold Medal Russian Acad. Sci. 1995, O'Ceallaigh Medal 2001, "Triumph" Prize 2002, Nobel Prize in Physics 2003 "for pioneering contributions to the theory of superconductors and superfluids".

V L Ginzburg is productive theoretical physicist with broad interests. He had made a number of significant contributions to particle physics, plasma physics, condensed matter theory, astrophysics, contributions that range from his seminal work on the origin of cosmic rays, radioastronomy, and the propagation of electromagnetic waves in plasma, in the ionosphere, and in stellar corona to the prediction (with Frank) of transition radiation and pioneering work on new mechanisms that might lead to a substantial increase in the superconducting transition temperature. A major contribution with great impact on many areas of physics in the Ginzburg–Landau theory of superconductivity, in which the superconducting properties are described by a complex order parameter with amplitude and phase. First given on phenomenological grounds in 1950, the theory has since been confirmed by microscopic theory. It has been applied not only to superconductivity but to many other phase transitions including superfluidity. Relativistic generalizations are being used to account for the weak, electromagnetic, and strong interactions of high energy physics.

V L Ginzburg is the author of several hundreds papers and many books including 1979 *Theoretical Physics and Astrophysics* (Pergamon Press), 1983 *Waynflete Lectures of Physics* (Pergamon Press), 1964 *Origin of Cosmic Rays* (with S I Syrovatskii, Pergamon Press), 1970 *Propagation of Electromagnetic Waves in Plasma* (Pergamon Press), 1990 *Transition Radiation and Transition Scattering* (with V N Tsytovich, A Hilger), 1984 *Crystal Optics with Spatial Dispersion and Excitons* (with V M Agranovich, Springer), 1985 *Physics and Astrophysics. A Selection of Key Problems* (Pergamon Press), 1992 *On Physics and Astrophysics* (in Russian) (Moscow: Nauka); second Russian edition, 1995 (Moscow: Quantum).

The Physics of a Lifetime: Reflections on the Problems and Personalities of 20th Century Physics

(published by Springer-Verlag in 2001, 514 pp)

Contents

Part III